[2.1]	x^n	the nth power of x, or x to the nth power
	$P(x)$, $D(y)$, etc.	P of x, D of y, etc.
[2.8]	a^0	1
	a^{-n}	the reciprocal of a^n
	$a^{m/n}$	the mth power of the nth root of a
	$a^{1/n}$	the nth real root of a for $a \in R$, or the positive one if there are two
[2.9]	$\sqrt[n]{a}$	the nth real root of a for $a \in R$, or the positive one if there are two
[4.1]	(x, y)	the ordered pair of numbers whose first component is x and whose second component is y
	$A \times B$	the Cartesian product of A and B
	$R \times R$, or R^2	the Cartesian product of R and R
	f, g, h, F, etc.	names of functions
	$f(x)$	f of x, or value of f at x
	f^{-1}	the inverse function of f
[4.2]	d	distance between two points
	m	the slope of a line
[4.8]	(x, y, z)	the ordered triple of numbers whose first component is x, second component is y, and third component is z
	$(R \times R) \times R$, or R^3	the Cartesian product of $(R \times R)$ and R
[6.2]	$\log_b x$	the logarithm of x to the base b
[6.3]	$\text{antilog}_b x$	the antilogarithm of x to the base of b
	e	an irrational number, approximately equal to 2.7182818
[7.1]	$\cos x$	element in the range of the cosine function
	$\sin x$	element in the range of the sine function
[7.3]	$\bar{s}$ or $\bar{x}$	length of reference arc corresponding to arc of length s or x
[7.7]	$\tan x$	element in the range of the tangent function
	$\cot x$	element in the range of the cotangent function
	$\sec x$	element in the range of the secant function
	$\csc x$	element in the range of the cosecant function

(*continued inside back cover*)

Modern College Algebra and Trigonometry
Second Edition

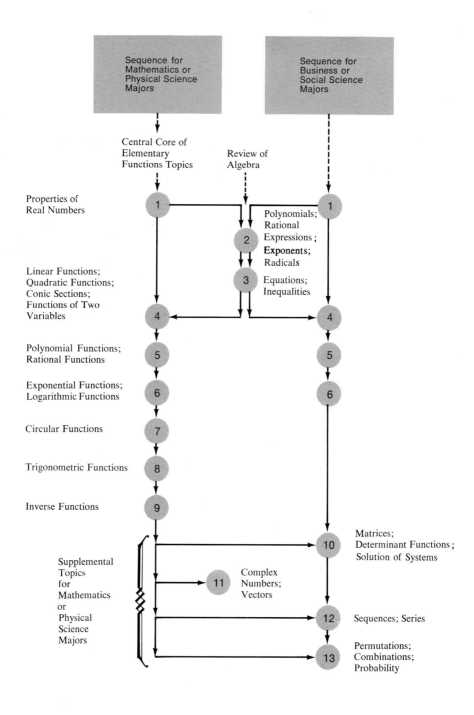

Sequence for Mathematics or Physical Science Majors

Sequence for Business or Social Science Majors

Central Core of Elementary Functions Topics

Review of Algebra

Properties of Real Numbers — 1

Polynomials; Rational Expressions; **Exponents**; Radicals — 2

Equations; Inequalities — 3

Linear Functions; Quadratic Functions; Conic Sections; Functions of Two Variables — 4

Polynomial Functions; Rational Functions — 5

Exponential Functions; Logarithmic Functions — 6

Circular Functions — 7

Trigonometric Functions — 8

Inverse Functions — 9

Supplemental Topics for Mathematics or Physical Science Majors

Matrices; Determinant Functions; Solution of Systems — 10

Complex Numbers; Vectors — 11

Sequences; Series — 12

Permutations; Combinations; Probability — 13

Modern College Algebra and Trigonometry
Second Edition

Edwin F. Beckenbach
University of California, Los Angeles

Irving Drooyan
Los Angeles Pierce College

Wadsworth Publishing Company, Inc., Belmont, California

To the Student

A partially programmed study guide based on the text is available from your local bookstore under the name A Semi-Programmed Study Guide for **Modern College Algebra and Trigonometry**, Second Edition, by Bernard Feldman.

ISBN: 0-534-00110-6
L.C. Cat. Card No.: 78-173312

Printed in the United States of America

3 4 5 6 7 8 9 10—76 75 74 73

Preface

Our primary purpose in revising *Modern College Algebra and Trigonometry* has been to reduce the level of rigor and abstraction in the presentation and to increase the number of worked-out examples in the exercise sets.

This book, like its predecessor, is designed for students who have completed the equivalent of one and one-half to two years of high school algebra and a year of geometry but need additional preparation for the study of calculus or other mathematically oriented fields, such as business. The topics covered and the spirit in which they are presented reflect the recommendations of various mathematics curriculum study groups. In particular, all of those topics recommended by the Committee on the Undergraduate Program of the Mathematical Association of America for the Level O course are covered. The organization of the material permits considerable flexibility in the kind of course for which the book can be used. Specific suggestions are charted on page ii.

Major revisions that make the material more accessible to the student are as follows: Chapter 1 has been rewritten to change the formal statement-reason format for proofs to an informal paragraph style. In Chapter 4, new sections covering parallel and perpendicular lines and loci in the plane have been added. The former Chapters 10 and 11 have been combined to reduce emphasis on transformations of the plane.

In addition, major rewriting has been done in the chapters on circular functions (7), complex numbers (11), sequences (12), and probability (13) in the interest of improving clarity.

The first three chapters, concerned with basic properties of real numbers and simple facts about algebraic expressions, equations, and inequalities, provide an extended review of the fundamental skills of algebra. The student is introduced to relations and functions in Chapter 4; from there on, discussions center around the function concept. After introduction of constant, linear, and quadratic algebraic functions, the study proceeds through polynomial, rational, exponential, and logarithmic functions, with heavy emphasis given to sketching graphs of these functions. The traditional topics of plane analytic geometry are covered in Chapters 4, 5, and 6 in sufficient depth that students can proceed directly from this text to a study of calculus. The periodic circular and trigonometric functions become an integral part of the presentation in Chapters 7 through 9. Chapter 10, on matrices, includes a section on linear systems, and the discussion of complex numbers in Chapter 11 includes a treatment of the trigonometric form for a complex number as well as a section on De Moivre's theorem.

The last three chapters are essentially independent of each other, and any or all may be omitted for shorter courses. The degree of rigor in the course can be varied by the assignment or nonassignment of some problems occurring at the ends of exercise sets, in particular those problems that ask for proofs omitted from the text.

A second color is used throughout to highlight key procedures in routine manipulations, to stress basic concepts, and to focus attention on key elements of figures. Marginal annotations direct the reader's attention to important ideas. An exercise set is included for each section, and answers are given for the odd-numbered problems. A solutions manual with detailed solutions of all problems and also a manual containing the even-numbered answers, a set of test items for each chapter, and a final examination are available for instructors. A semi-programmed study guide based on the text is available to students and instructors for supplemental work.

We wish to express our appreciation to Professor William Wooton, our coauthor in *Essentials of College Algebra*, for his permission to use material from that text. Our special thanks go to Dr. Edwin S. Beckenbach for assistance in the preparation of this edition.

Edwin F. Beckenbach

Irving Drooyan

Contents

4 Relations and Functions

5 Polynomial and Rational Functions

6 Exponential and Logarithmic Functions

7 Circular Functions

8　Trigonometric Functions

9　Inverse Functions and Conditional Equations

10　Matrices and Determinants

11　Complex Numbers and Vectors

1

Properties
of real numbers

1.1

Definitions and symbols

A **set** is simply a collection of some kind. It may be a collection of people, colors, numbers, or anything else. In algebra, we are interested in sets of numbers of various sorts and in their relations to sets of points or lines in a plane or in space. Any one of the collection of things in a set is called a **member** or **element** of the set, and is said to be **contained in** or **included in** (or, sometimes, just **in**) the set. For example, the counting numbers 1, 2, 3, ... (where the dots indicate that the sequence continues indefinitely) are the elements of the set we call the set of **natural numbers**.

Set notation Sets are usually designated by means of capital letters, A, B, C, etc. They are identified by means of **braces**, { }, with the members either listed or described. For example, the elements might be listed as in {1, 2, 3}, or described as in {first three natural numbers}. The expression "{1, 2, 3}" is read "the set whose elements are one, two, and three"; "{first three natural numbers}" is read "the set whose elements are the first three natural numbers."

Using the undefined notion of set membership, we can be more specific about some other terms we shall be using.

Definition 1.1 *Two sets A and B are equal, $A = B$, if and only if they have the same members—that is, if and only if every member of each is a member of the other.*

Thus, if A denotes {1, 2, 3}, B denotes {3, 2, 1}, C denotes {2, 3, 4}, and D denotes {natural numbers between 1 and 5}, then $A = B$ and $C = D$. The phrase "if and only if" used in this definition is simply the mathematician's way of making two statements at once. Definition 1.1 means: "Two sets are equal if they have the same members. Two sets are equal only if they have the same members." The second of these statements is logically equivalent to: "Two sets have the same members if they are equal."

Definition 1.2 *If the elements of a set A can be paired with the elements of a set B in such fashion that each element of A is paired with one and only one element of B, and conversely, then such a pairing is called a one-to-one correspondence between A and B.*

For example, if $A = \{a, b, c\}$ and $B = \{1, 2, 3\}$, then the sets A and B can be put into one-to-one correspondence in six different ways, two of which are shown:

$$\{a, b, c\} \qquad \{a, b, c\}$$
$$\updownarrow \; \updownarrow \; \updownarrow \qquad \updownarrow \; \updownarrow \; \updownarrow$$
$$\{1, 2, 3\} \qquad \{2, 1, 3\}$$

Definition 1.3 *Two sets are equivalent if and only if a one-to-one correspondence exists between them.*

Equivalence symbol The symbol $\sim$ is used to denote equivalence. Thus, $A \sim B$ is read "A is equivalent to B." Intuitively, equivalent sets are sets that contain the same number of members. Clearly, if two sets are equal, they are equivalent, but the converse is not necessarily true—equality of sets requires that the members be identical, not merely that the sets be in one-to-one correspondence.

Definition 1.4 *If every member of a set A is a member of a set B, then A is a subset of B. If, in addition, B contains at least one member not in A, then A is a proper subset of B.*

Subset symbols The symbol $\subset$ (read "is a subset of" or "is contained in") will be used to denote the subset relationship. Thus

$$\{1, 2, 3\} \subset \{1, 2, 3, 4\} \quad \text{and} \quad \{1, 2, 3\} \subset \{1, 2, 3\}.$$

Notice that, by definition, every set is a subset of itself.

The set that contains no elements is called the **empty set**, or **null set**, and is denoted by the symbol $\emptyset$ (read "the empty set" or "the null set"); $\emptyset$ is a subset of every set, and it is a proper subset of every set except itself. If a set S is the null set or is equivalent to the set $\{1, 2, 3, \ldots, n\}$ for some fixed natural number n, then S is said to be **finite**. A set that is not finite is said to be **infinite**. For example, the set of *all* natural numbers, $\{1, 2, 3, \ldots\}$, is an infinite set.

Definition 1.5   *Two sets A and B are disjoint if and only if A and B contain no member in common.*

For example, if $A = \{1, 2, 3\}$ and $B = \{5, 6, 7\}$, then A and B are disjoint.

Set-membership notation   The symbol $\in$ (read "is a member of" or "is an element of") is used to denote membership in a set. Thus,

$$2 \in \{1, 2, 3\}.$$

Note that we write

$$\{2\} \subset \{1, 2, 3\} \quad \text{and} \quad 2 \in \{1, 2, 3\},$$

since $\{2\}$ is a *subset*, whereas 2 is an *element*, of $\{1, 2, 3\}$.

When discussing an individual but unspecified element of a set containing more than one member, we usually denote the element by a lowercase italic letter (for example, a, d, s, x), or sometimes by a letter from the Greek alphabet: α (alpha), β (beta), γ (gamma), and so on. Symbols used in this way are called **variables**.

Definition 1.6   *A variable is a symbol representing an unspecified element of a given set containing more than one element.*

The given set is called the **replacement set**, or **domain**, of the variable. If the domain is a set of numbers, then the variable represents a number. Thus

$$x \in A$$

means that the variable x represents an (unspecified) element of the set A.

The members of the replacement set are called the **values** of the variable. If the set has just one member, this one value is a **constant**.

When discussing sets, it is often helpful to have in mind some general set from which the elements of all of the sets under consideration are drawn. For example, if we wish to talk about sets of college students, we may want to consider all college students in this country, or all students in general; or, taking a larger view, we may want to consider students as a special kind of human being—say, all those human beings who are consciously striving to increase their knowledge. Thus, we can draw sets of college students from any one of a number of different general sets. Such a general set is called the **universe of discourse**, or the **universal set**, and we shall usually denote it by the capital letter U. It follows that any set in a particular discussion is a subset of U for that discussion.

Negation symbol   The slant bar, $/$, drawn through certain symbols of relation, is used to indicate negation. Thus $\neq$ is read "is not equal to," $\not\subset$ is read "is not a subset of," and $\notin$ is read "is not an element of." For example,

$$\{1, 2\} \neq \{1, 2, 3\}, \quad \{1, 2, 3\} \not\subset \{1, 2\}, \quad \text{and} \quad 3 \notin \{1, 2\}.$$

Set-builder   Another symbolism useful in discussing sets is illustrated by
notation

$$\{x \mid x \in A \text{ and } x \notin B\}$$

(read "the set of all x such that x is a member of A and is not a member of B").
This symbolism, called **set-builder notation**, is used extensively in this book. What
it does is specify a variable (in this case, x) and, at the same time, state a condition
on the variable (in this case, that x is contained in the set A and is not contained
in the set B).

Exercise 1.1

Designate each of the following sets by using braces and listing the members.

Example   {natural numbers between 8 and 12}

Solution   {9, 10, 11}

1. {natural numbers between 2 and 7}

2. {natural numbers between 13 and 20}

3. {natural numbers between 88 and 89}

4. {natural numbers between 100 and 102}

5. {days in the week}

6. {months in the year}

Replace the colored comma with either $=$ or $\neq$ to make a true statement.

7. {natural numbers less than 3}, {1, 2}

8. {integers between -3 and 1}, {-2, -1}

9. {2}, {-2}

10. {4}, {7}

11. $\emptyset$, {0}

12. {5, 7, 9}, {7, 9, 5}

Replace the colored comma with either ∈ or ∉ to make a true statement.

13. 3, {2, 3, 4}   14. 15, {2, 4, 6, ...}

15. {2}, {2, 3, 4}   16. ∅, {2, 3, 4}

Replace the colored comma with either ⊂ or ⊄ to make a true statement.

17. 5, {4, 5, 6}   18. {5, 4, 6}, {4, 5, 6}

19. ∅, {4, 5, 6}   20. {3, 4}, {4, 5, 6}

21. Let $U = \{5, 6, 7\}$. List the subsets of U that contain

    a. three members   b. two members

    c. one member   d. no members

22. Let $U = \{1, 2, 3, 4\}$. List the subsets of U that contain

    a. four members   b. three members   c. two members

    d. one member   e. no members   f. 2 and one other member

23. Let $U = \{1, 2, 3, 4, 5, 6, 7, 8, 9\}$, $A = \{1, 2, 3, 4\}$, $B = \{4, 5, 6, 7\}$, and $C = \{6, 7\}$. Replace the comma with either ⊂ or ⊄ to make a true statement.

    a. A, U b. C, A c. A, B d. C, B

24. Let $U = \{\text{natural numbers}\}$, $A = \{\text{even natural numbers}\}$, $B = \{\text{odd natural numbers}\}$, $C = \{x \mid x \text{ is a natural number between 1 and 10}\}$, and $D = \{x \mid x \text{ is a natural number less than 9}\}$. Which of the following statements are true?

    a. $A \sim D$ b. $C = D$ c. $C \subset D$

    d. $A \sim B$ e. $A = B$ f. $D \subset C$

    g. $A \sim U$ h. A and B are disjoint.   i. $A \subset U$

    j. $C \sim D$ k. $C \subset A$ l. C and B are disjoint.

Designate each of the following sets by using set-builder notation.

Example   {even natural numbers}

Solution $\{x \mid x = 2n, n \text{ a natural number}\}$

25. {odd natural numbers}   26. {natural numbers}

27. {solutions of $2^x = 5$}   28. {solutions of $x^x = 5$}

29. {elements not in set A}   30. {elements in set B}

1.2

Operations on sets

Ideas involving universal sets, subsets thereof, and certain operations on sets can be depicted using plane geometric figures called **Venn diagrams**. Figure 1.1 shows such a diagram representing a universe having as its elements all points of the rectangle and its interior. It also shows a number of subsets of the universe, denoted by circles and their interiors. In this figure, sets A, B, and C are disjoint, D is a subset of C, and E is neither a subset of C nor disjoint from C.

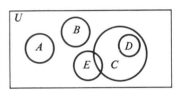

Figure 1.1

There are several mathematically important operations on the subsets of a given universe U. One such operation is defined as follows.

Definition 1.7   *The union of two subsets A and B of U is the set of all elements of U that belong either to A or to B or to both*

Set-union symbol   The symbol $\cup$ is used to denote the union of sets. Thus $A \cup B$ (read "the union of A and B" or, sometimes, "A cup B") is the set of all elements that are in either A or B or both.

Example   If $A = \{1, 2, 3, 4, 5\}$ and $B = \{2, 3, 4, 5, 6\}$,

then $A \cup B = \{1, 2, 3, 4, 5, 6\}$.

Notice that each element in $A \cup B$ is listed only once in this example, since repetition would be redundant. Figure 1.2 is a Venn diagram in which the shaded region depicts $A \cup B$.

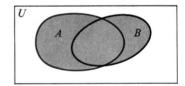

Figure 1.2

A second set operation of interest is defined as follows.

Definition 1.8   *The intersection of two subsets A and B of U is the set of all elements of U that belong to both A and B.*

Set-intersection symbol   The symbol $\cap$ is used to denote intersection. Thus $A \cap B$ (read "the intersection of A and B" or, sometimes, "A cap B") denotes the set of all elements of U that are in both A and B.

Example   If $A = \{1, 2, 3, 4, 5\}$ and $B = \{2, 3, 4, 5, 6\}$,

then $A \cap B = \{2, 3, 4, 5\}$.

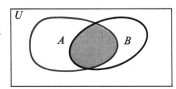

In the Venn diagram of Figure 1.3, the shaded region depicts $A \cap B$.

Figure 1.3

Both union and intersection are applied to two sets in relation to each other. The following operation on sets, however, applies to only one set in relation to U.

Definition 1.9 *The complement of a set A in U is the set of all elements of U that do not belong to A.*

The symbol A' (or, sometimes, $\bar{A}$, $\sim A$, or $\tilde{A}$) denotes the complement of A in U.

Example   If $U = \{1, 2, 3, 4, 5\}$ and $A = \{2, 4\}$,

then $A' = \{1, 3, 5\}$.

The shaded part of the Venn diagram in Figure 1.4 represents A'.

Figure 1.4

Exercise 1.2

Let $U = \{1, 2, 3, 4, 5, 6, 7, 8, 9, 10\}$, $A = \{2, 4, 6, 8, 10\}$, $B = \{1, 2, 3, 4, 5\}$, *and* $C = \{1, 3, 5, 7, 9\}$. *List the members of each of the following sets.*

1. A' 2. B' 3. C' 4. $A \cap B$

5. $A \cup B$ 6. $A \cup C$ 7. $A \cap C$ 8. $A' \cap B'$

9. $A' \cup C'$ 10. $(A \cap B)'$ 11. $A' \cup C$ 12. $C' \cap B$

For each of the Problems 13–24, *copy the Venn diagram shown here on a sheet of paper.*
Shade the part of the diagram corresponding to each of the following sets.

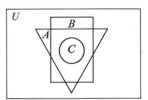

13. $A \cap B$ 14. $C \cap A$ 15. $C \cap B$

16. $A \cap C'$ 17. $B \cap C'$ 18. $B' \cup A'$

19. $B' \cap A'$ 20. $B' \cap A$ 21. $A' \cap B$

22. $(C \cap B)'$ 23. $(C' \cap A)'$ 24. $(A' \cap B)'$

Using Venn diagrams if necessary, complete each of the following equations.

25. $(A')' =$ 26. $A \cap A' =$ 27. $A \cap U =$ 28. $A \cup A' =$

29. $A \cup U =$ 30. $A \cap \emptyset =$ 31. $A \cup \emptyset =$ 32. $A \cap A =$

33. $\emptyset' \cap \emptyset =$ 34. $A \cup A =$ 35. $\emptyset \cup \emptyset =$ 36. $U \cap U =$

37. Under what conditions would each of the following statements be true?

  a. $A \cup B = \emptyset$ b. $A \cup \emptyset = \emptyset$ c. $A \cap U = U$

  d. $A \cup B = A$ e. $A \cup \emptyset = U$ f. $A' \cap U = U$

  g. $A \cap B = A$ h. $A' \cup \emptyset = \emptyset$ i. $A \cup B = A \cap B$

1.3

The field postulates

You should already be somewhat familiar with the seven sets of numbers described below.

1. The set N of **natural numbers**, whose elements are the counting numbers:

$$N = \{1, 2, 3, \ldots\}.$$

2. The set J of **integers**, whose elements are the counting numbers, their negatives,
and zero:

$$J = \{\ldots, -2, -1, 0, 1, 2, \ldots\}.$$

3. The set Q of **rational numbers**, whose elements are all those numbers that can
be represented as the quotient of two integers $\dfrac{a}{b}$ (or a/b, or $a \div b$), where b is
not 0. Among the elements of Q are such numbers as $-3/4$, $18/27$, $3/1$, and
$-6/1$. In symbols,

$$Q = \left\{ x \,\middle|\, x = \frac{a}{b}, \quad a, b \in J, b \neq 0 \right\}.$$

Included in the set of rational numbers are numbers with terminating or repeating decimal representations.

4. The set H of **irrational numbers**, whose elements are the numbers with decimal representations that are nonterminating and nonrepeating. Among the elements of this set are such numbers as $\sqrt{2}$, π, and $-\sqrt{7}$. An irrational number cannot be represented in the form a/b, where a and b are integers. In symbols,

$$H = \{\text{irrational numbers}\}.$$

5. The set R of **real numbers**, which is the union of the set of all rational numbers and the set of all irrational numbers:

$$R = \{x \mid x \in (Q \cup H)\}.$$

6. The set I of **imaginary numbers**, whose members can be represented in the form $x + yi$, where x and y are real numbers, $y \neq 0$, and $i^2 = -1$:

$$I = \{x + yi \mid x, y \in R, \ y \neq 0, \ i^2 = -1\}.$$

If $x = 0$, then the imaginary number $x + yi$ is written yi, with $y \in R$, $y \neq 0$. Such a number is called a **pure imaginary number**.

7. The set C of **complex numbers**, whose members can be represented in the form $x + yi$, where x and y are real numbers and $i^2 = -1$:

$$C = \{x + yi \mid x, y \in R, \ i^2 = -1\}.$$

Relations between sets of numbers   The foregoing sets of numbers are related as indicated in Figure 1.5. We can characterize each set by stating properties that we assume it to have. In mathematics, when we make

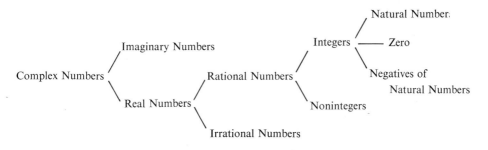

Figure 1.5

formal assumptions about members of a set or about their properties, we call the assumptions **axioms**, or **postulates**. The words **property**, **law**, and **principle** are sometimes used in referring to assumptions, although these words may also be applied to certain consequences thereof.

The first assumptions to be considered here have to do with *equality*. An **equality**, or an " is equal to " assertion, is simply a mathematical statement that two symbols or words, or two groups of symbols or words, are names for the same thing.

Equality
postulates

We postulate that, for any members a, b, and c of any set S, the equality ($=$) relationship satisfies the following laws:

E-1 $a = a$.   *Reflexive law for equality.*
E-2   If $a = b$, then $b = a$.   *Symmetric law for equality.*
E-3   If $a = b$ and $b = c$, then $a = c$.   *Transitive law for equality.*
E-4   If $a = b$, then a may be replaced by b *Substitution law for equality.*
      and b by a in any mathematical state-
      ment without altering the truth or
      falsity of the statement.†

The equality axioms indicate how we use symbols and warn that we must neither change the meaning of a symbol in the middle of a discussion nor use the same symbol for two different things in the same context. Other axioms, somewhat different conceptually, are used to characterize mathematical systems. They are assumptions made about the behavior of elements of sets under **binary operations.**

Definition 1.10   *A binary operation in a set A is an operation that assigns to each pair a and b of elements of A, taken in a definite order (a first and b second), a corresponding element of A.*

Addition and
multiplication
postulates

Since the set R of real numbers is the set of greatest interest to us at the moment, we postulate the basic laws for the elements of R in relation to the binary operations of addition and multiplication. You are probably familiar with these laws from your earlier work. They are listed on page 11 for reference. Parentheses are used in stating some of the relations to indicate that symbols within parentheses are to be viewed as representing a single entity.

To give meaning to expressions $a + b + c$, $a \times b \times c$, $a + b + c + d$, and so on, let us make the following agreement.

Definition 1.11   *If $a, b, c, d, \ldots \in R$, then*

$$a + b + c = a + (b + c), \qquad a + b + c + d = a + (b + c + d), \ldots,$$

and

$$a \times b \times c = a \times (b \times c), \qquad a \times b \times c \times d = a \times (b \times c \times d), \ldots.$$

---

† We have adopted a very powerful postulate in E-4, one that subsumes E-2 and E-3 as special cases. By wording E-4 as we have, we can eliminate a great deal of fussiness in later arguments, and, because E-2 and E-3 are fundamental properties of the equality relation, we have elected to retain them as postulates.

Field postulates of the set R of real numbers

Let a, b, c be arbitrary elements of R.

F-1 $a + b$ is a unique element of R.   *Closure law for addition.*

F-2 $(a + b) + c = a + (b + c)$.   *Associative law for addition.*

F-3   There exists an element $0 \in R$ (called   *Additive identity law.*
the **identity element for addition**) with
the property

$$a + 0 = a \quad \text{and} \quad 0 + a = a$$

for each $a \in R$.

F-4   For each $a \in R$, there exists an   *Additive-inverse law.*
element $^-a \in R$ (called the **additive
inverse**, or **negative**, of a) with the
property

$$a + (^-a) = 0 \quad \text{and} \quad (^-a) + a = 0.$$

F-5 $a + b = b + a$.   *Commutative law for addition.*

F-6 $a \times b$ is a unique element of R.   *Closure law for multiplication.*

F-7 $(a \times b) \times c = a \times (b \times c)$.   *Associative law for multiplication.*

F-8 $a \times (b + c) = a \times b + a \times c$ and   *Distributive law.*
$(b + c) \times a = b \times a + c \times a$.

F-9   There exists an element $1 \in R$ (called   *Multiplicative-identity law.*
the **identity element for multiplication**),
$1 \neq 0$, with the property

$$a \times 1 = a \quad \text{and} \quad 1 \times a = a$$

for each $a \in R$.

F-10 $a \times b = b \times a$.   *Commutative law for multiplication.*

F-11 For each element $a \in R$, $a \neq 0$, there   *Multiplicative-inverse law.*
exists an element $a^{-1} \in R$ (called the
multiplicative inverse, or **reciprocal**,
of a) with the property

$$a \times (a^{-1}) = 1 \quad \text{and} \quad (a^{-1}) \times a = 1.$$

If, for a given set and a given pair of binary operations (not necessarily the set of real numbers or ordinary addition and multiplication) Postulates F-1 through F-11 are satisfied by the elements of the set, then the set is called a **field** under these operations. Thus, we speak of the **field** R **of real numbers**.

Inverse operations In terms of the operations of addition and multiplication, let us now define two new operations on real numbers.

Definition 1.12 *The difference of elements $a \in R$ and $b \in R$, denoted by $a - b$, is given by*

$$a - b = a + (^-b).$$

The operation of finding a difference is called **subtraction**. This definition explains why we ordinarily write $-b$ for ^-b (the negative of b).

Definition 1.13 *The quotient of elements $a \in R$ and $b \in R$, $b \neq 0$, denoted by $\dfrac{a}{b}$, a/b, or $a \div b$, is given by*

$$\frac{a}{b} = a \times (b^{-1}).$$

The operation of finding a quotient is called **division**. In particular, for $a = 1$ we have, by the substitution law for equality,

$$\frac{1}{b} = 1 \times b^{-1},$$

from which we obtain

$$\frac{1}{b} = b^{-1}.$$

In all that follows, we shall ordinarily write $-a$ for the additive inverse of a and $\dfrac{1}{b}$ or $1/b$ for the multiplicative inverse of b. We shall also follow the customary practice of writing ab or $a \cdot b$ for $a \times b$.

Exercise 1.3

Let $N = \{natural\ numbers\}$, $J = \{integers\}$, $Q = \{rational\ numbers\}$, $H = \{irrational\ numbers\}$, $R = \{real\ numbers\}$, $I = \{imaginary\ numbers\}$, and $C = \{complex\ numbers\}$. State whether each statement is true or false.

1. $-5 \in N$ 2. $0 \in Q$ 3. $\sqrt{3} \in R$

4. $\pi \in C$ 5. $-3 \in R$ 6. $-7 \in H$

7. $\{-1, 1\} \subset N$ 8. $\{-1, 1\} \subset J$ 9. $\{-1, 1\} \subset Q$

10. $\{-1, 1\} \subset H$ 11. $\{-1, 1\} \subset R$ 12. $\{-1, 1\} \subset C$

Represent each set, listing members if finite, and using ellipsis dots, $\ldots$, if infinite.

Examples   a. {natural numbers between 5 and 9}   b. {integers greater than 3}

Solutions   a. {6, 7, 8}   b. {4, 5, 6, $\ldots$}

13. {first five natural numbers}   14. {integers between -4 and 5}

15. {natural numbers less than 7}   16. {integers greater than -5}

17. {integers between -10 and -5}   18. {nonnegative integers}

Use set-builder notation, $\{x \mid$ condition on $x\}$, to represent each set.

Example   {natural numbers greater than 12}

Solution $\{x \mid x \in N$ and x is greater than 12$\}$

19. {natural numbers}   20. {integers}

21. {real numbers}   22. {integers less than -6}

23. {real numbers between -4 and 3}   24. {nonnegative real numbers}

25. Let $A = \{4, -2, 2/5, \sqrt{-7}, 0, -3/4, \sqrt{2}, \sqrt{7}, \sqrt{-1}\}$. Designate each of the following sets by using braces and listing the members.

   a. {natural numbers in A}   b. {integers in A}

   c. {rational numbers in A}   d. {real numbers in A}

26. Let $B = \{-6, 3, \sqrt{5}, -3/4, \sqrt{-2}, 0, 5, -1, \sqrt{3}\}$. Designate each set by using braces and listing the members.

   a. {natural numbers in B}   b. {integers in B}

   c. {irrational numbers in B}   d. {real numbers in B}

Each variable in Problems 27–50 denotes a real number. Each of the statements 27–32 is an application of one of the postulates E-1 through E-4. Justify each statement by citing an appropriate postulate. (There may be more than one correct justification.)

Example   If $3 = a$ then $a = 3$.

Solution   Symmetric law for equality

27. If $a + 3 = b$ and $b = 7$, then $a + 3 = 7$.   28. If $x = 5$ and $y = x + 2$, then $y = 5 + 2$.

29. If $2 + y = 6$, then $6 = 2 + y$.   30. If $a = 2c$ and $c = 6$, then $a = 2 \cdot 6$.

31. If $y = 7 + x$ and $7 + x = z$, then $y = z$.   32. $x + 8 = x + 8$.

Each of the statements 33–50 is an application of one of the Postulates F-1 through F-11. Justify each statement by citing the appropriate postulate.

Example $2(3 + 1) = 2 \cdot 3 + 2 \cdot 1$

Solution   Distributive law

33. ab is a real number

34. $7 + 0 = 7$

35. $(2 \cdot 3) \cdot 4 = 2 \cdot (3 \cdot 4)$

36. $(5 + 4) + 1 = (4 + 5) + 1$

37. $3 + (-3) = 0$

38. $5 + (-2) = (-2) + 5$

39. $a\left(\dfrac{1}{a}\right) = \left(\dfrac{1}{a}\right)a \quad (a \neq 0)$

40. $\dfrac{1}{c}(a + b) = \dfrac{1}{c} \cdot a + \dfrac{1}{c} \cdot b \quad (c \neq 0)$

41. $(a + b) + c = c + (a + b)$

42. $(a + b) + c = (b + a) + c$

43. $a + (b + c)d = a + bd + cd$

44. $a + (b + c)d = a + d(b + c)$

45. $a + (b + c)d = (b + c)d + a$

46. $(a + b) + [-(a + b)] = 0$

47. $a(b + c) = (b + c)a$

48. $a[b + (c + d)] = ab + a(c + d)$

49. $ab + a(c + d) = ab + ac + ad$

50. $ab + ac = ba + ac$

Which of the sets are closed under the stated operation?

Example   {integers}, division

Solution   Not closed. The set is not closed, because the quotient of two integers is not always an integer; for example, the quotient $-2/3 \notin J$.

51. {even natural numbers}, addition

52. {odd natural numbers}, addition

53. {odd natural numbers}, multiplication

54. {even natural numbers}, multiplication

55. {0, 1}, multiplication   56. {0, 1}, addition

57. {natural numbers}, division   58. {natural numbers}, subtraction

1.4
Field properties

The field postulates together with the postulates for equality imply other properties of the real numbers. Such implications are generally stated as **theorems**. A theorem is simply an assertion of a fact that follows logically from the postulates (or axioms) and other theorems. We shall list some of the theorems ordinarily encountered in lower-level algebra courses. We have numbered these theorems to provide an efficient way to refer to them later and have also named those that have commonly accepted names.

Addition law for equality   First, consider the following result, which reaffirms the uniqueness of the sum of two real numbers.

Theorem 1.1   *If a, b, $c \in R$ and $a = b$, then*

$$a + c = b + c \quad and \quad c + a = c + b.$$

This result follows directly from the reflexive and substitution laws of equality. Since $a = b$, and, by the reflexive law,

$$a + c = a + c \quad and \quad c + a = c + a,$$

we can substitute b for a in the right-hand member of each of the latter equations to obtain

$$a + c = b + c \quad and \quad c + a = c + b.$$

Both of these equations were included in the conclusion of Theorem 1.1, because applications to further results are sometimes immediate from one form and sometimes from the other.

Multiplication law for equality   A theorem closely analogous to Theorem 1.1 can be stated as follows.

Theorem 1.2   *If a, b, $c \in R$ and $a = b$, then*

$$ac = bc \quad and \quad ca = cb.$$

The proof of Theorem 1.2 exactly parallels that of Theorem 1.1, and is left as an exercise.

Additional
properties

The next theorem asserts that the additive inverse of a real
number is unique—that is, that a given real number has only
one additive inverse.

Theorem 1.3 *If $a, b \in R$ and $a + b = 0$, then*

$$b = -a \quad and \quad a = -b.$$

To prove this, we can begin by observing that if $a + b = 0$, then, by Theorem 1.1,

$$(-a) + (a + b) = (-a) + 0.$$

Next, the associative law of addition and the additive-inverse law imply

$$[(-a) + a] + b = (-a) + 0 \quad and \quad 0 + b = -a + 0.$$

Finally, by the additive-identity law, we have

$$b = -a.$$

A similar argument establishes also that $a = -b$.

In the last step of the foregoing proof, we did not state that $0 + b = b$ and
$(-a) + 0 = -a$ and then argue that $b = -a$ by the substitution law; this amount
of detail is superfluous. The degree of rigorous detail desirable in proofs of this
kind is purely relative. If a line of argument is clear and valid, most reasonable
persons will not insist on dotting all i's and crossing all t's. What is important is
that you understand the argument.

A result analogous to Theorem 1.3 holds for the multiplicative inverse and is left
as an exercise, as are the proofs of the succeeding four theorems, which follow
directly from the previous results.

Theorem 1.4 *If $a, b, c \in R$ and $a + c = b + c$, then $a = b$.*

Theorem 1.5 *If $a, b, c \in R$, $c \neq 0$, and $ac = bc$, then $a = b$.*

Theorem 1.6 *For every $a \in R$, $a \cdot 0 = 0$.*

Theorem 1.7 *If $a, b \in R$ and $a \cdot b = 0$, then either $a = 0$ or $b = 0$ or both.*

Combining Theorems 1.6 and 1.7, we see that for $a, b \in R$ we have $ab = 0$ *if and
only if* at least one of the factors is 0.

The following theorem concerns the familiar " laws of signs " for operating with
real numbers.

Theorem 1.8 *If a, b ∈ R, then*

$$\text{I} \quad -(-a) = a, \qquad\qquad \text{II} \quad (-a) + (-b) = -(a+b),$$

$$\text{III} \quad (-a)(b) = -(ab), \qquad\qquad \text{IV} \quad (-a)(-b) = ab,$$

$$\text{V} \quad \frac{-a}{b} = \frac{a}{-b} = -\frac{a}{b} \quad (b \neq 0), \qquad \text{VI} \quad \frac{-a}{-b} = \frac{a}{b} \quad (b \neq 0).$$

Part I of this theorem is an immediate consequence of the uniqueness of the additive inverse. That is, the symbol "$-(-a)$" denotes the additive inverse of $-a$, as does the symbol "a." Hence, $-(-a) = a$.

The proofs of the remaining parts of Theorem 1.8 are left as exercises, as are the proofs of Theorems 1.9 to 1.11, which are concerned with the familiar properties of quotients.

Theorem 1.9 *If a, b, c ∈ R, then*

$$\frac{a}{b} = \frac{c}{d} \quad (b, d \neq 0) \quad \textit{if and only if} \quad ad = bc.$$

As a direct consequence of this characterization of equal quotients, we have a theorem that is sometimes referred to as the *fundamental principle of fractions*.

Theorem 1.10 *If a, b, c ∈ R, then*

$$\frac{ac}{bc} = \frac{a}{b} \quad (b, c \neq 0).$$

Next, let us group a number of assertions about quotients into a single theorem.

Theorem 1.11 *If a, b, c, d ∈ R, then*

$$\text{I} \quad \frac{1}{a} \cdot \frac{1}{b} = \frac{1}{ab} \quad (a, b \neq 0), \qquad \text{II} \quad \frac{a}{b} \cdot \frac{c}{d} = \frac{ac}{bd} \quad (b, d \neq 0),$$

$$\text{III} \quad \frac{a}{c} + \frac{b}{c} = \frac{a+b}{c} \quad (c \neq 0), \qquad \text{IV} \quad \frac{a}{b} + \frac{c}{d} = \frac{ad+bc}{bd} \quad (b, d \neq 0),$$

$$\text{V} \quad \frac{a}{b} - \frac{c}{d} = \frac{ad-bc}{bd} \quad (b, d \neq 0), \qquad \text{VI} \quad \frac{1}{\dfrac{a}{b}} = \frac{b}{a} \quad (a, b \neq 0),$$

$$\text{VII} \quad \frac{\dfrac{a}{b}}{\dfrac{c}{d}} = \frac{ad}{bc} \quad (b, c, d \neq 0).$$

Exercise 1.4

In Problems 1–20, each statement is justified by one part of Theorems 1.1–1.11. Cite an appropriate justification. All variables denote elements of the set R of real numbers.

Example   If $p + q + 3 = 4 + 3$, then $p + q = 4$.

Solution   Theorem 1.4

1. If $x = 3$, then $x + 9 = 3 + 9$.

2. $-2 - q = -(2 + q)$

3. $\dfrac{-2}{5} = -\dfrac{2}{5}$

4. $\dfrac{4(x + y)}{6} = \dfrac{2(x + y)}{3}$

5. $\dfrac{x}{3} + \dfrac{y + z}{3} = \dfrac{x + y + z}{3}$

6. $\dfrac{p}{3} \cdot \dfrac{q}{4} = \dfrac{pq}{12}$

7. If $4p = 0$, then $p = 0$.

8. $(-5)(-6) = 5 \cdot 6$

9. If $3x = 2y$, then $\dfrac{3}{2} = \dfrac{y}{x}$.

10. $\dfrac{-x}{-3} = \dfrac{x}{3}$

11. $\dfrac{\frac{2}{5}}{\frac{3}{7}} = \dfrac{2 \cdot 7}{5 \cdot 3}$

12. $\dfrac{\frac{1}{2}}{\frac{3}{2}} = \dfrac{3}{2}$

13. $(-3)(p) = -(3p)$

14. $x - (-y) = x + y$

15. $\dfrac{x}{3} + \dfrac{y}{2} = \dfrac{2x + 3y}{3 \cdot 2}$

16. $\dfrac{x}{-3} = -\dfrac{x}{3}$

17. If $p + q = 7$, then $4(p + q) = 4 \cdot 7$.

18. If $(r + s) + 7 = 0$, then $r + s = -7$.

19. If $6(x - y) = 3z$, then $2(x - y) = z$.

20. $\dfrac{x + 1}{4} - \dfrac{y + 3}{3} = \dfrac{3(x + 1) - 4(y + 3)}{4 \cdot 3}$

Prove each of the following statements. All variables denote elements of the set R of real numbers.

21. If $a = b$, then $ac = bc$ and $ca = cb$.

22. $\dfrac{1}{a}$ $(a \neq 0)$ is unique; that is, if $a \cdot b = 1$, then $b = \dfrac{1}{a}$.

23. 0 is unique; that is, if $a + b = a$, then $b = 0$.

24. 1 is unique; that is, if $a \neq 0$ and $a \cdot b = a$, then $b = 1$.

25. If $a + c = b + c$, then $a = b$.

26. If $ac = bc$ and $c \neq 0$, then $a = b$.

27. If $a \in R$, then $a \cdot 0 = 0$.

28. If $a \cdot b = 0$, then either $a = 0$ or $b = 0$ or both.

29. $(-a) + (-b) = -(a + b)$

30. $(-a)(b) = -(ab)$

31. $(-a)(-b) = ab$

32. $\dfrac{-a}{b} = \dfrac{a}{-b} = -\dfrac{a}{b}$ $(b \neq 0)$

33. $\dfrac{-a}{-b} = \dfrac{a}{b}$ $(b \neq 0)$

34. $\dfrac{a}{b} = \dfrac{c}{d}$ $(b, d \neq 0)$ if and only if $ad = bc$.

35. $\dfrac{ac}{bc} = \dfrac{a}{b}$ $(b, c \neq 0)$

36. $\dfrac{1}{a} \cdot \dfrac{1}{b} = \dfrac{1}{ab}$ $(a, b \neq 0)$

37. $\dfrac{a}{b} \cdot \dfrac{c}{d} = \dfrac{ac}{bd}$ $(b, d \neq 0)$

38. $\dfrac{a}{c} + \dfrac{b}{c} = \dfrac{a + b}{c}$ $(c \neq 0)$

39. $\dfrac{a}{b} + \dfrac{c}{d} = \dfrac{ad + bc}{bd}$ $(b, d \neq 0)$

40. $\dfrac{a}{b} - \dfrac{c}{d} = \dfrac{ad - bc}{bd}$ $(b, d \neq 0)$

41. $\dfrac{\dfrac{1}{a}}{\dfrac{b}{a}} = \dfrac{b}{a}$ $(a, b \neq 0)$

42. $\dfrac{\dfrac{a}{b}}{\dfrac{c}{d}} = \dfrac{ad}{bc}$ $(b, c, d \neq 0)$

43. $\dfrac{a}{b} = q$ if and only if $a = bq$ $(b \neq 0)$.

44. Derive the second equation,

$$(b + c)a = ba + ca,$$

in Field Postulate F-8 from the other postulates and the first equation,

$$a(b + c) = ab + ac.$$

1.5

Order and completeness

As you probably recall from earlier courses, there is a one-to-one correspondence between the real numbers and the points on a geometric line (to each real number there corresponds one and only one point on the line, and vice versa). To illustrate this, we imagine the line scaled in convenient units, with the positive direction (from 0 toward 1) denoted by an arrowhead. The line is then called a **number line**;

the real number corresponding to a point on the line is called the **coordinate** of the point, and the point is called the **graph** of the number. For example, a number-line representation of {1, 3, 5} is shown in Figure 1.6.

Figure 1.6

A horizontal number-line graph directed to the right can be used to illustrate the separation of the real numbers into three disjoint subsets: {negative real numbers}, {0}, {positive real numbers}. The point associated with 0 is called the **origin**. The set of numbers whose elements are associated with the points on the right-hand side of the origin belong to the set R_+ of **positive real numbers**, and the set whose elements are associated with the points on the other side belong to the set R_- of **negative real numbers**.

Notice that the word "negative" has now been used in two ways. In one case, we refer to the *negative of a number*, as in Postulate F-4, whereas, in the other, we refer to a *negative number*, which is the negative of a positive number.

Order postulates It is possible to categorize the set of positive real numbers without recourse to geometric considerations, though of course we shall continue to find it convenient to refer also to the number line. With this in mind, let us state two more postulates that apply to real numbers.

O-1 If a is a real number, then exactly one   *Trichotomy law.*
of the following statements is true:
a is positive, a is zero, or $-a$ is positive.

O-2 If a and b are positive real numbers,   *Closure law for positive numbers.*
then $a + b$ is positive and ab is positive.

The first of these postulates asserts that every real number belongs to one of the sets R_+, {0}, or R_-, but to only one of them. The second asserts that the set R_+ of positive real numbers is closed with respect to the binary operations of addition and multiplication.

Since the set R of real numbers satisfies Postulates O-1 and O-2 as well as Postulates F-1 through F-11, we say that R is an **ordered field**. Similarly, the set Q of rational numbers is an ordered field. But, as we shall see in Chapter 11, the set C of complex numbers is not an ordered field even though it is a field.

By Postulate O-2, if a and b are positive real numbers, then ab is a positive real number. This fact, together with Parts III and IV of Theorem 1.8, is sufficient to establish that the product of a positive real number and a negative real number is a negative real number, while the product of two negative real numbers is a positive real number.

Less than and greater than   The addition of a positive real number d to a real number a can be visualized on a number-line graph as the process of locating the point corresponding to a on the line and then moving along the line d units to the right to arrive at the point corresponding to $a + d$ (Figure 1.7). With this idea in mind, we define what is meant by "less than."

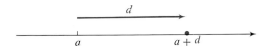

Figure 1.7

Definition 1.14   *If $a, b \in R$, then a is less than b if and only if there exists a positive real number d such that $a + d = b$.*

Since d is positive, this implies for any two real numbers a and b, that if the graph of a lies to the left of the graph of b, then a is less than b. The inequality symbol $<$ is used to denote the phrase "is less than," and $a < b$ is read "a is less than b." The inequality symbol $>$ means "is greater than." The statements $a < b$ and $b > a$ are taken as equivalent.

Definition 1.14 and the field postulates have the following implications, which we shall not prove.

Theorem 1.12   *For any $a, b, c \in R$:*

I   *If $a < b$ and $b < c$, then $a < c$.*   II   *If $a < b$, then $a + c < b + c$.*

III   *If $a < b$ and $c > 0$, then $ac < bc$.*   IV   *If $a < b$ and $c < 0$, then $ac > bc$.*

As you can see, Theorem 1.12-I is comparable to the transitive law for equality. That is, we can say that "less than" is a transitive relationship.

Each of the symbols $\leq$ and $\geq$ (read "is less than or equal to" and "is greater than or equal to," respectively) is a contraction for two symbols, one of equality and one of inequality, connected by the word "or." For example, $x \leq 7$ is the statement that x is less than 7 *or* x equals 7.

We often write together two inequalities that express a transitive relationship. Thus $a < b$ and $b < c$ are written together as $a < b < c$ and read "a is less than b and b is less than c." Similarly, $0 < x$ and $x \leq 1$ are written together as $0 < x \leq 1$.

Absolute value   The graphs of the numbers a and $-a$ on a number line lie the same distance from the origin, but on opposite sides of it. If we wish to refer to the *distance* of the graph of a number from the origin, and not to the side of the origin on which it is located, then we use the term **absolute value**. Thus, the absolute value of a and the absolute value of $-a$ are the same nonnegative number. The symbol $|a|$ is used to denote the absolute value of a. We formalize the definition as follows:

Definition 1.15   *If $a \in R$, then the absolute value of a is given by*

$$|a| = \begin{cases} a, & if\ a \geq 0, \\ -a, & if\ a < 0. \end{cases}$$

For example, $|-3| = -(-3) = 3$, $|7| = 7$,   and $|0| = 0$.

Completeness   On page 19 we mentioned in passing a property of the set R
postulate   of real numbers that is of fundamental importance and necessary for establishing the existence of irrational numbers. Let us formalize this property by assuming the following **completeness property**.

O-3   There is a one-to-one correspondence between the set of real numbers and the set of points on a geometric line.

Although this property might be asserted more precisely, for our purposes this formulation is quite sufficient. Because of the completeness property, the set R of real numbers is said to be a **complete ordered field**. Note, however, that while the set Q of rational numbers is an ordered field, and between any two rational numbers there are infinitely many other rational numbers, Q is not a complete ordered field—because between any two rational numbers there are also infinitely many irrational numbers; and the set of points corresponding to Q does not completely "fill" the number line.

Exercise 1.5

For $x, y \in R$, justify each statement by citing one part of Theorem 1.12.

Example   If $2 < 3x$, then $-4 > -6x$.

Solution   Part IV. Each member of $2 < 3x$ is multiplied by -2.

1. If $x < 3$ and $y < x$, then $y < 3$.   2. If $x + 1 < 0$, then $x < -1$.

3. If $y < 8$, then $3y < 24$.   4. If $y < 4$, then $y + 2 < 6$.

5. If $x < 7$, then $x - 2 < 5$.   6. If $x < 9$, then $-2x > -18$.

7. If $-6x < 12$, then $x > -2$.   8. If $x - 3 < 5$, then $x < 8$.

Express each statement by means of symbols.

Examples   a. 5 is not greater than 7.   b. x is between 5 and 8.

Solutions   a. $5 \ngtr 7$, or $5 \leq 7$ b. $5 < x < 8$

9. 7 is greater than 3.   10. 2 is less than 5.

11. -4 is less than -3.   12. -4 is greater than -7.

13. x is between -1 and 1, inclusive.   14. x is negative.

15. x is positive.   16. x is nonpositive.

17. x is nonnegative.   18. $2x$ is less than or equal to 8.

For $x, y \in R$, rewrite each of the following expressions without using absolute-value notation.

Examples   a. $|-11|$ b. $|x - 6|$

Solutions   a. 11   b. $x - 6$, if $x - 6 \geq 0$;
                        $-(x - 6)$, if $x - 6 < 0$.

19. $|-3|$ 20. $|-5|$ 21. $|7|$ 22. $|4|$

23. $-|-2|$ 24. $-|-5|$ 25. $|3x|$ 26. $|2y|$

27. $|x + 1|$ 28. $|x + 2|$ 29. $|y - 3|$ 30. $|y - 4|$

Replace the comma with an appropriate order symbol to form a true statement.

31. $|-2|, |-5|$ 32. $|3|, |-4|$ 33. $-7, |-1|$

34. $5, |-2|$ 35. $|-3|, 0$ 36. $-|-4|, 0$

For $x, y \in R$, write an equivalent relation without using the negation symbol, $/$.

37. $2 \ngtr 5$ 38. $-1 \nleq -2$ 39. $7 \ngeq 8$

40. $|x| \nless 3$ 41. $|x| \ngeq 3$ 42. $x \ngtr |y|$

2

Algebraic expressions

The fundamental algebraic ideas reviewed in this chapter serve as a basis for the remainder of the course.

2.1

Definitions; sums of polynomials

Any grouping of constants and variables generated by applying a finite number of the elementary operations—addition, subtraction, multiplication, division, or the extraction of roots—is called an **algebraic expression**. For example,

$$\frac{3x^2 + \sqrt{2x - 1}}{3} \quad \text{and} \quad xy + 3x^2z - \sqrt[5]{z}$$

are algebraic expressions.

If two algebraic expressions have equal values for all values of the variable(s) for which both expressions are defined, then we say that on the set of these values of the variable(s) the expressions are **equivalent.**

You should recall that an expression of the form x^n is called a **power** of x, where x is the **base** of the power and n is the **exponent** of the power.

Definition 2.1 *If $n \in N$ and $x \in R$, then*

$$x^n = x \cdot x \cdot x \cdots x \quad (n \text{ factors}).$$

In any algebraic expression of the form $A + B + C + \cdots$, where $A, B, C, \ldots$ are algebraic expressions, $A, B, C, \ldots$ are called **terms** of the expression. If an algebraic expression contains no variable in a denominator and contains only positive integral powers of each variable, then the expression is a **polynomial.** For example,

$$5x, \quad \frac{3x^2}{2} - \frac{7x}{2}, \quad 0, \quad 2x^2 - 3x + 4, \quad \text{and} \quad \frac{x}{4} - \frac{\sqrt{7}}{4}$$

are polynomials, whereas

$$\frac{3}{x}, \quad 3 + \sqrt{x}, \quad \text{and} \quad \frac{2x - 1}{2x + 1}$$

are algebraic expressions, but not polynomials, in the variable x.

Polynomials consisting of one, two, or three terms are also called **monomials**, **binomials**, and **trinomials**, respectively. Thus $3x^2y$ is a monomial, $x + 4x^2$ is a binomial, and $x + y + z$ is a trinomial. In a monomial, the constant factor is called the **coefficient** of the remaining factors. In a monomial such as x^2, where no constant factor appears, we shall agree that the coefficient is 1.

The **degree** of a monomial is given by the exponent of the variable in the monomial. Thus, 5 is of degree zero, $2x$ is of first degree, and $3x^4$ is of fourth degree; but no degree is assigned to the special monomial 0. If a monomial contains more than one variable, its degree is given by the sum of the exponents on the variables; $3x^2y^3z$ is of sixth degree in x, y, and z. It can also be described as being of second degree in x, third degree in y, fifth degree in x and y, and so on. The **degree of a polynomial** is the same as the degree of its term of greatest degree. Since no degree is assigned to the monomial 0, no degree is assigned to the zero polynomial 0, either.

Fundamental operations on polynomials   Because $a - b$ is defined to be $a + (-b)$, we shall view the signs in any polynomial with numerical coefficients as signs denoting positive or negative coefficients, and the operation involved to be addition. Thus

$$3x - 5x + 4x = (3x) + (-5x) + (4x),$$

and its terms are $3x$, $-5x$, and $4x$. Also, as regards multiplication and division, since

$$\frac{a}{b} = a\left(\frac{1}{b}\right),$$

we can view division by a constant as multiplication by its reciprocal (multiplicative inverse), and, for example, write

$$\frac{3x^2}{4} + \frac{x}{2} \quad \text{as} \quad \frac{3}{4}x^2 + \frac{1}{2}x.$$

Accordingly, since a polynomial can be considered to involve only the operations of addition and multiplication, and since the set R of real numbers is closed with respect to these operations, it follows that, for any specific real value of x, a polynomial with real coefficients represents a real number. Therefore, the postulates for the real numbers are applicable to the terms in such polynomials and to the polynomials themselves.

Simplification of polynomials
By applying the commutative, associative, and distributive laws in various ways, we can frequently rewrite polynomials and their sums in what might be termed "simpler" form.

Example

$$2x^2 + 3x + 5 + 2x + 6x^2 + 7 = 2x^2 + 6x^2 + 3x + 2x + 5 + 7$$
$$= (2 + 6)x^2 + (3 + 2)x + (5 + 7)$$
$$= 8x^2 + 5x + 12$$

Here we have reduced the number of terms from six to three.

Representation of polynomials
We shall often be concerned with polynomials in one variable. A polynomial of degree n, $n \geq 0$, in x can be represented—when its terms are rearranged, if need be—by an expression of the form

$$a_0 x^n + a_1 x^{n-1} + a_2 x^{n-2} + \cdots + a_{n-1}x + a_n \qquad (a_0 \neq 0),$$

where it is understood that the a's are the (constant) coefficients of the powers of x in the polynomial.

The term $a_0 x^n$ is called the **leading term**, and the coefficient a_0 is called the **leading coefficient**, in the polynomial. It is often convenient to have the leading term of the form x^n. In this case (that is, when $a_0 = 1$), the polynomial is said to be **monic**.

If the coefficients in a polynomial are real numbers, then the polynomial is called a **polynomial over the real-number field**, or simply a **polynomial over** R. If the variable x is restricted to represent only real numbers, then the polynomial is said to be a **polynomial in the real variable** x. If both the coefficients and the variable are restricted to real values, we say that the polynomial is a **real polynomial**. In this chapter we shall deal only with real polynomials.

Symbols for polynomials
Polynomials are frequently represented by symbols such as

$$P(x), \quad D(y), \quad \text{and} \quad Q(z)$$

(read "P of x," "D of y," and "Q of z"), where the symbols in the parentheses designate the variables. Thus, we might write

$$P(x) = 2x^3 - 3x + 2,$$
$$D(y) = y^6 - 2y^2 + 3y - 2,$$
$$Q(z) = 8z^4 + 3z^3 - 2z^2 + z - 1.$$

The notation $P(x)$ can be used to denote values of the polynomial for specific values of x. Thus, $P(2)$ means the value of the polynomial $P(x)$ when x is replaced by 2. For example, if

$$P(x) = x^2 - 2x + 1,$$

then

$$P(2) = (2)^2 - 2(2) + 1 = 1$$

and

$$P(-4) = (-4)^2 - 2(-4) + 1 = 25.$$

In some applications, the notation $P(x)|_a^b$ denotes $P(b) - P(a)$. Thus if

$$P(x) = \frac{x^2}{2} - 4x,$$

$$P(x)\Big|_2^3 = P(3) - P(2) = \left[\frac{3^2}{2} - 4(3)\right] - \left[\frac{2^2}{2} - 4(2)\right] = -\frac{3}{2}.$$

Exercise 2.1

Give the degree of each polynomial. If the expression is not a polynomial, so state.

Example $x^3y^2 + y^4 + x$

Solution   Fifth degree in x and y; fourth degree in y; third degree in x.

1. $y^3 + 6y + 4$ 2. $x^2 - x$ 3. $x^3y - xy^2 + x^2$

4. $4 - \dfrac{2}{x^2}$ 5. $\dfrac{x^2 + 3}{x^3}$ 6. $x^4 - x^3y^2 - y^3$

Example   If $P(x) = 2x^2 - x + 3$, find $P(3)$, $P(-3)$, $P(0)$, and $P(a)$.

Solution $P(3) = 2(3)^2 - (3) + 3 = 18$

$P(-3) = 2(-3)^2 - (-3) + 3 = 24$

$P(0) = 2(0)^2 - (0) + 3 = 3$

$P(a) = 2a^2 - a + 3$

7. If $P(x) = x^3 - 3x^2 + x + 1$, find $P(2)$, $P(0)$, $P(x)|_0^2$, and $P(x)|_{-2}^2$.

8. If $P(x) = 2x^3 + x^2 - 3x + 4$, find $P(3)$, $P(-3)$, $P(0)$, $P(x)|_{-3}^0$, and $P(x)|_0^3$.

9. If $P(x) = x^{12}$, find $P(1)$, $P(-1)$, $P(0)$, $P(x)|_0^1$, and $P(x)|_{-1}^1$.

10. If $P(x) = x^{13}$, find $P(1)$, $P(-1)$, $P(0)$, $P(x)|_0^1$, and $P(x)|_{-1}^1$.

Example   If $P(x) = x - 4$ and $Q(x) = x + 2$, find $P(Q(2))$ and $P(Q(x))$.

Solution $Q(2) = 2 + 2 = 4$, so $P(Q(2)) = P(4) = 4 - 4 = 0$

$P(Q(x)) = (x + 2) - 4 = x - 2$

11. If $P(x) = x + 2$ and $Q(x) = x - 3$, find $P(Q(2))$ and $Q(P(2))$.

12. If $P(x) = 2x + 1$ and $Q(x) = \dfrac{1}{2}(x - 1)$, find $P(Q(x))$ and $Q(P(x))$.

13. If $P(x) = x^2 + 3$ and $Q(x) = 6$, find $P(Q(2))$ and $Q(P(2))$.

14. If $P(x) = 2x^2 - 3x$ and $Q(x) = x^2 + 1$, find $P(Q(0))$ and $Q(P(0))$.

Express as a polynomial   **a.** $P(x) + Q(x)$ and   **b.** $P(x) - Q(x)$.

15. $P(x) = 3x - 2$ and $Q(x) = 3 - x$

16. $P(x) = x^2 + 3x - 2$ and $Q(x) = 2x^2 + x - 2$

17. $P(x) = x^2 - 2x + 3$ and $Q(x) = 2x^2 - 2x - 1$

18. $P(x) = 2x^3 - 3x^2 + x - 1$ and $Q(x) = x^3 + 3x - 2$

Given that $P(x) = 2x^2 - 3x + 2$, $Q(x) = 3 - 2x + x^2$, *and* $S(x) = -2x^2 + 3x - 5$, *write each expression as an equivalent polynomial.*

19. $P(x) + Q(x)$ 20. $P(x) + [Q(x) - S(x)]$

21. $P(x) - [Q(x) + S(x)]$ 22. $P(x) - [Q(x) - S(x)]$

23. $[Q(x) - P(x)] - S(x)$ 24. $S(x) - [-P(x) - Q(x)]$

25. If $P(x)$ is of degree n and $Q(x)$ is of degree $n - 2$, what is the degree of $P(x) + Q(x)$? Of $P(x) - Q(x)$?

26. If $P(x)$ and $Q(x)$ are polynomials, with $P(0) = 4$ and $Q(0) = 3$, what is the value of $P(x) + Q(x)$ for $x = 0$? Of $P(x) - Q(x)$ for $x = 0$?

2.2
Products of polynomials

By Definition 2.1, for $x \in R$ and $n \in N$, we have

$$x^n = x \cdot x \cdot x \cdots x \quad (n \text{ factors})$$

Now, consider the product $x^m \cdot x^n$, where m and n are natural numbers. Since

$$x^m \cdot x^n = \underbrace{(x \cdot x \cdot x \cdots x)}_{m \text{ factors}}\underbrace{(x \cdot x \cdot x \cdots x)}_{n \text{ factors}},$$

it follows that

$$x^m \cdot x^n = \underbrace{x \cdot x \cdot x \cdots x}_{(m+n) \text{ factors}} = x^{m+n}.$$

In the same way, we have

$$(x^m)^n = \underbrace{\underbrace{(x \cdot x \cdot x \cdots x)}_{m \text{ factors}}\underbrace{(x \cdot x \cdot x \cdots x)}_{m \text{ factors}} \cdots \underbrace{(x \cdot x \cdot x \cdots x)}_{m \text{ factors}}}_{n \text{ factors}} = x^{mn}$$

and, for $y \in R$,

$$(xy)^n = \underbrace{(xy)(xy)(xy) \cdots (xy)}_{n \text{ factors}} = \underbrace{(x \cdot x \cdot x \cdots x)}_{n \text{ factors}}\underbrace{(y \cdot y \cdot y \cdots y)}_{n \text{ factors}} = x^n y^n.$$

We can state the above results formally:

Theorem 2.1 *If $x, y \in R$ and $m, n \in N$, then*

$$\text{I} \quad x^m x^n = x^{m+n}, \qquad \text{II} \quad (x^m)^n = x^{mn}, \qquad \text{III} \quad (xy)^n = x^n y^n.$$

Examples   a. $x^2 x^3 = x^5$ b. $(x^3)^5 = x^{15}$ c. $(xy)^4 = x^4 y^4$

Products of monomials   We can use the commutative and associative laws and Theorem 2.1 to rewrite an expression for the product of any two monomials.

Examples   a. $(3x^2 y)(2xy^2) = 6x^3 y^3$ b. $(-xy^3)(4xyz)(2yz) = -8x^2 y^5 z^2$

The **generalized distributive law**,

$$a(b_1 + b_2 + \cdots + b_n) = ab_1 + ab_2 + \cdots + ab_n,$$

can be applied to write as a polynomial the product of a monomial and a polynomial containing more than one term.

Example $3x(x + y + z) = 3xx + 3xy + 3xz = 3x^2 + 3xy + 3xz$

In its complete form, the generalized distributive law requires mathematical induction for its proof, but its validity as applied in our example here—or in any similar example—can readily be verified; thus we have

$$3x(x + y + z) = 3x[(x + y) + z] = 3x(x + y) + 3xz$$
$$= 3x^2 + 3xy + 3xz.$$

Binomial products   The distributive law can be applied successively to the product of polynomials containing more than one term.

Example $(3x + 2y)(x - y) = 3x(x - y) + 2y(x - y)$
$$= 3x^2 - 3xy + 2xy - 2y^2$$
$$= 3x^2 - xy - 2y^2$$

The following binomial products are types so frequently encountered that you should learn to recognize them on sight:

$$(x + a)(x + b) = x^2 + (a + b)x + ab,$$
$$(x + a)^2 = x^2 + 2ax + a^2,$$
$$(x + a)(x - a) = x^2 - a^2.$$

Exercise 2.2

Write each product in polynomial form in which constants and powers of each variable in each term are combined.

Examples   a. $(-2x^2)(3xy)(y^2)$ b. $2(x^2 - x - 1)$ c. $(x - 3)(x + 5)$

Solutions   a. $-6x^3y^3$ b. $2x^2 - 2x - 2$ c. $x^2 + 5x - 3x - 15$
$x^2 + 2x - 15$

1. $(-3x^2)(-2xy)(-y^3)$ 2. $(a^3)(-2ab^2)(-b^3)$

3. $-4x(2x^2 - x + 1)$ 4. $3y(2y^2 - 6y - 3)$

5. $abc(a - b + 2c)$ 6. $-ab(2a - b + 3c)$

7. $(x + 2)(x + 5)$ 8. $(x - 3)(x + 2)$

9. $(x - 2y)^2$ 10. $(2x - y)^2$

11. $(5x + 1)(2x + 3)$ 12. $(2x + 3)(x - 5)$

13. $(3a + 2b)(3a - 2b)$ 14. $(5x - y)(5x + y)$

15. $(x + 4)(x^2 + 2x - 1)$ 16. $(x - 2)(x^2 - x + 3)$

17. $2(x + 1)(x + 3)$ 18. $3(x - 1)(x + 2)$

19. $-(2a - b)(c - 3d)$ 20. $-(3a - b)(c + 3d)$

21. $a(a - b)(a^2 + ab + b^2)$ 22. $b(a + b)(a^2 - ab + b^2)$

23. $2\{a - [a - 2(a + 1) + 1] + 1\}$ 24. $-\{4 - [3 - 2(a - 1) + a] + a\}$

25. $2x\{x + 3[2(2x - 1) - x + 1] + 5\}$ 26. $-x\{4 - 2[1 - 2(x + 3)] - x\}$

27. If $P(x) = x^2 - 3x + 7$, find $P(x - 1)$ and $P(2 - x)$.

28. If $P(x) = x^2 + 2x + 1$, find $P(x + h)$ and $P(x - h)$.

29. If $P(x) = x^2 - 3x$, find $[P(-x)]^2$ and $[P(a^2)]^2 - P(a^2)$.

30. If $P(x) = 3 - x^2$, find $[P(3)]^2$ and $[P(3)]^2 - P(3^2)$.

31. Simplify the difference $(a + b)^2 - (a^2 + b^2)$. What are the conditions on a and b for $a^2 + b^2$ to be greater than $(a + b)^2$? For $a^2 + b^2$ to be less than $(a + b)^2$?

2.3

Factoring polynomials

Factors in a domain   What do we mean when we speak of *factoring* an integer or a polynomial? It is true, for example, that

$$. = 4\left(\frac{1}{2}\right),$$

but we would not ordinarily say that 4 and 1/2 are factors of 2. On the other hand, since

$$10 = (2)(5),$$

we do say that 2 and 5 are factors of 10 in the domain (or universe) J of integers.
  Now consider the polynomial

$$2x^2 - 10.$$

As we shall presently see, in the domain of polynomials in x having *integral* coefficients—that is, coefficients that are integers—its complete factorization is given by

$$2x^2 - 10 = 2(x^2 - 5),$$

but in the domain of polynomials in x having real numbers as coefficients, its complete factorization is

$$2x^2 - 10 = 2(x - \sqrt{5})(x + \sqrt{5}).$$

Thus the result of a factorization depends in part on the domain under consideration. The choice of order of factors is arbitrary; the arrangement that seems most "natural" should be used, although this arrangement is admittedly not always easy to determine. For instance, the forms

$$2(x - \sqrt{5})(x + \sqrt{5}) \quad \text{and} \quad 2(x + \sqrt{5})(x - \sqrt{5})$$

are equivalent, but it is difficult to affirm one as more "natural" than the other. Ordinarily, however, we write monomial factors first, and order the terms within a factor according to descending degree of the variable.

We say that a polynomial having integral coefficients is **prime** if it is not the product of two polynomials having integral coefficients with no common integral factor other than 1 or -1. For example,

$$x^2 - 5$$

is a prime polynomial. Note, however, that in the polynomial

$$2x^2 + x - 3,$$

the coefficients 2, 1, and -3 have no common integral factor other than 1 or -1, but the polynomial is not prime; instead, it is the product of two prime polynomials, since

$$2x^2 + x - 3 = (2x + 3)(x - 1).$$

Factors of polynomials   We say that a polynomial other than 0 or 1 with integral coefficients is **completely factored** if it is written equivalently as a product of prime polynomials. One very common type of factoring is that involving quadratic (second-degree) binomials or trinomials with integral coefficients. Recall, from Section 2.2, that

$$(x + a)(x + b) = x^2 + (a + b)x + ab, \tag{1}$$

$$(x + a)^2 = x^2 + 2ax + a^2, \tag{2}$$

$$(x + a)(x - a) = x^2 - a^2. \tag{3}$$

These three forms are those most commonly encountered in the chapters that follow. In this section, we are interested in viewing these relationships from right to left—that is, from polynomial to factored form.

A few other polynomials occur frequently enough to justify a study of their factorization. In particular, the forms

$$(a + b)(x + y) = ax + ay + bx + by, \tag{4}$$

$$(x + a)(x^2 - ax + a^2) = x^3 + a^3, \tag{5}$$

$$(x - a)(x^2 + ax + a^2) = x^3 - a^3 \tag{6}$$

are often encountered in one or another part of mathematics. We are again interested in viewing these relationships from right to left. Expressions such as the right-hand member of form (4) are factorable by grouping. For example, to factor

$$3x^2y + 2y + 3xy^2 + 2x,$$

we write it in the form

$$3x^2y + 2x + 3xy^2 + 2y$$

and factor the common monomial x from the first group of two terms and y from the second group of two terms, obtaining

$$x(3xy + 2) + y(3xy + 2).$$

If now we factor the common binomial $(3xy + 2)$ from each term, we have

$$(3xy + 2)(x + y),$$

in which both factors are prime.

The application of forms (5) and (6) is direct.

Example $8a^3 + b^3 = (2a)^3 + b^3$

$$= (2a + b)[(2a)^2 - 2ab + b^2]$$

$$= (2a + b)(4a^2 - 2ab + b^2)$$

Exercise 2.3

Factor completely in the domain of polynomials with integer coefficients.

Examples   a. $18x^2y - 24xy^2 + 6xy$ b. $4a^3 - 5a^2 + a$

Solutions   a. $(2)(3)xy(3x - 4y + 1)$ b. $a(4a^2 - 5a + 1)$
                                                                  $a(4a - 1)(a - 1)$

1. $xy^2 + x^2y$

2. $4x^2y^3 - 2xy^2$

3. $9x^5y - 3x^4y + 6x^3y$

4. $x^2y^2z^2 + 2xyz - xz$

5. $x^2 - 8x + 12$

6. $a^2 - a - 6$

7. $x^2 - 25$

8. $4x^2 - 36$

9. $12x^2 - 5x - 3$

10. $4x^2 + 12x + 9$

11. $3x^2 + 12x + 12$

12. $x^4y^2 - x^2y^2$

13. $y^4 + 3y^2 + 2$

14. $x^4 - 5x^2 + 4$

15. $2a^4 - a^2 - 1$

16. $3z^4 - 11z^2 - 4$

17. $x^4 - (y - 2x)^4$

18. $ax^2 + x + ax + 1$

Examples   a. $by - ay + bx - ax$ b. $8x^3 - y^3$

Solutions   a. $y(b - a) + x(b - a)$ b. $(2x)^3 - y^3$
                     $(b - a)(y + x)$ $(2x - y)(4x^2 + 2xy + y^2)$

19. $x^2 + ax + xy + ay$

20. $3x + y - 6x^2 - 2xy$

21. $a^3 + 2ab^2 - 4b^3 - 2a^2b$

22. $6x^3 - 4x^2 + 3x - 2$

23. $y^3 - 27x^3$

24. $8 + x^3y^3$

25. $x^3 + (x - y)^3$

26. $(x + y)^3 - z^3$

In Problems 27–34, assume that all variables in exponents represent natural numbers.

Examples   a. $a^{2n} - 9$ b. $x^{4n} - 3x^{2n} - 4$

Solutions   a. $(a^n - 3)(a^n + 3)$ b. $(x^{2n} - 4)(x^{2n} + 1)$
                                                       $(x^n - 2)(x^n + 2)(x^{2n} + 1)$

27. $a^{2n} - 4$

28. $x^{2n} - y^{2n}$

29. $x^{4n} - y^{4n}$

30. $x^{4n} - 2x^{2n} + 1$

31. $3x^{4n} - 10x^{2n} + 3$

32. $6y^{2n} + 30y^n - 900$

33. $2y^{2n} - 12y^n - 1440$

34. $2x^{2n} - 23x^ny^n - 39y^{2n}$

35. Show that $ac - ad + bd - bc$ can be factored both as $(a - b)(c - d)$ and as $(b - a)(d - c)$.

36. Show that $a^2 - b^2 - c^2 + 2bc$ can be factored as $(a - b + c)(a + b - c)$.

37. Consider the polynomial $x^4 + x^2y^2 + 25y^4$. If $9x^2y^2$ is both added to and subtracted from this expression (thus producing an equivalent expression), we have

$$x^4 + x^2y^2 + 25y^4 + 9x^2y^2 - 9x^2y^2,$$
$$(x^4 + 10x^2y^2 + 25y^4) - 9x^2y^2,$$
$$(x^2 + 5y^2)^2 - 9x^2y^2,$$
$$[(x^2 + 5y^2) - 3xy][(x^2 + 5y^2) + 3xy],$$
$$(x^2 - 3xy + 5y^2)(x^2 + 3xy + 5y^2).$$

By adding and subtracting an appropriate monomial, factor $x^4 + x^2y^2 + y^4$.

38. Use the method of Problem 37 to factor $x^4 - 3x^2y^2 + y^4$.

2.4
Quotients of polynomials

It is easy to show that the set of integers is not closed with respect to division; $2/3$, $1/5$, $8/9$, and $-3/4$ are all examples of noninteger quotients of integers. Similarly, we can see that the set of polynomials is not closed with respect to division, because $1/x$ is a counterexample. That is, $1/x$ is the quotient of the polynomials 1 and x, but is not, itself, a polynomial. In Exercise 1.4, Problem 43, you were asked to show that if a, b, and q are real numbers, with $b \neq 0$, then

$$\frac{a}{b} = q \quad \text{if and only if} \quad a = bq.$$

When we observe that the closure laws guarantee the applicability of the real-number axioms to real polynomials, this problem suffices to validate the following theorem.

Theorem 2.2 *If A, D, and Q are real polynomials, then for values of the variables for which $D \neq 0$,*

$$\frac{A}{D} = Q \quad \text{if and only if} \quad A = DQ.$$

If a polynomial Q exists such that $A = DQ$, $D \neq 0$, then A is said to be **exactly divisible** by D. If A is not exactly divisible by D, then the quotient A/D cannot be written equivalently as a polynomial.

Quotients of powers   Let us examine some ways in which we can rewrite quotients of polynomials, even if the resulting expressions are not always, themselves, polynomials. We begin with the simplest case, that in which A and D are monomials and their quotient is a polynomial.

Consider

$$\frac{x^m}{x^n} \quad (x \in R;\ x \neq 0;\ m,\ n \in N;\ \text{and}\ m > n).$$

We have

$$\frac{x^m}{x^n} = x^m \cdot \frac{1}{x^n} = (x^{m-n} \cdot x^n) \cdot \frac{1}{x^n}$$

$$= x^{m-n} \cdot \left(x^n \cdot \frac{1}{x^n}\right) = x^{m-n} \cdot 1,$$

from which we obtain

$$\frac{x^m}{x^n} = x^{m-n}.$$

This establishes the first part of the following result. A proof of the second part would be similar to the proof of Theorem 2.1-III.

Theorem 2.3 *If* $x,\ y \in R,\ x \neq 0,\ m,\ n \in N,\ \text{and}\ m > n,\ \text{then}$

$$\text{I} \quad \frac{x^m}{x^n} = x^{m-n} \qquad \text{II} \quad \left(\frac{y}{x}\right)^n = \frac{y^n}{x^n}.$$

The theorem enables us to write, for example,

$$\frac{12a^5b^3}{4a^2b^2} = \frac{12}{4} \cdot a^{5-2}b^{3-2} = 3a^3b \quad (a,\ b \neq 0).$$

Note that a and b are not permitted to take the value 0, because if they were, $3a^3b$ would represent a real number, 0, although $12a^5b^3/4a^2b^2$ would not be defined.

Theorem 1.11-III (together with the closure laws) permits us to rewrite quotients of polynomials of the form $(A + B)/C$, $C \neq 0$, in the form $A/C + B/C$, and then to do such further rewriting as seems indicated. For example,

$$\frac{2x^3 + 4x^2 + 8x}{2x} = \frac{2x^3}{2x} + \frac{4x^2}{2x} + \frac{8x}{2x},$$

and, by an application of Theorem 2.3-I, the right-hand member can then be denoted by the expression

$$x^2 + 2x + \frac{8x}{2x} \quad (x \neq 0).$$

Though Theorem 2.3 is not applicable to the variable factors in expressions such as $8x/2x$, by Theorem 1.11-II and the fact that $x/x = 1$ for $x \neq 0$, we can write

$$\frac{8x}{2x} = \frac{8}{2} \cdot \frac{x}{x} = \frac{8}{2} \cdot 1 = 4,$$

for every $x \neq 0$. Thus,

$$\frac{2x^3 + 4x^2 + 8x}{2x} = x^2 + 2x + 4 \quad (x \neq 0).$$

As mentioned at the start of this section, quotients of polynomials cannot always be represented by polynomials. For example, we have

$$\frac{2x^3 + 4x + 1}{x} = \frac{2x^3}{x} + \frac{4x}{x} + \frac{1}{x}$$

$$= 2x^2 + 4 + \frac{1}{x} \quad (x \neq 0),$$

where the resulting expression is not a polynomial.

Long division   If the divisor of a quotient contains more than one term, the familiar **long-division algorithm** involving successive subtractions can be used to rewrite the quotient. For example, the computation

$$
\begin{array}{r}
x - 3 \\
x^2 + 2x - 1 \overline{\smash{\big)}\ x^3 - x^2 - 7x + 3} \\
\underline{x^3 + 2x^2 - x} \\
-3x^2 - 6x + 3 \\
\underline{-3x^2 - 6x + 3} \\
0
\end{array}
$$

shows that, for $x^2 + 2x - 1 \neq 0$,

$$\frac{x^3 - x^2 - 7x + 3}{x^2 + 2x - 1} = x - 3.$$

It is most convenient to arrange the dividend and the divisor in descending powers of the variable before using the division algorithm and to leave an appropriate space for any missing terms (terms with coefficient 0) in the dividend.

When the divisor is not a factor of the dividend, the division process will produce a nonzero remainder. For example, from

$$
\begin{array}{r}
x^3 - 3x^2 + 10x - 28 \\
x + 3 \overline{\smash{\big)}\ x^4 + x^2 + 2x - 1} \\
\underline{x^4 + 3x^3} \\
-3x^3 + x^2 \\
\underline{-3x^3 - 9x^2} \\
10x^2 + 2x \\
\underline{10x^2 + 30x} \\
-28x - 1 \\
\underline{-28x - 84} \\
83 \quad \text{(remainder)}
\end{array}
$$

we see that

$$\frac{x^4 + x^2 + 2x - 1}{x + 3} = x^3 - 3x^2 + 10x - 28 + \frac{83}{x + 3} \qquad (x \neq -3).$$

Observe in the foregoing example that a space is left in the dividend for a term involving x^3, even though the dividend contains no such term.

Synthetic division If the divisor is of the form $x + c$, the process of dividing one polynomial by another can be simplified by a process called **synthetic division**. Consider the foregoing example. If we omit writing the variables and write only the coefficients of the terms, and use zero for the coefficient of any missing power, we have

$$
\begin{array}{r}
1 - 3 + 10 - 28 \\
1 + 3 \,\overline{)\,1 + 0 + 1\ \ +2\ \ -1} \\
\underline{1 + 3} \\
-3 + (1) \\
\underline{-3 - 9} \\
10 + (2) \\
\underline{10 + 30} \\
-28 - (1) \\
\underline{-28 - 84} \\
83 \quad \text{(remainder).}
\end{array}
$$

Now, observe that the numerals shown in color are repetitions of the numerals written immediately above and are also repetitions of the coefficients of the associated variable in the quotient; the numbers in parentheses are repetitions of the coefficients of the dividend. Therefore, the whole process can be written in compact form as

(1)	$3\,\rfloor\,1$	0	1	2	-1
(2)		3	-9	30	-84
(3)	1	-3	10	-28	83 (remainder: 83)

where the repetitions are omitted and where 1, the coefficient of x in the divisor, has also been omitted.

The entries in line (3), which are the coefficients of the variables in the quotient and the remainder, have been obtained by *subtracting* the **detached coefficients** in line (2) from the detached coefficients of terms of the same degree in line (1). We could obtain the same result by replacing 3 with -3 in the divisor and *adding* instead of subtracting at each step, and this is what is done in the *synthetic-division* process. The final form then appears as

$$
\begin{array}{r}
(1) \quad -3\,\vert\,1 \quad 0 \quad 1 \quad 2 \quad -1 \\
(2) \quad\quad\quad\quad -3 \quad 9 \quad -30 \quad 84 \\
\hline
(3) \quad\quad\quad 1 \quad -3 \quad 10 \quad -28 \quad 83 \quad \text{(remainder: 83).}
\end{array}
$$

Comparing the results using synthetic division with those obtained by the same process using long division, we observe that the entries in line (3) are the coefficients of the polynomial $x^3 - 3x^2 + 10x - 28$ and that there is a remainder of 83.

Example   Write $\dfrac{3x^3 - 4x - 1}{x - 2}$ in the form $Q + \dfrac{r}{D}$, where r is a constant.

Solution   Using synthetic division, we first write

$$
2\,\vert\; 3 \quad 0 \quad -4 \quad -1,
$$

where 0 has been inserted in the position that would be occupied by the coefficient of a second-degree term if such a term were present in the dividend. The divisor, $x - 2$, is indicated by the negative of -2, or 2. Continuing the synthetic division, we write,

$$
\begin{array}{r}
(1) \quad 2\,\vert\; 3 \quad 0 \quad -4 \quad -1 \\
(2) \quad\quad\quad 6 \quad 12 \quad 16 \\
\hline
(3) \quad\quad 3 \quad 6 \quad 8 \quad 15 \quad \text{(remainder: 15).}
\end{array}
$$

This process employs these steps:

1. 3 is "brought down" from line (1) to line (3).
2. 6, the product of 2 and 3, is written in the next position on line (2).
3. 6, the sum of 0 and 6, is written on line (3).
4. 12, the product of 2 and 6, is written in the next position on line (2).
5. 8, the sum of -4 and 12, is written on line (3).
6. 16, the product of 2 and 8, is written in the next position in line (2).
7. 15, the sum of -1 and 16, is written on line (3).

We can use the first three entries in line (3) as coefficients to write a polynomial of degree one less than the degree of the dividend. This polynomial is the quotient lacking the remainder. The last number is the remainder. Thus, for $x - 2 \neq 0$, the quotient of $3x^3 - 4x - 1$ divided by $x - 2$ is

$$
3x^2 + 6x + 8,
$$

with a remainder of 15; that is,

$$
\frac{3x^3 - 4x - 1}{x - 2} = 3x^2 + 6x + 8 + \frac{15}{x - 2} \quad (x \neq 2).
$$

The foregoing example illustrates a theorem that we shall state without proof.

Theorem 2.4 *If $P(x)$ is a real polynomial of degree $n \geq 1$ and c is any real number, then there exist a unique real polynomial $Q(x)$ of degree $n - 1$ and a unique real number r, such that*

$$P(x) = (x - c)Q(x) + r.$$

Although this theorem does not directly involve the quotient $\dfrac{P(x)}{x - c}$, it does assure us that, for $x \neq c$,

$$\frac{P(x)}{x - c} = Q(x) + \frac{r}{x - c}.$$

Remainder   Now consider the following important result.
Theorem

Theorem 2.5 *If $P(x)$ is a real polynomial, then for every real number c there exists a unique polynomial $Q(x)$ such that*

$$P(x) = (x - c)Q(x) + P(c).$$

Proof From Theorem 2.4, we know that, for every real number c, there exists a real polynomial $Q(x)$ and a real number r such that

$$P(x) = (x - c)Q(x) + r.$$

Since this is true for all $x \in R$, it must be true for $x = c$. Thus we have

$$P(c) = (c - c)Q(c) + r$$
$$= 0 \cdot Q(c) + r$$
$$= r,$$

and the theorem is proved.

This theorem is called the **remainder theorem** because it asserts that the remainder, when $P(x)$ is divided by $x - c$, is the value of P at c, that is, $P(c)$. Since synthetic division offers a quick means of obtaining this remainder, we can usually find values $P(c)$ more rapidly by synthetic division than by direct substitution.

Example   If $P(x) = x^3 - x^2 + 3$, find $P(3)$ by the remainder theorem.

Solution   Synthetically dividing $x^3 - x^2 + 3$ by $x - 3$, we have

$$
\begin{array}{r|rrrr}
3 & 1 & -1 & 0 & 3 \\
  &   & 3 & 6 & 18 \\
\hline
  & 1 & 2 & 6 & 21
\end{array}
$$

and, by inspection, $r = P(3) = 21$.

Exercise 2.4

Write each quotient as a polynomial.

Examples   a. $\dfrac{6x^2y^3}{2xy}$ b. $\dfrac{2y^3 - 6y^2 + y}{y}$

Solutions   a. $\dfrac{6}{2} \cdot x^{2-1}y^{3-1}$ b. $\dfrac{2y^3}{y} - \dfrac{6y^2}{y} + \dfrac{y}{y}$

$\qquad\qquad 3xy^2 \quad (x,\, y \neq 0)$ $\qquad 2y^2 - 6y + 1 \quad (y \neq 0)$

1. $\dfrac{8a^3y^5}{2a^2y^3}$ 2. $\dfrac{38a^3b^5}{19ab^3}$ 3. $\dfrac{x^2y^3z}{xy^2}$

4. $\dfrac{8x^3yz^4}{2xyz^2}$ 5. $\dfrac{6x^2 + 3x}{3x}$ 6. $\dfrac{12xy^2 - 4xy}{2xy}$

7. $\dfrac{8a^3 + 4a^2 + 4a}{2a}$ 8. $\dfrac{12x^3 - 8x^2 + 36x}{4x}$

9. $\dfrac{x^3 - 4x^2 - 3x}{x}$ 10. $\dfrac{8a^2x^2 - 4ax^2 + ax}{ax}$

Write each quotient P/D in the form $Q + R/D$, where the degree of R is less than the degree of D.

Examples   a. $\dfrac{2y^3 - 6y^2 + 2y - 4}{y}$ b. $\dfrac{y^3 + y^2 - 5y + 2}{y^2 - 2y}$

Solutions   a. $\dfrac{2y^3}{y} - \dfrac{6y^2}{y} + \dfrac{2y}{y} - \dfrac{4}{y}$

b. $\quad y^2 - 2y\,\overline{\smash{\big)}\,y^3 + y^2 - 5y + 2}$

$\qquad\qquad 2y^2 - 6y + 2 - \dfrac{4}{y} \quad (y \neq 0)$

$\qquad\qquad\qquad\quad \dfrac{y^3 - 2y^2}{3y^2 - 5y}$

$\qquad\qquad\qquad\qquad \dfrac{3y^2 - 6y}{y + 2}$

$\qquad\qquad\qquad y + 3 + \dfrac{y + 2}{y^2 - 2y} \quad (y \neq 0,\, 2)$

11. $\dfrac{15x^3y - 10x^2y + 3y}{5xy}$ 12. $\dfrac{38a^3b^2 - 19a^2b^3 + 3}{19ab}$

13. $\dfrac{4y^3 + 12y + 5}{2y + 1}$ 14. $\dfrac{2x^4 + 13x^3 - 7}{2x - 1}$

15. $\dfrac{4y^5 - 4y^2 - 5y + 1}{2y^2 + y + 1}$

16. $\dfrac{2x^3 - 3x^2 - 15x - 1}{x^2 + 5}$

Use synthetic division to write each quotient $P(x)/D(x)$ in the form $Q(x) + r/D(x)$ where $Q(x)$ is a polynomial and r is a constant.

Examples   a. $\dfrac{2x^4 + x^3 - 1}{x + 2}$ b. $\dfrac{x^3 - 1}{x - 1}$

Solutions   a.

$$
\begin{array}{r|rrrrr}
-2 & 2 & 1 & 0 & 0 & -1 \\
   &   & -4 & 6 & -12 & 24 \\
\hline
   & 2 & -3 & 6 & -12 & 23
\end{array}
$$

b.
$$
\begin{array}{r|rrrr}
1 & 1 & 0 & 0 & -1 \\
  &   & 1 & 1 & 1 \\
\hline
  & 1 & 1 & 1 & 0
\end{array}
$$

$$2x^3 - 3x^2 + 6x - 12 + \frac{23}{x + 2} \quad (x \neq -2) \qquad x^2 + x + 1 \quad (x \neq 1)$$

17. $\dfrac{x^4 - 3x^3 + 2x^2 - 1}{x - 2}$

18. $\dfrac{x^4 + 2x^2 - 3x + 5}{x - 3}$

19. $\dfrac{2x^3 + x - 5}{x + 1}$

20. $\dfrac{3x^3 + x^2 - 7}{x + 2}$

21. $\dfrac{2x^4 - x + 6}{x - 5}$

22. $\dfrac{3x^4 - x^2 + 1}{x - 4}$

23. $\dfrac{x^3 + 4x^2 + x - 2}{x + 2}$

24. $\dfrac{x^3 - 7x^2 - x + 3}{x + 3}$

25. $\dfrac{x^6 + x^4 - x}{x - 1}$

26. $\dfrac{x^6 + 3x^3 - 2x - 1}{x - 2}$

27. $\dfrac{x^5 - 1}{x - 1}$

28. $\dfrac{x^5 + 1}{x + 1}$

Use synthetic division in Problems 29–34.

29. $P(x) = x^3 + 2x^2 + x - 1$; find $P(1)$, $(P(2)$, and $P(3)$.

30. $P(x) = x^3 - 3x^2 - x + 3$; find $P(1)$, $P(2)$, and $P(3)$.

31. $P(x) = 2x^4 - 3x^3 + x + 2$; find $P(-2)$, $P(2)$, and $P(4)$.

32. $P(x) = 3x^4 + 3x^2 - x + 3$; find $P(-2)$, $P(2)$, and $(P(4)$.

33. $P(x) = 3x^5 - x^3 + 2x^2 - 1$; find $P(-3)$, $P(2)$, and $P(3)$.

34. $P(x) = 2x^6 - x^4 + 3x^3 + 1$; find $P(-3)$, $P(2)$, and $P(3)$.

2.5
Equivalent rational expressions

    A fraction is an expression denoting a quotient. If the numerator (dividend) and the denominator (divisor) are polynomials, then the fraction is said to be a

rational expression. Trivially, any polynomial can be considered as being a rational expression, since it is the quotient of itself and 1. For each replacement of the variable(s) for which the denominator is not zero, a rational expression represents a real number. Of course, for any value(s) of the variable(s) for which the denominator vanishes (is equal to zero), the fraction does not represent a real number; it is then said to be undefined.

Since a rational expression represents a real number for each replacement of the variable(s) for which its denominator is not zero, some theorems for rational expressions follow directly from Theorems 1.10 and 1.11 and the laws of closure.

Theorem 2.6 *If A, B, C, and D represent polynomials, then for values of the variables for which the denominators do not vanish,*

I $\dfrac{A}{B} = \dfrac{C}{D}$ *if and only if* $AD = BC$,

II $-\dfrac{A}{B} = \dfrac{-A}{B} = \dfrac{A}{-B} = -\dfrac{-A}{-B}$,

III $\dfrac{A}{B} = \dfrac{-A}{-B} = -\dfrac{-A}{B} = -\dfrac{A}{-B}$,

IV $\dfrac{AC}{BC} = \dfrac{A}{B}$ (*fundamental principle of fractions*).

A fraction is said to be in **lowest terms** when the numerator and denominator do not contain certain types of factors in common. The arithmetic fraction a/b, where a and b are integers and $b \neq 0$, is in lowest terms provided a and b are relatively prime—that is, provided they contain no common positive integral factors other than 1. If the numerator and denominator of a fraction are polynomials with integral coefficients, then the fraction is said to be in lowest terms if the numerator and denominator cannot be expressed as products of polynomials having integral coefficients with a common factor other than 1 or -1.

To express a given fraction in lowest terms (called *reducing* the fraction), we can factor the numerator and denominator and apply the fundamental principle of fractions.

Example $\dfrac{y}{y^2} = \dfrac{1 \cdot y}{y \cdot y} = \dfrac{1}{y}$ $(y \neq 0)$

Diagonal lines are sometimes used to abbreviate the foregoing procedure.

We may write

$$\frac{y}{y^2} = \frac{\overset{1}{\cancel{y}}}{\underset{y}{\cancel{y^2}}} = \frac{1}{y} \quad (y \neq 0),$$

using the diagonal lines instead of writing $\dfrac{1 \cdot y}{y \cdot y}$. Reducing a fraction to lowest terms should be accomplished mentally whenever convenient.

Reduction of rational expressions　To reduce fractions with polynomial numerators and denominators, we first write them in factored form when possible. Common factors are then evident by inspection.

Example
$$\frac{2x^2 + x - 15}{2x + 6} = \frac{(2x - 5)(x + 3)}{2(x + 3)}$$

$$= \frac{2x - 5}{2} \quad (x \neq -3)$$

Building to higher terms　We can also change fractions to equivalent fractions in higher terms by applying the fundamental principle in the form

$$\frac{A}{B} = \frac{AC}{BC} \quad (B, C \neq 0).$$

We might want to do this, for instance, in order to express two given fractions A/B and D/C as equivalent fractions with the same denominator BC.

In general, to change a fraction A/B to an equivalent fraction with BC as denominator, we can determine the factor C by inspection and then multiply the numerator and the denominator of the original fraction by this factor.

Examples　　a. $\dfrac{2}{3x} = \dfrac{?}{6x^2y}$　　　　b. $\dfrac{3}{x - 1} = \dfrac{?}{x^2 - 5x + 4}$

Solutions　　a. $\dfrac{2(2xy)}{3x(2xy)} = \dfrac{4xy}{6x^2y}$　　　b. $\dfrac{3}{x - 1} = \dfrac{?}{(x - 1)(x - 4)}$

$$= \frac{3(x - 4)}{(x - 1)(x - 4)} = \frac{3x - 12}{x^2 - 5x + 4}$$

In a situation such as that in (b) of the foregoing example, it is usually convenient to leave the result in the factored form $3(x - 4)/(x - 1)(x - 4)$.

Exercise 2.5

Reduce to lowest terms where possible. Specify restrictions on the variable(s) for which the reduction is not valid.

Examples   a. $\dfrac{a-b}{b^2-a^2}$ b. $\dfrac{a+b}{b}$ c. $\dfrac{8x^3-1}{2x-1}$

Solutions   a. $\dfrac{-(b-a)}{(b-a)(b+a)}$ b. Expression   c. $\dfrac{(2x)^3-1}{2x-1}$

$\dfrac{-1}{b+a}$ $(b \neq a, -a)$ is in lowest terms $(b \neq 0)$. $\dfrac{(2x-1)(4x^2+2x+1)}{2x-1}$

$4x^2+2x+1 \quad \left(x \neq \dfrac{1}{2}\right)$

1. $\dfrac{a^2bc}{ab^2c}$

2. $\dfrac{24x^4y^2z}{16x^3y^2z}$

3. $\dfrac{2x+2y}{x+y}$

4. $\dfrac{x^2+x}{x+1}$

5. $\dfrac{a-b}{b-a}$

6. $\dfrac{x^2-xy}{y-x}$

7. $\dfrac{x^2-1}{1-x}$

8. $\dfrac{x^2-16}{4-x}$

9. $\dfrac{2x^3-4x^2-3x}{2x}$

10. $\dfrac{8a^2x^2-4ax^2+ax}{2ax}$

11. $\dfrac{y^2+5y-14}{y-2}$

12. $\dfrac{x^2+5x+6}{x+3}$

13. $\dfrac{4y^2+8y-5}{1-2y}$

14. $\dfrac{2x^2+13x-7}{1-2x}$

15. $\dfrac{x^3-y^3}{x^2-y^2}$

16. $\dfrac{x^4-y^4}{x^2+y^2}$

17. $\dfrac{y^4-16}{y^4-y^2-12}$

18. $\dfrac{a^3-a^2+b^2+b^3}{3a^2+6ab+3b^2}$

19. $\dfrac{x^2-2xy+y^2+x-y}{y^2-x^2}$

20. $\dfrac{8x^2-8xy-x^3-y^3+8y^2}{x^4+x^2y^2+y^4}$

Express each of the given fractions as an equivalent fraction with the given denominator. Specify values of the variables for which the fractions are not equivalent.

21. $\dfrac{3}{4}; \dfrac{}{12}$

22. $\dfrac{1}{5}; \dfrac{}{10}$

23. $\dfrac{b}{a}; \dfrac{}{a^2b}$

24. $\dfrac{b}{2a}; \dfrac{}{6a^3b^2}$

25. $\dfrac{3}{y+2}; \dfrac{}{y^2-y-6}$

26. $\dfrac{2}{x+3}; \dfrac{}{x^2+x-6}$

27. $\dfrac{3}{a+3}; \dfrac{}{a^3+27}$

28. $\dfrac{-2}{x^2+y^2}; \dfrac{}{x^4-y^4}$

29. Is the fraction $\dfrac{x-2}{1+x^2}$ defined for all values of $x \in R$? For what value(s) of x does the fraction equal zero?

30. Write three equivalent forms of the fraction $\dfrac{1}{a-b}$ $(a \neq b)$ by changing the sign or signs of the numerator, denominator, or fraction itself.

31. What are the conditions on a and b for the fraction $\dfrac{-1}{a-b}$ $(a \neq b)$ to represent a positive number? A negative number?

2.6

Sums of rational expressions

Since rational expressions represent real numbers for each replacement of the variable(s) for which the denominators are not zero, the following theorem is a direct consequence of Theorem 1.11-III and the laws of closure.

Theorem 2.7 *If A, B, C are polynomials, then for values of the variables for which $C \neq 0$,*

$$\frac{A}{C} + \frac{B}{C} = \frac{A+B}{C}.$$

This principle, of course, extends to the sum of any number of fractions. If the fractions in a sum have unlike denominators, we can replace the fractions with equivalent fractions having common denominators and then write the sum as a single fraction.

Example

$$\frac{3x^2-5x}{2} + \frac{x^2+2}{3} = \frac{(3x^2-5x)(3)}{2(3)} + \frac{(x^2+2)(2)}{3(2)}$$

$$= \frac{(9x^2-15x)+(2x^2+4)}{6}$$

$$= \frac{11x^2-15x+4}{6}$$

Differences can be viewed as sums and then expressed as a single fraction by an application of Theorem 2.7.

Least common
multiple of
polynomials

In rewriting fractions in a sum so that they share a common denominator, any such denominator may be used. If the **least common multiple** of the denominators (called the **least common denominator**) is used, however, the resulting fraction will be in simpler form than if any other common denominator is employed. The least common multiple of two or more natural numbers is the least natural number that is exactly divisible by each of the given numbers.

The notion of a least common multiple among several polynomial expressions is generally meaningless without further specification of what is desired. We can, however, define the least common multiple of a set of polynomials with integral coefficients to be (1) the polynomial of lowest degree with integral coefficients yielding a polynomial quotient upon division by each of the given polynomials, and (2) among all such polynomials, the one having the least possible positive leading coefficient.

Very often, the least common multiple of a set of natural numbers or polynomials can be determined by inspection. When inspection fails, we can find the least common multiple of a set of polynomials with integer coefficients as follows:

1. Express each polynomial in completely factored form.

2. Write as factors of a product each *different* factor occurring in any of the polynomials, including each factor the greatest number of times it occurs in any one of the given polynomials.

Example

Find the least common multiple of 12, 15, and 18.

Solution

12	15	18
$2 \cdot 2 \cdot 3$	$3 \cdot 5$	$3 \cdot 3 \cdot 2$

The least common multiple is $2^2 \cdot 3^2 \cdot 5$, or 180.

Example

Find the least common multiple of x^2, $x^2 - 9$, and $x^3 - x^2 - 6x$.

Solution

x^2	$x^2 - 9$	$x^3 - x^2 - 6x$
$x \cdot x$	$(x - 3)(x + 3)$	$x(x - 3)(x + 2)$

The least common multiple is $x^2(x + 2)(x + 3)(x - 3)$.

Addition of
rational
expressions

To simplify sums of fractions having different denominators, we can ascertain the least common denominator of the fractions, determine the factor necessary to express each of the fractions as a fraction having this common denominator, write the fractions accordingly, and then express the sum as a single fraction.

Example   Simplify $\dfrac{3}{x} + \dfrac{2}{x^2} + \dfrac{3}{xy}$.

Solution   The least common denominator of the fractions is x^2y.

$$\frac{3}{x} + \frac{2}{x^2} + \frac{3}{xy} = \frac{3(xy)}{x(xy)} + \frac{2(y)}{x^2(y)} + \frac{3(x)}{xy(x)}$$

$$= \frac{3xy}{x^2y} + \frac{2y}{x^2y} + \frac{3x}{x^2y}$$

$$= \frac{3xy + 2y + 3x}{x^2y} \quad (x, y \neq 0)$$

Exercise 2.6

Write each sum or difference as a single fraction in lowest terms. Assume that no variable in a denominator takes a value for which the denominator equals zero.

Example $\dfrac{5}{x^2 - 9} - \dfrac{2}{x - 3}$

Solution $\dfrac{5}{x^2 - 9} - \dfrac{2}{x - 3} = \dfrac{5}{(x - 3)(x + 3)} - \dfrac{2(x + 3)}{(x - 3)(x + 3)}$

$$= \frac{5 - 2(x + 3)}{(x - 3)(x + 3)}$$

$$= \frac{5 - 2x - 6}{x^2 - 9}$$

$$= \frac{-2x - 1}{x^2 - 9}$$

1. $\dfrac{x - 1}{y} + \dfrac{x}{y}$

2. $\dfrac{y + 1}{x} + \dfrac{y - 1}{x}$

3. $\dfrac{2a - b}{a} - \dfrac{a - b}{a}$

4. $\dfrac{3a - 1}{b} - \dfrac{2 - a}{b}$

5. $\dfrac{a + 2}{3} - \dfrac{a - 3}{9}$

6. $\dfrac{a - 2}{9} - \dfrac{a + 1}{3}$

7. $\dfrac{x + 1}{6} - \dfrac{2x - 1}{15}$

8. $\dfrac{x - 3}{10x} - \dfrac{x}{8x}$

9. $\dfrac{2}{a + b} + \dfrac{1}{2a + 2b}$

10. $\dfrac{7}{5x - 10} + \dfrac{5}{3x - 6}$

11. $\dfrac{1}{2y - 6} - \dfrac{3}{3y - 9}$

12. $\dfrac{2}{3x + 3} - \dfrac{4}{5x + 5}$

13. $\dfrac{2}{3-x} - \dfrac{1}{x-3}$

14. $\dfrac{7}{y-3} + \dfrac{3}{3-y}$

15. $\dfrac{a+1}{a+2} - \dfrac{a+2}{a+3}$

16. $\dfrac{5x-y}{3x+y} - \dfrac{6x-5y}{2x-y}$

17. $\dfrac{x+2y}{2x-y} - \dfrac{2x+y}{x-2y}$

18. $\dfrac{x-2y}{x+y} - \dfrac{2x-y}{x-y}$

19. $\dfrac{y}{y^2-16} - \dfrac{y+1}{y^2-5y+4} + \dfrac{1}{y+4}$

20. $\dfrac{1}{b^2-1} - \dfrac{1}{b^2+2b+1} + \dfrac{1}{b+1}$

21. $x + \dfrac{1}{x-1} - \dfrac{1}{(x-1)^2}$

22. $y - \dfrac{2y}{y^2-1} + \dfrac{3}{y+1}$

23. $x - 1 + \dfrac{3}{2x-1} - \dfrac{x}{4x^2-1}$

24. $2y - 3 - \dfrac{1}{y^2+2y+1} + \dfrac{3}{y+1}$

25. $\dfrac{3x}{x^2-5x+4} + \dfrac{2x}{x^2-2x+1}$

26. $\dfrac{2y}{y^2-y-2} + \dfrac{y}{y^2+3y-10}$

27. $\dfrac{y+3}{3y^2+7y+4} - \dfrac{y-7}{3y^2+13y+12}$

28. $\dfrac{a+b}{a^2+2ab-3b^2} - \dfrac{2a+b}{a^2+4ab+3b^2}$

29. $\dfrac{xy}{(z-x)(x-y)} + \dfrac{yz}{(z-y)(x-z)} + \dfrac{xz}{(y-x)(y-z)}$

30. $\dfrac{1}{(a-b)(b-c)} + \dfrac{1}{(b-c)(c-a)} + \dfrac{1}{(c-a)(a-b)}$

2.7

Products and quotients of rational expressions

By Parts II and VII of Theorem 1.11 and the laws of closure, we have the following result.

Theorem 2.8 *If A, B, C, and D are real polynomials, then for values of the variables for which the denominators do not equal zero,*

$$\text{I} \quad \frac{A}{B} \cdot \frac{C}{D} = \frac{AC}{BD}, \qquad \text{II} \quad \frac{A}{B} \div \frac{C}{D} = \frac{AD}{BC}.$$

We can use Theorem 2.8 to rewrite a product or quotient of fractions as a single fraction *in* lowest terms.

Example

$$\frac{x^2 - 2x + 1}{x^2 + 2x - 3} \cdot \frac{x^2 + 3x}{x^2 + 2x} = \frac{(x-1)(x-1)}{(x+3)(x-1)} \cdot \frac{x(x+3)}{x(x+2)}$$

$$= \frac{(x-1)[(x-1)(x+3)x]}{(x+2)[(x-1)(x+3)x]}$$

$$= \frac{x-1}{x+2} \quad (x \neq -3, -2, 0, 1)$$

Since the factors of the numerator and denominator of the product of two fractions are just the respective factors of the numerators and denominators of the fractions, we can divide common factors out of the numerators and denominators before writing the product as a single fraction. Thus, in the example above, we could write

$$\frac{x^2 - 2x + 1}{x^2 + 2x - 3} \cdot \frac{x^2 + 3x}{x^2 + 2x} = \frac{\overset{1}{\cancel{(x-1)}}(x-1)}{\cancel{(x+3)}\cancel{(x-1)}} \cdot \frac{\overset{1}{\cancel{x}}\overset{1}{\cancel{(x+3)}}}{\cancel{x}(x+2)}$$

$$= \frac{x-1}{x+2} \quad (x \neq -3, -2, 0, 1).$$

Example

$$\frac{x^3 - 8}{x^3 + 8} \div \frac{(x+1)(x^2 + 2x + 4)}{(x-1)(x^2 - 2x + 4)} = \frac{x^3 - 8}{x^3 + 8} \cdot \frac{(x-1)(x^2 - 2x + 4)}{(x+1)(x^2 + 2x + 4)}$$

$$= \frac{(x-2)\overset{1}{\cancel{(x^2 + 2x + 4)}}}{(x+2)\cancel{(x^2 - 2x + 4)}} \cdot \frac{(x-1)\overset{1}{\cancel{(x^2 - 2x + 4)}}}{(x+1)\cancel{(x^2 + 2x + 4)}}$$

$$= \frac{(x-2)(x-1)}{(x+2)(x+1)} \quad (x \neq -2, -1, 1)$$

When the quotient of two fractions is given in the form of a **complex fraction** (one containing a fraction in either the numerator or the denominator or both), we have a choice of procedures for writing the quotient in the form of a simple (not complex) fraction.

Example

Write $\dfrac{x + \dfrac{3}{4}}{x - \dfrac{1}{2}}$ as a simple fraction in lowest terms.

Solution 1

We can apply the fundamental principle of fractions to multiply numerator and denominator by the least common denominator of the simple fractions involved. Thus, we have

$$\frac{\left(x+\dfrac{3}{4}\right)4}{\left(x-\dfrac{1}{2}\right)4}=\frac{4x+3}{4x-2}\quad\left(x\neq\frac{1}{2}\right).$$

Solution 2   Alternatively, we can rewrite the complex fraction as follows.

$$\frac{x+\dfrac{3}{4}}{x-\dfrac{1}{2}}=\frac{\dfrac{4x+3}{4}}{\dfrac{2x-1}{2}}$$

$$=\frac{4x+3}{4}\cdot\frac{2}{2x-1}$$

$$=\frac{4x+3}{\underset{2}{\cancel{4}}}\cdot\frac{\overset{1}{\cancel{2}}}{2x-1}=\frac{4x+3}{4x-2}\quad\left(x\neq\frac{1}{2}\right)$$

In the event you have more complicated expressions involving complex fractions, you can rewrite the expression by attacking small parts of it at a time. An example is given before Problem 25 in the following exercise set.

Exercise 2.7

Write each product or quotient as a single fraction in lowest terms. Assume that no variable in a denominator takes a value for which the denominator equals zero.

Examples

a. $\dfrac{4y^2-1}{y^2-4}\cdot\dfrac{y^2+2y}{4y+2}$

b. $\dfrac{3x^2-3}{2x+2}\div\dfrac{5x-5}{2}$

Solutions

a. $\dfrac{(2y-1)(2y+1)}{(y-2)(y+2)}\cdot\dfrac{y(y+2)}{2(2y+1)}$

b. $\dfrac{3(x-1)(x+1)}{2(x+1)}\cdot\dfrac{2}{5(x-1)}$

$\dfrac{y(2y-1)(2y+1)(y+2)}{2(y-2)(2y+1)(y+2)}$

$\dfrac{3\cdot 2(x-1)(x+1)}{5\cdot 2(x-1)(x+1)}$

$\dfrac{y(2y-1)}{2(y-2)}$

$\dfrac{3}{5}$

1. $\dfrac{-12a^2b}{5c}\cdot\dfrac{10b^2c}{24a^3b}$

2. $\dfrac{a^2}{xy}\cdot\dfrac{3x^3y}{4a}$

3. $\dfrac{xy}{a^2b} \div \dfrac{x^3y^2}{ab}$

4. $\dfrac{24a^3b}{-6xy^2} \div \dfrac{3a^2b}{12x}$

5. $\dfrac{5x+25}{2x} \cdot \dfrac{4x}{2x+10}$

6. $\dfrac{3x}{4x-6x^2} \cdot \dfrac{2-3x}{12}$

7. $\dfrac{a^2-ab}{ab^2} \div \dfrac{2a-2b}{a^2b}$

8. $\dfrac{a^2+2ab}{ab^2} \div \dfrac{a+2b}{b^3}$

9. $\dfrac{x^2-1}{2x+6} \div \dfrac{x^2-2x+1}{x^2-9}$

10. $\dfrac{2y^2-y-3}{y^2-1} \div \dfrac{4y^2-9}{y+1}$

11. $\dfrac{x^2-x-20}{x^2+7x+12} \cdot \dfrac{2x^2+6x}{x^2-25}$

12. $\dfrac{4x^2+8x+3}{2x^2-5x+3} \cdot \dfrac{6x^2-9x}{1-4x^2}$

13. $\dfrac{25a^2b^2-16}{4ab+1} \div \dfrac{5ab+4}{16a^2b^2+16ab+3}$

14. $\dfrac{a^2-25}{a^2-16} \div \dfrac{a^2+2a-15}{a^2+a-12}$

15. $\dfrac{x^2-y^2}{x^2} \cdot \dfrac{x^2-xy+y^2}{x^2} \div \dfrac{x^3+y^3}{x^4}$

16. $\dfrac{a^3-b^3}{ab} \cdot \dfrac{a^2b}{a^2-b^2} \div \dfrac{ab^2}{a+b}$

17. $\left(1+\dfrac{1}{x}\right) \cdot \left(1-\dfrac{1}{x}\right)$

18. $\left(x-\dfrac{1}{x}\right) \div \left(x+\dfrac{1}{x}\right)$

19. $\left[\dfrac{3}{x-1}-\dfrac{2}{x+1}\right] \cdot \dfrac{x-1}{x}$

20. $\left[\dfrac{x}{x^2-9}+\dfrac{2}{x-3}\right] \cdot \dfrac{x-1}{x}$

21. $\left[\dfrac{2y}{2y-1}-\dfrac{3}{y}\right] \div \dfrac{3}{2y^2-y}$

22. $\left[\dfrac{y}{y^2-1}-\dfrac{y}{y^2-2y+1}\right] \div \dfrac{y}{y-1}$

23. $\dfrac{\dfrac{2}{a}+\dfrac{3}{2a}}{5+\dfrac{1}{a}}$

24. $\dfrac{1+\dfrac{1}{x}}{1-\dfrac{1}{x}}$

Example Write $\dfrac{1}{x+\dfrac{1}{x+\dfrac{1}{x}}}$ as a simple fraction in lowest terms.

Solution We can begin by simplifying the lower right-hand expression, $\dfrac{1}{x+\dfrac{1}{x}}$.

We have
$$\dfrac{1}{x+\dfrac{1}{x}} = \dfrac{(1)x}{\left(x+\dfrac{1}{x}\right)x} = \dfrac{x}{x^2+1}.$$

Thus,

$$\cfrac{1}{x + \cfrac{1}{x + \cfrac{1}{x}}} = \cfrac{1}{x + \cfrac{x}{x^2 + 1}}.$$

From this point, we can apply either of the methods shown in the example on pages 50 and 51 to the right-hand member above. Using the first method, we have

$$\cfrac{1}{x + \cfrac{1}{x + \cfrac{1}{x}}} = \frac{1(x^2+1)}{\left(x + \cfrac{x}{x^2+1}\right)(x^2+1)} = \frac{x^2+1}{x^3+x+x} = \frac{x^2+1}{x^3+2x} \quad (x \neq 0).$$

25. $a - \cfrac{a}{a + \cfrac{1}{4}}$

26. $x - \cfrac{x}{1 - \cfrac{x}{1-x}}$

27. $1 - \cfrac{1}{1 - \cfrac{1}{y-2}}$

28. $2y + \cfrac{3}{3 - \cfrac{2y}{y-1}}$

29. $\cfrac{1 + \cfrac{1}{1 - \cfrac{a}{b}}}{1 - \cfrac{3}{1 - \cfrac{a}{b}}}$

30. $\cfrac{1 - \cfrac{1}{\cfrac{a}{b} + 2}}{1 + \cfrac{3}{\cfrac{a}{2b} + 1}}$

2.8

Roots and exponents

Thus far, powers of real numbers have been defined for natural-number exponents, and some simple properties of products and quotients of powers have been examined. Powers with integral and rational exponents can now be defined in a manner consistent with these properties.

Reason for defining a^0 to be 1

By Theorem 2.3-I, for $a \in R$, $a \neq 0$, and $m, n \in N$, $m > n$, we have

$$\frac{a^m}{a^n} = a^{m-n}. \tag{1}$$

If (1) is to hold also for $m = n$, then we must have

$$\frac{a^n}{a^n} = a^{n-n} = a^0.$$

Since $a^n/a^n = 1$, we therefore make the following definition.

Definition 2.2   *If $a \in R$, $a \neq 0$, then*

$$a^0 = 1.$$

In the same way, if (1) is also to hold for $m = 0$, then we must have

$$\frac{a^0}{a^n} = a^{0-n} = a^{-n}.$$

Reason for
defining a^{-n}
to be $1/a^n$ Since $a^0 = 1$, we therefore make the following definition.

Definition 2.3   *If $a \in R$, $a \neq 0$, and $n \in N$, then*

$$a^{-n} = \frac{1}{a^n}.$$

Next, by Theorem 2.1-II, for $a \in R$, and $m, n \in N$, we have

$$(a^m)^n = a^{mn}. \tag{2}$$

If (2) is to hold for $m = 1/n$, and $a^{1/n}$ is a real number, then we must have

$$(a^{1/n})^n = a^{(1/n)(n)} = a^{n/n} = a^1 = a,$$

so that the nth power of $a^{1/n}$ must be a. A number having a as its nth power is called an **nth root** of a. In particular, for $n = 2$ or 3, an nth root is called a **square root** or a **cube root**, respectively.

Number of   For n odd, each $a \in R$ has just one real nth root. Thus
nth roots

$$(-2)^3 = -8 \quad \text{and} \quad 2^3 = 8,$$

so that -2 is the real cube root of -8, and 2 is the real cube root of 8. For n even and $a > 0$, a has two real nth roots. Thus

$$(-2)^4 = 16 \quad \text{and} \quad 2^4 = 16,$$

so that -2 and 2 are both fourth roots of 16. For n even and $a < 0$, a has no real nth root. Thus -1 has no real square root, since the square of each real number is nonnegative. If $a = 0$, then a has exactly one nth root, namely 0. We therefore make the following definition.

Definition 2.4 *If $a \in R$, $n \in N$, then $a^{1/n}$ is the real number if exactly one exists, and is the positive one if two exist, such that*

$$(a^{1/n})^n = a.$$

Examples   a. $25^{1/2} = 5$ b. $-25^{1/2} = -5$

             c. $(-25)^{1/2}$ is not a real number.   d. $27^{1/3} = 3$

             e. $-27^{1/3} = -3$ f. $(-27)^{1/3} = -3$

Order of taking power and root   The two theorems that follow will enable us to generalize from rational exponents of the form $1/n$, for $n \in N$, to rational exponents of the form m/n, for $m \in J$, $n \in N$.

Theorem 2.9 *If $a^{1/n} \in R$, $m \in J$, and $n \in N$, then*

$$(a^{1/n})^m = (a^m)^{1/n}.$$

Proof   Since $n \in N$, we can write

$$\underbrace{(a^{1/n})^m \cdot (a^{1/n})^m \cdot (a^{1/n})^m \cdots (a^{1/n})^m}_{n \text{ factors}} = [(a^{1/n})^m]^n = (a^{1/n})^{mn}$$
$$= (a^{1/n})^{nm} = [(a^{1/n})^n]^m$$
$$= a^m.$$

Therefore $[(a^{1/n})^m]^n = a^m$, and accordingly $(a^{1/n})^m$ is an nth root of a^m. Hence, by Definition 2.4, $(a^{1/n})^m = (a^m)^{1/n}$.

Observe that Theorem 2.9 requires that $a^{1/n}$ be a real number. This requirement is not satisfied if n is even and a is negative. If, for instance, $a = -3$, $m = 2$, and $n = 2$, then $(a^{1/n})^m = [(-3)^{1/2}]^2$ is not a real number, because $(-3)^{1/2}$ is not a real number.

The proof of the following result is similar to that of Theorem 2.9.

Theorem 2.10 *If $a^{1/np} \in R$, $m \in J$, $n \in N$, and $p \in N$, then*

$$(a^{1/np})^{mp} = (a^{1/n})^m.$$

Theorems 2.9 and 2.10 show that, if $a^{1/np}$ is a real number, then $a^{m/n}$ can be considered equally well as $(a^{1/n})^m$, $(a^m)^{1/n}$, $(a^{1/np})^{mp}$, or $(a^{mp})^{1/np}$. For example,

$$(16^{1/2})^3 = 4^3 = 64, \qquad (16^3)^{1/2} = 4096^{1/2} = 64,$$
$$(16^{1/4})^6 = 2^6 = 64, \qquad (16^6)^{1/4} = 16,777,216^{1/4} = 64.$$

We choose the following definition.

Definition 2.5 *If $a^{1/n} \in R$, $m \in J$, and $n \in N$, then*

$$a^{m/n} = (a^{1/n})^m.$$

Observe that requiring $n \in N$ does not alter the fact that m/n can represent every rational number, since all that is done is to restrict the denominator of the fraction representing the rational number to be positive, which is always possible in light of V and VI of Theorem 1.8.

Because we define $a^{1/n}$ to be the positive nth root of a for a positive and n an even natural number, and since a^m is positive for a negative and m an even natural number, it follows that, for m and n *even* natural numbers, and a *any* real number,

$$(a^m)^{1/n} = |a|^{m/n}.$$

For the special case, $m = n$ (m and n even),

$$(a^n)^{1/n} = |a|.$$

Properties of
rational powers

Powers with rational exponents have the same fundamental properties as powers with natural-number exponents, as long as the powers are real numbers. Recall from Theorems 2.1 and 2.3 that powers with natural-number exponents exhibit the properties set forth in the following theorem for powers with rational exponents. We state this theorem without proof.

Theorem 2.11 *If a^m, a^n, $b^n \in R$, $a, b \neq 0$,   and $m, n \in Q$, then*

$$\text{I} \quad a^m \cdot a^n = a^{m+n}, \qquad \text{II} \quad (a^m)^n = a^{mn},$$

$$\text{III} \quad (ab)^n = a^n b^n, \qquad \text{IV} \quad \frac{a^m}{a^n} = a^{m-n},$$

$$\text{V} \quad \left(\frac{a}{b}\right)^n = \frac{a^n}{b^n}.$$

Let us look at a few applications of this theorem.

Examples

a. $\dfrac{x^{2/3}}{x^{1/3}} = x^{2/3 - 1/3} = x^{1/3} \quad (x \neq 0)$

b. $\left(\dfrac{a^3 b^6}{c^{12}}\right)^{2/3} = \dfrac{(a^3)^{2/3}(b^6)^{2/3}}{(c^{12})^{2/3}} = \dfrac{a^2 b^4}{c^8} \quad (c \neq 0)$

c. $(a^6)^{1/2} = (|a|^6)^{1/2} = |a|^3$

Observe that in the last example it was necessary to use absolute-value notation because the expression has been defined to be positive for m and n even. For example, if $a = -2$, we have

$$(a^6)^{1/2} = [(-2)^6]^{1/2} = 64^{1/2} = 8 = |a|^3,$$

whereas

$$a^3 = (-2)^3 = -8 = -|a|^3 \neq |a|^3.$$

Exercise 2.8

Write each of the following as a power with exponent 1.

Examples   a. $64^{1/2}$ b. $\left(\dfrac{8}{27}\right)^{-2/3}$

Solutions   a. 8 b. $\left[\left(\dfrac{8}{27}\right)^{1/3}\right]^{-2} = \left(\dfrac{2}{3}\right)^{-2} = \dfrac{9}{4}$

1. $(32)^{1/5}$ 2. $(-27)^{1/3}$ 3. $(81)^{-3/4}$ 4. $(81)^{-1/2}$

5. $\left(\dfrac{1}{8}\right)^{-5/3}$ 6. $\left(\dfrac{1}{8}\right)^{5/3}$ 7. $\left(\dfrac{4}{9}\right)^{3/2}$ 8. $\left(\dfrac{4}{9}\right)^{-3/2}$

Write each expression as a product or quotient of powers in which each variable occurs but once and in which all exponents are positive. Assume all variable bases are positive and all variable exponents are natural numbers.

Examples   a. $\dfrac{(x^{1/2}y^2)^2}{(x^{2/3}y)^3}$ b. $(y^{2n}y^{n/2})^4$

Solutions   a. $\dfrac{xy^4}{x^2y^3} = \dfrac{y}{x}$ b. $y^{8n}y^{2n} = y^{10n}$

9. $x^{1/3}x^{5/3}$ 10. $x^{4/3}x^{1/2}$ 11. $a^{2/3}a^{3/4}$ 12. $x^{1/2}x^{5/6}$

13. $\dfrac{x^{5/6}}{x^{1/2}}$ 14. $\dfrac{x^{1/2}}{x^{1/3}}$ 15. $\dfrac{x^{-2/5}}{x^{2/3}}$ 16. $\left(\dfrac{a^6}{c^3}\right)^{-2/3}$

17. $\left(\dfrac{y^4}{x^2}\right)^{1/2}$ 18. $\left(\dfrac{16}{ab^2}\right)^{1/4}$ 19. $\left(\dfrac{x^5y^8}{y^{13}}\right)^{1/4}$ 20. $\left(\dfrac{125x^3y^4}{27x^{-6}y}\right)^{1/3}$

21. $(x^2)^{n/2}(y^{2n})^{2/n}$

22. $(x^{n/2})^2(y^n)^{5/n}$

23. $\dfrac{x^{2n}}{x^{n/2}}$

24. $\left(\dfrac{a^n}{b}\right)^{1/2} \cdot \left(\dfrac{b}{a^{2n}}\right)^{3/2}$

25. $\dfrac{x^{3n}y^{2m-1}}{(x^n y^m)^{1/2}}$

26. $\left(\dfrac{x^{2n^2}}{y^{4n}}\right)^{1/n}$

Apply the distributive law to write each product as a sum.

Examples   a. $x^{1/2}(x^{1/2} - x)$

b. $(x^{1/3} - x)(x^{1/3} + x)$

Solutions   a. $x^{1/2}x^{1/2} - x^{1/2}x$
                  $x - x^{3/2}$

b. $x^{1/3}x^{1/3} - x^2$
   $x^{2/3} - x^2$

27. $x^{1/3}(x^{2/3} - x^{1/3})$

28. $y^{2/3}(y^{2/3} + y^{1/3})$

29. $(x^{1/2} - y^{-1/2})(x^{1/2} - y^{-1/2})$

30. $(x^{1/2} + y^{1/2})(x^{1/2} - y^{1/2})$

31. $(x + y)^{1/2}[(x + y)^{1/2} - (x + y)]$

32. $(a - b)^{2/3}[(a - b)^{-1/3} + (a - b)]$

33. $(x^{1/3} + y^{1/3})(x^{2/3} - x^{1/3}y^{1/3} + y^{2/3})$

34. $(a^{1/3} - b^{1/3})(a^{2/3} + a^{1/3}b^{1/3} + b^{2/3})$

Factor as indicated.

Examples   a. $y^{-1/2} + y^{1/2} = y^{-1/2}(\ ? \ )$ b. $x^{3/2} - x^{-1/2} = x^{-1/2}(\ ? \ )$

Solutions   a. $y^{-1/2}(1 + y)$ b. $x^{-1/2}(x^2 - 1)$

35. $x^{3/2} + x = x(\ ? \ )$

36. $y - y^{2/3} = y^{1/3}(\ ? \ )$

37. $x^{-3/2} + x^{-1/2} = x^{-1/2}(\ ? \ )$

38. $y^{1/2} + y = y^{1/2}(\ ? \ )$

39. $y^{-3/2} - y = y^{-3/2}(\ ? \ )$

40. $x^{1/3} - x^{-2/3} = x^{-2/3}(\ ? \ )$

In the previous problems, the variables were restricted to represent positive numbers. In Problems 41–46, consider variable bases to denote any *elements of the set of real numbers and simplify.*

Examples   a. $[(-3)^2]^{1/2}$ b. $[u^2(u + 5)]^{1/2}$

Solutions   a. $|-3| = 3$ b. $|u|(u + 5)^{1/2}$

41. $[(-5)^2]^{1/2}$

42. $[(-3)^{12}]^{1/4}$

43. $[4x^2]^{1/2}$

44. $[x^2(x - 1)]^{1/2}$

45. $\dfrac{2}{[x^2(x + 1)]^{1/2}}$

46. $\left[\dfrac{9}{x^6(x^2 + 1)}\right]^{1/2}$

2.9

Radical expressions

Powers of real numbers with rational numbers for exponents are frequently denoted by symbols involving the use of the radical sign, $\sqrt{}$.

Definition 2.6 *If $a^{1/n} \in R$ and $n \in N$, then*

$$\sqrt[n]{a} = a^{1/n}.$$

Naturally, the radical expression on the left is not defined if the power on the right is not. In the symbolism $\sqrt[n]{a}$, a is called the **radicand** and n the **index** of the radical, and the expression is called a **radical expression of order n**. If no index is shown with a radical expression, as, for example, in the case $\sqrt{a}$, then the index 2 is understood to apply. The symbol $\sqrt{a}$ denotes the nonnegative square root of a, where, of course, a cannot be negative. The symbol $\sqrt{x^2}$, where $x \in R$, therefore provides us with an alternative means of writing $|x|$. That is, $\sqrt{x^2} = |x|$.

Properties of
radicals
An immediate consequence of the foregoing definition and the theorems pertaining to exponents in Section 2.8 follows.

Theorem 2.12 *For real values of a and b for which all the radical expressions in the equation denote real numbers,*

$$\text{I} \quad \sqrt[n]{a^n} = \begin{cases} a & (n \text{ an odd natural number}), \\ |a| & (n \text{ an even natural number}), \end{cases}$$

$$\text{II} \quad \sqrt[n]{a^m} = \sqrt[n]{a^m} \quad (n \in N, m \in J),$$

$$\text{III} \quad \sqrt[n]{a} \cdot \sqrt[n]{b} = \sqrt[n]{ab} \quad (n \in N),$$

$$\text{IV} \quad \frac{\sqrt[n]{a}}{\sqrt[n]{b}} = \sqrt[n]{\frac{a}{b}} \quad (b \neq 0, n \in N).$$

It might be noted in III and IV that if $a, b < 0$, and n is even, then the radicals in the left-hand member are not defined, even though the radical in the right-hand member is.

The several parts of this theorem can be used to rewrite radical expressions in various ways, and, in particular, to write them in what is called "simplest" form.

A radical expression is said to be in **simplest form** if:
a. the radicand contains no polynomial factor raised to a power equal to or greater than the index of the radical,
b. the radicand contains no fractions,
c. no radical expressions are contained in denominators of fractions, and
d. the index of the radical and the power in the radicand have no common factor other than 1.

Examples

a. $\sqrt[3]{24x^3y^2} = \sqrt[3]{8x}\,\sqrt[3]{3y^2} = 2x\sqrt[3]{3y^2}$

b. $\sqrt[6]{49x^2} = \sqrt[3\cdot2]{7^2x^2} = \sqrt[3]{7x}$ $(x \geq 0)$

c. $\dfrac{\sqrt[3]{4a^2}}{\sqrt[3]{b}} = \dfrac{\sqrt[3]{4a^2}\sqrt[3]{b^2}}{\sqrt[3]{b}\,\sqrt[3]{b^2}} = \dfrac{\sqrt[3]{4a^2b^2}}{\sqrt[3]{b^3}} = \dfrac{\sqrt[3]{4a^2b^2}}{b}$ $(b \neq 0)$

The process used to simplify the expression in part c in the foregoing example is called "rationalizing the denominator," because the result is a fraction with denominator free of radicals. This does not exclude the possibility that the denominator is an irrational number.

Sums and products

Since we have defined our radical expressions so that they represent real numbers, the properties of the real numbers can be applied. For example, the distributive law permits us to express certain sums as products and certain products as sums.

Examples

a. $2\sqrt{5} + 4\sqrt{5} = (2+4)\sqrt{5} = 6\sqrt{5}$

b. $(\sqrt{x} + 2)(\sqrt{x} - 1) = \sqrt{x}(\sqrt{x} - 1) + 2(\sqrt{x} - 1)$

$= x - \sqrt{x} + 2\sqrt{x} - 2 = x + \sqrt{x} - 2$

Rationalization of binomial denominators

The distributive law also enables us to rationalize denominators of fractions in which radicals occur in one or both of two terms. To accomplish this, we first recall that

$$(a - b)(a + b) = a^2 - b^2,$$

where the expression in the right-hand member contains no linear term. Each of the two factors of a product exhibiting this property is said to be the **conjugate** of the other. Now consider a fraction of the form

$$\frac{a}{b + \sqrt{c}},$$

where c is positive and $b^2 \neq c$. If we multiply the numerator and denominator of this fraction by the conjugate of the denominator, then the denominator of the resulting fraction will contain no term involving $\sqrt{c}$ and hence will be free of radicals. That is,

$$\frac{a}{b + \sqrt{c}} = \frac{a(b - \sqrt{c})}{(b + \sqrt{c})(b - \sqrt{c})} = \frac{ab - a\sqrt{c}}{b^2 - c} \quad (b^2 \neq c),$$

where the denominator has been rationalized. This process is equally applicable to radical fractions of the form

$$\frac{a}{\sqrt{b} + \sqrt{c}},$$

since

$$\frac{a}{\sqrt{b} + \sqrt{c}} = \frac{a(\sqrt{b} - \sqrt{c})}{(\sqrt{b} + \sqrt{c})(\sqrt{b} - \sqrt{c})} = \frac{a\sqrt{b} - a\sqrt{c}}{b - c}.$$

It should be noted, though, that it is not *always* preferable, in working with fractions, to rationalize their denominators. Sometimes, in fact, it is desirable to rationalize the numerator. Thus, for example,

$$\frac{\sqrt{b} + \sqrt{c}}{a} = \frac{(\sqrt{b} + \sqrt{c})(\sqrt{b} - \sqrt{c})}{a(\sqrt{b} - \sqrt{c})} = \frac{b - c}{a(\sqrt{b} - \sqrt{c})}.$$

Exercise 2.9

In all exercises in this set, assume that all variables represent positive real numbers and that all radicands are positive unless otherwise specified.

Write in radical form.

Examples   a. $5^{1/2}$ b. $xy^{2/3}$ c. $(x - y^2)^{-1/2}$

Solutions   a. $\sqrt{5}$ b. $x\sqrt[3]{y^2}$ c. $\dfrac{1}{\sqrt{x - y^2}}$

1. $a^{2/3}$ 2. $x^{3/2}$ 3. $3x^{1/3}$ 4. $-6xy^{1/2}$

5. $-6(xy)^{2/3}$ 6. $x^{1/5}y^{3/5}$ 7. $(x - y)^{4/7}$ 8. $(24 - 3x)^{-3/4}$

Write an equivalent expression, using positive fractional exponents in lowest terms.

Examples   a. $\sqrt{2^3}$ b. $7\sqrt[3]{a^2}$ c. $\dfrac{1}{\sqrt{a-b}}$

Solutions   a. $2^{3/2}$ b. $7a^{2/3}$ c. $\dfrac{1}{(a-b)^{1/2}}$

9. $\sqrt[3]{x^2}$ 10. $\sqrt[5]{xy}$ 11. $\sqrt[5]{2xy^2}$ 12. $a\sqrt[5]{x^2y^3}$

13. $-3\sqrt[4]{a^3b}$ 14. $7\sqrt[7]{x^5}$ 15. $3\sqrt[3]{x^2-y}$ 16. $-a\sqrt[5]{x^4-y^4}$

Find the root indicated.

Examples   a. $\sqrt[5]{-32}$ b. $\sqrt[3]{x^6y^3}$ c. $\sqrt{x^2y^6}$

Solutions   a. -2 b. x^2y c. xy^3

17. $\sqrt{144}$ 18. $-\sqrt{169}$ 19. $\sqrt[3]{-27}$ 20. $-\sqrt[6]{64}$

21. $\sqrt{x^4y^2}$ 22. $\sqrt[3]{8y^6}$ 23. $\sqrt{\dfrac{4}{9}x^6y^{10}}$ 24. $\sqrt[3]{\dfrac{-8x^3}{125}}$

Write in simplest form.

Examples   a. $\sqrt[3]{2x^7y^3}$ b. $\sqrt{2xy}\sqrt{8x}$ c. $\dfrac{\sqrt{6a}\sqrt{5a}}{\sqrt{15}}$

Solutions   a. $\sqrt[3]{x^6y^3}\sqrt[3]{2x}$ b. $\sqrt{16x^2}\sqrt{y}$ c. $\sqrt{\dfrac{30a^2}{15}}$

               $x^2y\sqrt[3]{2x}$ $4x\sqrt{y}$ $a\sqrt{2}$

25. $\sqrt{4x^5}$ 26. $\sqrt{16y^3}$ 27. $\sqrt[4]{3x^5y^5}$

28. $\sqrt[3]{-8x^6}$ 29. $\sqrt[3]{9}\sqrt[4]{27}$ 30. $\sqrt[3]{a^4}\sqrt[3]{a^7}$

31. $\dfrac{\sqrt{6x}}{\sqrt{2x}}$ 32. $\dfrac{\sqrt{a^5b^3}}{\sqrt{ab}}$ 33. $\dfrac{\sqrt[3]{2a^2b^3}}{\sqrt[3]{a^2b^2}}$ 34. $\dfrac{\sqrt[3]{42x^3}}{\sqrt[3]{6x^2}}$

35. $\dfrac{\sqrt{x}\sqrt{xy^3}}{\sqrt{y}}$ 36. $\dfrac{\sqrt{ab}\sqrt{ab^4}}{\sqrt{b}}$ 37. $\dfrac{\sqrt[3]{ab}\sqrt[3]{b^2}}{\sqrt[3]{a}}$ 38. $\dfrac{\sqrt[3]{4ab^2}\sqrt[3]{2a}}{\sqrt[3]{4ab}}$

Rationalize the denominator of each of the following expressions.

Examples   a. $\dfrac{1}{\sqrt{2}}$ b. $\sqrt{\dfrac{3x}{7y}}$ c. $\sqrt[3]{\dfrac{2}{y}}$

Solutions   a. $\dfrac{1}{\sqrt{2}}\dfrac{\sqrt{2}}{\sqrt{2}}$ b. $\dfrac{\sqrt{3x}\sqrt{7y}}{\sqrt{7y}\sqrt{7y}}$ c. $\dfrac{\sqrt[3]{2}\sqrt[3]{y}\sqrt[3]{y}}{\sqrt[3]{y}\sqrt[3]{y}\sqrt[3]{y}}$

               $\dfrac{\sqrt{2}}{2}$ $\dfrac{\sqrt{21xy}}{7y}$ $\dfrac{\sqrt[3]{2y^2}}{y}$

39. $\dfrac{1}{\sqrt{3}}$ 40. $\dfrac{3}{\sqrt{5}}$ 41. $\sqrt{\dfrac{x}{2}}$

42. $\sqrt{\dfrac{3}{y}}$ 43. $\sqrt[3]{\dfrac{3}{x^2}}$ 44. $\sqrt[3]{\dfrac{5}{x}}$

Rationalize the numerator of each expression.

45. $\dfrac{\sqrt{7}}{7}$ 46. $\dfrac{\sqrt{5}}{10}$ 47. $\dfrac{\sqrt{x}}{\sqrt{y}}$ 48. $\dfrac{\sqrt{xy}}{x}$

Write each sum as a product.

Examples   a. $4\sqrt{2}+3\sqrt{2}-\sqrt{2}$ b. $2\sqrt{3}+4\sqrt{12}$

Solutions   a. $(4+3-1)\sqrt{2}$ b. $2\sqrt{3}+4\cdot2\sqrt{3}$

               $6\sqrt{2}$ $10\sqrt{3}$

49. $\sqrt{3}+2\sqrt{3}$ 50. $3\sqrt{5}-6\sqrt{5}$ 51. $\sqrt{8}-\sqrt{50}-\sqrt{2}$

52. $\sqrt{50}+2\sqrt{32}-\sqrt{2}$ 53. $3\sqrt[3]{16}-\sqrt[3]{2}$ 54. $\sqrt[3]{54}+2\sqrt[3]{128}$

Multiply factors and write all radicals in the result in simplest form.

Examples   a. $\sqrt{x}(\sqrt{2x}-\sqrt{x})$ b. $(\sqrt{x}-2\sqrt{y})(2\sqrt{x}+\sqrt{y})$

Solutions   a. $x\sqrt{2}-x$ b. $2x-3\sqrt{xy}-2y$

55. $\sqrt{2}(3+\sqrt{3})$ 56. $\sqrt{6}(3+\sqrt{6})$ 57. $(3+\sqrt{5})(2-\sqrt{5})$

58. $(5-\sqrt{3})(5+\sqrt{3})$ 59. $(\sqrt{x}-2\sqrt{3})(\sqrt{x}+\sqrt{3})$ 60. $(2-\sqrt[3]{4})(2+\sqrt[3]{4})$

Rationalize denominators.

Examples   a. $\dfrac{3}{\sqrt{2}-1}$ b. $\dfrac{1}{\sqrt{x}-\sqrt{y}}$

Solutions   a. $\dfrac{3}{(\sqrt{2}-1)}\dfrac{(\sqrt{2}+1)}{(\sqrt{2}+1)}$ b. $\dfrac{1}{(\sqrt{x}-\sqrt{y})}\dfrac{(\sqrt{x}+\sqrt{y})}{(\sqrt{x}+\sqrt{y})}$

$\qquad\qquad\qquad \dfrac{3\sqrt{2}+3}{2-1}$ $\dfrac{\sqrt{x}+\sqrt{y}}{x-y}$

$\qquad\qquad\qquad 3\sqrt{2}+3$

61. $\dfrac{-4}{1+\sqrt{3}}$ 62. $\dfrac{1}{2-\sqrt{2}}$ 63. $\dfrac{x}{\sqrt{x}-3}$ 64. $\dfrac{\sqrt{x}}{\sqrt{x}-\sqrt{y}}$

Rationalize numerators.

65. $\dfrac{1-\sqrt{2}}{2}$ 66. $\dfrac{1-\sqrt{x+a}}{\sqrt{x+a}}$

In the previous problems, the variables were restricted to represent positive numbers. In Problems 67–70, consider variable bases to denote any elements of the set of real numbers and simplify. Hint: See Problems 41–46 on page 58.

67. $\sqrt{4x^2}$ 68. $\sqrt{9x^2y^4}$ 69. $\sqrt{9(x^3-x^2)}$ 70. $\sqrt{18(x^6-x^7)}$

3

Equations and inequalities in one variable

3.1

Equivalent equations; first-degree equations

We come now to one of the more important tasks in the study of elementary algebra—that of solving equations and inequalities over an ordered field—in particular, over the ordered field R of real numbers. In this chapter, we review a few of the procedures involved.

Equations and inequalities involving variables and constants are called **sentences**, while those involving only constants are called **statements**. Sentences can be classified as either **conditional sentences** or **identities**. An identity is a sentence that is true for every value in the replacement set of any variable or variables involved, while a conditional sentence is false for at least one value in the replacement set. For example, $x + 2 = 3$ is a conditional sentence for $x \in R$, while $x(x + 2) = x^2 + 2x$ is an identity over R.

If we replace the variable x in $P(x) = Q(x)$ with an element from its replacement set U and the resulting statement is true, then the element is a **root**, or **solution**, of the equation and is said to **satisfy** the equation. The subset of U consisting of all solutions of an equation is called the **solution set** of the equation in U. If two equations have the same solution set in U, then they are said to be **equivalent**.

Elementary transformations   To solve an equation over a given set, we usually determine the members of the solution set by inspection, or else we generate a sequence of equivalent equations until we arrive at one with an obvious solution set. The following theorem, which follows directly from the addition and multiplication laws for real numbers, is frequently used in generating equivalent equations over the set of real numbers.

Theorem 3.1   *If $P(x)$, $Q(x)$, and $R(x)$ are expressions, then for all values of x for which $P(x)$, $Q(x)$, and $R(x)$ are real numbers,*

$$P(x) = Q(x)$$

is equivalent to each of the following equations:

   I $P(x) + R(x) = Q(x) + R(x)$,

   II $P(x) \cdot R(x) = Q(x) \cdot R(x)$ *for $x \in \{x \mid R(x) \neq 0\}$.*

Any application of Theorem 3.1 is called an **elementary transformation**.

Solution of a
linear equation

The equation

$$ax + b = 0, \tag{1}$$

where $a, b \in R$ and $a \neq 0$, is a **first-degree**, or **linear**, equation over R. Any equation that can be reduced to this form by elementary transformations is therefore equivalent to a first-degree equation. We can show that such an equation has one and only one solution in R. By Theorem 3.1-I, Equation (1) is equivalent to

$$ax = -b; \tag{2}$$

and, by Theorem 3.1-II, Equation (2) is equivalent to

$$x = -\frac{b}{a}, \tag{3}$$

which, of course, has the unique solution $-b/a$. Since (1), (2), and (3) are equivalent, Equation (1) has the unique solution $-b/a$.

An equation containing more than one variable, or containing symbols such as a, b, and c, representing constants, can often be solved for one of the symbols in terms of the remaining symbols by applying elementary transformations until the desired symbol is obtained by itself as one member of an equation.

Example

Solve $cx = c - x$, for x.

Solution

We generate the following sequence of equivalent equations.

$$cx + x = c$$
$$x(c + 1) = c$$
$$x = \frac{c}{c + 1} \quad (c \neq -1)$$

Word problems

Equations can be used to express quantitative relationships in word problems symbolically. The following set of exercises includes problems that lead to linear equations.

Exercise 3.1

Solve. Consider R to be the replacement set of the variable.

1. $-3[x - (2x + 3) - 2x] = -9$

2. $4 - (x - 3)(x + 2) = 10 - x^2$

3. $6 + 3x - x^2 = 4 - (x - 2)(x + 3)$

4. $\dfrac{y}{2} + \dfrac{y}{3} - \dfrac{y}{4} = 7$

5. $\dfrac{2x - 1}{5} = \dfrac{x + 1}{2}$

6. $\dfrac{3}{5} = \dfrac{x}{x + 2}$

7. $\dfrac{2}{x - 9} = \dfrac{9}{x + 12}$

8. $\dfrac{x}{x - 2} = \dfrac{2}{x - 2} + 7$

9. $\dfrac{5}{x - 3} = \dfrac{x + 2}{x - 3} + 3$

10. $\dfrac{2}{y + 1} + \dfrac{1}{3y + 3} = \dfrac{1}{6}$

11. $\dfrac{y}{y + 2} - \dfrac{3}{y - 2} = \dfrac{y^2 + 8}{y^2 - 4}$

12. $\dfrac{4}{2x - 3} + \dfrac{4x}{4x^2 - 9} = \dfrac{1}{2x + 3}$

Solve. Assume that all constants are real numbers and that the replacement set of all variables is R. Leave the results in the form of an equation equivalent to the given equation. State any restrictions on the variables.

13. $v = k + gt$,   for k

14. $v = k + gt$,   for t

15. $A = \dfrac{h}{2}(b + c)$,   for c

16. $S = \dfrac{a}{1 - r}$,   for r

17. $l = a + (n - 1)d$,   for n

18. $\dfrac{1}{r} = \dfrac{1}{r_1} + \dfrac{1}{r_2}$,   for r

19. $x^2 y' - 3x - 2y^3 y' = 1$,   for y'

20. $2xy' - 3y' + x^2 = 0$,   for y'

21. $x_1 x_2 - 2x_1 x_3 = x_4$,   for x_1

22. $3x_1 x_3 + x_1 x_2 = x_4$,   for x_1

23. $\dfrac{y - y_1}{x - x_1} = 6$,   for y

24. $\dfrac{y - y_1}{x - x_1} = 2$,   for x

25. For what value of k does the equation $2x - 3 = \dfrac{4 + x}{k}$ have as its solution set $\{-1\}$?

26. Find a value of k in $3x - 1 = k$ so that the equation is equivalent to $2x + 5 = 1$.

27. When the length of each side of a square is increased by five inches, the area is increased by 85 square inches. Find the length of a side of the original square.

28. How much pure alcohol should be added to 12 ounces of a 45% solution to obtain a 60% solution?

29. A sum of $2000 is invested at simple interest, part at 5% and the remainder at 7%. Find the amount invested at each rate if the yearly income from the two investments is $112.

30. An airplane travels 1260 miles in the same time than an automobile travels 420 miles. If the rate of the airplane is 120 miles per hour greater than the rate of the automobile, find the rate of each.

31. Two cars start together and travel in the same direction, one going twice as fast as the other. At the end of three hours they are 96 miles apart. How fast is each traveling?

32. Clerk A can process 50 applications in four hours, and clerk B can process 50 applications in eight hours. How long will it take both clerks working together to process 100 applications?

3.2

Second-degree equations

The equation

$$ax^2 + bx + c = 0 \quad (a \neq 0)$$

is a **second-degree**, or **quadratic**, equation. Any equation that can be reduced to this form by elementary transformations is therefore equivalent to a quadratic equation. We shall designate the form shown as the **standard form** for such equations.

Solution by factoring   Theorem 3.2 will help us find solutions of quadratic equations.

Theorem 3.2 *If $P(x)$ and $Q(x)$ are expressions, then for all values of r for which $P(r)$ and $Q(r)$ are real numbers, r is a solution of $P(x) \cdot Q(x) = 0$ if and only if $P(r) = 0$, or $Q(r) = 0$, or both.*

The proof of this Theorem follows directly from Theorems 1.6 and 1.7.

Example   Find the solution set in R of $x^2 + 2x - 15 = 0$.

Solution   Since $x^2 + 2x - 15 = (x + 5)(x - 3)$ is an identity over R, we have $(x + 5)(x - 3) = 0$. But, by Theorem 3.2, $(x + 5)(x - 3) = 0$ is true if and only if $x + 5 = 0$ or $x - 3 = 0$, and we can see by inspection that the solutions to these equations are -5 and 3. Hence, the solution set we seek is $\{-5, 3\}$.

Number of solutions   In general, the solution set of a quadratic equation *over the real numbers* may contain two, one, or no real numbers as elements. The equation in the foregoing example has two real solutions. Now, consider the equation

$$x^2 - 2x + 1 = 0.$$

Since $x^2 - 2x + 1 = 0$ is equivalent to

$$(x - 1)^2 = 0,$$

and since the only value of x for which $(x - 1)^2 = 0$ is 1, this is the only member of the solution set.

Before exhibiting an example of a quadratic equation over the real numbers in the real variable x having an empty solution set, let us consider the special case of a quadratic equation, $x^2 - a = 0$, where $a > 0$. Since $x^2 - a = 0$ is equivalent to $x^2 = a$, and since $x^2 = a$ implies that x must be a square root of a, we have as the solution set $\{\sqrt{a}, -\sqrt{a}\}$. It is now easy to exhibit a quadratic equation having no real solutions, for example, $x^2 + 1 = 0$. Since there exists no number $x \in R$ such that its square is negative, the solution set in R for the given equation is $\emptyset$. In Chapter 11 when we discuss the set C of complex numbers, you will see that *every* quadratic equation over this set has a nonempty solution set in C.

Quadratic equations of the form

$$(x - a)^2 = b \quad (b \geq 0)$$

can be solved by observing that $x - a$ must be one of the square roots of b. That is, if $(x - a)^2 = b$, then either

$$x - a = \sqrt{b} \quad \text{or} \quad x - a = -\sqrt{b},$$

and conversely. Thus it is evident that the solution set is

$$\{a + \sqrt{b}, a - \sqrt{b}\}.$$

General solution   Once we know how to find solution sets for quadratic equations of the form $(x - a)^2 = b$, we can find the solution set of any quadratic equation. Let us first consider the general quadratic equation in standard form

$$ax^2 + bx + c = 0 \quad (a \neq 0), \quad a, b, c \in R,$$

for the special case in which $a = 1$, so that

$$x^2 + bx + c = 0. \tag{1}$$

We can write the equation in the equivalent form used above,

$$(x - p)^2 = q.$$

We begin by adding $-c$ to each member of (1). This yields

$$x^2 + bx \qquad = -c. \tag{2}$$

If we then add $(b/2)^2$ to each member of (2), we obtain

$$x^2 + bx + \left(\frac{b}{2}\right)^2 = -c + \left(\frac{b}{2}\right)^2, \tag{3}$$

in which the left-hand member is equal to $(x + b/2)^2$, and we have

$$\left(x + \frac{b}{2}\right)^2 = -c + \frac{b^2}{4}. \tag{4}$$

Since we have performed only elementary transformations, (4) is equivalent to (1), and we can solve (4) by the method given on page 69, provided that

$$-c + \frac{b^2}{4} \geq 0.$$

The technique used to obtain Equations (3) and (4) is called **completing the square**. We can determine the term necessary to complete the square in (2) by dividing the coefficient b of the first-degree term by the number 2 and squaring the result. The expression obtained, $x^2 + bx + (b/2)^2$, is called a **perfect square** and may be written in the form $(x + b/2)^2$.

Because the general quadratic equation

$$ax^2 + bx + c = 0 \quad (a \neq 0)$$

can be written equivalently in the form

$$x^2 + \frac{b}{a}x + \frac{c}{a} = 0,$$

the foregoing process can be applied to obtain the **quadratic formula**,

$$x = \frac{-b \pm \sqrt{b^2 - 4ac}}{2a},$$

where the roots of the general quadratic equation are expressed in terms of the coefficients. The symbol $\pm$ (read "plus or minus") is used to condense the writing of the two equations

$$x = \frac{-b + \sqrt{b^2 - 4ac}}{2a} \quad \text{and} \quad x = \frac{-b - \sqrt{b^2 - 4ac}}{2a}$$

into a single equation. We need only substitute the coefficients a, b, and c of a given quadratic equation in the formula to find the solution set for the equation.

Methods for
solving

We now have the following methods available to solve quadratic equations:

1. Factoring when convenient and using Theorem 3.2.
2. Extracting roots when the member containing the variable is a perfect square, completing the square if necessary.
3. Applying the quadratic formula, which is simply the end product of completing the square in the general case.

Determination
of number of
real solutions

An examination of the quadratic formula,

$$x = \frac{-b \pm \sqrt{b^2 - 4ac}}{2a},$$

suffices to show that if $ax^2 + bx + c = 0$ is to have a nonempty solution set in the set of real numbers, then $\sqrt{b^2 - 4ac}$ must be real. This in turn implies that only those quadratic equations for which $b^2 - 4ac \geq 0$ have real solutions. The number represented by $b^2 - 4ac$ is called the **discriminant** of the quadratic equation $ax^2 + bx + c = 0$. It yields the following information about the solution set of the equation for $a, b, c \in R$:

1. If $b^2 - 4ac = 0$, then there is precisely one real solution.
2. If $b^2 - 4ac < 0$, then there are no real solutions.
3. If $b^2 - 4ac > 0$, then there are two real solutions.

Solutions of
applied problems

In some cases, the mathematical model we obtain for a physical situation is a quadratic equation that has two real solutions. It may be that one but not both of the solutions to the equation fits the physical situation. For example, if we were asked to find two consecutive *natural numbers* having the product 72, we would write the equation

$$x(x + 1) = 72$$

as our model. Solving this equation, we have

$$x^2 + x - 72 = 0,$$

$$(x + 9)(x - 8) = 0,$$

with solution set $\{8, -9\}$. Since -9 is not a natural number, we must reject it as a possible solution of our original problem; the solution 8, however, leads to the consecutive natural numbers 8 and 9. For additional examples, observe that we would not accept -6 feet as the height of a man, or 27/4 for the number of persons in a room.

A quadratic equation used as a model for a physical situation may have two, one, or no meaningful solutions—meaningful, that is, in a physical sense. Answers to word problems should always be checked against the universal set of meaningful numbers for the original problem.

Exercise 3.2

Solve by factoring.

Example $x^2 + x = 30$

Solution   Write an equivalent equation in standard form and factor the left-hand member.

$$x^2 + x - 30 = 0$$

$$(x + 6)(x - 5) = 0$$

Determine solutions by inspection or set each factor equal to zero and solve each linear equation.

$$x + 6 = 0 \qquad x - 5 = 0$$

$$x = -6 \qquad x = 5$$

The solution set is $\{-6, 5\}$.

1. $x^2 + 2x = 0$

2. $x^2 - x = 5x$

3. $x^2 + 5x - 14 = 0$

4. $3x^2 - 6x = -3$

5. $x(2x - 3) = -1$

6. $(x - 2)(x + 1) = 4$

7. $3 = \dfrac{10}{x^2} - \dfrac{7}{x}$

8. $\dfrac{2}{x - 3} - \dfrac{6}{x - 8} = -1$

Solve for x by the extraction of roots.

Example $(x + 3)^2 = 7$

Solution   Set $x + 3$ equal to each square root of 7.

$$x + 3 = \sqrt{7} \qquad x + 3 = -\sqrt{7}$$

$$x = -3 + \sqrt{7} \qquad x = -3 - \sqrt{7}$$

The solution set is $\{-3 + \sqrt{7}, -3 - \sqrt{7}\}$.

9. $x^2 = 4$ 10. $9x^2 - 100 = 0$ 11. $x^2 = 5$

12. $(x - 1)^2 = 4$ 13. $(x - 6)^2 = 5$ 14. $(x - a)^2 = 4$

Solve by completing the square.

Example $2x^2 + x - 1 = 0$

Solution   Write an equivalent equation with the constant term as the right-hand member and the coefficient of x^2 equal to 1.

$$x^2 + \frac{1}{2}x = \frac{1}{2}$$

Add the square of one-half of the coefficient of the first-degree term to each member.

$$x^2 + \frac{1}{2}x + \frac{1}{16} = \frac{1}{2} + \frac{1}{16}$$

Rewrite the left-hand member as the square of an expression.

$$\left(x + \frac{1}{4}\right)^2 = \frac{9}{16}$$

Set $x + \frac{1}{4}$ equal to each square root of $\frac{9}{16}$.

$$x + \frac{1}{4} = \frac{3}{4} \qquad x + \frac{1}{4} = -\frac{3}{4}$$

$$x = \frac{1}{2} \qquad x = -1$$

The solution set is $\left\{\frac{1}{2}, -1\right\}$.

15. $x^2 + 4x - 12 = 0$ 16. $x^2 - 2x + 1 = 0$

17. $x^2 + 9x + 20 = 0$ 18. $x^2 - 2x - 1 = 0$

19. $2x^2 = 2 - 3x$ 20. $2x^2 + 4x = -1$

Reduce each of the following equations to equivalent equations of the form

$$(x - h)^2 + (y - k)^2 = r^2$$

by completing the squares in x and y.

Example $x^2 + y^2 - 4x + 6y = 5$

Solution   Write an equivalent equation in the form

$$[x^2 - 4x + (\ \ )] + [y^2 + 6y + (\ \ )] = 5 + (\ \ ) + (\ \ ).$$

Complete the squares in x and y.

$$[x^2 - 4x + 4] + [y^2 + 6y + 9] = 5 + 4 + 9$$

$$(x - 2)^2 + (y + 3)^2 = 18, \quad \text{or} \quad (x - 2)^2 + [y - (-3)]^2 = (\sqrt{18})^2$$

21. $x^2 + y^2 - 4x - 4y - 17 = 0$ 22. $x^2 + y^2 + 6x - 6y + 18 = 0$

23. $x^2 + y^2 + 6x - 2y + 6 = 0$ 24. $x^2 + y^2 - 2x + 4y + 2 = 0$

25. $4x^2 + 4y^2 - 4x + 8y = 11$ 26. $16x^2 + 16y^2 - 8x + 16y = 59$

Solve for x using the quadratic formula.

Example $\dfrac{x^2}{4} + \dfrac{x}{4} = 3$

Solution   Write an equivalent equation in standard form.

$$x^2 + x - 12 = 0$$

Substitute 1 for a, 1 for b, and -12 for c in the quadratic formula, and simplify.

$$x = \frac{-1 \pm \sqrt{1 + 48}}{2} = \frac{-1 \pm 7}{2}$$

The solution set is $\{3, -4\}$.

27. $x^2 - 3x + 2 = 0$ 28. $x^2 + 4x + 4 = 0$

29. $2x^2 = 7x - 6$ 30. $3x^2 = 5x - 1$

31. $\dfrac{x^2}{3} = \dfrac{1}{2}x + \dfrac{3}{2}$ 32. $\dfrac{x^2 - 3}{2} + \dfrac{x}{4} = 1$

33. $x^2 - 2\sqrt{5}x + 5 = 0$ 34. $x^2 + 2\sqrt{2}x + 2 = 0$

35. $2x^2 - \sqrt{3}x - 3 = 0$ 36. $2x^2 + \sqrt{7}x - 7 = 0$

37. $x^2 - kx - 2k^2 = 0$ 38. $2x^2 - kx + 3 = 0$

39. $ax^2 - x + c = 0$ 40. $x^2 + 2x + c + 3 = 0$

41. Determine k so that the roots of $kx^2 + 4x + 1 = 0$ will be equal. *Hint:* Use the discriminant.

42. Determine k so that $x^2 - kx + 9 = 0$ will have just one real root.

43. Determine k so that the roots of $x^2 + 2x + k + 3 = 0$ will be two real numbers.

44. Determine k so that the roots of $x^2 + 9x + k = 2$ will be two real numbers.

45. Find two consecutive natural numbers such that the sum of their squares is 85.

46. Two airplanes with their lines of flight at right angles pass each other (at slightly different altitudes) at noon. One is flying at 140 miles per hour and one at 180 miles per hour. How far apart are they at 12:30 PM?

47. A box without a top is to be made from a square piece of tin by cutting a two-inch square from each corner and folding up the sides. If the box is to hold 128 cubic inches, what should be the length of each side of the original square?

48. Suppose a ball thrown upward reaches a height h in feet given by the equation $h = 32t - 8t^2$, where t is the time in seconds after the throw. How long will it take the ball to reach a height of 24 feet on its way up? How long after the throw will the ball return to the ground?

49. The distance s a body falls in a vacuum is given by $s = v_0 t + \frac{1}{2}gt^2$, where s is measured in feet, t is measured in seconds, v_0 is the initial velocity in feet per second, and g is the constant of acceleration due to gravity (approximately 32 ft/sec/sec). How long will it take a body to fall 150 feet if v_0 is 20 feet per second? If the body starts from rest?

50. A man sailed a boat across a lake and back in two and a half hours. If his rate returning was two miles per hour less than his rate going, and if the distance each way was six miles, find his rate each way.

3.3

Equations containing radical expressions

Equations involving radicals   In order to find solution sets for equations containing radical expressions, we shall need the following result.

Theorem 3.3 *If $U(x)$ and $V(x)$ are expressions in x, then the solution set of $U(x) = V(x)$ is a subset of the solution set of $[U(x)]^n = [V(x)]^n$, for each natural number n.*

This theorem, which follows simply from the fact that products of equal numbers are equal numbers, permits us to raise both members of an equation to the same natural-number power with the assurance that we do not lose any solutions of the original equation in the process. On the other hand, it does not assert that the resulting equation will be equivalent to the original equation, and indeed it will not always be. The equation $[U(x)]^n = [V(x)]^n$ may have additional solutions, called

extraneous solutions, that are not solutions of $U(x) = V(x)$. Thus, if $a = b$, then $a^4 = b^4$, but the converse does not necessarily hold. That is, a^4 and b^4 may be equal, but $a \neq b$. For example, $(3)^4 = (-3)^4$, but $3 \neq -3$. The solution set of the equation $x^4 = 81$, obtained from $x = 3$ by raising each member to the fourth power, contains -3 as an extraneous real solution, since -3 does not satisfy the original equation even though it does satisfy $x^4 = 81$.

Necessity of checking solutions   Because the result of applying the foregoing process is not always an equivalent equation, each solution obtained through its use *must* be substituted for the variable in the original equation to check its validity.

Example   Find the solution set of $\sqrt[3]{x-1} = -1$.

Solution   If we raise each member of $\sqrt[3]{x-1} = -1$ to the third power, we obtain

$$(\sqrt[3]{x-1})^3 = (-1)^3,$$

$$x - 1 = -1,$$

which is equivalent to

$$x = 0.$$

Since $\sqrt[3]{0-1} = -1$, 0 is a solution of the original equation. Moreover, 0 is the only real solution, since Theorem 3.3 guarantees that the solution set of $\sqrt[3]{x-1} = -1$ is a subset of the solution set of $x = 0$.

Example   Find the solution set of $\sqrt{x+2} + 4 = x$.

Solution   We first write the equivalent equation,

$$\sqrt{x+2} = x - 4,$$

and then apply Theorem 3.3. We obtain

$$(\sqrt{x+2})^2 = (x-4)^2,$$

$$x + 2 = x^2 - 8x + 16.$$

This last equation is equivalent to

$$x^2 - 9x + 14 = 0,$$

$$(x-2)(x-7) = 0,$$

which clearly has solutions 2 and 7. Upon replacing x with 2 in the original equation, however, we obtain

$$\sqrt{2+2} + 4 = 2,$$

or

$$6 = 2,$$

which is false. Hence, 2 is not a solution of the original equation; it is an extraneous root. On the other hand, 7 does satisfy the original equation, so the solution set we seek is {7}.

It is sometimes necessary to apply Theorem 3.3 more than once in solving certain equations, such as those in Problems 7–10 of Exercise 3.3.

Substitution of variables   Some equations that are not polynomial equations can nevertheless be solved through the solutions of related polynomial equations.

Example   Find the solution set of $y + 2\sqrt{y} - 8 = 0$.

Solution   If we set $p = \sqrt{y}$ and substitute p for $\sqrt{y}$ and p^2 for y in the given equation, we have

$$p^2 + 2p - 8 = 0,$$
$$(p + 4)(p - 2) = 0,$$

which has -4 and 2 as solutions. Since $-4 < 0$, we must reject it as a source for solutions, because $p = \sqrt{y}$ and $\sqrt{y}$ is always nonnegative. The other value, $p = 2$, leads to $\sqrt{y} = 2$ and hence $y = 4$. The solution set we seek is {4}.

The technique of substituting one variable for another—or, more generally, a variable for an expression—is not limited to cases involving radicals, but is useful in any situation in which an equation is polynomial in form.

Exercise 3.3

Solve and check. If there is no solution, so state.

1. $\sqrt{x} = 8$
2. $\sqrt{y + 8} = 1$
3. $\sqrt[3]{2 - y} = 3$
4. $\sqrt[5]{7 - x} = 2$
5. $2x - 3 = \sqrt{7x - 3}$
6. $\sqrt{x + 3}\sqrt{x - 9} = 8$
7. $\sqrt{y + 4} = \sqrt{y + 20} - 2$
8. $\sqrt{x} + \sqrt{2} = \sqrt{x + 2}$
9. $\sqrt{5 + \sqrt{x}} = \sqrt{x} - 1$
10. $\sqrt{13 + \sqrt{x}} = \sqrt{x} + 1$
11. $(5 + x)^{1/2} + x^{1/2} = 5$
12. $(y + 7)^{1/2} + (y + 4)^{1/2} = 3$
13. $(y^2 - 3y + 5)^{1/2} - (y + 2)^{1/2} = 0$
14. $(z - 3)^{1/2} + (z + 5)^{1/2} = 4$

Solve. Leave the results in the form of an equation. Assume denominators not zero.

15. $r = \sqrt{\dfrac{A}{\pi}}$,   for A

16. $t = \sqrt{\dfrac{2v}{g}}$,   for g

17. $x\sqrt{xy} = 1$,   for y

18. $P = \pi\sqrt{\dfrac{l}{g}}$,   for g

19. $x = \sqrt{a^2 - y^2}$,   for y

20. $y = \dfrac{1}{\sqrt{1-x}}$,   for x

Solve for x, y, or z.

Example $x^4 - 10x^2 + 9 = 0$

Solution   Set $p = x^2$, substitute p for x^2 and p^2 for x^4, and solve for p.

$$p^2 - 10p + 9 = 0$$
$$(p - 9)(p - 1) = 0$$
$$p = 9 \quad \text{or} \quad p = 1$$

Set each value of $p = x^2$ and solve for x.

$$x^2 = 9 \qquad\qquad x^2 = 1$$
$$x = 3 \text{ or } -3 \qquad x = 1 \text{ or } -1$$

The solution set is $\{3,\ -3,\ 1,\ -1\}$.

21. $x - 2\sqrt{x} - 15 = 0$

22. $x^4 - 5x^2 + 4 = 0$

23. $2x^4 + 17x^2 - 9 = 0$

24. $z^4 - 2z^2 - 24 = 0$

25. $(y^2 + 5y)^2 - 8(y^2 + 5y) - 84 = 0$

26. $y^2 - 5 - 5\sqrt{y^2 - 5} + 6 = 0$

27. $y^{2/3} - 2y^{1/3} - 8 = 0$

28. $z^{2/3} - 2z^{1/3} - 35 = 0$

29. $y^{-2} - y^{-1} - 12 = 0$

30. $z^{-2} + 9z^{-1} - 10 = 0$

3.4
Solution of inequalities

Sentences such as

$$x + 3 \geq 10 \tag{1}$$

and

$$\frac{-2y - 3}{3} < 5 \tag{2}$$

are called **inequalities**. For appropriate values of the variable, one member of an inequality represents a real number that is less than ($<$), less than or equal to ($\leq$), greater than or equal to ($\geq$), or greater than ($>$) the real number represented by the other member.

Any element of the replacement set of the variable for which an inequality is valid is called a **solution**, and the set of all solutions of an inequality is called the **solution set** of the inequality. Inequalities that are true for every element in the replacement set of the variable—such as $x^2 + 1 > 0$, $x \in R$—are called **absolute inequalities**, or **unconditional inequalities**. Inequalities that are not true for every element of the replacement set are called **conditional inequalities**—for example, (1) and (2) above.

Elementary transformations   As in the case with equations, we shall solve a given inequality by generating a series of **equivalent inequalities** (inequalities having the same solution set), by means of elementary transformations, until we arrive at one with an obvious solution set. To do this, we shall need the following theorem applicable to inequalities. The proof of this theorem follows directly from the properties of order for the set of real numbers.

Theorem 3.4 *If $P(x)$, $Q(x)$, and $R(x)$ are expressions, then for all values of x for which $P(x)$, $Q(x)$, and $R(x)$ are real numbers,*

$$P(x) < Q(x)$$

is equivalent to each of the following sentences:

I $P(x) + R(x) < Q(x) + R(x)$.

II $P(x) \cdot R(x) < Q(x) \cdot R(x)$ *for* $x \in \{x \mid R(x) > 0\}$.

III $P(x) \cdot R(x) > Q(x) \cdot R(x)$ *for* $x \in \{x \mid R(x) < 0\}$.

Similarly,

$$P(x) \leq Q(x)$$

is equivalent to sentences of the form I–III, *with* $<$ *(or* $>$*) replaced by* $\leq$ *(or* $\geq$*) under the same conditions,* $R(x) > 0$ *and* $R(x) < 0$, *as above.*

Note that Theorem 3.4 does not permit multiplying by zero, and variables in multipliers are restricted from values for which the expression vanishes. The result of applying any part of this theorem is an elementary transformation.

Solution of a linear inequality   Theorem 3.4 can be applied to solve inequalities in the same way that the theorems of equality are applied to solve equations.

Example   Solve $\dfrac{x-3}{4} < \dfrac{2}{3}$.

Solution   Multiplying each member by 12, we have

$$3(x-3) < 8, \quad \text{or} \quad 3x - 9 < 8.$$

Adding 9 to each member gives

$$3x < 17.$$

Finally, multiplying each member by $\dfrac{1}{3}$, we obtain $x < \dfrac{17}{3}$,

and the solution set is

$$S = \left\{ x \,\middle|\, x < \frac{17}{3} \right\}.$$

Graphical representation of inequality solution set   The solution set in the foregoing example can be pictured on a number-line graph as shown in Figure 3.1. The heavy line indicates points with coordinates in the solution set.

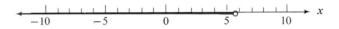

Figure 3.1

Inequalities sometimes appear in a form such as

$$-6 < 3x \le 15, \tag{3}$$

where an expression is bracketed between two inequality symbols. As observed in Section 1.5, this means $-6 < 3x$ *and* $3x \le 15$. The solution set of such an inequality is obtained in the same manner as the solution set of any other inequality. In (3) above, each expression may be multiplied by $1/3$ to obtain

$$-2 < x \le 5.$$

The solution set,

$$S = \{x \,|\, -2 < x \le 5\},$$

is shown on the number-line graph in Figure 3.2. Note that the open dot at the

Figure 3.2

left-hand endpoint of the interval indicates that -2 *is not* a member of the solution set, whereas the solid dot at the other end indicates that 5 *is* a member of the solution set.

Solution of a quadratic inequality
    Quadratic inequalities offer somewhat different problems. For example, consider the inequality

$$x^2 + 4x < 5.$$

To determine values of x for which this condition holds, we might first rewrite the sentence equivalently as

$$x^2 + 4x - 5 < 0,$$
$$(x + 5)(x - 1) < 0.$$

It is clear here that only those values of x for which the factors $x + 5$ and $x - 1$ are opposite in sign will be in the solution set, which can be determined analytically by noting that $(x + 5)(x - 1) < 0$ implies that either

$$x + 5 < 0 \quad \text{and} \quad x - 1 > 0$$

or

$$x + 5 > 0 \quad \text{and} \quad x - 1 < 0.$$

Each of these two cases can be considered separately.

First, $x + 5 < 0$ and $x - 1 > 0$ imply $x < -5$ and $x > 1$, conditions which are not satisfied by any values of x. But $x + 5 > 0$ and $x - 1 < 0$ imply $x > -5$ and $x < 1$, which lead to the solution set,

$$S = \{x \mid -5 < x < 1\}.$$

An alternative set notation for this solution set is

$$S = \{x \mid x > -5\} \cap \{x \mid x < 1\}.$$

Use of sign graphs in solving inequalities
    One relatively easy way to visualize the solution set of a quadratic inequality is to indicate on a number line the signs associated with each factor for number replacements for the variable. Figure 3.3 on page 82 shows such an arrangement, or **sign graph**, for the example above; it is constructed by first showing on the top number line the places where $x + 5$ is positive ($x > -5$) and the places where it is negative ($x < -5$), and then showing on a second line those places where $x - 1$ is positive ($x > 1$) and those where it is negative ($x < 1$). The bottom line can then be marked by observing those parts of the first two lines where the signs are alike and those parts where the signs are opposite. Since it is desired that the product $(x + 5)(x - 1)$ be negative, the third line shows clearly that this occurs where $-5 < x < 1$, so that the solution set of the inequality is $\{x \mid -5 < x < 1\}$.

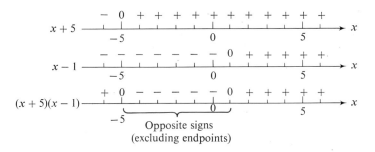

Figure 3.3

Multiplication of an inequality by a variable   Inequalities involving fractions have to be approached with care if any fraction contains a variable in the denominator. If Theorem 3.4 is invoked to multiply each member by an expression containing the variable, we have to be careful either to distinguish between those values of the variable for which the expression denotes a positive and negative number, respectively, or else make sure that the expression by which we multiply is always positive. An alternative approach using a sign graph is shown in an example in the following exercise set.

Exercise 3.4

Solve and represent the solution set on a line graph.

1. $x + 7 > 8$

2. $x - 5 \leq 7$

3. $3x - 2 > 1 + 2x$

4. $2x + 3 \leq x - 1$

5. $\dfrac{2x - 3}{2} \leq 5$

6. $\dfrac{3x + 4}{3} > 12$

Graph each of the following sets and rewrite in simpler set notation.

Example $\{x \mid x + 2 \geq 0\} \cap \{x \mid x - 3 < 1\}$

Solution   Solve each inequality and graph. Indicate the region where the graphs overlap.

$$\{x \mid -2 \leq x < 4 \mid\}$$

7. $\{x\,|\,x - 2 < 3\} \cap \{x\,|\,x + 4 > 2\}$

8. $\left\{x\,\middle|\,\dfrac{1+x}{2} \leq 3\right\} \cap \{x\,|\,x \leq 6\}$

9. $\{x\,|\,2x - 1 > 5\} \cap \left\{x\,\middle|\,\dfrac{x-1}{3} \geq 4\right\}$

10. $\{x\,|\,x > 2\} \cap \left\{x\,\middle|\,\dfrac{2x+5}{2} < 0\right\}$

Find the solution set of each inequality.

11. $(x + 1)(x - 2) > 0$

12. $(x + 2)(x + 5) < 0$

13. $x^2 - 3x - 4 > 0$

14. $x^2 - 5x - 6 \geq 0$

15. $x^2 < 5$

16. $4x^2 + 1 < 0$

Example

$$\frac{x}{x - 2} \geq 5$$

Solution

We can approach this directly by means of a sign graph. We first rewrite the given inequality equivalently as

$$\frac{x}{x - 2} - 5 \geq 0,$$

from which we obtain

$$\frac{-4x + 10}{x - 2} \geq 0.$$

For this to be valid, $x - 2$ must not be 0, and the numerator and denominator must be of like sign. The sign graph shows that the quotient $(-4x + 10)/(x - 2)$ is positive or zero for x between 2 and 5/2, including 5/2 but excluding 2, a value of x for which the denominator is 0. The desired solution set is therefore $\{x\,|\,2 < x \leq 5/2\}$, with graph as shown on the last line of the sign graph.

Like signs
(excluding left endpoint)

17. $\dfrac{2}{x} \leq 4$

18. $\dfrac{3}{x-6} > 8$

19. $\dfrac{x}{x+2} > 4$

20. $\dfrac{x+2}{x-2} \geq 6$

21. $\dfrac{2}{x-2} \geq \dfrac{4}{x}$

22. $\dfrac{3}{4x+1} > \dfrac{2}{x-5}$

23. $x(x-2)(x+3) > 0$

24. $x^3 - 4x \leq 0$

25. A student must have an average of at least 80%, but less than 90%, on five tests in a course to receive a B. His grades on the first four tests were 98%, 76%, 86%, and 92%. What grade on the fifth test would give him a B in the course?

26. Fahrenheit and centigrade temperatures are related by $C = \dfrac{5}{9}(F - 32)$. Within what range must the temperature be in Fahrenheit degrees for the temperature in centigrade degrees to lie between $-10°$ and $20°$, inclusive?

3.5
Sentences involving absolute values

Equations involving absolute values   In Section 1.5, we defined the absolute value of a real number by

$$|x| = \begin{cases} x, & \text{if } x \geq 0, \\ -x, & \text{if } x < 0. \end{cases}$$

More generally, then, the expression $|x - a|$ satisfies

$$|x - a| = \begin{cases} x - a, & \text{if } x - a \geq 0, \text{ or, equivalently, if } x \geq a, \\ -(x - a), & \text{if } x - a < 0, \text{ or, equivalently, if } x < a, \end{cases}$$

and can be interpreted on a number line as denoting the distance the graph of x is located from the graph of a, as shown in Figure 3.4.

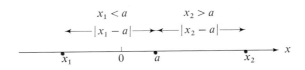

Figure 3.4

Example   Find the solution set of $|x - 3| = 5$.

Solution 1   This can be solved by inspection. Since $x - 3$ represents the distance the graph of x is located from the graph of 3, and since by the equation this distance is 5, the two solutions of the equation are $3 + 5$, or 8, and $3 - 5$, or -2. Thus, the solution set is $\{-2, 8\}$.

Solution 2   By definition, $|x - 3| = 5$ implies that

$$x - 3 = 5$$

or

$$-(x - 3) = 5.$$

The equation $x - 3 = 5$ is equivalent to $x = 8$, and $-(x - 3) = 5$ is equivalent to $x = -2$. Hence the solution set is given by

$$S = \{x \mid x = 8\} \cup \{x \mid x = -2\} = \{-2, 8\}.$$

Inequalities involving absolute values   Inequalities involving absolute-value notation require some additional discussion. For example, consider the inequality

$$|x + 1| > 3.$$

By definition, this inequality is equivalent to

$$x + 1 > 3 \quad \text{for} \quad x + 1 \geq 0,$$

and to

$$-(x + 1) > 3 \quad \text{for} \quad x + 1 < 0,$$

so that the solution set is given by

$$S = \{x \mid x > 2 \text{ or } x < -4\}. \tag{1}$$

Alternatively, we could also write

$$S = \{x \mid x > 2\} \cup \{x \mid x < -4\},$$

where the union gives a precise meaning to the word "or" used in (1). The graph of the solution set is shown in Figure 3.5.

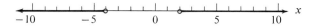

Figure 3.5

Exercise 3.5

Solve.

Example $|x + 5| = 8$

Solution   Write the equation as two first-degree equations and solve each of them.

$$x + 5 = 8 \qquad -(x + 5) = 8$$
$$x = 3 \qquad x = -13$$

The solution set is {3, −13}.

1. $|x| = 6$

2. $|x| = 3$

3. $|x - 1| = 4$

4. $|x - 6| = 3$

5. $\left| x - \dfrac{2}{3} \right| = \dfrac{1}{3}$

6. $\left| x - \dfrac{3}{4} \right| = \dfrac{1}{2}$

Solve and graph each solution set on a number line.

Example $|3x - 6| < 9$

Solution   By definition, the inequality is equivalent to

$$3x - 6 < 9 \quad \text{for} \quad 3x - 6 \geq 0, \quad \text{or} \quad x \geq 2,$$

and to

$$-(3x - 6) < 9 \quad \text{for} \quad 3x - 6 < 0, \quad \text{or} \quad x < 2,$$

Hence the solution set is given by

$$[\{x \,|\, x < 5\} \cap \{x \,|\, x \geq 2\}] \cup [\{x \,|\, x > -1\} \cap \{x \,|\, x < 2\}],$$

or alternatively,

$$\{x \,|\, -1 < x < 5\}.$$

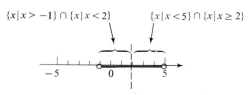

7. $|x| < 2$

8. $|x - 1| > 2$

9. $|x + 3| \geq 4$

10. $|x + 1| \leq 8$

11. $|2x - 5| \geq 3$

12. $|2x + 4| < -1$

Replace each of the following inequalities with a single inequality involving an absolute-value symbol.

Example $-3 < x < 7$

Solution   Since the average of 7 and -3 is $[7 + (-3)]/2 = 4/2 = 2$, the values of x are centered about 2. Subtracting 2 from each member, we have:

$$-5 < x - 2 < 5, \qquad |x - 2| < 5.$$

13. $1 < x < 3$ 14. $-5 \leq x \leq 9$ 15. $-9 \leq x \leq -7$

16. $5 < x < 13$ 17. $-7 \leq 2x \leq 12$ 18. $-5 < 3x < 10$

In Problems 19–22, consider $a, b \in R$.

19. Show that $|-a| = |a|$. *Hint:* Consider two possible cases, a nonnegative and a negative.

20. Show that $|a - b| = |b - a|$. *Hint:* Consider two possible cases, $a - b \geq 0$ and $a - b < 0$.

21. Show that $|a^2| = |a|^2 = a^2$.

22. Show that $|ab| = |a| \cdot |b|$. *Hint:* Consider four possible cases, with a nonnegative and negative, and b nonnegative and negative.

4

Relations and functions

4.1

Ordered pairs of real numbers

When the order in which the numbers of a number pair are to be considered is specified, the pair is called an **ordered pair**, and the pair is denoted by a symbol such as (3, 2), (2, 3), (−1, 5), or (0, 3). Each of the two numbers in an ordered pair is called a **component** of the ordered pair, the first and second being called the **first component** and the **second component**, respectively. Having established what is meant by an ordered pair, we are ready to define a set operation involving such pairs.

Definition 4.1 *The Cartesian product of two sets A and B, denoted by A × B, is the set of all ordered pairs (x, y) such that x ∈ A and y ∈ B.*

$R \times R$, or R^2, and the geometric plane

The most important Cartesian product with which we shall be concerned is that formed from the set R of real numbers. The product $R \times R$, which is often denoted by R^2, is the set of all possible ordered pairs of real numbers. The fact that each member of R^2 corresponds to a point in the geometric plane, and the coordinates of each point in the geometric plane are the components of a member of R^2, is the basis for all plane graphing. As you probably recall from your earlier study of algebra, the correspondence between points in the plane and ordered pairs of real numbers is usually established through a **Cartesian (or rectangular) coordinate system**, as shown in Figure 4.1.

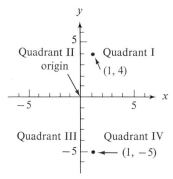

Figure 4.1

Solutions of an
equation in
two variables

Equations in two variables, such as

$$3x + 2y = 12, \quad x^2y + 3x = y^5, \quad \text{and} \quad \sqrt{xy} = y^2 - 5,$$

with $x, y \in R$, have, as solutions, ordered pairs of numbers. For example, if the components of (2, 3) are substituted for the variables x and y, in that order, in the equation

$$3x + 2y = 12, \tag{1}$$

the result is

$$3(2) + 2(3) = 12,$$

which is true. Hence, (2, 3) is a solution of Equation (1). In this book, the first component of an ordered pair is a value for the **abscissa** x, and the second component a value for the **ordinate** y. Since many equations (and inequalities) in two variables have an infinite number of solutions, we shall use the set-builder notation

$$\{(x, y) \mid \text{condition on } x \text{ and } y\}$$

to represent the set of all solutions.

Subsets
of $R \times R$

Any condition on x and y—that is, any sentence in two variables, x and y—expresses a relationship between elements in the replacement sets of the two variables, and this relationship is precisely represented by the solution set of the sentence in two variables. For $x, y \in R$, the solution set is always a subset of $R \times R$. This leads us to the following definition.

Definition 4.2   *Any subset of $R \times R$ is a relation in R.*

The relation is said to be in R because the components of the ordered pairs in the relation are elements of R. Alternatively, we frequently refer to the relationship as being in $R \times R$.

The set of all first components in the ordered pairs in a relation is called the **domain** of the relation, and the set of all second components is called the **range** of the relation. Thus

elements in the domain
$\{(2, 5), (3, 10), (4, 15)\}$
elements in the range

is a relation with domain $\{2, 3, 4\}$ and range $\{5, 10, 15\}$.

If a relation is defined by an equation and the domain is not specified, we shall understand that the domain is *the set of all real numbers for which a real number exists in the range* (such relations are called real-valued relations of a real variable). For example, the domain of

$$S = \left\{(x, y) \,\Big|\, y = \frac{1}{x - 2}\right\}$$

is $\{x \,|\, x \in R, x \neq 2\}$, because for every real number x except 2, the expression $1/(x - 2)$ represents a real number.

A special kind of relation that is very important in mathematics is called a *function*.

Definition 4.3   *A function is a relation in which no two ordered pairs have the same first component and different second components.*

Graphical characterization of a function   A function, therefore, associates each element in its domain with one and only one element in its range. In a graphical sense, this implies that no two of the ordered pairs in a function graph into points on the same vertical line.

As you should recall from your earlier study of algebra, graphs in R^2, that is, in $R \times R$, are often continuous lines and curves. Figure 4.2 shows three such graphs.

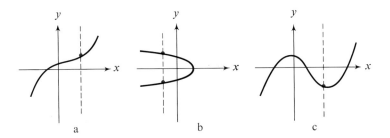

Figure 4.2

Imagine a vertical line moving across each of these from left to right. If the line at any position cuts the graph of the relation in more than one point, then the relation is not a function. Thus, although Figures 4.2-a and 4.2-c show the graphs of relations that are functions, Figure 4.2-b shows the graph of a relation that is not a function, because the vertical line shown in the figure meets the graph in two places. This means that, for the particular value of x involved, the relation associates two distinct values for y.

Algebraic characterization of a function | When a relation is defined by an equation, one way to test whether or not the relation is a function is to solve the equation explicitly for the variable y representing an element in the range. This will show whether or not more than one value of y is associated with any single value of x.

Example | Is the relation $\{(x, y) | y^2 = 1 + x^2\}$ in R^2 a function?

Solution | Since $y^2 = 1 + x^2$ implies either $y = \sqrt{1 + x^2}$ or $y = -\sqrt{1 + x^2}$, the assignment of a real value to x will result in two different values for y, and hence the relation is not a function.

Function notation | In Section 2.1 you became acquainted with a notation that is widely used in discussing functions. In general, functions are denoted by single symbols; for example, f, g, h, and F might designate functions. The symbol for a function can be used in conjunction with the variable representing an element in the domain to represent the associated element in the range. Thus $f(x)$, read "f of x" or "the value of f at x," is the element in the range of f associated with the element x in the domain.

Suppose

$$f = \{(x, y) | y = x + 3\}.$$

The alternative notation

$$f = \{(x, f(x)) | f(x) = x + 3\}$$

can be used, where $f(x)$ plays the same role as y.

Inverse functions | By Definition 4.3, a function is a set of ordered pairs (x, y) such that no two have the same first component and different second components. When the components of every ordered pair in a function f are interchanged, another set of ordered pairs results, a set of ordered pairs which may or may not be a function. For example, if

$$(1, 5), (2, 5), \text{ and } (3, 6)$$

are ordered pairs in f, then

$$(5, 1), (5, 2), \text{ and } (6, 3)$$

will be members of the relation formed by interchanging the components of the ordered pairs in f. Clearly these latter pairs cannot be members of a function, since two of them, $(5, 1)$ and $(5, 2)$, have the same first component and different second components. If, however, a function f is *one-to-one*—that is, if no two ordered pairs in f have the same second components (same values for y)—then the relation obtained by interchanging the first and second components of every pair in the function will also be a function. This function is called the *inverse function* of *f*. The notation f^{-1} (read "*f* inverse") is frequently used to denote the inverse function of *f*.

Definition 4.4 *If the function f is such that no two of its ordered pairs with different first components have the same second component, then the inverse function f^{-1} is the set of ordered pairs obtained from f by interchanging the first and second components of each ordered pair in f.*

It is evident from this definition that the domain and range of f^{-1} are just the range and domain, respectively, of *f*. If $y = f(x)$ defines a function *f*, and if *f* is one-to-one, then an interchange of the variables *x* and *y* yields the equation $x = f(y)$, which defines the inverse of *f*. For example, the inverse of the function defined by

$$y = 3x + 2$$

is defined by

$$x = 3y + 2,$$

or, when *y* is expressed in terms of *x*, by

$$y = \frac{1}{3}x - \frac{2}{3}.$$

Exercise 4.1

Supply the missing components so that the ordered pairs

a. (0,)   b. (1,)   c. (2,)   d. (−3,)   e. $\left(\frac{2}{3},\ \ \right)$

are solutions of the given equations.

1. $2x + y = 6$ 2. $y = 9 - x^2$ 3. $y = \dfrac{3x}{x^2 - 2}$

4. $y = 0$ 5. $y = \sqrt{3x + 11}$ 6. $y = |x - 1|$

a. *Specify the domain and the range of each relation.*
b. *State whether or not each relation is a function.*

Example $\{(3, 5), (4, 8), (4, 9), (5, 10)\}$

Solution
    a. Domain (the set of first components): $\{3, 4, 5\}$
        Range (the set of second components): $\{5, 8, 9, 10\}$
    b. The relation is not a function, because two ordered pairs, $(4, 8)$ and $(4, 9)$, have the same first components.

7. $\{(2, 3), (5, 7), (7, 8)\}$ 8. $\{(-1, 6), (0, 2), (3, 3)\}$

9. $\{(2, -1), (3, 4), (3, 6)\}$ 10. $\{(-4, 7), (-4, 8), (3, 2)\}$

11. $\{(5, 5), (6, 6), (7, 7)\}$ 12. $\{(0, 0), (2, 4), (4, 2)\}$

Specify the maximum domain of the real numbers that would yield real numbers y for elements in the range of the relation defined by each equation.

Examples   a. $y = \sqrt{16 - x^2}$ b. $y = \dfrac{1}{x(x + 2)}$

Solutions
    a. For what values of x is   b. For what values of x is
       $16 - x^2 \geq 0$? $x(x + 2) \neq 0$?

       The domain is   The domain is
       $\{x \mid -4 \leq x \leq 4\}$. $\{x \mid x \neq 0, -2\}$.

13. $y = x + 7$ 14. $y = 2x - 3$ 15. $y = x^2$

16. $y = \dfrac{1}{x}$ 17. $y = \dfrac{1}{x - 2}$ 18. $y = \dfrac{1}{x^2 + 1}$

19. $y = \sqrt{x}$ 20. $y = \sqrt{4 - x}$ 21. $y = \sqrt{4 - x^2}$

22. $y = \sqrt{x^2 - 9}$ 23. $y = \dfrac{4}{x(x - 1)}$ 24. $y = \dfrac{x}{(x - 1)(x + 2)}$

State whether or not the given equation defines a function.

Examples   a. $x^2 y = 3$ b. $x^2 + y^2 = 36$

Solutions   Solve explicitly for y.

       a. $y = \dfrac{3}{x^2}$ b. $y = \pm\sqrt{36 - x^2}$

       Yes. There is only one value   No. There are two values of y
       of y associated with each   associated with values of x
       value of x $(x \neq 0)$.   satisfying $|x| < 6$.

25. $x + y = 3$ 26. $y = -x^2$ 27. $y = \sqrt{x^2 - 5}$

28. $y = \sqrt{16 - x^2}$ 29. $x^2 + y^2 = 16$ 30. $y = \pm\sqrt{x^2}$

31. $y^2 = x^3$ 32. $y = ax^n$

If $f(x) = x + 2$, find the given element in the range.

Example $f(3)$

Solution   Substitute 3 for x.

$$f(3) = 3 + 2 = 5$$

The element is 5.

33. $f(0)$ 34. $f(1)$ 35. $f(-3)$ 36. $f(a)$

If $g(x) = x^2 - 2x + 1$, find the given element in the range.

37. $g(-2)$ 38. $g(0)$ 39. $g(a + 1)$ 40. $g(a - 1)$

If $f(x) = x + 2$ defines a function, find the element in the domain of f associated with the given element in the range.

Example $f(x) = 5$

Solution   Replacing $f(x)$ with $x + 2$, we have

$$x + 2 = 5,$$

$$x = 3.$$

The element is 3.

41. $f(x) = 3$ 42. $f(x) = -2$ 43. $f(x) = a$ 44. $f(x) = a + 2$

If $g(x) = x^2 - 1$, find all elements in the domain of g associated with the given element in the range.

45. $g(x) = 0$ 46. $g(x) = 3$ 47. $g(x) = 8$ 48. $g(x) = 5$

49. Suppose $f(x) = x + 2$ and $g(x) = x - 2$. Find each of the following.

     a. $f(0)$ b. $g(2)$ c. $f(g(2))$ d. $f(g(x))$

50. If $f(x) = x^2 - x + 1$, find each of the following.

a. $f(x + h) - f(x)$

b. $\dfrac{f(x + h) - f(x)}{h}$

Any function satisfying the condition that $f(-x) = f(x)$ for all x in the domain is called an **even function**. *Any function satisfying the condition that $f(-x) = -f(x)$ for all x in the domain is an* **odd** **function**. *Which of the following functions are even and which are odd?*

51. a. $\{(x, f(x)) | f(x) = x^2\}$ b. $\{(x, f(x)) | f(x) = x^3\}$

52. a. $\{(x, f(x)) | f(x) = x^4 - x^2\}$ b. $\{(x, f(x)) | f(x) = x^3 - x\}$

Each of the following equations defines a function. Write an equation defining the inverse relation and state whether or not the inverse relation is a function.

Example $y = 2x - 8$

Solution An equation defining the inverse is obtained by interchanging the variables. Hence,

$$x = 2y - 8$$

is an equation of the inverse. Solving explicitly for y, we obtain the equivalent equation

$$y = \frac{x + 8}{2}.$$

Since with each x this equation pairs only one y, the inverse is also a function.

53. $y = x + 2$ 54. $y = 3x - 1$ 55. $2x - 3y = 6$

56. $x + 4y = 0$ 57. $y = x^2 - 3$ 58. $x^2 - 2y = 4$

4.2
Linear functions

A **first-degree equation**, or **linear equation**, in x and y is an equation that can be written equivalently in the form

$$Ax + By + C = 0 \quad (A \text{ and } B \text{ not both } 0). \tag{1}$$

Graphs of first-degree equations   We shall call (1) the **standard form** for a linear equation. The graph of any such equation (technically, of its solution set) in R^2 is a straight line, although we do not prove this here.

For $B \neq 0$, such an equation defines a **linear function** with domain the set of real numbers x. Since any two distinct points determine a straight line, it is evident that we need find only two solutions of such an equation to determine its graph, that is, the graph of the solution set of the equation. In practice, the two solutions easiest to find are usually those whose first and second components, respectively, are zero, that is, the solutions $(0, y)$ and $(x, 0)$. The x-coordinate of the point at which the graph crosses the x-axis is called the **x-intercept**, and the y-coordinate of the point at which the graph crosses the y-axis is called the **y-intercept**. As an example, consider the function

$$f = \{(x, y) \mid 3x + 4y = 12\}. \qquad (2)$$

If $y = 0$, we have $x = 4$, and the x-intercept is 4. If $x = 0$, then $y = 3$, and the y-intercept is 3. Thus the graph of (2) appears as in Figure 4.3.

If the graph intersects both axes at or near the origin, either the intercepts do not represent two separate points, or the points are too close together to be of much use in drawing the graph. It is then necessary to plot at least one other point at a distance far enough removed from the origin to establish the line accurately.

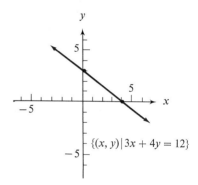

Figure 4.3

Equations of horizontal lines   There are two special cases of linear equations worth noting. First, an equation such as

$$y - 4 = 0$$

may be considered an equation in two variables in R^2,

$$0x + y = 4.$$

For each x, this equation assigns $y = 4$. That is, any ordered pair of the form $(x, 4)$ is a solution of the equation. For instance,

$$(1, 4), (2, 4), (3, 4)$$

are all solutions of the equation. If we graph these points and connect them with a straight line, we have Figure 4.4-a.

Since the equation

$$y - 4 = 0$$

assigns to each x the same value for y, the function defined by this equation is called a **constant function**.

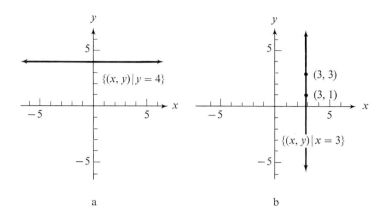

Figure 4.4

Equations of vertical lines   The other special case of the linear equation is of the type

$$x - 3 = 0,$$

which may be looked upon in R^2 as

$$x + 0y = 3.$$

Here, only one value is permissible for x, namely 3, whereas any value may be assigned to y. That is, any ordered pair of the form $(3, y)$ is a solution of this equation. If we choose two solutions, say $(3, 1)$ and $(3, 3)$, and complete the graph, we have Figure 4.4-b. It is clear from the graph that the equation does not define a function. This fact accounts for our restriction $B \neq 0$ on the standard form of a first-degree equation in two variables, $Ax + By + C = 0$, in order that this equation should define a function.

Distance between two points   Any two distinct points in the plane are the endpoints of a line segment. Two fundamental properties of a line segment are its **length** and its **inclination** with respect to the x-axis.

Figure 4.5 on page 98 shows the line segment joining points $P_1(x_1, y_1)$ and $P_2(x_2, y_2)$. If a line parallel to the x-axis is constructed through P_1, and a line parallel to the y-axis through P_2, then these lines will intersect at a point P_3 with coordinates (x_2, y_1). The distance from P_1 to P_3 is then $|x_2 - x_1|$, and the distance from P_2 to P_3 is $|y_2 - y_1|$. The Pythagorean theorem applied to these distances yields the length d of the line segment joining P_1 and P_2, namely,

$$d = \sqrt{|x_2 - x_1|^2 + |y_2 - y_1|^2}.$$

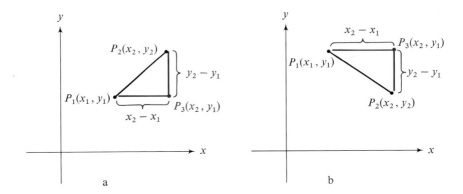

Figure 4.5

Since $|x_2 - x_1|^2 = (x_2 - x_1)^2$ and $|y_2 - y_1|^2 = (y_2 - y_1)^2$, we accordingly have

$$d = \sqrt{(x_2 - x_1)^2 + (y_2 - y_1)^2}.$$

This is known as the **distance formula**.

Slope of a line   The inclination of the line segment joining P_1 and P_2 is measured by forming the ratio of the differences $y_2 - y_1$ and
$x_2 - x_1$, and is called the **slope** m of the line segment. Thus

$$m = \frac{y_2 - y_1}{x_2 - x_1} \quad (x_2 - x_1 \neq 0).$$

For the segment in Figure 4.5-a the slope is positive, while for the one in Figure
4.5-b the slope is negative. If a line segment is parallel to the x-axis, then $y_2 - y_1 = 0$
and the line segment has slope 0; but if it is parallel to the y-axis, then $x_2 - x_1 = 0$
and its slope is not defined (see Figure 4.6).

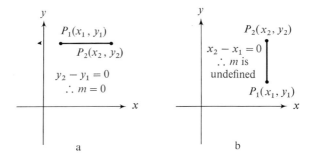

Figure 4.6

It can be shown (by similar triangles, say) that the slopes of any two line segments contained in the same line are equal, and hence we can say that the slope of a line is equal to the slope of any of its segments.

Forms of linear equations   Using the slope concept, we can rewrite the defining equation for a linear function in several useful forms. Consider a line having slope m and passing through a given point (x_1, y_1), as shown in Figure 4.7. If we choose *any other* point on the line and assign to it the coordinates (x, y), it is evident that the slope of the line is given by

$$\frac{y - y_1}{x - x_1} = m,$$

from which

$$y - y_1 = m(x - x_1). \qquad (3)$$

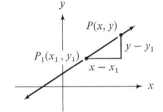

Figure 4.7

Note that (3) is satisfied also by $(x, y) = (x_1, y_1)$. Since now x and y are the co-ordinates of *any* point on the line, (3) is an equation of the line passing through (x_1, y_1) with slope m. This is called the **point-slope form** for a linear equation.

Now consider an equation of the line with slope m that passes through a given point on the y-axis having coordinates $(0, b)$, as shown in Figure 4.8. Substituting $(0, b)$ in the point-slope form of a linear equation,

$$y - y_1 = m(x - x_1),$$

we obtain

$$y - b = m(x - 0),$$

from which

$$y = mx + b. \qquad (4)$$

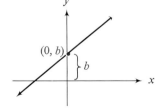

Figure 4.8

Equation (4) is called the **slope-intercept form** for a linear equation. Any linear equation in the standard form

$$Ax + By + C = 0$$

can be written equivalently in the slope-intercept form by solving for y in terms of x if $B \neq 0$. For example,

$$2x + 3y - 6 = 0$$

can be written equivalently as

$$y = -\frac{2}{3}x + 2.$$

The slope of the line, $-2/3$, and the y-intercept, 2, can now be read directly from the last form of the equation.

Graph of the inverse of a linear function In Section 4.1 we defined the inverse, f^{-1}, of a function f as the set of ordered pairs obtained from f by interchanging the first and second components of each ordered pair in f.

The graphs of inverse relations are related in an interesting way. To see this, we first observe that the graphs of the ordered pairs (a, b) and (b, a) are always located symmetrically with respect to the graph of $y = x$ in Figure 4.9. Therefore, because for every ordered pair (a, b) in Q the ordered pair (b, a) is in Q^{-1}, the graphs of $y = Q^{-1}(x)$ and $y = Q(x)$ are reflections of each other about the graph of $y = x$.

Figure 4.10 shows the graphs of the linear function

$$Q = \{(x, y) \mid y = 4x - 3\}$$

and its inverse

$$Q^{-1} = \{(x, y) \mid x = 4y - 3\} = \left\{(x, y) \mid y = \frac{1}{4}(x + 3)\right\}$$

together with the graph of $y = x$.

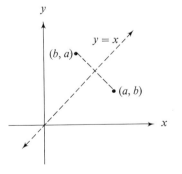

Figure 4.9

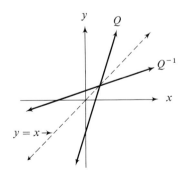

Figure 4.10

Exercise 4.2

Graph each of the following equations.

Example $2x - 3y = 6$

Solution   Find the intercepts of the graph.
When $x = 0$ we have $y = -2$, and
when $y = 0$ we have $x = 3$. Hence
the y-intercept is -2 and the
x-intercept is 3. Graph the points
$(0, -2)$ and $(3, 0)$ and draw a line
through the points.

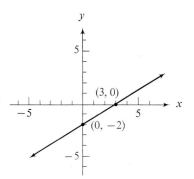

1. $y = x + 5$ 2. $y = x - 7$

3. $2x + 5y = 10$ 4. $6x - y = 12$

5. $2x - 6y = -12$ 6. $3x + y = -6$

7. $x - 5 = 0$ 8. $x + 2 = 0$

9. $y + 4 = 0$ 10. $x + 3 = 0$

Find the distance between each of the given pairs of points, and find the slope of the line segment joining them.

Example $(3, -5), (2, 4)$

Solution   Consider $(3, -5)$ as P_1 and $(2, 4)$ as P_2.

$$d = \sqrt{(x_2 - x_1)^2 + (y_2 - y_1)^2}$$ $$m = \frac{y_2 - y_1}{x_2 - x_1}$$

$$= \sqrt{(2 - 3)^2 + [4 - (-5)]^2} = \sqrt{1 + 81}$$ $$= \frac{4 - (-5)}{2 - 3} = \frac{9}{-1}$$

Distance, $\sqrt{82}$; slope, -9.

11. $(1, 1), (4, 5)$ 12. $(-1, 1), (5, 9)$

13. $(-3, 2), (2, 14)$ 14. $(-4, -3), (1, 9)$

15. $(2, 1), (1, 0)$ 16. $(-3, 2), (0, 0)$

Find the lengths of the sides of the triangle having vertices as given.

17. $(10, 1)$, $(3, 1)$, $(5, 9)$ 18. $(0, 6)$, $(9, -6)$, $(-3, 0)$

19. $(5, 6)$, $(11, -2)$, $(-10, -2)$ 20. $(-1, 5)$, $(8, -7)$, $(4, 1)$

Find an equation, in standard form, of the line through each of the given points and having the given slope.

Example $(3, -5)$, $m = -2$

Solution   Substitute given values in the point-slope form of the linear equation.

$$y - y_1 = m(x - x_1)$$
$$y - (-5) = -2(x - 3)$$
$$y + 5 = -2x + 6$$
$$2x + y - 1 = 0$$

21. $(-3, -2)$, $m = \dfrac{1}{2}$ 22. $(0, 0)$, $m = 3$ 23. $(-1, 0)$, $m = 1$

24. $(2, -3)$, $m = 0$ 25. $(-4, 2)$, $m = 0$ 26. $(-1, -2)$, parallel to y-axis

Write each equation in slope-intercept form; specify the slope and y-intercept of the line.

Example $2x - 3y = 5$

Solution   Solving explicitly for y, we obtain

$$y = \frac{2}{3}x - \frac{5}{3}.$$

Compare with the general slope-intercept form

$$y = mx + b.$$

Slope, $\dfrac{2}{3}$; y-intercept, $-\dfrac{5}{3}$.

27. $3x + 2y = 1$ 28. $3x - y = 7$ 29. $x - 3y = 2$

30. $2x - 3y = 0$ 31. $8x - 3y = 0$ 32. $-x = 2y - 5$

33. Write an equation, in standard form, of the line with the same slope as $x - 2y = 5$ and passing through the origin. Sketch the graph of this equation.

34. Write an equation, in standard form, of the line through $(0, 5)$ with the same slope as $2y - 3x = 5$. Sketch the graph of this equation.

35. Show that, for $x_2 \neq x_1$,

$$y - y_1 = \left(\frac{y_2 - y_1}{x_2 - x_1}\right)(x - x_1)$$

is an equation of the line joining the points (x_1, y_1) and (x_2, y_2). This is the **two-point form** of the linear equation.

36. Using the general equation in Problem 35, find an equation of the lines through the given points.

    a. $(2, 1)$ and $(-1, 3)$ b. $(3, 0)$ and $(5, 0)$

    c. $(-2, 1)$ and $(3, -2)$ d. $(-1, -1)$ and $(1, 1)$

37. Consider the linear function

$$F = \{(x, y) \mid y = F(x)\}.$$

If $(2, 3)$ and $(-1, 4)$ are known to be in F, find $F(x)$ in terms of x.

38. Show that if a and b are nonzero numbers denoting the x- and y-intercepts of a straight line, then $\dfrac{x}{a} + \dfrac{y}{b} = 1$ is an equation for the line. This is called the **intercept form** of the equation.

39. Using the general equation in Problem 38, find an equation of the line having the given intercepts.

    a. x-intercept 3, y-intercept -1 b. x-intercept -2, y-intercept 2

40. Show that the triangle with vertices $(0, 6)$, $(9, -6)$, and $(-3, 0)$ is a right triangle. *Hint:* Use the converse of the Pythagorean theorem; that is, if $c^2 = a^2 + b^2$, then the triangle is a right triangle.

41. Show by similar triangles that the coordinates of the midpoint of the line segment joining the points $P_1(x_1, y_1)$ and $P_2(x_2, y_2)$ are given by

$$x = \frac{x_1 + x_2}{2} \quad \text{and} \quad y = \frac{y_1 + y_2}{2}.$$

42. Using the results of Problem 41, find the coordinates of the midpoint of the line segment joining the points whose coordinates are given.

    a. $(2, 4)$ and $(6, 8)$ b. $(-4, 6)$ and $(6, -10)$

Graph each given function f and its inverse f^{-1}, using the same set of axes.

43. $f = \{(x, y) \mid y = 2x + 6\}$ 44. $f = \{(x, y) \mid y = 4 - 2x\}$

45. $f = \{(x, y) \mid 3x - 4y = 12\}$ 46. $f = \{(x, y) \mid x - 6y = 6\}$

4.3

Parallel and perpendicular lines

Parallel lines
Triangles can be used to show that line segments (and lines) not perpendicular to the *x*-axis are parallel if and only if they have equal slopes. Since the proof of the following theorem depends on familiar geometric properties, it is quite simple and is left as an exercise.

Theorem 4.1 *Two lines with slopes m_1 and m_2 are parallel if and only if $m_1 = m_2$. Two lines perpendicular to the x-axis are parallel to each other.*

Example
Find an equation of the line through $(-1, 2)$ parallel to $3x - 2y = 6$.

Solution
Make a sketch. The given equation can be written equivalently in slope-intercept form as

$$y = \frac{3}{2}x - 3,$$

and, by inspection, we see that the slope of its graph is 3/2. Then, using the point-slope form for a linear equation, with $m = 3/2$, $x_1 = -1$, and $y_1 = 2$, we have

$$(y - 2) = \frac{3}{2}(x + 1),$$

which reduces to the standard form $3x - 2y + 7 = 0$.

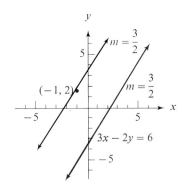

Perpendicular lines
The slopes of two perpendicular lines are related to each other in an interesting way.

Theorem 4.2 *Two lines with slopes m_1 and m_2 are perpendicular if and only if $m_1 \cdot m_2 = -1$. Two lines that are parallel to the x- and y-axes, respectively, are also perpendicular to each other.*

The proof of this theorem also follows from a familiar geometric property and is left as an exercise.

Example   Find an equation of the line passing through $(3, -2)$ and perpendicular to the graph of $2x + 5y = 10$.

Solution   Make a sketch. The given equation can be written equivalently in slope-intercept form as

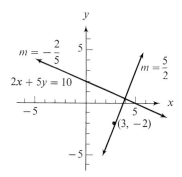

$$y = -\frac{2}{5}x + 2,$$

and, by inspection, the slope of its graph is $-2/5$. Hence, from Theorem 4.2, the slope of any line perpendicular to this graph must be $5/2$. Using the point-slope form for a linear equation with $m = 5/2$, $x_1 = 3$, and $y_1 = -2$, we have

$$y - (-2) = \frac{5}{2}(x - 3),$$

which reduces to the standard form

$$5x - 2y - 19 = 0.$$

Exercise 4.3

Find an equation of the line which is (a) *parallel and* (b) *perpendicular to the graph of the given equation, and which passes through the given point.*

1. $3x + y = 6$; $(5, 1)$
2. $2x - y = -3$; $(-2, 1)$
3. $3x - 2y = 5$; $(0, 0)$
4. $4x - 2y = -5$; $(0, 0)$
5. $4x + 6y = 3$; $(-2, -1)$
6. $5x - 3y = 2$; $(-4, -1)$
7. $2x + 4y = 9$; $(-3, 4)$
8. $3x - 9y = 4$; $(5, -7)$

Find an equation of the line that is the perpendicular bisector of the segment whose endpoints are given.

Example $(7, -4)$ and $(-5, -9)$

Solution   The slope of the segment is

$$m = \frac{y_2 - y_1}{x_2 - x_1} = \frac{-9 - (-4)}{-5 - 7} = \frac{-5}{-12} = \frac{5}{12}.$$

(Solution continued.)

From Theorem 4.2, it follows that the slope of the desired bisector is $-12/5$. From Exercise 41, page 103, we see that the coordinates of the midpoint of the given segment are

$$x = \frac{x_1 + x_2}{2} = \frac{7 + (-5)}{2} = \frac{2}{2} = 1,$$

$$y = \frac{y_1 + y_2}{2} = \frac{-4 + (-9)}{2} = -\frac{13}{2}.$$

Using the point-slope form for a linear equation, with $m = -12/5$ and given point $(1, -13/2)$, we have the desired equation

$$y + \frac{13}{2} = -\frac{12}{5}(x - 1),$$

which simplifies to the standard form

$$24x + 10y + 41 = 0.$$

9. $(5, 3)$ and $(9, 7)$ 10. $(9, 1)$ and $(-3, -5)$

11. $(5, -5)$ and $(-9, 1)$ 12. $(9, 4)$ and $(-3, 4)$

13. $(a, 0)$ and $(0, b)$ 14. (a, b) and $(a + k_1, b + k_2)$

15. Use slopes to show that the triangle with vertices at $A(0, 8)$, $B(6, 2)$, and $C(-4, 4)$ is a right triangle.

16. Use slopes to show that the triangle with vertices $D(2, 5)$, $E(5, 2)$, and $F(10, 7)$ is a right triangle.

17. Use slopes to show that the quadrilateral with vertices at $P(-1, 2)$, $Q(5, 4)$, $R(8, 2)$, and $S(2, 0)$ is a parallelogram.

18. Show that the quadrilateral with vertices at $A(-5, -1)$, $B(0, 0)$, $C(1, -5)$, and $D(-4, -6)$ is a square.

19. Recall from geometry that if two parallel lines are cut by a transversal, then corresponding angles are congruent. Use this fact to show, in the figure at the right, where L_1 is parallel to L_2, that $\triangle P_1 Q_1 P_3$ is similar to $\triangle P_2 Q_2 P_4$ and, hence, that the slopes, m_1 and m_2, of the lines are equal.

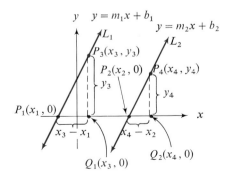

20. In the figure for Problem 19, assume that $m_1 = m_2$. Deduce that $\triangle P_1 Q_1 P_3$ is similar to $\triangle P_2 Q_2 P_4$ and, hence, that L_1 is parallel to L_2.

21. In the adjoining figure, where L_1 is perpendicular to L_2, use the fact that $\Delta P_1 Q P_2$ is similar to $\Delta P_3 Q P_1$, so that

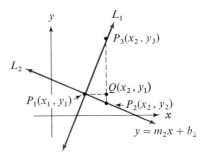

$$\frac{y_1 - y_2}{x_2 - x_1} = \frac{x_2 - x_1}{y_3 - y_1}$$

to deduce that $m_1 m_2 = -1$.

22. Use the figure in Problem 21 to argue that, because

$$(m_1 - m_2)^2 = m_1^2 - 2m_1 m_2 + m_2^2$$

and because the converse of the Pythagorean theorem is true, the relation $m_1 m_2 = -1$ implies that L_1 is perpendicular to L_2.

4.4

Quadratic functions

Graph of a quadratic function   Consider the quadratic equation in two variables

$$y = x^2 - 4. \tag{1}$$

As with linear equations in two variables, solutions of this equation must be ordered pairs (x, y). We need replacements for both x and y in order to obtain a statement we can adjudge to be true or false. As before, such ordered pairs can be found by arbitrarily assigning values to x and computing related values for y. For instance, assigning the value -3 to x in Equation (1), we obtain

$$y = (-3)^2 - 4,$$

$$y = 5,$$

and $(-3, 5)$ is a solution. Similarly, we find that

$$(-2, 0), (-1, -3), (0, -4), (1, -3), (2, 0), \text{ and } (3, 5)$$

are also solutions of (1). Plotting the corresponding points on the plane, we have the graph in Figure 4.11-a on page 108. Clearly, these points do not lie on a straight line, and we may reasonably inquire whether the graph of the solution set of (1),

$$S = \{(x, y) \mid y = x^2 - 4\},$$

forms any kind of meaningful pattern on the plane. By plotting additional solutions of (1)—solutions with x-components between those already found—we may be able to obtain a clearer picture. Thus we find the solutions

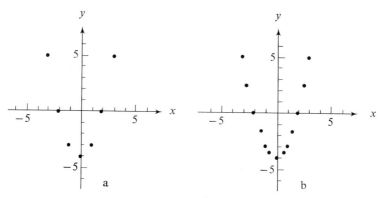

Figure 4.11

$$\left(\frac{-5}{2},\frac{9}{4}\right), \left(\frac{-3}{2},\frac{-7}{4}\right), \left(\frac{-1}{2},\frac{-15}{4}\right), \left(\frac{1}{2},\frac{-15}{4}\right), \left(\frac{3}{2},\frac{-7}{4}\right), \left(\frac{5}{2},\frac{9}{4}\right),$$

and by plotting these points in addition to those found earlier, we have the graph in Figure 4.11-b. It now appears reasonable to connect these points in sequence, say from left to right, by a smooth curve as in Figure 4.12, and to assume that the resulting curve is a good approxima-tion to the graph of (1). (You should realize, of course, that regardless of how many individual points are plotted, we have no absolute assurance that the smooth curve is a good approximation to the true graph; more information is needed—for example, in this case, that for $|x_2| > |x_1|$ we have correspond-ingly $y_2 > y_1$.) This curve is an example of a **parabola**.

More generally, the graph of the solu-tion set of any quadratic equation of the form

$$y = ax^2 + bx + c, \qquad (2)$$

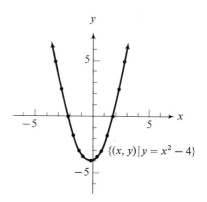

Figure 4.12

where a, b, and c are real and $a \neq 0$, is a parabola. Since for each x an equation of the form (2) will determine only one y, such an equation defines a function having as domain the entire set of real numbers and as range some subset of the reals. For example, we observe from the graph in Figure 4.12 that the range of the function defined by (1) is the set of real numbers

$$\{y \mid y \geq -4\}.$$

The parabola that is the graph of an equation of the form (2) will have a lowest (minimum) point or a highest (maximum) point, depending on whether $a > 0$ or $a < 0$, respectively. Such a point is called the **vertex** of the parabola. The line through the vertex and parallel to the y-axis is called the **axis of symmetry**, or, simply, the **axis** of the parabola; it separates the parabola into two parts, each the mirror image of the other in the axis. If we observe that the graphs of

$$y = ax^2 + bx + c \qquad (3)$$

and

$$y = ax^2 + bx \qquad (4)$$

have the same axis (Figure 4.13), we can find an equation for the axis of (3) by inspecting

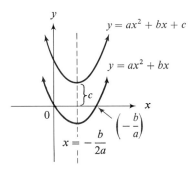

Figure 4.13

Equation (4). Factoring the right-hand member of $y = ax^2 + bx$ yields

$$y = x(ax + b),$$

and thus we can see that 0 and $-b/a$ are the x-intercepts of the graph. Since the axis of symmetry bisects the segment with these endpoints, an equation for the axis of symmetry is

$$x = -\frac{b}{2a}.$$

Example   Find an equation for the axis of symmetry of the graph of

$$\{(x, y) \mid y = 2x^2 - 5x + 7\}.$$

Solution   By comparing the given equation to $y = ax^2 + bx + c$, we see that $a = 2$ and $b = -5$. Hence, we have

$$x = -\frac{b}{2a} = -\frac{-5}{2(2)} = \frac{5}{4},$$

and $x = 5/4$ is the desired equation for the axis.

Since the vertex of a parabola lies on its axis, obtaining an equation for this axis will give us the x-coordinate of the vertex. The y-coordinate can then easily be obtained by substitution in the equation for the parabola.

Graphing parabolas   When graphing a quadratic equation in two variables, it is desirable to select first components for the ordered pairs that ensure that the more significant parts of the graph are displayed. For a parabola, these parts include the intercepts, if they exist, and the maximum or minimum point on the curve.

Example   Graph $y = x^2 - 3x - 4$.

Solution   By inspection, the y-intercept is -4. Setting $y = 0$, we have

$$0 = x^2 - 3x - 4 = (x - 4)(x + 1),$$

and the x-intercepts are 4 and -1. Finally, the x-coordinate of the minimum point is

$$x = -\frac{b}{2a} = -\frac{-3}{2(1)} = \frac{3}{2}.$$

By substituting $3/2$ for x in $y = x^2 - 3x - 4$, we obtain

$$y = \left(\frac{3}{2}\right)^2 - 3\left(\frac{3}{2}\right) - 4$$

$$= \frac{9}{4} - \frac{9}{2} - 4 = -\frac{25}{4}.$$

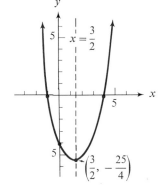

Graphing the intercepts and the coordinates of the minimum point and then sketching the curve produce the graph shown at the right above.

Solutions, zeros, and x-intercepts   Consider the graph of the function

$$S = \{(x, f(x)) \mid f(x) = ax^2 + bx + c\}, \tag{5}$$

for $a \neq 0$, and the solution set of the equation

$$ax^2 + bx + c = 0. \tag{6}$$

Any value of x for which $f(x) = 0$ in (5) will be a solution of (6). Since any point on the x-axis has y-coordinate zero (i.e., $f(x) = 0$), the x-intercepts of the graph of (2) are the real solutions of (6). Since values of x for which $f(x) = 0$ are also called the **zeros of the function**, we have three different names for the same set of values:

1. The *elements of the solution set* of the equation $ax^2 + bx + c = 0$.
2. The *zeros of the function* defined by $f(x) = ax^2 + bx + c$.
3. The *x-intercepts* of the graph of the equation $y = ax^2 + bx + c$.

Inverse of a
quadratic function
In Section 4.2 we observed that the graph of the inverse of a linear function is also a straight line and hence the inverse is also a linear function if its graph is not vertical. Because every function is a relation, every function has an inverse, but the inverse is not always a function. For example, the graphs of

$$F = \{(x, y)\,|\,y = x^2\}$$

and its inverse,

$$F^{-1} = \{(x, y)\,|\,x = y^2\} = \{(x, y)\,|\,y = \pm\sqrt{x}\},$$

are shown in Figure 4.14. Since F^{-1} associates two different y's with each x for all but one value in its domain, this inverse is not a function.

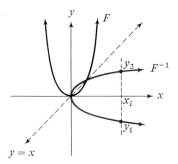

Figure 4.14

Exercise 4.4

Graph. (Obtain analytically the x-intercepts and the maximum or minimum point, and then sketch the rest of the curve.)

Example $\{(x, y)\,|\,y = x^2 - 7x + 6\}$

Solution   The y-intercept is clearly 6. Since the solutions of

$$x^2 - 7x + 6 = (x - 1)(x - 6) = 0$$

are 1 and 6, these are the x-intercepts. The axis of symmetry has equation

$$x = -\frac{b}{2a} = \frac{7}{2}.$$

Substituting 7/2 for x in $y = x^2 - 7x + 6$, we obtain

$$y = -\frac{25}{4}.$$

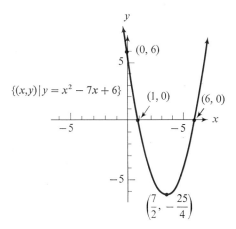

Hence the minimum point is (7/2, −25/4). Using these points, we can sketch the graph as shown.

1. $\{(x, y) \mid y = x^2 - 5x + 4\}$

2. $\{(x, g(x)) \mid g(x) = x^2 - 3x + 2\}$

3. $\{(x, y) \mid y = x^2 - 6x - 7\}$

4. $\{(x, y) \mid y = x^2 - 3x + 2\}$

5. $\{(x, f(x)) \mid f(x) = -x^2 + 5x - 4\}$

6. $\{(x, f(x)) \mid f(x) = -x^2 - 8x + 9\}$

7. $\left\{(x, g(x)) \mid g(x) = \dfrac{1}{2}x^2 + 2\right\}$

8. $\left\{(x, y) \mid y = -\dfrac{1}{2}x^2 - 2x\right\}$

9. Graph $\{(x, f(x)) \mid f(x) = x^2 + 1\}$. Represent $f(0)$ and $f(4)$ by drawing line segments from $(0, 0)$ to $(0, f(0))$ and from $(4, 0)$ to $(4, f(4))$.

10. Graph $\{(x, g(x)) \mid g(x) = x^2 + 1\}$. Represent $g(-3)$ and $g(2)$ by drawing line segments from $(-3, 0)$ to $(-3, g(-3))$ and from $(2, 0)$ to $(2, g(2))$.

Solve Problems 11 *and* 12 *by completing the square.*

11. Find two numbers having sum 8 and product as great as possible.

12. Find the maximum possible area of a rectangle with perimeter 100 inches.

Graph each given function f and f^{-1}, using the same set of axes.

13. $f = \{(x, y) \mid y = x^2 - 4\}$

14. $f = \{(x, y) \mid y = x^2 + 4\}$

15. $f = \{(x, y) \mid y = x^2 - 4x + 4\}$

16. $f = \{(x, y) \mid y = x^2 - 2x - 3\}$

17. On a single set of axes, sketch the family of four curves that are the graphs of

$$y = x^2 + k \quad (k = -2, 0, 2, 4).$$

What effect does varying k have on the graph?

18. On a single set of axes, sketch the family of six curves that are the graphs of

$$y = kx^2 \quad \left(k = \frac{1}{2}, 1, 2, -\frac{1}{2}, -1, -2\right).$$

What effect does varying k have on the graph?

4.5

Conic sections

In Section 4.4, we discussed quadratic equations of the form

$$y = ax^2 + bx + c \quad (a \neq 0), \tag{1}$$

whose graphs are parabolas, opening upward if $a > 0$ and downward if $a < 0$. Similarly, an equation of the form

$$x = ay^2 + by + c \quad (a \neq 0) \tag{2}$$

also has a graph that is a parabola, opening to the right if $a > 0$ and to the left if $a < 0$.

In addition to (1) and (2), above, there are other types of second-degree equations in two variables. Their graphs are referred to as **conic sections**, or **conics**, because such curves result from the intersection of a plane and either a right circular cone, as shown in Figure 4.15, or a right circular cylinder.

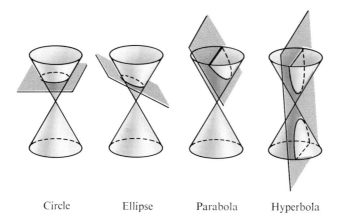

Circle   Ellipse   Parabola   Hyperbola

Figure 4.15

Classification
of central conics
The graphs of relations defined by equations of the form

$$Ax^2 + By^2 = C \quad (A^2 + B^2 \neq 0) \tag{3}$$

are symmetric *with respect to the origin*. That is, if the point with coordinates (x_1, y_1) is on the graph, then so is the point with coordinates $(-x_1, -y_1)$. For this reason, their graphs are called **central conics**. There exist the following possibilities: The graph is

(a) a circle if $A = B$ and A, B, and C have like signs;

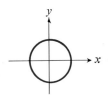

(b) an ellipse if $A \neq B$ and A, B, and C have like signs;

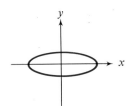 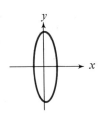

(c) a hyperbola if A and B are opposite in sign and $C \neq 0$;

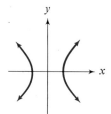 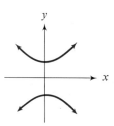

(d) two distinct lines through the origin if A and B are opposite in sign and $C = 0$;

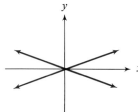

(e) two distinct parallel lines if one of A and $B = 0$ and the other has the same sign as C;

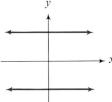 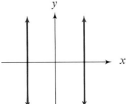

(f) two coincident parallel lines (one line) through the origin if one of A and $B = 0$ and also $C = 0$;

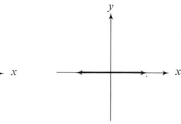

(g) a point if A and B are both ≥ 0 or both ≤ 0 and $C = 0$;

(h) the null set if A and B are both ≥ 0 and $C < 0$, or if A and B are both ≤ 0 and $C > 0$.

Graph of a central conic   After we recognize the general form of a central conic, the location of a few points should suffice to sketch the complete graph. The intercepts, for instance, are always easy to identify.

Example   Graph $\{(x, y) \mid 4y^2 = 8 - x^2\}$.

Solution   By comparing the defining equation in standard form,

$$x^2 + 4y^2 - 8 = 0,$$

with (b) above, we note immediately that its graph is an ellipse.

If $y = 0$, then $x = \pm\sqrt{8}$; and if $x = 0$, then $y = \pm\sqrt{2}$. We can accordingly sketch the graph of the relation.

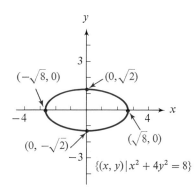

Example   Graph $\{(x, y) \mid x^2 - y^2 = 3\}$.   (4)

Solution   By comparing the defining equation with (c), we see that its graph is a
                 hyperbola. If $y = 0$, then $x = \pm\sqrt{3}$. If, however, $x = 0$, then y^2 would
                 have to be negative, which is an impossibility in the field of real numbers.
                 Thus the graph does not cross the y-axis. By assigning a few other
                 arbitrary values to one of the variables, say x—for example, (4,   )
                 and $(-4,$ $)$—we can find additional ordered pairs,

$$(4, \sqrt{13}), (4, -\sqrt{13}), (-4, \sqrt{13}), (-4, -\sqrt{13}),$$

                 satisfying (5). The graph can then be sketched as shown.

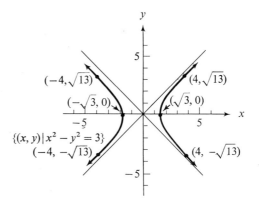

Asymptotes   The thin black lines in the figure of the foregoing example were
                 used to help sketch the curve. These are called **asymptotes** of the
                 graph. They comprise the graph of $x^2 - y^2 = 0$. In general,
         for $A, B, C > 0$, the graph of $Ax^2 - By^2 = C$ has as asymptotes the two straight
         lines that form the graph of $Ax^2 - By^2 = 0$ (see Problem 16).

Exercise 4.5

Name and sketch the graph of each of the following relations if the graphs exist.

1. $\{(x, y) \mid 2x^2 + 2y^2 = 8\}$ 2. $\{(x, y) \mid 4x^2 + 4y^2 = 36\}$

3. $\{(x, y) \mid x^2 + 2y^2 = 8\}$ 4. $\{(x, y) \mid 4x^2 + y^2 = 16\}$

5. $\{(x, y) \mid x^2 - 9y^2 = 36\}$ 6. $\{(x, y) \mid 2x^2 - y^2 = 8\}$

7. $\{(x, y) \mid x^2 - 9y^2 = 0\}$ 8. $\{(x, y) \mid 4x^2 - y^2 = 0\}$

9. $\{(x, y) \mid x^2 + 9y^2 = 0\}$ 10. $\{(x, y) \mid 4x^2 + y^2 = 0\}$

11. $\{(x, y)\,|\,x^2 + 2y^2 = -4\}$ 12. $\{(x, y)\,|\,4x^2 + y^2 = -8\}$

13. $\{(x, y)\,|\,x^2 = 16\}$ 14. $\{(x, y)\,|\,y^2 = 9\}$

15. Graph $\{(x, y)\,|\,x^2 + y^2 = 25\}$ and $\{(x, y)\,|\,4x^2 + y^2 = 36\}$ on the same set of axes. What is the significance of the coordinates of the points of intersection, $\{(x, y)\,|\,x^2 + y^2 = 25\} \cap \{(x, y)\,|\,4x^2 + y^2 = 36\}$?

16. By solving $Ax^2 - By^2 = C$ $(A, B, C > 0)$ for y, obtain the expression

$$y = \pm \sqrt{\frac{A}{B}}\,|x|\left(\sqrt{1 - \frac{C}{Ax^2}}\right)$$

and argue that the graph of $Ax^2 - By^2 = C$ approaches the graphs of

$$y = \pm \sqrt{\frac{A}{B}}\,|x|$$

as $|x|$ increases.

4.6

Relations whose graphs are given

Any set of points in the plane can be called a graph or a **locus** (plural **loci**). However, the words *graph* and *locus* are ordinarily used to refer to the set of points meeting a predesignated condition or conditions. For example, the set of points on a circle with radius r can be called the locus of points in the plane located a distance r from a given point (the center of the circle). A fundamental mathematical problem is that of finding an equation (or inequality) of a given locus.

Example   Find an equation of the circle having center at $P_1(3, -2)$ and radius 7.

Solution   It is best first to make a sketch showing a representative point in the locus. By the distance formula, the distance d from $P_1(3, -2)$ to any point $P(x, y)$ on the circle is

$$d = \sqrt{(x - 3)^2 + (y + 2)^2}.$$

Since this distance is given to be 7, we have

$$\sqrt{(x - 3)^2 + (y + 2)^2} = 7.$$

Squaring both members produces

$$(x - 3)^2 + (y + 2)^2 = 49,$$

$$x^2 - 6x + 9 + y^2 + 4y + 4 = 49,$$

or finally

$$x^2 + y^2 - 6x + 4y - 36 = 0. \tag{1}$$

(Solution continued overleaf.)

Conversely, since the steps are reversible, any point whose coordinates satisfy (1) is on the circle. Hence (1) is an equation of the circle. Observe, however, that it is not an equation of the form discussed in Section 4.5, because the center is not at the origin.

Example

Find an equation of the locus of all points located twice as far from the point $(3, -2)$ as they are from $(-2, 1)$.

Solution

Make a sketch. By the distance formula, any point $P(x, y)$ in the plane is located at a distance d_1 from $(-2, 1)$ given by

$$d_1 = \sqrt{(x + 2)^2 + (y - 1)^2}$$

and at a distance d_2 from $(3, -2)$ given by

$$d_2 = \sqrt{(x - 3)^2 + (y + 2)^2}.$$

In order for $P(x, y)$ to lie in the described locus, it is necessary that

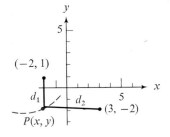

$$d_2 = 2d_1,$$

$$\sqrt{(x - 3)^2 + (y + 2)^2} = 2\sqrt{(x + 2)^2 + (y - 1)^2}.$$

Then, equating squares of both members, we have

$$(x - 3)^2 + (y + 2)^2 = 4[(x + 2)^2 + (y - 1)^2],$$

$$x^2 - 6x + 9 + y^2 + 4y + 4 = 4x^2 + 16x + 16 + 4y^2 - 8y + 4,$$

from which we obtain the simplified equation

$$3x^2 + 3y^2 + 22x - 12y + 7 = 0. \qquad (2)$$

Again the steps are reversible, so that (2) is an equation of the given locus.

Loci as paths

Another way of considering loci in an intuitive way is to think of a set of points as having been traced out in the plane by a single moving particle.

Example

Find an equation of a particle that moves in the plane so that it is always at the same distance from the line $x = 3$ as it is from the point $(8, 0)$.

Solution

Make a sketch. The x-coordinate of P_1 is 3. Since P_1P is parallel to the x-axis, the distance d_1 from $P_1(3, y)$ to $P(x, y)$ is given by

$$d_1 = |x - 3|,$$

while d_2, the distance from $P(x, y)$ to $P_2(8, 0)$, is given by

$$d_2 = \sqrt{(x - 8)^2 + (y - 0)^2}.$$

The condition imposed on $P(x, y)$ is that

$$d_1 = d_2.$$

Hence, the condition can be expressed as

$$|x - 3| = \sqrt{(x - 8)^2 + (y - 0)^2}.$$

Squaring each member, we have

$$(x - 3)^2 = (x - 8)^2 + y^2,$$

$$x^2 - 6x + 9 = x^2 - 16x + 64 + y^2,$$

from which we obtain the simplified equation

$$y^2 - 10x + 55 = 0. \tag{3}$$

The steps are reversible, so that (3) is an equation of the given locus.

Exercise 4.6

Find an equation of the circle with the given center and radius.

1. $(1, 0)$; $r = 1$ 2. $(4, 2)$; $r = 3$ 3. $(-3, 1)$; $r = 2$

4. $(4, -5)$; $r = 8$ 5. (h, k); $r = r$ 6. $(-h, -k)$; $r = r$

7. Find an equation of the circle which has center $(3, 5)$ and contains the point $(6, 9)$.

8. Find an equation of the circle which has center $(-2, 3)$ and contains the point $(3, -9)$.

Find an equation of the locus of all points that are located so that:

9. They are equidistant from $(3, 2)$ and $(5, 0)$.

10. They are equidistant from $(-3, -4)$ and $(1, 2)$.

11. Their distance from $(1, -4)$ is twice their distance from $(6, -1)$.

12. Their distance from $(5, 3)$ is three times their distance from $(5, 9)$.

13. Their distance from the line with equation $x = -2$ is equal to their distance from $(2, 0)$.

14. Their distance from the line with equation $y = 4$ is equal to their distance from $(0, 0)$.

15. They are equidistant from the x-axis and $(-3, -2)$.

16. They are equidistant from the y-axis and $(4, 3)$.

17. The sum of their distances from $(2, 0)$ and $(-2, 0)$ is 8.

18. The sum of their distances from $(0, 4)$ and $(0, -4)$ is 12.

19. The sum of their distances from $(4, 8)$ and $(4, 2)$ is 10.

20. The sum of their distances from $(3, -5)$ and $(-1, -5)$ is 6.

21. The difference of their distances from $(2, 0)$ and $(-2, 0)$ is 2.

22. The difference of their distances from $(0, 4)$ and $(0, -4)$ is 6.

23. The difference of their distances from $(2, -5)$ and $(2, 5)$ is 6.

24. The difference of their distances from $(-3, 4)$ and $(-10, 4)$ is 5.

25. Their distance from $(0, p)$ equals their distance from the x-axis.

26. Their distance from $(p, 0)$ equals their distance from the y-axis.

4.7
Graphs of inequalities

Graph of a
linear inequality

Given A and B not both zero, a sentence of the form

$$Ax + By + C \le 0 \quad \text{or} \quad Ax + By + C < 0$$

is an inequality of the first degree that defines the relation

$$\{(x, y)\,|\,Ax + By + C \le 0\} \quad \text{or} \quad \{(x, y)\,|\,Ax + By + C < 0\},$$

respectively. Such relations in R^2 can be represented on the plane; the graph will be a region of the plane (a half-plane). For example, consider the relation

$$S = \{(x, y)\,|\,2x + y - 3 < 0\}. \tag{1}$$

When the defining inequality is rewritten in the equivalent form

$$y < -2x + 3, \tag{2}$$

we see that for every x in a solution (x, y), y is less than $-2x + 3$. The graph of the equation

$$y = -2x + 3 \tag{3}$$

is simply a straight line, as illustrated in Figure 4.16-a. To graph the relation S, we

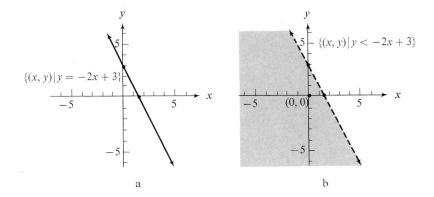

Figure 4.16

need only observe that any point below this line has a y-coordinate that satisfies (2). Consequently the solution set of (2), which is S, corresponds to the entire region below the line. The region is indicated on the graph with shading, as in Figure 4.16-b. That the line itself is not in the graph is indicated by means of a broken line. If the defining inequality were

$$2x + y - 3 \leq 0,$$

the line would be a part of the graph and would appear as a solid line.

Determination of the half-plane   To determine which half-plane to shade in constructing graphs of first-degree relations, we select any point in either half-plane and find out whether or not the coordinates of the point satisfy the defining sentence. If so, the half-plane containing the selected point is shaded; if not, the opposite half-plane is shaded. A very convenient point to use in this process is the origin. Thus, in the foregoing example, if x and y in $2x + y - 3 < 0$ are replaced by 0 we have

$$2(0) + 0 - 3 < 0,$$

which is true, and hence the half-plane containing the origin is shaded.

Graph of a
quadratic
inequality

Relations of the form

$$\{(x, y)|y < ax^2 + bx + c\} \tag{4}$$

or

$$\{(x, y)|y > ax^2 + bx + c\} \tag{5}$$

can be graphed by the same method used to graph relations defined by linear inequalities in two variables. We first graph the relation defined by the equation having the same members as the defining inequality and then shade an appropriate region as required. For instance, to graph

$$\{(x, y)|y < x^2 + 2\}, \tag{6}$$

we first graph

$$\{(x, y)|y = x^2 + 2\}. \tag{7}$$

Next, we substitute 0 for x and y in (6) to obtain

$$0 < 0^2 + 2,$$

which is true. Hence, the region below the curve, which contains the origin, is shaded, as shown in Figure 4.17. Since the graph of (7) is not part of the graph of (6), a broken curve is used.

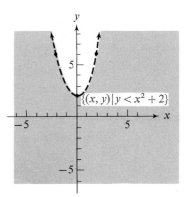

Figure 4.17

Inequalities do not ordinarily define functions according to the definition of a function in Section 4.1, because it usually is not true that each element of the domain is associated with a unique element in the range.

Exercise 4.7

Graph the relation.

Example

$$\{(x, y)|2x + y \geq 4\}$$

Solution

Graph the equality $2x + y = 4$.

Substitute 0 for x and y and determine that $2(0) + 0 \geq 4$ is false, so the origin is not in the graph.

Shade the region above the graph of $2x+y=4$. The line is included in the graph of the inequality.

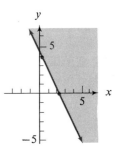

1. $\{(x, y)\,|\,y < x\}$ 2. $\{(x, y)\,|\,y > x\}$ 3. $\{(x, y)\,|\,y \leq x + 2\}$

4. $\{(x, y)\,|\,y \geq x - 2\}$ 5. $\{(x, y)\,|\,x + y < 5\}$ 6. $\{(x, y)\,|\,2x + y < 2\}$

7. $\{(x, y)\,|\,3 \geq 2x - 2y\}$ 8. $\{(x, y)\,|\,0 \geq x + y\}$ 9. $\{(x, y)\,|\,x > 0\}$

10. $\{(x, y)\,|\,y < 0\}$ 11. $\{(x, y)\,|\,x < 0\}$ 12. $\{(x, y)\,|\,x < -2\}$

13. $\{(x, y)\,|\,-1 < x < 5\}$ 14. $\{(x, y)\,|\,0 \leq y \leq 1\}$ 15. $\{(x, y)\,|\,|x| < 3\}$

16. $\{(x, y)\,|\,|y| > 1\}$

Example $\{(x, y)\,|\,y \geq x^2 + 2x\}$

Solution   Graph $\{(x, y)\,|\,y = x^2 + 2x\}$.

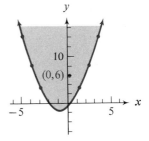

Substitute the coordinates of any point not on the curve for x and y in $y \geq x^2 + 2x$. We shall use $(0, 6)$. Since

$$6 \geq 0^2 + 2(0)$$

is true, the point $(0, 6)$ is in the graph of the inequality and the region above the graph of $y = x^2 + 2x$ is shaded. The curve is included in the graph of the inequality.

17. $\{(x, y)\,|\,y > x^2\}$ 18. $\{(x, y)\,|\,y < x^2\}$

19. $\{(x, y)\,|\,y \geq x^2 + 3\}$ 20. $\{(x, y)\,|\,y \leq x^2 + 3\}$

21. $\{(x, y)\,|\,y < 3x^2 + 2x\}$ 22. $\{(x, y)\,|\,y > 3x^2 + 2x\}$

23. $\{(x, y)\,|\,y \leq x^2 + 3x + 2\}$ 24. $\{(x, y)\,|\,y \geq x^2 + 3x + 2\}$

25. $\{(x, y)\,|\,y \geq 2x^2 - 5x + 2\}$ 26. $\{(x, y)\,|\,y \leq 2x^2 - 5x + 2\}$

27. $\{(x, y)\,|\,|x| + |y| \leq 1\}$ 28. $\{(x, y)\,|\,|x| + |y| \geq 1\}$

Hint: In Problems 27 and 28, consider the graphs in each quadrant separately: $x, y \geq 0$; $x \leq 0, y \geq 0$; $x, y \leq 0$; and $x \geq 0, y \leq 0$.

4.8

Equations in three variables

If each ordered pair of real numbers (x, y) is itself paired with a real number z, the pairing can be represented by the ordered triple (x, y, z). The universe for the set of all such ordered triples is sometimes represented by the symbol R^3. This denotes the Cartesian product of $R \times R$, or R^2, with R, that is, $(R \times R) \times R$.

Solutions of
equations
    An equation in three variables, such as

$$x^2 + 2y^2 - 3z - 4 = 0, \tag{1}$$

has **ordered triples** of real numbers as solutions, just as an equation in two variables has ordered pairs of real numbers as solutions. For example, $(1, 0, -1)$ is a solution of (1) because if x, y, and z are replaced with 1, 0, and -1, respectively, the result is

$$1^2 + 2(0)^2 - 3(-1) - 4 = 0,$$

$$1 + 0 + 3 - 4 = 0,$$

$$0 = 0,$$

which is a true statement. On the other hand, $(1, 1, 1)$ is not a solution of (1), because

$$1^2 + 2(1)^2 - 3(1) - 4 \neq 0.$$

Example   Find a second solution in R^3 of (1) above.

Solution   Select any arbitrary real-number replacements for *any two* variables, say 2 for x and 1 for y, and determine the value of the third variable, in this case z. Thus,

$$2^2 + 2(1)^2 - 3z - 4 = 0,$$

$$z = \frac{2}{3},$$

and a solution is $\left(2, 1, \dfrac{2}{3}\right)$.

Example   Find the solution of (1) above with first and second components zero.

Solution   Substituting 0 for x and 0 for y, we have

$$0 + 0 - 3z - 4 = 0,$$

$$z = -\frac{4}{3},$$

and the solution is $\left(0, 0, -\dfrac{4}{3}\right)$.

A solution in R^3 of an equation such as $x = 3$ is any ordered triple of the form $(3, y, z)$, where y and z may be any real numbers, just as a solution in R^2 is any ordered pair of the form $(3, y)$, where y may be any real number. Similarly, solutions in R^3 of an equation such as $y = 4$ are of the form $(x, 4, z)$, and solutions in

R^3 of an equation such as $z = 5$ are of the form $(x, y, 5)$. Furthermore, solutions in R^3 of an equation such as $x^2 + y^2 = 25$ are ordered triples (x, y, z) such that $x^2 + y^2 = 25$ and z is any real number.

R^3 and   Each member of R^3 can be paired with a point in space by
geometric space   using an extension of a standard Cartesian coordinate system
of the plane. If each of three mutually perpendicular lines in
space intersecting at a point is scaled (or coordinatized) with
origin at the point of intersection, each line then becomes an **axis** of a three-dimensional coordinate system, as suggested by Figure 4.18. The planes determined by the axes taken in pairs are called **coordinate planes**, and we identify them by using the letters associated with the two axes they contain; that is, we refer to them as the xy-plane, the xz-plane, and the yz-plane.

The orientation of the axes in Figure 4.18 constitutes a **right-hand system**. Interchanging the x- and y-axes produces a **left-hand system**.

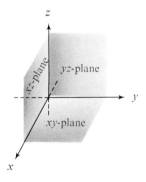

Figure 4.18

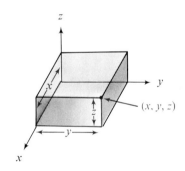

Figure 4.19

By associating each component of an ordered triple (x, y, z) with the directed (perpendicular) distance from a coordinate plane to a point in space, a one-to-one correspondence can be established between R^3 and the set of all points in space. One way to visualize the pairing of an ordered triple (x, y, z) with its graph in space is suggested by Figure 4.19. The rectangular prism with one vertex at the origin, as shown, will have the point paired with (x, y, z) as vertex opposite the origin. For example, the graphs of $(2, 3, 5)$, $(3, 4, -2)$, and $(1, -5, 2)$ are shown in Figure 4.20 on page 126. Of course it is not necessary to sketch the entire rectangular prism in order to locate a point.

The coordinate planes separate space into eight regions called **octants**. The region in which all the coordinates of each point are positive numbers is called the **first octant**. The remaining octants are not ordinarily assigned numbers.

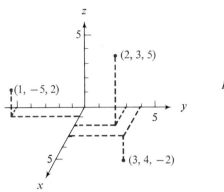

Figure 4.20   Figure 4.21

Distance between points in space   The formula for the distance between two points in a plane (see page 98) has a natural extension to three dimensions. Figure 4.21 shows points $P(x_1, y_1, z_1)$ and $Q(x_2, y_2, z_2)$ as opposite vertices of a rectangular prism with faces *parallel* to the coordinate planes. Vertices R and S in the bottom face are also shown, together with their coordinates. Since $P, R,$ and S are coplanar points in a plane parallel to the xy-plane, the distance from P to S is given by

$$d_1 = \sqrt{(x_2 - x_1)^2 + (y_2 - y_1)^2}.$$

Then, because $P, Q,$ and S are coplanar points in a plane parallel to the z-axis, we have

$$d_2 = |z_2 - z_1|.$$

From the Pythagorean theorem, we obtain

$$d^2 = d_1^2 + d_2^2,$$

$$d^2 = (x_2 - x_1)^2 + (y_2 - y_1)^2 + (z_2 - z_1)^2,$$

from which

$$d = \sqrt{(x_2 - x_1)^2 + (y_2 - y_1)^2 + (z_2 - z_1)^2}.$$

This is known as the **distance formula** in R^3.

Example   Find the distance from $(6, 3, -6)$ to $(10, 0, 6)$.

Solution   Setting $(x_2, y_2, z_2) = (6, 3, -6)$ and $(x_1, y_1, z_1) = (10, 0, 6)$, we have

$$d = \sqrt{(6 - 10)^2 + (3 - 0)^2 + (-6 - 6)^2}$$

$$= \sqrt{(-4)^2 + 3^2 + (-12)^2}$$

$$= \sqrt{16 + 9 + 144} = \sqrt{169} = 13.$$

Exercise 4.8

Find the solution(s) with the given components of each equation in R^3.

Example $x^2 + 2y^2 - z^2 = 4;$ $(0, 3, ?)$

Solution   Substituting 0 for x and 3 for y yields

$$(0)^2 + 2(3)^2 - z^2 = 4,$$
$$-z^2 = -14,$$
$$z = \sqrt{14} \quad \text{or} \quad z = -\sqrt{14}.$$

Hence, the solutions are $(0, 3, \sqrt{14})$ and $(0, 3, -\sqrt{14})$.

1. $x - y + z = 2;$ a. $(2, 1, ?)$,   b. $(-1, ?, 3)$,   c. $(?, 2, -2)$

2. $x^2 - 3y - z = 4;$ a. $(1, -1, ?)$,   b. $(4, ?, 2)$,   c. $(?, -2, 0)$

3. $x^2 + 2y^2 + z = 0;$ a. $(4, 0, ?)$,   b. $(0, ?, -4)$,   c. $(?, 0, 5)$

4. $3x^2 - y^2 + 2z^2 = 2;$ a. $(0, -1, ?)$,   b. $(2, ?, 0)$,   c. $(?, 3, 0)$

In Problems 5 and 6, find the value of the constant k if the given ordered triple is a solution of each equation in R^3.

Example $x + 4y - kz = 3;$ $(1, 2, 3)$

Solution $1 + 4(2) - 3k = 3;$ hence $k = 2$.

5. $x - ky + z = 4;$ a. $(2, -1, 3)$ b. $(0, 1, 0)$ c. $(2, 2, -2)$

6. $z = x^2 - ky^2;$ a. $(4, 1, 7)$ b. $(-1, 2, 3)$ c. $(0, 4, 0)$

Find the solutions of each equation in R^3 where
a. the first and second components are zero,
b. the first and third components are zero, and
c. the second and third components are zero.

Example $x^2 + 3y - z = 4$

Solution   a. $(0)^2 + 3(0) - z = 4;$ hence $z = -4,$ and $(0, 0, -4)$ is a solution.

b. $(0)^2 + 3y - (0) = 4;$ hence $y = \dfrac{4}{3},$ and $\left(0, \dfrac{4}{3}, 0\right)$ is a solution.

c. $x^2 + 3(0) - (0) = 4;$ hence $x = 2$ or $x = -2,$ and $(2, 0, 0)$ and $(-2, 0, 0)$ are solutions.

7. $x + 2y + 3z = 6$ 8. $x - 3y + 4z = 12$

9. $2x^2 + y^2 + z^2 = 4$ 10. $x^2 + 3y^2 - z = 12$

11. $x^2 - y - z^2 = 9$ 12. $4x^2 - 2y^2 - z = 8$

13. $z = 4x^2 - y^2$ 14. $z = x^2 + 2y^2$

Specify the form *of a solution for each equation in R^3 with the given component.*

Example $x^2 + y = 4$, where $x = 3$

Solution   Substituting 3 for x, we have

$$(3)^2 + y = 4,$$

$$y = -5.$$

Hence, a solution in R^3 is of the form $(3, -5, z)$, where z can be any real number.

15. $x^2 - y = 6$, where $x = 2$ 16. $y^2 + 3z = 12$, where $y = 3$

17. $x^2 - z^2 = 9$, where $z = 2$ 18. $x^2 + y^2 = 4$, where $x = 0$

19. $y = 3$ 20. $z = -4$ 21. $x = -2$ 22. $y = 0$

Graph each set of ordered triples.

Example $\{(4, 5, 2), (2, -6, 3), (0, 0, 4)\}$

Solution   The graph is shown at the right.

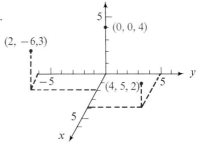

23. $\{(0, 0, 2), (3, 0, 0), (0, 4, 0)\}$

24. $\{(0, 0, -2), (-3, 0, 0), (0, -5, 0)\}$

25. $\{(2, 3, 0), (3, 0, 4), (0, 2, 5)\}$

26. $\{(-3, 1, 0), (4, 0, -2), (0, -3, 1)\}$

27. $\{(4, 3, 4), (7, 6, 1), (1, 5, 3)\}$

28. $\{(3, 3, 6), (2, -3, 3), (2, 4, -3)\}$

29. $\{(-5, 3, 1), (-5, -4, 2), (1, 2, -6)\}$

30. $\{(5, 5, 5,), (5, 5, -5), (5, -5, -5)\}$

Find the distance between the points with given coordinates.

Example $(2, -1, 4)$ and $(-3, 4, 5)$

Solution   From the distance formula in R^3, we have

$$d = \sqrt{(x_2 - x_1)^2 + (y_2 - y_1)^2 + (z_2 - z_1)^2}$$

$$= \sqrt{(-3 - 2)^2 + [4 - (-1)]^2 + (5 - 4)^2}$$

$$= \sqrt{25 + 25 + 1} = \sqrt{51}.$$

31. $(4, 2, -1)$ and $(5, -1, 2)$ 32. $(3, -3, 0)$ and $(0, 2, -1)$

33. $(3, 3, -5)$ and $(1, 4, -2)$ 34. $(-6, 1, 3)$ and $(-4, 4, 2)$

35. $(0, -2, 4)$ and $(-3, 1, 2)$ 36. $(-5, 4, 6)$ and $(2, -7, -2)$

4.9

Graphs of functions of two variables

If the equation $2x + y + z - 6 = 0$ is solved for z in terms of x and y, we obtain

$$z = -2x - y + 6.$$

In this form, the equation apparently serves to pair every ordered pair (x, y) of real numbers with exactly one real number z. This pairing constitutes a function, with domain the set R^2. In set notation, we can represent the function as

$$\{((x, y), z) \mid z = -2x - y + 6\}. \tag{1}$$

Such a function is said to be a function of the two variables x and y.

Just as the solution set of an equation in two variables in R^2 has a graph that is generally a line or a curve in the plane, the graph of the solution set of an equation in three variables in R^3 is generally a plane or a surface in space. For example, Figure 4.22 shows the part of the plane that is the graph of the first-degree function of two variables specified in (1) above.

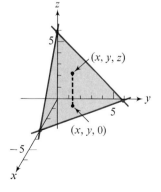

Figure 4.22

Because pictures of surfaces in three dimensions are difficult to draw, we shall focus our attention mainly on **sections of surfaces**, which are the curves formed where a surface intersects a plane. In particular, we shall examine sections in planes parallel to the coordinate planes and in the coordinate planes themselves.

Planes parallel to coordinate planes   In the universe R^3, equations of the form

$$x = k, \quad y = k, \quad \text{and} \quad z = k,$$

where k is a constant, determine planes parallel to coordinate planes, as shown in Figures 4.23-a, b, and c. This is simply a reflection of the fact

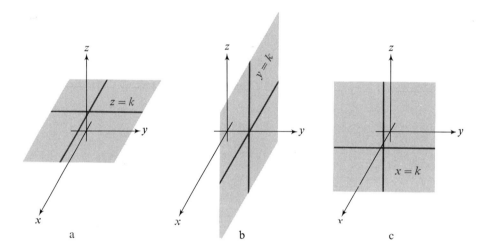

Figure 4.23

that ordered triples of the form

$$(k, y, z), \quad (x, k, z), \quad \text{and} \quad (x, y, k)$$

have graphs located at a fixed distance $|k|$ on one side or the other of one of the coordinate planes, with the side depending on the sign of k.

Graphs of traces   We can use planes of this kind to help sketch surfaces in space by sketching sections of the surface determined by the planes.
          In many cases, sections parallel to a particular one of the coordinate planes offer a better picture than sections parallel to either of the other coordinate planes. This suggests that some care should be given to selecting the most appropriate planes to use. Of course, in sketching any surface the first thing to

do is draw the sections in the coordinate planes. These particular sections are
called **traces**. To identify the traces of a surface, we look at the equation of the
surface when *each of the variables has, in turn, the value* 0.

For example, to sketch the graph of

$$z = -2x - y + 6$$

shown in Figure 4.22, we can find the traces in the coordinate planes as follows:

> In x, y plane, set $z = 0$ to obtain $\quad 2x + y = 6$.
> In y, z plane, set $x = 0$ to obtain $\quad\;\; y + z = 6$.
> In x, z plane, set $y = 0$ to obtain $\quad 2x + z = 6$.

Then each of these traces can easily be sketched using their intercepts with the
coordinate axes.

As another example, consider the equation

$$2x^2 + 2y^2 + z^2 = 72. \tag{2}$$

To find an equation for the trace in the xy-plane, we set z equal to 0. Thus this trace
has equation $2x^2 + 2y^2 = 72$, with graph a circle, as shown in Figure 4.24-a.

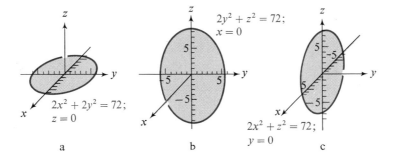

Figure 4.24

To find the trace in the yz-plane, we set x equal to 0. This trace has equation
$2y^2 + z^2 = 72$, with graph an ellipse, as shown in Figure 4.24-b.

To find the trace in the xz-plane, we set y equal to 0. This trace has equation
$2x^2 + z^2 = 72$, with graph an ellipse, as shown in Figure 4.24-c.

Now several additional sections for the surface associated with (2) will suggest
the appearance of the entire surface. Rewriting (2) equivalently as

$$2x^2 + 2y^2 = 72 - z^2,$$

we observe that for values $|z| < \sqrt{72}$, the equation has the form

$$x^2 + y^2 = k, \quad k > 0,$$

the graph of which is a circle for all such k. This means that sections which are graphs of such equations can aid us in completing the sketch of the entire surface. First let us take z equal to 2 and 6, for example, in turn and obtain the resulting equations and their respective sections as shown in Figure 4.25-a. Similar sections can be obtained below the xy-plane for z equal to -2 and -6.

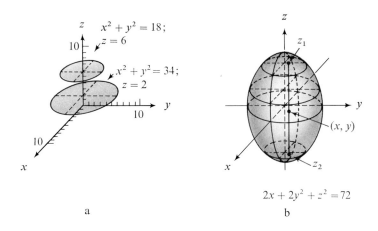

$$2x + 2y^2 + z^2 = 72$$

a   b

Figure 4.25

Using all the information available from the traces in the coordinate planes and the sections obtained for selected values of z, we can sketch the entire surface, as shown in Figure 4.25-b.

Because the equation $2x^2 + 2y^2 + z^2 = 72$ pairs each ordered pair in $\{(x, y) \mid 2x^2 + 2y^2 < 72\}$ with *more than one value* z, the set of ordered triples determined by the equation is not a function. By solving for z, however, we have

$$z = \sqrt{72 - 2x^2 - 2y^2} \quad \text{or} \quad z = -\sqrt{72 - 2x^2 - 2y^2},$$

and either of these equations defines a function with domain

$$\{(x, y) \mid 2x^2 + 2y^2 \le 72\}$$

and with range either

$$\{z \mid 0 \le z \le \sqrt{72}\} \quad \text{or} \quad \{z \mid -\sqrt{72} \le z \le 0\}.$$

Now consider, as a third example, the graph of

$$x^2 + y^2 = 4z.$$

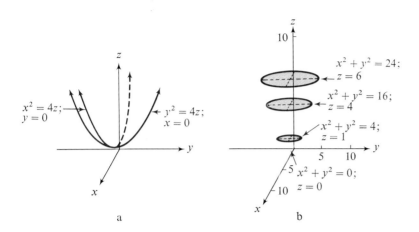

Figure 4.26

We can first sketch the traces of the surface in the coordinate planes, as shown in Figure 4.26-a. Notice that the trace in the xy-plane, with equation $x^2 + y^2 = 0$, is the point at the origin. By inspection, we observe that the sections that will best delineate the surface are those parallel to (and above) the xy-plane. For $z < 0$, we have $x^2 + y^2 < 0$, which has no solution with real-number components. We can therefore let $z = 1$, $z = 4$, and $z = 6$, for example, in turn and sketch the associated sections.

The equation of the section in the plane $z = 1$ is

$$x^2 + y^2 = 4,$$

the equation for the section in the plane $z = 4$ is

$$x^2 + y^2 = 16,$$

and the equation for the section in the plane $z = 6$ is

$$x^2 + y^2 = 24.$$

The sections are shown in Figure 4.26-b.

Using all the information obtained from the traces in the coordinate planes and sections in the planes corresponding to $z = 1$, $z = 4$, and $z = 6$, we complete the graph of the entire surface, as shown in Figure 4.27.

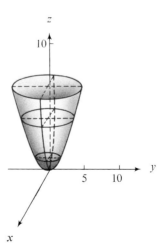

Figure 4.27

Exercise 4.9

As suggested by the example on page 131, *the graph in* R^3 *of a first-degree equation, of the form* $ax + by + cz = d$, *is a plane. Find the intercepts of the graph of each first-degree equation on the coordinate axes and show the graph in one octant.*

Example $3x - y + 2z = 6$

Solution   The ordered triples corresponding to the points of intersection of the plane and the coordinate axes are shown in the figure.

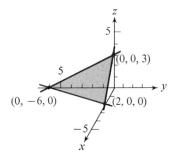

1. $x + 3y + 2z = 6$ 　　　　　　 2. $x + 4y + 2z = 8$

3. $2x + 4y + 3z = 12$ 　　　　　 4. $3x + y + 2z = 6$

5. $3x - y + 3z = 6$ 　　　　　　 6. $2x - 2y + z = 4$

7. $2x + y - 2z = 8$ 　　　　　　 8. $-3x + 2y + z = 6$

Graph each set of ordered triples or each equation in R^3.

Example $\{(x, y, z) \,|\, z = 3\}$

Solution   The graph is shown at the right.

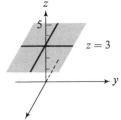

 9. $\{(x, y, z) \,|\, x = 3\}$

10. $\{(x, y, z) \,|\, y = 4\}$

11. $\{(x, y, z) \,|\, z = 5\}$

12. $\{(x, y, z) \,|\, x = -2\}$

13. $y = -6$

14. $z = -3$

15. $x = 5$

16. $y = 0$

Graph.

17. $\{(x, y, z) \,|\, 2x + y = 6\}$ 　　　　　 18. $\{(x, y, z) \,|\, x^2 + y^2 = 25\}$

19. $\{(x, y, z) \,|\, 4y^2 + z^2 = 16\}$ 　　　　 20. $\{(x, y, z) \,|\, 4x^2 - y = 0\}$

Graph the solution set in R³ of each equation.

21. $x + 3y = 6$

22. $x^2 + z^2 = 16$

23. $y^2 + z = 4$

24. $x^2 - 4y^2 = 16$

Graph the trace on the specified coordinate plane for each equation if such trace exists.

Example $4x^2 - y^2 - z^2 = -16$; xy-plane

Solution   On the xy-plane, $z = 0$.

For $z = 0$,

$4x^2 - y^2 = -16$,

$y^2 - 4x^2 = 16$.

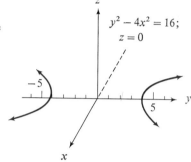

25. $3x + 2y + 4z = 12$; xy-plane

26. $x - 4y + 2z = 8$; yz-plane

27. $x^2 + 2y^2 + 4z^2 = 16$; xz-plane

28. $x^2 + y + 2z^2 = 4$; yz-plane

29. $4x^2 - y^2 + z = 36$; xy-plane

30. $x^2 + 3y^2 - z^2 = 16$; xz-plane

31. $-4x^2 + y^2 + 2z^2 = -4$; yz-plane

32. $-x + y - 6z^2 = 12$; xz-plane

Graph the section of each surface on the specified plane if such section exists.

Example $y = x^2 + z^2$; $y = 4$

Solution   The graph is shown at the right.

For $y = 4$,

$x^2 + z^2 = 4$.

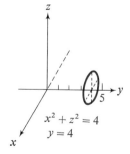

33. $4x + 2y + 3z = 13$; $z = 3$

34. $x + 3y + 2z = -6$; $y = -4$

35. $x^2 + y^2 + z^2 = 25$; $x = 3$

36. $x^2 + 4y^2 - 2z^2 = 16$; $z = 3$

37. $x - 4y^2 + z^2 = 0$; $y = 1$

38. $3x^2 + y^2 - z = 4$; $x = 2$

39. $4x^2 - 9y^2 - 2z^2 = 28$; $z = 2$

40. $2x^2 + y - z^2 = -12$; $y = -4$

Using the appropriate traces and sections, graph each equation in R^3.

41. $x^2 + y^2 + z^2 = 9$

42. $2x^2 + 4y^2 + z^2 = 16$

43. $3x - y + 3z = 6$

44. $2x + y + z = 4$

45. $x^2 + 9y^2 + 9z^2 = 36$

46. $x^2 - 2y^2 + z^2 = 16$

47. $x^2 -- y + z^2 = 0$

48. $4x^2 + 2y^2 - z = 8$

5

Polynomial and rational functions

5.1

Polynomial functions

Graphs of polynomial functions In Section 4.2, we graphed linear functions

$$\{(x, f(x)) \mid f(x) = a_0 x + a_1, \ a_0 \neq 0\},$$

and in Section 4.4, we graphed quadratic functions

$$\{(x, f(x)) \mid f(x) = a_0 x^2 + a_1 x + a_2, \ a_0 \neq 0\}.$$

We can graph any real polynomial function

$$\{(x, P(x)) \mid P(x) = a_0 x^n + a_1 x^{n-1} + \cdots + a_n, \ a_0 \neq 0\}$$

by similar methods—that is, by combining the plotting of points with a consideration of certain general properties of the defining equations. In the case of the general polynomial equation, we shall lean more heavily on the use of plotted points. There is one fact about polynomials, however, that can be useful. This involves *turning points*, or local maximum and minimum values of y. Thus, for example, in Figure 5.1-a there is one local maximum as well as one local minimum, or a total of two turning points; in Figure 5.1-c are one local maximum and two local

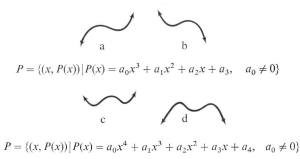

$$P = \{(x, P(x)) \mid P(x) = a_0 x^3 + a_1 x^2 + a_2 x + a_3, \quad a_0 \neq 0\}$$

$$P = \{(x, P(x)) \mid P(x) = a_0 x^4 + a_1 x^3 + a_2 x^2 + a_3 x + a_4, \quad a_0 \neq 0\}$$

Figure 5.1

minima, for a total of three turning points. In general, we have the following theorem.

Theorem 5.1   *If* $P(x) = a_0 x^n + a_1 x^{n-1} + \cdots + a_n$ *is a real polynomial equation of degree n, then the graph of*

$$\{(x, P(x)) \mid P(x) = a_0 x^n + a_1 x^{n-1} + \cdots + a_n, \ a_0 \neq 0\}$$

is a smooth curve that has at most $n - 1$ *turning points.*

The proof of this theorem is omitted because it involves ideas we shall not discuss herein. In accordance with this theorem, the graphs of third- and fourth-degree polynomial functions might appear as in a, b, and c, d, respectively, of Figure 5.1.

General form of a polynomial graph   To ascertain whether or not the graph of a polynomial function ultimately goes up to the right, taking the general form a or c, we can examine the leading coefficient, a_0, of the right-hand member of the defining equation; if $a_0 > 0$, then we can look for a form similar to a or c. But if $a_0 < 0$, we can expect something similar to b or d, where the ultimate direction is down. In other words, the graph ultimately goes up or down to the right according as $a_0 > 0$ or $a_0 < 0$.

If $a_0 > 0$, then the graph ultimately goes down to the left as in a and d, or up to the left as in b and c, according as n is odd or even. If $a_0 < 0$, the situation is reversed.

For the actual graphing process, we can obtain ordered pairs $(x, f(x))$ for any polynomial function either by direct substitution or by using the remainder theorem.

Example   Graph $\{(x, P(x)) \mid P(x) = 2x^3 + 13x^2 + 6x\}$.

Solution   Since P is defined by a cubic polynomial with positive leading coefficient, we expect a graph having a form similar to Figure 5.1-a. Now, to find points $(x, P(x))$ lying on the graph, we shall use the process of synthetic division and find $P(x)$ by the remainder theorem. Since we know nothing about where to look for turning points, let us start with $x = 0$. By inspection, $P(0) = 0$, so that the graph includes the origin. For $x = 1$, we have

$$
\begin{array}{r|rrrr}
1 & 2 & 13 & 6 & 0 \\
  &   & 2 & 15 & 21 \\
\hline
  & 2 & 15 & 21 & 21 \\
\end{array}
$$

and $P(1) = 21$, so that $(1, 21)$ is on the graph. For $x = 2$, we have

$$
\begin{array}{r|rrrr}
2 & 2 & 13 & 6 & 0 \\
 & & 4 & 34 & 80 \\
\hline
 & 2 & 17 & 40 & 80
\end{array}
$$

and $(2, 80)$ is on the graph. Since the signs involved at each step in the last row of the division process here are positive, it is evident that for values $x > 2$, $P(x)$ is positive and will grow increasingly large; consequently, let us turn our attention to negative values of x. For $x = -1$, we have

$$
\begin{array}{r|rrrr}
-1 & 2 & 13 & 6 & 0 \\
 & & -2 & -11 & 5 \\
\hline
 & 2 & 11 & -5 & 5
\end{array}
$$

and $(-1, 5)$ is on the graph. Similarly we find that the following points lie on the graph of P:

$$(-2, 24), (-3, 45), (-4, 56), (-5, 45), \text{ and } (-6, 0).$$

The graphs of the nine ordered pairs are shown in the figure on the left below. These points make the general appearance of the graph clear, and there remains only the question of whether or not there is a zero for the function between -1 and 0. If to x we assign values $-3/4$, $-1/4$, and $-1/2$, we obtain the additional pairs $(-3/4, 63/32)$, $(-1/4, -23/32)$, and $(-1/2, 0)$. For $-1/2 < x < 0$, we have $P(x) < 0$, and the graph can be sketched as shown in the right-hand figure.

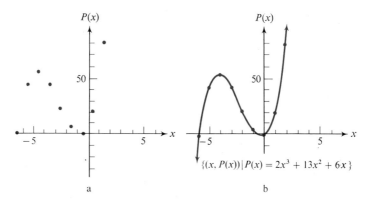

$$\{(x, P(x)) \mid P(x) = 2x^3 + 13x^2 + 6x\}$$

a   b

Exercise 5.1

Graph. Include approximations to all turning points.

1. $\{(x, P(x)) \mid P(x) = x^3 - 4x^2 + 3x\}$ 2. $\{x, P(x)) \mid P(x) = x^3 - 2x^2 + 1\}$
3. $\{(x, P(x)) \mid P(x) = 2x^3 + 9x^2 + 7x - 6\}$ 4. $\{(x, P(x)) \mid P(x) = 3x^3 + 2x^2 - x + 1\}$

5. $\{(x, P(x)) | P(x) = x^4\}$　　　　　　6. $\{(x, P(x)) | P(x) = -x^4 + x\}$

7. $\{(x, P(x)) | P(x) = x^4 - x^3 - 2x^2 + 3x - 3\}$　8. $\{(x, P(x)) | P(x) = x^4 - 4x^3 - 4x + 12\}$

5.2
Real zeros of polynomial functions

Continuity　In Section 5.1, we graphed polynomial functions by plotting a number of points and using the degree of the polynomial involved to help us deduce a basic pattern for the curve. We also implicitly assumed the following theorem on continuity in the process.

The proofs of this and the succeeding theorems in this section are beyond the scope of this book and are omitted.

Theorem 5.2　*If $P(x) = a_0 x^n + a_1 x^{n-1} + \cdots + a_n$, $a_i \in R$, and k is a real number between $P(x_1)$ and $P(x_2)$, then there exists at least one $c \in R$ between x_1 and x_2 such that $P(c) = k$.*

Essentially, $P(x)$ must assume all values between any two of its values, because of the fact that the graph of P must be a continuous curve. This continuity property of the values of a polynomial, together with the theorems given below, enables us to deduce some facts about real zeros of a polynomial function without having to graph the function.

Number of zeros　To begin with, a polynomial function of degree $n \geq 1$ over R can have at most n real zeros, and may have none. In Chapter 11, you will find that, with an agreement on how to count zeros, every such function has exactly n complex zeros (zeros that are complex numbers, real or imaginary) and that when imaginary zeros (zeros that are imaginary complex numbers) exist for a polynomial over R, there is always an even number of them.

Turning now to real zeros only, let us agree that a **variation in sign** occurs in a polynomial with real coefficients if, as the polynomial is viewed from left to right, successive coefficients are opposite in sign. For example, in the polynomial

$$P(x) = 3x^5 - 2x^4 - 2x^2 + x - 1,$$

there are three variations in sign, and in

$$P(-x) = -3x^5 - 2x^4 - 2x^2 - x - 1,$$

there are no variations in sign. We have the following result, called **Descartes' rule of signs**.

Theorem 5.3   *If $P(x)$ is a polynomial over the field R of real numbers, then the number of positive real zeros of $P(x)$ either is equal to the number of variations in sign occurring in the coefficients of $P(x)$ or else is less than this number by an even natural number. Moreover, the number of negative real zeros of $P(x)$ either is equal to the number of variations in sign occurring in $P(-x)$ or else is less than this number by an even natural number.*

Example   Find an upper bound on the number of real positive zeros and real negative zeros of

$$\{(x, P(x))\,|\,P(x) = 3x^4 + 3x^3 - 2x^2 + x + 1\}.$$

Solution   Since $P(x) = 3x^4 + 3x^3 - 2x^2 + x + 1$ has but two variations in sign, $P(x)$ can have no more than two positive real zeros. Since

$$P(-x) = 3(-x)^4 + 3(-x)^3 - 2(-x)^2 + (-x) + 1$$
$$= 3x^4 - 3x^3 - 2x^2 - x + 1$$

has two variations in sign, $P(x)$ can have at most two negative real zeros.

Bounds on sets   The following theorem is sometimes helpful in isolating real
of zeros   zeros of a polynomial function with real coefficients.

Theorem 5.4   *Let $P(x)$ be a polynomial over the field R of real numbers.*

I   *If $c_1 \geq 0$ and the coefficients of the terms in $Q(x)$ and the term $P(c_1)$ are all of the same sign in the right-hand member of*

$$P(x) = (x - c_1)Q(x) + P(c_1),$$

then $P(x)$ can have no zero greater than c_1.

II   *If $c_2 \leq 0$ and the coefficients of the terms in $Q(x)$ and the term $P(c_2)$ alternate in sign (zero suitably denoted by $+0$ or -0) in the right-hand member of*

$$P(x) = (x - c_2)Q(x) + P(c_2),$$

then $P(x)$ can have no zero less than c_2.

   This theorem permits us to place upper and lower bounds on the set of zeros of the polynomial function

$$P = \{(x, P(x))\,|\,P(x) = a_0 x^n + \cdots + a_n\},$$

that is, on the members of the solution set of $P(x) = 0$.

Example

Show that 2 and -2 are upper and lower bounds, respectively, for the set of zeros of

$$\{(x, P(x)) \mid P(x) = 18x^3 - 12x^2 - 11x + 10\}.$$

Solution

To find $Q(x)$, we use synthetic division to divide $P(x)$ by $(x - 2)$:

$$
\begin{array}{r|rrrr}
2 & 18 & -12 & -11 & 10 \\
  &   & 36 & 48 & 74 \\
\hline
  & 18 & 24 & 37 & 84 \\
\end{array}
$$

Since $Q(x) = 18x^2 + 24x + 37$ and $P(2) = 84 > 0$, from Theorem 5.4-I it follows that 2 is an upper bound for the set of zeros of P. Next dividing $P(x)$ by $(x + 2)$, we have

$$
\begin{array}{r|rrrr}
-2 & 18 & -12 & -11 & 10 \\
   &   & -36 & 96 & -170 \\
\hline
   & 18 & -48 & 85 & -160 \\
\end{array}
$$

Since $Q(x) = 18x^2 - 48x + 85$ and $P(-2) = -160$, it follows from Theorem 5.4-II that -2 is a lower bound for the set of zeros of P.

Example

Find the least nonnegative integer and the greatest nonpositive integer that are, by Theorem 5.4, upper and lower bounds, respectively, for the set of zeros of

$$\{(x, P(x)) \mid P(x) = x^4 - x^3 - 10x^2 - 2x + 12\}.$$

Solution

We shall first seek an upper bound by dividing $P(x)$ successively by $(x - 1)$, $(x - 2)$, and so on. Each row after the first in the following array is the bottom row in the synthetic division involved.

$$
\begin{array}{r|rrrrr}
  & 1 & -1 & -10 & -2 & 12 \\
\hline
1 & 1 & 0 & -10 & -12 & 0 \\
2 & 1 & 1 & -8 & -18 & -24 \\
3 & 1 & 2 & -4 & -14 & -30 \\
4 & 1 & 3 & 2   & 6   & 36 \\
\end{array}
$$

Since the numbers in the last row are all positive, 4 is an upper bound. Next, we divide by $(x + 1), (x + 2)$, and so on, in search of a lower bound.

$$
\begin{array}{r|rrrrr}
   & 1 & -1 & -10 & -2 & 12 \\
\hline
-1 & 1 & -2 & -8 & 6 & 6 \\
-2 & 1 & -3 & -4 & 6 & 0 \\
-3 & 1 & -4 & 2   & -8 & 36 \\
\end{array}
$$

Since the signs alternate in the last row, -3 is a lower bound. Had the numbers in the row been $1, 0, 2, -8, 36$, then the sign " $-$ " could arbitrarily have been assigned to 0 to give the desired pattern of alternating signs.

To narrow the search for real zeros of a polynomial function still further, we have the following **location theorem**, which follows directly from Theorem 5.2.

Theorem 5.5 *Let $P(x)$ be a polynomial over the field R of real numbers. If $x_1, x_2 \in R$, with $x_1 < x_2$, and $P(x_1)$ and $P(x_2)$ are opposite in sign, then there exists at least one $c \in R$, $x_1 < c < x_2$, such that $P(c) = 0$.*

This theorem expresses the fact that if the graphs of $(x_1, P(x_1))$ and $(x_2, P(x_2))$ are on opposite sides of the x-axis, then the graph of $y = P(x)$ must cross the x-axis at some (at least one) point with abscissa c between x_1 and x_2.

Example

Verify that $\{(x, f(x)) | f(x) = 2x^3 - 3x^2 + 4x - 6\}$ has a zero between 1 and 2.

Solution

We can apply synthetic division to find $f(1)$ and $f(2)$.

		2	−3	4	−6
1		2	−1	3	−3
2		2	1	6	6

Since $f(1) = -3$ and $f(2) = 6$, Theorem 5.5 assures us that f has a zero between 1 and 2.

Exercise 5.2

Use Theorem 5.3 to find bounds on the number of positive zeros and the number of negative zeros of the function defined by each equation.

1. $P(x) = x^4 - 2x^3 + 2x + 1$ 2. $R(x) = 3x^4 + 3x^3 + 2x^2 - x + 1$

3. $Q(x) = 2x^5 + 3x^3 + 2x + 1$ 4. $P(x) = 4x^5 - 2x^3 - 3x - 2$

5. $S(x) = 3x^4 + 1$ 6. $Q(x) = 2x^5 - 1$

Find the least nonnegative integer and the greatest nonpositive integer that are, by Theorem 5.4, an upper bound and a lower bound, respectively, for the set of real zeros of the function defined by each equation.

7. $P(x) = x^3 + 2x^2 - 7x - 8$ 8. $Q(x) = x^3 - 8x + 5$

9. $Q(x) = x^4 - 2x^3 - 7x^2 + 10x + 10$ 10. $R(x) = x^3 - 4x^2 - 4x + 12$

11. $S(x) = x^5 - 3x^3 + 24$ 12. $R(x) = x^5 - 3x^4 - 1$

13. $P(x) = 2x^5 + x^4 - 2x - 1$ 14. $G(x) = 2x^5 - 2x^2 + x - 2$

Use Theorem 5.5 to verify each statement.

15. $\{(x, f(x))|f(x) = x^3 - 3x + 1\}$ has a zero between 0 and 1.

16. $\{(x, f(x))|f(x) = 2x^3 + 7x^2 + 2x - 6\}$ has a zero between -2 and -1.

17. $\{(x, g(x))|g(x) = x^4 - 2x^2 + 12x - 17\}$ has a zero between -3 and -2.

18. $\{(x, g(x))|g(x) = 2x^4 + 3x^3 - 14x^2 - 15x + 9\}$ has a zero between -2 and -1.

19. $\{(x, P(x))|P(x) = 2x^2 + 4x - 4\}$ has one zero between -3 and -2, and one between 0 and 1.

20. $\{(x, P(x))|P(x) = x^3 - x^2 - 2x + 1\}$ has one zero between -2 and -1, one between 0 and 1, and one between 1 and 2.

Use information from Theorems 5.1 to 5.5 and sketch the graph of the function given in each specified problem.

21. Problem 15 above   22. Problem 16

23. Problem 17   24. Problem 18

25. Problem 19   26. Problem 20

5.3

Rational zeros of polynomial functions

Possible rational zeros   If all the coefficients of the defining equation

$$P(x) = a_0 x^n + a_1 x^{n-1} + \cdots + a_n, \quad a_0 \neq 0,$$

of a polynomial function P are integers, then we can identify all possible rational zeros of P by means of the following theorem.

Theorem 5.6   *If the rational number in lowest terms p/q is a solution of*

$$P(x) = a_0 x^n + a_1 x^{n-1} + \cdots + a_{n-1} x + a_n = 0,$$

where $a_i \in J$ $(a_0 \neq 0)$, then p is an integral factor of a_n and q is an integral factor of a_0.

Proof   Since p/q is a solution of $P(x) = 0$, we have

$$a_0 \left(\frac{p}{q}\right)^n + a_1 \left(\frac{p}{q}\right)^{n-1} + \cdots + a_{n-1} \left(\frac{p}{q}\right) + a_n = 0,$$

and we can multiply each member here by q^n to obtain

$$a_0 p^n + a_1 p^{n-1} q + \cdots + a_{n-1} pq^{n-1} + a_n q^n = 0.$$

Adding $-a_n q^n$ to each member and factoring p from each term in the left-hand member of the resulting equation, we have

$$p(a_0 p^{n-1} + a_1 p^{n-2}q + \cdots + a_{n-1}q^{n-1}) = -a_n q^n.$$

Since the set of integers is closed under addition and multiplication, the expression in parentheses in the left-hand member here represents an integer, say r, so that we have

$$pr = -a_n q^n,$$

where pr is an integer having p as a factor. Hence, p is a factor of $-a_n q^n$. But p and q^n have no factor in common, because, by hypothesis, p/q is in lowest terms; hence p must be a factor of a_n. In a similar manner, by writing the equation

$$a_0 p^n + a_1 p^{n-1}q + \cdots + a_{n-1}pq^{n-1} + a_n q^n = 0$$

in the form

$$-a_0 p^n = a_1 p^{n-1}q + \cdots + a_{n-1}pq^{n-1} + a_n q^n,$$

we can factor q from each term in the right-hand member and show that q must be a factor of a_0.

Example

List all possible rational zeros of

$$\{(x, P(x)) \mid P(x) = 2x^3 - 4x^2 + 3x + 9\}.$$

Solution

Rational zeros p/q must, by Theorem 5.6, be such that p is an integral factor of 9 and q is an integral factor of 2. Hence

$$p \in \{-9, -3, -1, 1, 3, 9\}, \qquad q \in \{-2, -1, 1, 2\},$$

and the set of possible rational zeros of P is

$$\left\{-9, -\frac{9}{2}, -3, -\frac{3}{2}, -1, -\frac{1}{2}, \frac{1}{2}, 1, \frac{3}{2}, 3, \frac{9}{2}, 9\right\}.$$

Test for
rational zeros

It is important to observe that Theorem 5.6 does not assure us that a polynomial function with integral coefficients indeed has a rational zero; it simply enables us to identify possibilities for rational zeros. These can then be checked by synthetic division. The identification of the zeros of P in the previous example is left as an exercise.

As a special case of Theorem 5.6, it is evident that if a function P is defined by

$$P(x) = x^n + a_1 x^{n-1} + \cdots + a_n,$$

in which $a_i \in J$ and the coefficient of x^n is 1, then any rational zero of P must be an integer and, moreover, an integral factor of a_n.

Example   Find all rational zeros of

$$\{(x, P(x)) \mid P(x) = x^3 - 4x^2 + x + 6\}.$$

Solution   The only possible rational zeros of P are $-6, -3, -2, -1, 1, 2, 3$, and
6. Using synthetic division, we set up the following array.

	1	-4	1	6
-6	1	-10	61	-360
-3	1	-7	22	-60
-2	1	-6	13	-20
-1	1	-5	6	0

We can end our trials with -1, since, by the remainder theorem, -1 is
a zero of P, and the remaining zeros can be obtained from the depressed
equation $x^2 - 5x + 6 = 0$ by writing this as $(x - 2)(x - 3) = 0$ and
observing that 2 and 3 are also zeros. Had the trial process begun with
-1, 2, or 3, rather than -6, a single trial would have rendered the
rational zeros immediately evident.

Exercise 5.3

Find all integral zeros of each function.

1. $\{(x, f(x)) \mid f(x) = x^3 - x^2 - 4x + 4\}$

2. $\{(x, f(x)) \mid f(x) = x^3 - 4x^2 + x + 6\}$

3. $\{(x, f(x)) \mid f(x) = 2x^3 - 3x^2 - 2x + 3\}$

4. $\{(x, f(x)) \mid f(x) = 5x^3 + 11x^2 - 2x - 8\}$

5. $\{(x, P(x)) \mid P(x) = x^4 + 2x^3 - 9x^2 - 2x + 8\}$

6. $\{(x, P(x)) \mid P(x) = x^4 + 3x^3 - 2x^2 - 12x - 8\}$

Find all rational zeros of each function.

7. $\{(x, f(x)) \mid f(x) = 4x^3 - 16x^2 + 11x + 10\}$

8. $\{(x, f(x)) \mid f(x) = 10x^3 + 19x^2 - 5x - 6\}$

9. $\{(x, P(x)) \mid P(x) = 12x^3 - 28x^2 - 9x + 10\}$

10. $\{(x, P(x)) \mid P(x) = 6x^4 - 29x^3 + 30x^2 + 11x - 6\}$

11. $\{(x, Q(x)) | Q(x) = 4x^4 - 4x^3 - 25x^2 + x + 6\}$

12. $\{(x, Q(x)) | Q(x) = 6x^4 + x^3 - 23x^2 + 4x + 12\}$

Use information from Theorems 5.1 to 5.6 and sketch the graph of the function given in each specified problem.

13. Problem 1 above   14. Problem 2

15. Problem 5   16. Problem 6

17. Problem 11   18. Problem 12

5.4

Irrational zeros of polynomial functions

Theorem 5.5 can often be used to isolate some real zeros of polynomial functions on intervals of the domain. Once we have isolated such zeros, various means exist for obtaining closer approximations to them, in particular to zeros that are irrational. We shall be concerned with only one such means herein—namely, linear interpolation.

Isolation
of a zero   Consider the function

$$P = \{(x, P(x)) | P(x) = x^3 - 3x^2 - 2x + 5\}.$$

By means of synthetic division and the remainder theorem, we can establish that $(1, 1)$ and $(2, -3)$ are in the function, and hence, by the location theorem, there is at least one zero between 1 and 2. Actually, since $P(x)$ is positive for x large and positive, and $P(x)$ is negative for x large in absolute value and negative, we can see that one of the remaining two roots must be greater than 2 and the other must be less than 1; therefore, there is exactly one zero between 1 and 2. Since the only rational zero would have to be an integer (the leading coefficient is 1), any zeros between 1 and 2 must be irrational. Figure 5.2-a on page 148 shows the two points $P_1(1, 1)$ and $P_2(2, -3)$, and a line segment joining them.

The dashed lines show possibilities for the graph of P on the interval $1 < x < 2$; but since we are uncertain of the curvature, we cannot be sure on which side of the line segment the graph actually lies. In either case, however, the point where the segment intersects the x-axis clearly is close (in some sense) to the point where the graph of P intersects this axis. In Figure 5.2-b we show the same segment, this time with an additional detail. If we can find a value for the coordinate of the x-intercept

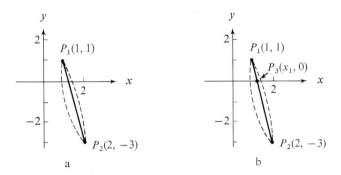

Figure 5.2

of the line segment P_1P_2, then we will have a first approximation to a zero for P. Since the slope of P_1P_3 is the same as the slope for P_1P_2, we have

$$\frac{0-1}{x_1-1} = \frac{-3-1}{2-1},$$

from which

$$x_1 = \frac{5}{4}.$$

To find $P(5/4)$, we divide $P(x)$ synthetically by 1.25, as follows.

$$
\begin{array}{r|rrrr}
1.25 & 1 & -3 & -2 & 5 \\
     &   & 1.25 & -2.1875 & -5.234375 \\
\hline
     & 1 & -1.75 & -4.1875 & -0.234375
\end{array}
$$

Thus, $P(5/4) \approx -0.2344$. Figure 5.3-a shows our present situation, from which it is

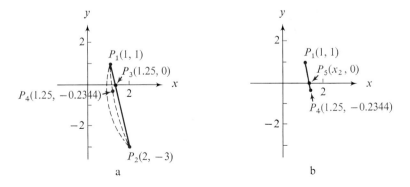

Figure 5.3

clear that the graph of P crosses the x-axis to the left of P_3, so that, at least insofar as this point is concerned, the graph of P is concave upward on this interval. We can now repeat the linear-interpolation process, using P_1 and P_4, to obtain another approximation to the zero for P. Figure 5.3-b shows the necessary detail. Again, since the slope of P_1P_4 is the same as the slope of P_1P_5, we have

$$\frac{0-1}{x_2-1} = \frac{-0.2344-1}{1.25-1},$$

from which

$$x_2 \approx 1.2025.$$

Thus, a second approximation to the desired zero of P is 1.2025. This process can be continued as long as necessary to obtain any desired degree of accuracy. To three decimal places, the zero sought here is 1.202.

Exercise 5.4

Find to one decimal place the indicated real zero(s) of the function.

1. $\{(x, P(x)) \mid P(x) = x^3 - 3x + 1\}$; between 1 and 2
2. $\{(x, P(x)) \mid P(x) = 2x^3 - x^2 + 3x + 1\}$; between 0 and -1
3. $\{(x, P(x)) \mid P(x) = x^3 - 2x - 5\}$; between 2 and 3
4. $\{(x, P(x)) \mid P(x) = x^3 + 2x^2 - 1\}$; between 0 and 1
5. $\{(x, P(x)) \mid P(x) = x^3 + 3x^2 - 6x - 3\}$; the greatest positive
6. $\{(x, P(x)) \mid P(x) = 2x^3 - 5x^2 - x + 5\}$; the least positive
7. $\{(x, P(x)) \mid P(x) = x^3 + x - 1\}$; all
8. $\{(x, P(x)) \mid P(x) = x^4 - 4x^3 - 4x + 12\}$; all
9. Find to two decimal places an approximation for $\sqrt[3]{5}$. *Hint*: Consider the equation $x^3 - 5 = 0$.
10. Find to two decimal places an approximation for $\sqrt[5]{2}$.

5.5

Rational functions

A function defined by an equation of the form

$$y = \frac{P(x)}{Q(x)}, \tag{1}$$

where $P(x)$ and $Q(x)$ are polynomials in x, and $Q(x)$ is not the zero polynomial, is called a **rational function**. Rational functions with real coefficients (that is, rational functions over R) are of importance in the calculus and provide some interesting problems with respect to their graphs. We shall consider them only briefly here.

Graph near vertical asymptotes
Since $P(x)/Q(x)$ is not defined for values of x for which $Q(x) = 0$, it is evident that we shall not be able to find points in $R \times R$ having such x-coordinates. We can, however, consider the graph for values of x as close as we please to a value, say x_0, for which $Q(x_0) = 0$, but still with $x \neq x_0$. This consideration is usually described by saying that x "approaches" x_0, "grows close" to x_0, etc., and correspondingly that $Q(x)$ approaches 0. Thus, the closer $Q(x)$ approaches 0, if $P(x)$ does not approach 0 at the same time, then the larger $|y|$ in (1) becomes. For example, Figure 5.4-a shows the behavior of

$$\left\{ (x, y) \,|\, y = \frac{2}{x - 2} \right\} \tag{2}$$

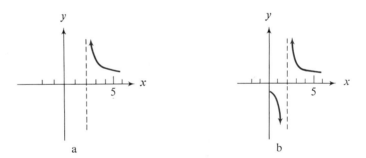

Figure 5.4

as x approaches 2 from the right. In situations such as this, the vertical line that the curve approaches is called a **vertical asymptote**.

Theorem 5.7 *The graph of the rational function over R defined by $y = P(x)/Q(x)$ has a vertical asymptote at $x = a$ for each value a at which $Q(x)$ vanishes and $P(x)$ does not vanish.*

In Figure 5.4-a we see the behavior of the function (2) as x approaches 2 from the right. We are also interested in its behavior as x approaches 2 from the left. Figure 5.4-b illustrates this. As long as $x > 2$, we have $x - 2 > 0$ and $[2/(x - 2)] > 0$; but if $x < 2$, then we have $x - 2 < 0$ and $[2/(x - 2)] < 0$.

Horizontal asymptotes   The graphs of some rational functions have **horizontal asymptotes**, which can, in general, be identified by using the following theorem.

Theorem 5.8   *The graph of the rational function over R defined by*

$$y = \frac{a_0 x^n + a_1 x^{n-1} + \cdots + a_n}{b_0 x^m + b_1 x^{m-1} + \cdots + b_m},$$

where a_0, $b_0 \neq 0$ and n, m are nonnegative integers, has

  I   *a horizontal asymptote at $y = 0$ if $n < m$,*

  II   *a horizontal asymptote at $y = a_0/b_0$ if $n = m$,*

 III   *no horizontal asymptotes if $n > m$.*

Though we shall not give a rigorous proof of this theorem here, we can certainly make the results plausible. If $n < m$, we can divide the numerator and denominator of the right-hand member of

$$y = \frac{a_0 x^n + a_1 x^{n-1} + \cdots + a_n}{b_0 x^m + b_1 x^{m-1} + \cdots + b_m}$$

by x^m to obtain, for $x \neq 0$,

$$y = \frac{\dfrac{a_0}{x^{m-n}} + \dfrac{a_1}{x^{m-n+1}} + \cdots + \dfrac{a_n}{x^m}}{b_0 + \dfrac{b_1}{x} + \cdots + \dfrac{b_m}{x^m}}.$$

Now, as $|x|$ grows larger and larger, each term containing an x in its denominator grows closer and closer to 0, and we find the expression on the right approaching $0/b_0$, so that y approaches 0. But if, as y grows close to 0, $|x|$ is increasing without

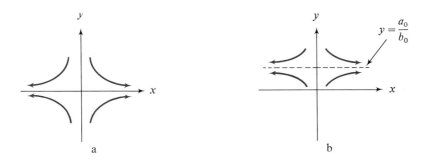

a   b

Figure 5.5

bound, then y is approaching the line $y = 0$ asymptotically, and actually must approach 0 from one of the four directions shown in Figure 5.5-a. A similar argument shows that if $n = m$, then, as $|x|$ increases without bound, y approaches the line $y = a_0/b_0$ from one of the directions shown in Figure 5.5-b. If $n > m$, then as $|x|$ becomes larger and larger, so does $|y|$.

Oblique
asymptotes

In particular, if $n = m + 1$, that is, if the numerator is of degree one greater than the denominator, we can argue that though the graph has no horizontal asymptote, it does have an **oblique asymptote**. We shall illustrate a special case only, but the technique involved is quite general.

Example

Find all asymptotes for the graph of $\left\{(x, y) \mid y = \dfrac{x^2 - 4}{x - 1}\right\}$.

Solution

First note that, by Theorem 5.7, there is a vertical asymptote at $x = 1$, and that, by Theorem 5.8, there are no horizontal asymptotes. If, however, we rewrite $y = (x^2 - 4)/(x - 1)$ by dividing $x^2 - 4$ by $x - 1$, we obtain

$$y = x + 1 - \frac{3}{x - 1}.$$

Now as $|x|$ grows larger and larger, $3/(x - 1)$ grows smaller and smaller, and the graph of

$$y = \frac{x^2 - 4}{x - 1}$$

approaches the graph of $y = x + 1$. Hence, the graph of $y = x + 1$ which is an oblique line, is an asymptote to the curve.

Helpful items
for graphing

Identifying asymptotes is one aid to the graphing of a rational function over R. Other helpful items are

1.   the zeros of the function, because these give us the x-intercepts;

2.   the domain and range, because these let us know where we can expect to find parts of the graph and where we cannot;

3.   some specific points on the graph, because these give us guidelines in sketching.

Example

Graph $\left\{(x, y) \mid y = \dfrac{x - 1}{x - 2}\right\}$.

Solution   We can begin by observing that the numerator of the right-hand
member will be equal to 0 when x is equal to 1. Therefore, when $x = 1$,
we have $y = 0$, and 1 is an x-intercept. Also, when $x = 0$, we have
$y = 1/2$, so that 1/2 is a y-intercept. Thus we can begin our graph as
shown in a. By inspection, $y = (x - 1)/(x - 2)$ is defined for all real

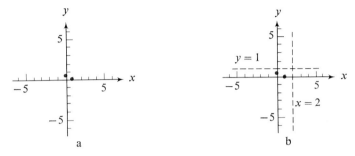

a   b

x except $x = 2$, so that $\{x \mid x \neq 2\}$ is the domain. Similarly, if we solve
the defining equation for x in terms of y, we have $x = (2y - 1)/(y - 1)$,
which is defined for all values of y except 1. Hence $\{y \mid y \neq 1\}$ is the
range of the function. From the defining equation and Theorem 5.7,
we see that there is a vertical asymptote at $x = 2$ and a horizontal
asymptote at $y = 1$. We can then add this information to our graph, as
indicated in b. Next we call our power of observation into play. That
the vertical asymptote, for example, is approached downward instead
of upward from the left can be confirmed by observing that if x is just
less than 2, say $2 - p$, $0 < p < 1/10$, then the denominator $x - 2$ in
the expression for y is $2 - p - 2 = -p$, which is barely negative,
whereas for $x - 1$, the numerator, we have $x - 1 = 2 - p - 1 = 1 - p$,
which is definitely positive; hence y is negative and $|y|$ large. The graph

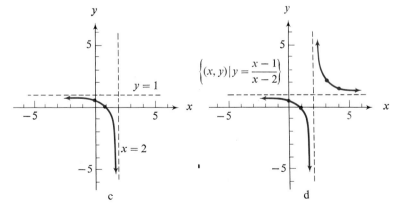

c   d

appears in c. To find the curve when $x > 2$, that is, to the right of the
vertical asymptote, we simply find one or two points associated with
such values of x. For example, we might choose 3 and 4. If $x = 3$, then

(Solution continued.)

$$y = \frac{3-1}{3-2} = 2,$$

and (3, 2) is on the curve. If $x = 4$, then

$$y = \frac{4-1}{4-2} = 3/2,$$

and (4, 3/2) is on the curve. Again, the knowledge that $x = 2$ and $y = 1$ are asymptotes, together with the location of the points (3, 2) and (4, 3/2), leads us to the complete graph of $\{(x, y) | y = (x-1)/(x-2)\}$, as shown in d.

Example   Graph $\left\{ (x, y) | y = \frac{x^2 - 4}{x - 1} \right\}.$

Solution   This is the same function we investigated in the example on page 152, where we found the vertical asymptote $x = 1$ and the oblique asymptote $y = x + 1$. By inspection, if $x = 0$, then $y = 4$, and if $y = 0$, then $x = 2$ or $x = -2$, so that there is a y-intercept at 4, as well as x-intercepts at 2 and -2. We therefore have the situation shown in a. Without any

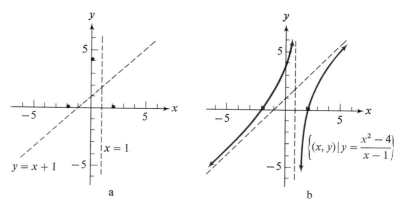

further information, we have strong reason to suspect that the graph will appear as shown in b. The plotting of a few check points, say $(-1, 3/2)$, $(1/2, 15/2)$, $(3/2, -7/2)$, and $(3, 5/2)$, would tend to confirm our conjecture.

Exercise 5.5

Determine the vertical asymptotes of the graph of each function.

Example $\{(x, y) | x^2 y - 4y = 1\}$

Solution   Express y explicitly in terms of x by means of an equivalent equation.

$$y(x^2 - 4) = 1$$

$$y = \frac{1}{x^2 - 4}$$

The denominator $x^2 - 4$ vanishes for $x = 2$, -2. Therefore, by Theorem 5.7, there are vertical asymptotes at $x = 2$ and $x = -2$.

1. $\left\{(x, y) \mid y = \dfrac{1}{x - 3}\right\}$ 2. $\left\{(x, y) \mid y = \dfrac{1}{x + 4}\right\}$

3. $\left\{(x, y) \mid y = \dfrac{4}{(x + 2)(x - 3)}\right\}$ 4. $\left\{(x, y) \mid y = \dfrac{8}{(x - 1)(x + 3)}\right\}$

5. $\left\{(x, y) \mid y = \dfrac{2x - 1}{x^2 + 5x + 4}\right\}$ 6. $\left\{(x, y) \mid y = \dfrac{x + 3}{2x^2 - 5x - 3}\right\}$

7. $\{(x, y) \mid xy + y = 4\}$ 8. $\{(x, y) \mid x^2y + xy = 3\}$

Graph.

9. $\left\{(x, y) \mid y = \dfrac{1}{x}\right\}$ 10. $\left\{(x, y) \mid y = \dfrac{1}{x + 4}\right\}$

11. $\left\{(x, y) \mid y = \dfrac{1}{x - 3}\right\}$ 12. $\left\{(x, y) \mid y = \dfrac{1}{x - 6}\right\}$

13. $\left\{(x, y) \mid y = \dfrac{4}{(x + 2)(x - 3)}\right\}$ 14. $\left\{(x, y) \mid y = \dfrac{8}{(x - 1)(x + 3)}\right\}$

15. $\left\{(x, y) \mid y = \dfrac{2}{(x - 3)^2}\right\}$ 16. $\left\{(x, y) \mid y = \dfrac{1}{(x + 4)^2}\right\}$

Determine any vertical, horizontal, or oblique asymptotes of the graphs of each of the following.

Example $\left\{(x, y) \mid y = \dfrac{6x^2 + 1}{2x^2 + 5x - 3}\right\}$

Solution   The defining equation can be written equivalently as

$$y = \frac{6x^2 + 1}{(2x - 1)(x + 3)}.$$

By Theorem 5.7, there are vertical asymptotes at $x = 1/2$ and $x = -3$.
By Theorem 5.8-II, there is a horizontal asymptote at $y = 6/2$, or 3.

17. $\left\{(x, y) \mid y = \dfrac{x}{x^2 - 4}\right\}$

18. $\left\{(x, y) \mid y = \dfrac{3x + 6}{x^2 + 3x + 2}\right\}$

19. $\left\{(x, y) \mid y = \dfrac{x^2 - 9}{x - 4}\right\}$

20. $\left\{(x, y) \mid y = \dfrac{x^3 - 27}{x^2 - 1}\right\}$

21. $\left\{(x, y) \mid y = \dfrac{x^2 - 3x + 2}{x^2 - 3x - 4}\right\}$

22. $\left\{(x, y) \mid y = \dfrac{x^2}{x^2 - x - 6}\right\}$

Graph. Use information concerning the zeros of each function, and any vertical, horizontal, and oblique asymptotes.

23. $\left\{(x, y) \mid y = \dfrac{x}{x - 2}\right\}$

24. $\left\{(x, y) \mid y = \dfrac{x - 1}{x + 3}\right\}$

25. $\left\{(x, y) \mid y = \dfrac{2x - 4}{x^2 - 9}\right\}$

26. $\left\{(x, y) \mid y = \dfrac{3x}{x^2 - 5x + 4}\right\}$

27. $\left\{(x, y) \mid y = \dfrac{x^2 - 4}{x^3}\right\}$

28. $\left\{(x, y) \mid y = \dfrac{x - 2}{x^2}\right\}$

29. $\left\{(x, y) \mid y = \dfrac{x^2 - 4x + 4}{x - 1}\right\}$

30. $\left\{(x, y) \mid y = \dfrac{x^2 + 4}{x - 2}\right\}$

31. $\left\{(x, y) \mid y = \dfrac{x + 1}{x(x^2 - 4)}\right\}$

32. $\left\{(x, y) \mid y = \dfrac{x^2 + x - 2}{x(x^2 - 9)}\right\}$

6

Exponential and logarithmic functions

6.1

Exponential functions

Powers of the form b^x, where $b \in R, b > 0$, and x denotes a *rational number*, were discussed in Chapter 2. These can be used to define functions. Notice that the base b is restricted here to positive real numbers to ensure that b^x be real for all rational numbers x.

Since we now want to define a function over R using b^x, we must be able to interpret powers with *irrational exponents*, such as

$$b^{\sqrt{2}}, \quad b^{-\sqrt{3}}, \quad \text{and} \quad b^{\pi},$$

to be real numbers. The following two theorems, the first one of which is presented without proof, will be useful in doing this. Recall that Q represents the set of rational numbers.

Theorem 6.1   *Let $x \in Q$ and let $x > 0$. Then*

$$b^x > 1 \text{ if } b > 1, \quad b^x = 1 \text{ if } b = 1, \quad \text{and} \quad 0 < b^x < 1 \text{ if } 0 < b < 1.$$

Theorem 6.2   *Let $x, y \in Q$, and let $x > y$. Then*

$$b^x > b^y \text{ if } b > 1, \quad b^x = b^y \text{ if } b = 1, \quad \text{and} \quad b^x < b^y \text{ if } 0 < b < 1.$$

Theorem 6.2 follows directly from Theorem 6.1. For example, if $b > 1$, then by Theorem 6.1, since $x - y > 0$, we have

$$\frac{b^x}{b^y} = b^{x-y} > 1,$$

from which

$$b^x > b^y.$$

You should recall that irrational numbers can be approximated by rational numbers to as great a degree of accuracy as desired. For example, $\sqrt{2} \approx 1.4$, or $\sqrt{2} \approx 1.414$, etc. Since b^x ($b > 0$) is defined for rational x, by Theorem 6.2 we can write the sequence of inequalities

$$2^1 < 2^2,$$
$$2^{1\cdot4} < 2^{1\cdot5},$$
$$2^{1\cdot41} < 2^{1\cdot42},$$
$$2^{1\cdot414} < 2^{1\cdot415},$$
$$\vdots$$

Powers with real-number exponents   As a result of Theorem 6.2, we can be sure that if $b > 1$, then b^x increases as x increases; and if $0 < b < 1$, then b^x decreases as x increases. This process can be continued indefinitely; thus, though we shall not prove it, in the foregoing sequence the difference between the number on the left and that on the right can be made as small as we please by going far enough in the sequence. This being the case, the completeness property O-3 of the real numbers guarantees that there is just one number, which we denote by $2^{\sqrt{2}}$, that lies between the number on the left and the number on the right, no matter how long this process of approximation is continued. Since we can produce the same type of argument for any irrational exponent x, we shall assume that b^x ($b > 0$) is defined in this way for all real values of x and that the laws governing rational exponents are valid for such powers. Thus Theorem 2.11, pertaining to powers with rational-number exponents, when appropriately reworded, is also applicable to powers with real-number exponents.

Since for each real x there is one and only one number b^x, the equation

$$f(x) = b^x \quad (b > 0) \tag{1}$$

defines a function. Because $1^x = 1$ for all $x \in R$, (1) defines a constant function if $b = 1$. If $b \neq 1$, we say that (1) defines an **exponential function**. Exponential functions can perhaps be visualized most clearly by considering their graphs. We illustrate two typical examples, in which $0 < b < 1$ and $b > 1$, respectively. Assigning values to x in the equations

$$f(x) = \left(\frac{1}{2}\right)^x \quad \text{and} \quad f(x) = 2^x,$$

we find some ordered pairs in each solution set and sketch the graphs of

$$\{(x, f(x)) \mid f(x) = (1/2)^x\} \quad \text{and} \quad \{(x, f(x)) \mid f(x) = 2^x\},$$

shown in Figures 6.1-a and 6.1-b, respectively.

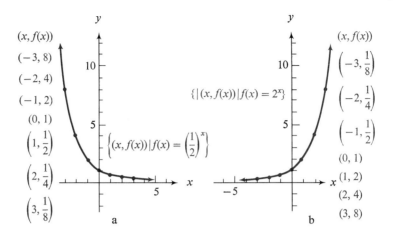

Figure 6.1

Increasing and decreasing functions

Notice that in accordance with Theorem 6.2, the graph of the function determined by $f(x) = (1/2)^x$ goes *down* to the right, and the graph of the function determined by $f(x) = 2^x$ goes *up* to the right. For this reason, we say that the former function is a *decreasing* function and the latter is an *increasing* function. In either case, the domain is the set of real numbers, and the range is the set of positive real numbers.

Exercise 6.1

Find the second component of each of the ordered pairs that makes the pair a solution of the corresponding equation.

1. $y = 3^x$;   (0,), (1,), (2,)

2. $y = -2^x$;   (0,), (1,), (2,)

3. $y = -5^x$;   (0,), (1,), (2,)

4. $y = 4^x$;   (0,), (1,), (2,)

5. $y = \left(\dfrac{1}{2}\right)^x$;   (−3,), (0,), (3,)

6. $y = \left(\dfrac{1}{3}\right)^x$;   (−3,), (0,), (3,)

7. $y = 10^x$;   (−2,), (1,), (0,)

8. $y = 10^{-x}$;   (0,), (1,), (2,)

Graph each function.

9. $\{(x, y)\,|\,y = 4^x\}$

10. $\{(x, y)\,|\,y = 5^x\}$

11. $\{(x, y)\,|\,y = 10^x\}$

12. $\{(x, y)\,|\,y = 10^{-x}\}$

13. $\{(x, y)|y = 2^{-x}\}$ 14. $\{(x, y)|y = 3^{-x}\}$

15. $\left\{(x, y)|y = \left(\dfrac{1}{3}\right)^x\right\}$ 16. $\left\{(x, y)|y = \left(\dfrac{1}{4}\right)^x\right\}$

17. $\left\{(x, y)|y = \left(\dfrac{1}{2}\right)^{-x}\right\}$ 18. $\left\{(x, y)|y = \left(\dfrac{1}{3}\right)^{-x}\right\}$

19. Graph $\{(x, f(x))|f(x) = 1^x\}$. Is this an exponential function? Name the function.

20. Graph $F = \{(x, y)|y = 10^x, x > 0\}$ and $F^{-1} = \{(x, y)|x = 10^y, x > 0\}$ on the same set of axes.

21. Solve for x by inspection.

   a. $10^x = \dfrac{1}{100}$ b. $\left(\dfrac{1}{2}\right)^x = 16$ c. $16^x = 8$

22. Determine an integer n such that $n < x < n + 1$.

   a. $3^x = 16.2$ b. $4^x = 87.1$ c. $10^x = 0.016$

6.2
Logarithmic functions

Inverse of an
exponential
function

In the exponential function

$$\{(x, y)|y = b^x \ (b > 0, b \neq 1)\}, \tag{1}$$

illustrated in Figure 6.1 for $b = 1/2$ and $b = 2$, there is only one x associated with each y. Thus by Definition 4.4 we have the inverse function

$$\{(x, y)|x = b^y \ (b > 0, b \neq 1)\}. \tag{2}$$

Observe that the relation $x > 0$ is implied by Equation (2), because there is no real number y for which b^y is not positive.

The graphs of functions of this form can be illustrated by the example

$$\{(x, y)|x = 10^y\}.$$

We consider x the variable denoting an element in the domain and, in the defining equation, we assign arbitrary values for x, say,

(0.01,   ), (0.1,   ), (1,   ), (10,   ), (100,   ),

to obtain the ordered pairs

$$(0.01, -2), (0.1, -1), (1, 0), (10, 1), (100, 2).$$

These can be graphed and connected with a smooth curve as shown in Figure 6.2.

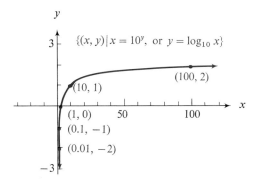

Figure 6.2

It is always useful to be able to express the variable denoting an element in the range explicitly in terms of the variable denoting an element in the domain. To do this in an equation such as that defining the function (2), we use the notation

$$y = \log_b x \quad (x > 0, b > 0, b \neq 1). \tag{3}$$

Here, $\log_b x$ is read "the logarithm to the base b of x," or "the logarithm of x to the base b." Functions defined by such equations are called **logarithmic functions**.

Properties of logarithmic functions   From the graph in Figure 6.2, we generalize from $\log_{10} x$ to $\log_b x$, and observe that for $b > 1$, a logarithmic function has the following properties:

1. The domain is the set of positive real numbers, and the range is the set of all real numbers.

2. $\log_b x < 0$ for $0 < x < 1$, $\log_b x = 0$ for $x = 1$, and $\log_b x > 0$ for $x > 1$.

Logarithmic and exponential statements   It should be recognized that the equation appearing in (2) and the one in (3) are different equations determining the same function, in the same way that $x = y + 4$ and $y = x - 4$ determine the same function, and we may use whichever equation suits our purpose. Thus, exponential statements may be written in logarithmic form, and logarithmic statements may be written in exponential form.

Examples   a. $5^2 = 25$ can be written as $\log_5 25 = 2$.

b. $8^{1/3} = 2$ can be written as $\log_8 2 = \frac{1}{3}$.

c. $3^{-2} = \frac{1}{9}$ can be written as $\log_3 \frac{1}{9} = -2$.

Examples

    a. $\log_{10} 100 = 2$ can be written as $10^2 = 100$.

    b. $\log_3 81 = 4$ can be written as $3^4 = 81$.

    c. $\log_2 \dfrac{1}{2} = -1$ can be written as $2^{-1} = \dfrac{1}{2}$.

The logarithmic function associates with each positive number x the exponent y such that the power b^y is equal to x. In other words, we can think of $\log_b x$ as an exponent on b. Thus

$$b^{\log_b x} = x.$$

Laws of logarithms   Since a logarithm is an exponent, the following theorem follows directly from the properties of powers with real-number exponents.

Theorem 6.3   *If $x_1, x_2 \in R$, $x_1 > 0$, $x_2 > 0$, $b > 0$, $b \neq 1$, $m \in R$, then*

    I $\log_b (x_1 x_2) = \log_b x_1 + \log_b x_2$,

    II $\log_b \dfrac{x_2}{x_1} = \log_b x_2 - \log_b x_1$,

    III $\log_b (x_1)^m = m \log_b x_1$.

The validity of I can be shown as follows: Since

$$x_1 = b^{\log_b x_1} \text{ and } x_2 = b^{\log_b x_2},$$

it follows that

$$x_1 x_2 = b^{\log_b x_1} \cdot b^{\log_b x_2}$$
$$= b^{\log_b x_1 + \log_b x_2},$$

and, by the definition of a logarithm,

$$\log_b (x_1 x_2) = \log_b x_1 + \log_b x_2.$$

The validity of II and III can also be established.

Exercise 6.2

Express in logarithmic notation.

1. $4^2 = 16$ 2. $5^3 = 125$ 3. $3^3 = 27$

4. $8^2 = 64$

5. $\left(\dfrac{1}{2}\right)^2 = \dfrac{1}{4}$

6. $\left(\dfrac{1}{3}\right)^2 = \dfrac{1}{9}$

7. $8^{-1/3} = \dfrac{1}{2}$

8. $64^{-1/6} = \dfrac{1}{2}$

9. $10^2 = 100$

10. $10^0 = 1$

11. $10^{-1} = 0.1$

12. $10^{-2} = 0.01$

Express in exponential notation.

13. $\log_2 64 = 6$

14. $\log_5 25 = 2$

15. $\log_3 9 = 2$

16. $\log_{16} 256 = 2$

17. $\log_{1/3} 9 = -2$

18. $\log_{1/2} 8 = -3$

19. $\log_{10} 1000 = 3$

20. $\log_{10} 1 = 0$

21. $\log_{10} 0.01 = -2$

Find the value of each expression.

22. $\log_5 5$

23. $\log_7 49$

24. $\log_2 32$

25. $\log_4 64$

26. $\log_5 \sqrt{5}$

27. $\log_3 \sqrt{3}$

28. $\log_3 \dfrac{1}{3}$

29. $\log_5 \dfrac{1}{5}$

30. $\log_3 3$

31. $\log_2 2$

32. $\log_{10} 10$

33. $\log_{10} 100$

34. $\log_{10} 1$

35. $\log_{10} 0.1$

36. $\log_{10} 0.01$

Solve for x, y, or b.

Examples   a. $\log_2 x = 3$ b. $\log_b 2 = \dfrac{1}{2}$

Solutions   Determine the solution by inspection or by first writing the equation equivalently in exponential form.

a. $2^3 = x$
$\quad\;\; x = 8$

b. $\quad b^{1/2} = 2$
$\quad (b^{1/2})^2 = (2)^2$
$\qquad\qquad b = 4$

37. $\log_3 9 = y$

38. $\log_5 125 = y$

39. $\log_b 8 = 3$

40. $\log_b 625 = 4$

41. $\log_4 x = 3$

42. $\log_{1/2} x = -5$

43. $\log_2 \dfrac{1}{8} = y$

44. $\log_5 5 = y$

45. $\log_b 10 = \dfrac{1}{2}$

46. $\log_b 0.1 = -1$

47. $\log_2 x = 2$

48. $\log_{10} x = -3$

49. Show that $\log_b 1 = 0$ for $b > 0$. *Hint*: Express the statement in exponential form.

50. Show that $\log_b b = 1$ for $b > 0$.

51. Show that $\log_b b^x = x$ for $b > 0$.

52. Graph $y = \log_2 x$. By examining the graph, what can you assert about $\log_2 a$ and $\log_2 b$ if $a < b$?

Express as the sum or difference of simpler logarithmic quantities.

Example $\quad\quad \log_b \left(\dfrac{xy}{z}\right)^{1/2}$

Solution $\quad\quad$ By Theorem 6.3-III,

$$\log_b \left(\frac{xy}{z}\right)^{1/2} = \frac{1}{2} \log_b \left(\frac{xy}{z}\right).$$

By Theorems 6.3-I and II,

$$\frac{1}{2} \log_b \left(\frac{xy}{z}\right) = \frac{1}{2} [\log_b x + \log_b y - \log_b z].$$

53. $\log_b (xy)$ $\quad\quad\quad\quad$ 54. $\log_b (xyz)$ $\quad\quad\quad\quad$ 55. $\log_b \left(\dfrac{x}{y}\right)$

56. $\log_b \left(\dfrac{xy}{z}\right)$ $\quad\quad\quad\quad$ 57. $\log_b x^5$ $\quad\quad\quad\quad$ 58. $\log_b x^{1/2}$

59. $\log_b \sqrt[3]{x}$ $\quad\quad\quad\quad$ 60. $\log_b \sqrt[3]{x^2}$ $\quad\quad\quad\quad$ 61. $\log_b \sqrt{\dfrac{x}{z}}$

62. $\log_b \sqrt{xy}$ $\quad\quad\quad\quad$ 63. $\log_{10} \sqrt[3]{\dfrac{xy^2}{z}}$ $\quad\quad\quad\quad$ 64. $\log_{10} \sqrt[5]{\dfrac{x^2 y}{z^3}}$

65. $\log_{10} \left(2\pi \sqrt{\dfrac{l}{g}}\right)$ $\quad\quad\quad\quad$ 66. $\log_{10} \sqrt{s(s-a)(s-b)(s-c)}$

Express as a single logarithm with coefficient 1.

Example $\quad\quad \dfrac{1}{2} (\log_b x - \log_b y)$

Solution $\quad\quad$ By Theorems 6.3-II and III,

$$\frac{1}{2} (\log_b x - \log_b y) = \frac{1}{2} \log_b \left(\frac{x}{y}\right) = \log_b \left(\frac{x}{y}\right)^{1/2}.$$

67. $\log_b x + \log_b y$ $\quad\quad\quad\quad\quad\quad$ 68. $\log_b x - \log_b y$

69. $2 \log_b x + 3 \log_b y$

70. $\dfrac{1}{4} \log_b x - \dfrac{3}{4} \log_b y$

71. $3 \log_b x + \log_b y - 2 \log_b z$

72. $\dfrac{1}{3} (\log_b x + \log_b y - 2 \log_b z)$

73. $\log_{10} (x - 2) + \log_{10} x - 2 \log_{10} z$

74. $\dfrac{1}{2} (\log_{10} x - 3 \log_{10} y - 5 \log_{10} z)$

75. Show that $\dfrac{1}{4} \log_{10} 8 + \dfrac{1}{4} \log_{10} 2 = \log_{10} 2$.

76. Show that $4 \log_{10} 3 - 2 \log_{10} 3 + 1 = \log_{10} 90$.

6.3
Logarithms to the base 10

There are two logarithmic functions of special interest; one is defined by

$$y = \log_{10} x, \tag{1}$$

and the other by

$$y = \log_e x, \tag{2}$$

where e is an irrational number with decimal approximation 2.7182818 to eight digits. Because these functions possess similar properties, and because we are more familiar with the number 10, we shall, for the present, confine our attention to (1).

Determination of $\log_{10} x$ Values for $\log_{10} x$ are called **logarithms to the base 10**, or **common logarithms**. From the definition of $\log_{10} x$,

$$10^{\log_{10} x} = x \quad (x > 0); \tag{3}$$

that is, $\log_{10} x$ is the exponent that must be placed on 10 so that the resulting power is x. Our concern in this section is that of finding $\log_{10} x$ for each positive x. First, $\log_{10} x$ can easily be determined for all values of x that are *integral* powers of 10 by inspection, directly from the definition of a logarithm:

$$\log_{10} 10 \quad = \log_{10} 10^1 = 1,$$
$$\log_{10} 100 \quad = \log_{10} 10^2 = 2,$$

and so forth; similarly

$$\log_{10} 1 \quad = \log_{10} 10^0 \quad = 0,$$
$$\log_{10} 0.1 \quad = \log_{10} 10^{-1} = -1,$$
$$\log_{10} 0.01 \quad = \log_{10} 10^{-2} = -2,$$
$$\log_{10} 0.001 = \log_{10} 10^{-3} = -3.$$

A table of logarithms is used to find $\log_{10} x$ for $1 \le x \le 10$ (see page 416). Consider the excerpt from this table shown in Figure 6.3. Each number in the

x	0	1	2	3	4	5	6	7	8	9
3.8	.5798	.5809	.5821	.5832	.5843	.5855	.5866	.5877	.5888	.5899
3.9	.5911	.5922	.5933	.5944	.5955	.5966	.5977	.5988	.5999	.6010
4.0	.6021	.6031	.6042	.6053	.6064	.6075	.6085	.6096	.6107	.6117
4.1	.6128	.6138	.6149	.6160	.6170	.6180	.6191	.6201	.6212	.6222
4.2	.6232	.6243	.6253	.6263	.6274	.6284	.6294	.6304	.6314	.6325
4.3	.6335	.6345	.6355	.6365	.6375	.6385	.6395	.6405	.6415	.6425
4.4	.6435	.6444	.6454	.6464	.6474	.6484	.6493	.6503	.6513	.6522
4.5	.6532	.6542	.6551	.6561	.6571	.6580	.6590	.6599	.6609	.6618
4.6	.6628	.6637	.6646	.6656	.6665	.6675	.6684	.6693	.6702	.6712

Figure 6.3

column headed by x represents the first two significant digits of x, while each number in the row opposite x contains the third significant digit of x. The digits located at the intersection of a row and a column form the logarithm of x. For example, to find $\log_{10} 4.25$, we look at the intersection of the row opposite 4.2 under x and the column headed by 5. Thus, we see that

$$\log_{10} 4.25 = 0.6284.$$

The equality sign is used here in an inexact sense; $\log_{10} 4.25 \approx 0.6284$ is more proper, because $\log_{10} 4.25$ is irrational and cannot be precisely represented by a rational number. We shall follow customary usage, however, writing $=$ instead of $\approx$, and leave the intent to the context.

Characteristic and mantissa Now suppose we wish to find $\log_{10} x$ for values of x outside the range of the table—that is, for $0 < x < 1$ or $x > 10$. This can be done quite readily by first representing the number in scientific notation—that is, as the product of a number between 1 and 10 and a power of 10—and then applying Theorem 6.3-I. For example,

$$\log_{10} 42.5 = \log_{10} (4.25 \times 10^1) = \log_{10} 4.25 + \log_{10} 10^1$$
$$= 0.6284 + 1 = 1.6284,$$

$$\log_{10} 425 = \log_{10} (4.25 \times 10^2) = \log_{10} 4.25 + \log_{10} 10^2$$
$$= 0.6284 + 2 = 2.6284.$$

Observe that the decimal portion of the logarithm is always 0.6284, and the integral portion is just the exponent on 10 when the number is written in scientific notation.

This process can be reduced to a mechanical one by considering $\log_{10} x$ to consist of two parts, an integral part (called the **characteristic**) and a nonnegative decimal fraction part (called the **mantissa**). Thus the table of values for $\log_{10} x$ for $1 < x < 10$ can be looked upon as a table of mantissas for $\log_{10} x$ for all $x > 0$.

For example, to find $\log_{10} 43700$, we first write

$$\log_{10} 43700 = \log_{10}(4.37 \times 10^4).$$

Upon examining the table of logarithms, we find that $\log_{10} 4.37 = 0.6405$, so that

$$\log_{10} 43700 = 4.6405,$$

where we have prefixed the characteristic 4.

Now consider an example of the form $\log_{10} x$ for $0 < x < 1$. To find $\log_{10} 0.00402$, we first write

$$\log_{10} 0.00402 = \log_{10}(4.02 \times 10^{-3}).$$

Examining the table, we find that $\log_{10} 4.02$ is 0.6042. Upon adding 0.6042 to the characteristic -3, we obtain

$$\log_{10} 0.00402 = -2.3958,$$

where the decimal portion of the logarithm is no longer 0.6042 as it is in the case of all numbers $x > 1$ for which the first three significant digits of x are 402. To circumvent this situation, it is customary to write the logarithm in a form in which the decimal part is positive. In the foregoing example, we write

$$\log_{10} 0.00402 = 0.6042 - 3$$
$$= 0.6042 + (7 - 10)$$
$$= 7.6042 - 10,$$

and the decimal part is positive. The logarithm

$$6.6042 - 9$$

is an equally valid representation, but $7.6042 - 10$, in which a multiple of 10 is subtracted, is customary in most cases.

Determination of antilog$_{10} x$ It is possible to reverse the process described in this section and, being given $\log_{10} x$, to find x. In this event, x is referred to as the **antilogarithm** (antilog$_{10}$) of $\log_{10} x$. For example, antilog$_{10} 1.6395$ can be obtained by locating the mantissa, 0.6395, in the body of the $\log_{10}$ tables and observing that the associated antilog$_{10}$ is 4.36. Thus

$$\text{antilog}_{10} 1.6395 = 4.36 \times 10^1 = 43.6.$$

Linear
interpolation If we seek the common logarithm of a number that is not an
entry in the table (for example, $\log_{10} 3712$), or if we seek x
when $\log_{10} x$ is not an entry in the table, it is customary to use
a procedure called **linear interpolation**. Observe that a table of common logarithms
is a set of ordered pairs. For each number x there is an associated number $\log_{10} x$,
and we have $(x, \log_{10} x)$ displayed in convenient tabular form. Because of space
limitations, only three digits for the number x and four for the number $\log_{10} x$
appear in Table II (pages 416 and 417). By means of linear interpolation, however,
the table can be used to find approximations to logarithms for four-digit numbers.

Let us examine geometrically the concepts involved. A portion of the graph of

$$y = \log_{10} x$$

is shown in Figure 6.4-a. The curvature is exaggerated to illustrate the principle

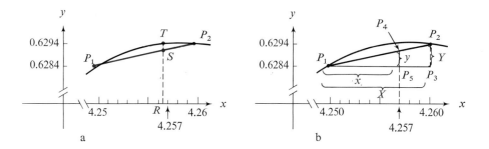

Figure 6.4

involved. We propose to use the line segment joining the points P_1 and P_2 as an
approximation to the curve passing through the points. If a large graph of
$y = \log_{10} x$ were available, the value of, say, $\log_{10} 4.257$ could be found by using
the value of the ordinate (RT) to the curve for $x = 4.257$. Since there is no way to
accomplish this with a table of values only, we shall instead use the value of the
ordinate (RS) to the line segment as an approximation to the ordinate of the curve.

This can be accomplished directly from the set of numbers available in the table
of logarithms. Consider Figure 6.4-b, where $P_2 P_3$ and $P_4 P_5$ are perpendicular to
$P_1 P_3$. From geometry, we have $\triangle P_1 P_4 P_5 \sim \triangle P_1 P_2 P_3$, where the lengths of the
corresponding sides are proportional, and hence

$$\frac{x}{X} = \frac{y}{Y}. \qquad (4)$$

If we know any three of these numbers, the fourth can be determined. For the
purpose of interpolation, we assume all of our numbers now have four-digit
numerals; that is, we consider 4.250 instead of 4.25, and 4.260 instead of 4.26. We

note that the point corresponding to 4.257 is located just $7/10$ of the distance be-tween the points corresponding to 4.250 and 4.260, respectively, and the value $Y (0.0010)$ is just the difference between the logarithms 0.6284 and 0.6294. It follows from (4) that

$$\frac{7}{10} = \frac{y}{0.0010},$$

from which

$$y = \frac{7}{10} (0.0010) = 0.0007.$$

We now add 0.0007 to 0.6284 and thus obtain a good approximation to the required logarithm. That is,

$$\log_{10} 4.257 = 0.6291.$$

An example in Exercise 6.3 shows a convenient arrangement for the calculations involved in the example presented here. The antilogarithm of a number can be found by a similar procedure. With practice, however, it is possible to interpolate mentally in both cases.

Exercise 6.3

Write the characteristic of each expression.

Examples   a. $\log_{10} 248$ b. $\log_{10} 0.0057$

Solutions   Represent the number in scientific notation.

a. $\log_{10} (2.48 \times 10^2)$ b. $\log_{10} (5.7 \times 10^{-3})$

The exponent on the base 10 is the characteristic.

          2 -3, or $7 - 10$

1. $\log_{10} 312$ 2. $\log_{10} 0.02$ 3. $\log_{10} 0.00851$

4. $\log_{10} 8.012$ 5. $\log_{10} 0.00031$ 6. $\log_{10} 0.0004$

7. $\log_{10} (15 \times 10^3)$ 8. $\log_{10} (820 \times 10^4)$

Find each logarithm.

Examples   a. $\log_{10} 16.8$ b. $\log_{10} 0.043$

Solutions   Represent the number in scientific notation.
                a. $\log_{10} (1.68 \times 10^1)$ b. $\log_{10} (4.3 \times 10^{-2})$

                Determine the mantissa from the table.
                    0.2253   0.6335

                Add the characteristic as determined by the exponent on the base 10.
                    1.2253 $8.6335 - 10$

9. $\log_{10} 6.73$ 10. $\log_{10} 891$
11. $\log_{10} 0.813$ 12. $\log_{10} 0.00214$
13. $\log_{10} 0.08$ 14. $\log_{10} 0.000413$
15. $\log_{10} (2.48 \times 10^2)$ 16. $\log_{10} (5.39 \times 10^{-3})$

Find each antilogarithm.

Example   antilog$_{10}$ 2.7364

Solution   Locate the mantissa in the body of the table of mantissas and determine
                the associated antilog$_{10}$ (a number between 1 and 10); write the charac-
                teristic as an exponent on the base 10.

                                $5.45 \times 10^2 = 545$

17. antilog$_{10}$ 0.6128   18. antilog$_{10}$ 0.2504
19. antilog$_{10}$ 0.5647   20. antilog$_{10}$ 3.9258
21. antilog$_{10}$ $(8.8075 - 10)$ 22. antilog$_{10}$ $(3.9722 - 5)$
23. antilog$_{10}$ 3.7388   24. antilog$_{10}$ 2.0086
25. antilog$_{10}$ $(6.8561 - 10)$ 26. antilog$_{10}$ $(1.8156 - 4)$

Find each logarithm.

Example $\log_{10} 4.257$

Solution   Interpolate mentally or use the following procedure.

$$
\begin{array}{cc}
x & \log_{10} x
\end{array}
$$

$$
10 \left\{ \, 7 \left\{ \begin{array}{c} 4.250 \\ 4.257 \\ 4.260 \end{array} \right. \left| \begin{array}{c} 0.6284 \\ ? \\ 0.6294 \end{array} \right\} y \, \right\} 0.0010
$$

Set up a proportion and solve for y.

$$\frac{7}{10} = \frac{y}{0.0010}$$

$$y = 0.0007$$

Add 0.0007, the value of y, to 0.6284.

$$\log_{10} 4.257 = 0.6284 + 0.0007 = 0.6291$$

27. $\log_{10} 4.213$ 28. $\log_{10} 8.184$ 29. $\log_{10} 1522$

30. $\log_{10} 203.4$ 31. $\log_{10} 37110$ 32. $\log_{10} 72.36$

33. $\log_{10} 0.5123$ 34. $\log_{10} 0.008351$

Find each antilogarithm.

Example   antilog$_{10}$ 0.6446

Solution   Interpolate mentally or use the following procedure.

$$x \qquad \text{antilog}_{10}\, x$$

$$0.0010 \left\{ 0.0002 \left\{ \begin{array}{c|c} 0.6444 & 4.410 \\ 0.6446 & ? \\ 0.6454 & 4.420 \end{array} \right\} y \right\} 0.010$$

Set up a proportion and solve for y.

$$\frac{0.0002}{0.0010} = \frac{y}{0.010}$$

$$y = 0.002$$

Add 0.002, the value of y, to 4.410.

$$\text{antilog}_{10}\, 0.6446 = 4.410 + 0.002 = 4.412$$

35. antilog$_{10}$ 0.5085   36. antilog$_{10}$ 0.8087

37. antilog$_{10}$ 1.0220   38. antilog$_{10}$ 3.0759

39. antilog$_{10}$ (8.7055 − 10)   40. antilog$_{10}$ (9.8742 − 10)

41. antilog$_{10}$ (2.8748 − 3)   42. antilog$_{10}$ (7.7397 − 10)

43. If we use linear interpolation to find $\log_{10}$ 3.751 and $\log_{10}$ 3.755, which of the resulting approximations should we expect to be more nearly correct? Why?

44. If we use linear interpolation to find $\log_{10} 1.025$ and $\log_{10} 9.025$, which of the resulting approximations should we expect to be more nearly correct? Why?

Find the value of each exponential expression by means of the table of logarithms to the base 10. Hint: Find the antilogarithm of the exponent $(10^x = \text{antilog}_{10} x)$.

45. $10^{0.9590}$ 46. $10^{0.8241}$ 47. $10^{3.6990}$

48. $10^{2.3874}$ 49. $10^{2.0531}$ 50. $10^{1.7396}$

6.4

Applications of logarithms

Computations   The use of the slide rule and the advent of high-speed computing devices have almost removed the need to perform routine numerical computations with pencil and paper by logarithms. Nevertheless, we introduce the techniques involved in making such computations because the writing of the logarithmic equations involved sheds light on the properties of the logarithmic function and on the usefulness of Theorem 6.3, which we reproduce here using the base 10.

If x_1 and x_2 are positive real numbers and $m \in R$, then

I $\log_{10} (x_1 x_2) = \log_{10} x_1 + \log_{10} x_2$,

II $\log_{10} \dfrac{x_1}{x_2} = \log_{10} x_1 - \log_{10} x_2$,

III $\log_{10} (x_1)^m = m \log_{10} x_1$.

Before illustrating the use of these laws, let us make two observations:

L-1   If $M = N$ $(M, N > 0)$, then $\log_b M = \log_b N$.

L-2   If $\log_b M = \log_b N$, then $M = N$.

L-1 is an application of the substitution law E-4, while L-2 follows from the fact that the variables in a logarithmic function are in one-to-one correspondence.

Example   Find the product of 3.825 and 0.00729, using logarithms and linear interpolation.

Solution   Let $N = (3.825)(0.00729)$. By Property L-1,

$$\log_{10} N = \log_{10} [(3.825)(0.00729)].$$

Now, by Theorem 6.3-I,

$$\log_{10} N = \log_{10} 3.825 + \log_{10} 0.00729,$$

and from the table we obtain

$$\log_{10} 3.825 = 0.5826, \quad \text{and} \quad \log_{10} 0.00729 = 7.8627 - 10,$$

so that

$$\log_{10} N = (0.5826) + (7.8627 - 10)$$
$$= 8.4453 - 10.$$

The computation is completed by referring to the table for

$$N = \text{antilog}_{10} (8.4453 - 10) = 2.788 \times 10^{-2}$$
$$= 0.02788.$$

Thus we have

$$N = (3.825)(0.00729) = 0.02788.$$

Actual computation shows the product to be 0.02788425, so that the result obtained by use of logarithms is correct to four significant digits. Some error should be expected when we employ a table of logarithms, because we are using approximations and linear interpolations.

Example   Compute $\dfrac{(8.21)^{1/2}(2.17)^{2/3}}{(3.14)^3}$.

Solution   Setting

$$N = \frac{(8.21)^{1/2}(2.17)^{2/3}}{(3.14)^3},$$

we obtain

$$\log_{10} N = \log_{10} \frac{(8.21)^{1/2}(2.17)^{2/3}}{(3.14)^3}$$

$$= \log_{10} (8.21)^{1/2} + \log_{10} (2.17)^{2/3} - \log_{10} (3.14)^3$$

$$= \frac{1}{2} \log_{10} 8.21 + \frac{2}{3} \log_{10} 2.17 - 3 \log_{10} 3.14.$$

The table provides values for the logarithms involved here, and the remainder of the computation is routine.

In computations involving numbers less than 1, it is sometimes convenient to use expressions other than such differences as 7.___ − 10 and 8.___ − 10 for negative characteristics.

Example   Compute $\sqrt[3]{0.0235}$.

Solution   Setting

$$N = \sqrt[3]{0.0235} = (0.0235)^{1/3},$$

we have

$$\log_{10} N = \log_{10} (0.0235)^{1/3} = \frac{1}{3} \log_{10} (0.0235).$$

From Table II, we see that the mantissa of this logarithm is 0.3711. Because we wish to multiply by 1/3, we select $7.3711 - 9$ instead of $8.3711 \quad 10$ for the logarithm. Thus

$$\log_{10} N = \frac{1}{3} (7.3711 \ -9) \approx 2.4570 - 3,$$

where we have avoided obtaining a nonintegral negative part of the logarithm. From the table,

$$N = \text{antilog}_{10} (2.4570 - 3) = 0.2864.$$

Thus,

$$\sqrt[3]{0.0235} \approx 0.2864.$$

Solving   An equation in one variable in which the variable occurs in an
exponential   exponent is called an **exponential equation**. Solution sets of
equations   some such equations can be found by means of logarithms.

Example   Find the solution set of $5^x = 7$.

Solution   Since $5^x > 0$ for all x, we can apply Property L-1 and write

$$\log_{10} 5^x = \log_{10} 7;$$

and from Theorem 6.3-III,

$$x \log_{10} 5 = \log_{10} 7.$$

Dividing each member by $\log_{10} 5$, we obtain

$$x = \frac{\log_{10} 7}{\log_{10} 5} = \frac{0.8451}{0.6990} \approx 1.209.$$

The solution set is

$$\left\{ \frac{\log_{10} 7}{\log_{10} 5} \right\},$$

and a decimal approximation for the single solution is 1.209. Note that in seeking this approximation, the logarithms are *divided*, not subtracted.

Exponential and logarithmic functions are of great importance in applied mathematics. Several applications are included in the following exercise set.

Exercise 6.4

Compute each of the indicated values by means of logarithms.

Example $\dfrac{(23.4)(0.681)}{4.13}$

Solution   Let $P = \dfrac{(23.4)(0.681)}{4.13}$.

Then

$$\log_{10} P = \log_{10} 23.4 + \log_{10} 0.681 - \log_{10} 4.13$$
$$= (1.3692) + (9.8331 - 10) - (0.6160)$$
$$= 0.5863,$$

and

$$\text{antilog}_{10}\ 0.5863 = 3.857.$$

1. $(2.32)(1.73)$

2. $(83.2)(6.12)$

3. $\dfrac{3.15}{1.37}$

4. $\dfrac{1.38}{2.52}$

5. $\dfrac{0.0149}{32.3}$

6. $\dfrac{0.00214}{3.17}$

7. $(2.3)^5$

8. $(4.62)^3$

9. $\sqrt[3]{8.12}$

10. $\sqrt[5]{75}$

11. $(0.0128)^5$

12. $(0.0021)^6$

13. $\sqrt{0.0021}$

14. $\sqrt[5]{0.0471}$

15. $\sqrt[3]{0.0214}$

16. $\sqrt[4]{0.0018}$

17. $\dfrac{(8.12)(8.74)}{7.19}$

18. $\dfrac{(0.412)^2(84.3)}{\sqrt{21.7}}$

19. $\dfrac{(6.49)^2\sqrt[3]{8.21}}{17.9}$

20. $\dfrac{(2.61)^2(4.32)}{\sqrt{7.83}}$

21. $\dfrac{(0.3498)(27.16)}{6.814}$

22. $\dfrac{(4.813)^2(20.14)}{3.612}$

23. $\sqrt{\dfrac{(4.71)(0.00481)}{(0.0432)^2}}$

24. $\sqrt{\dfrac{(2.85)^3(0.97)}{0.035}}$

Solve. Leave each solution in logarithmic form using the base 10.

Example $3^{x-2} = 16$

Solution   By Property L-1,
$$\log_{10} 3^{x-2} = \log_{10} 16.$$
By Theorem 6.3-III,
$$(x - 2) \log_{10} 3 = \log_{10} 16,$$
from which
$$x - 2 = \frac{\log_{10} 16}{\log_{10} 3},$$
$$x = \frac{\log_{10} 16}{\log_{10} 3} + 2.$$
The solution set is $\left\{ \dfrac{\log_{10} 16}{\log_{10} 3} + 2 \right\}$.

25. $2^x = 7$ 26. $3^x = 4$ 27. $3^{x+1} = 8$
28. $2^{x-1} = 9$ 29. $7^{2x-1} = 3$ 30. $3^{x+2} = 10$
31. $4^{x^2} = 15$ 32. $8^{x^2} = 21$ 33. $3^{-x} = 10$
34. $2.13^{-x} = 8.1$ 35. $3^{1-x} = 15$ 36. $4^{2-x} = 10$

Solve. Leave each result in the form of an equation equivalent to the given equation.

37. $y = x^n$,   for n 38. $y = Cx^{-n}$,   for n
39. $y = e^{kt}$,   for t 40. $y = Ce^{-kt}$,   for t

P dollars invested at an interest rate r compounded yearly yields an amount A after n years, given by $A = P(1 + r)^n$. *If the interest is compounded t times yearly, the amount is given by*

$$A = P\left(1 + \frac{r}{t}\right)^{tn}.$$

Solve for the variable n (nearest year), r (nearest $\frac{1}{2}\%$), or A (accuracy obtainable on four-place table of mantissas).

Example $(1 + r)^{12} = 1.127$

Solution   Equate $\log_{10}$ of each member and apply Theorem 6.3-III.
$$12 \log_{10} (1 + r) = \log_{10} 1.127 = 0.0519$$
Multiply each member by $1/12$.
$$\log_{10} (1 + r) = \frac{1}{12} (0.0519) = 0.0043$$

Determine $\text{antilog}_{10}\ 0.0043$ and solve for r.

$$\text{antilog}_{10}\ 0.0043 = 1.01$$
$$1 + r = 1.01$$
$$r = 0.01, \quad \text{or } 1\%$$

41. $(1 + 0.03)^{10} = A$ 42. $(1 + 0.04)^8 = A$

43. $(1 + r)^6 = 1.34$ 44. $(1 + r)^{10} = 1.48$

45. $100\left(1 + \dfrac{r}{2}\right)^{10} = 113$ 46. $40\left(1 + \dfrac{r}{4}\right)^{12} = 50.9$

Example $40(1 + 0.02)^n = 51.74$

Solution   Divide each member by 40; equate $\log_{10}$ of each member, and apply Theorem 6.3, Parts II and III.

$$n \log_{10}(1.02) = \log_{10} 51.74 - \log_{10} 40$$
$$n(0.0086) = 1.7138 - 1.6021$$
$$n = \frac{0.1117}{0.0086}$$
$$n = 13 \text{ years}$$

47. $(1 + 0.04)^n = 2.19$ 48. $(1 + 0.04)^n = 1.60$

49. $150(1 + 0.01)^{4n} = 240$ 50. $60(1 + 0.02)^{2n} = 116$

51. Find the amount of $5000 invested at 4% for 10 years when compounded annually. When compounded semi-annually.

52. Two men, A and B, each invested $10,000 at 4% for 20 years with a bank that computed interest quarterly. A withdrew his interest at the end of each three-month period but B let his investment be compounded. How much more did B earn than A from his investment over the period of 20 years?

Chemists define the pH (hydrogen potential) of a solution by

$$pH = \log_{10} \frac{1}{[H^+]}$$
$$= \log_{10} [H^+]^{-1}.$$

Therefore,

$$pH = -\log_{10} [H^+],$$

where $[H^+]$ is a numerical value for the concentration of hydrogen ions in aqueous solution in moles per liter.

Calculate the pH of a solution with given hydrogen-ion concentration.

53. $[H^+] = 10^{-7}$ 54. $[H^+] = 4.0 \times 10^{-5}$ 55. $[H^+] = 2.0 \times 10^{-8}$

56. $[H^+] = 8.5 \times 10^{-3}$ 57. $[H^+] = 6.3 \times 10^{-7}$ 58. $[H^+] = 5.7 \times 10^{-7}$

Calculate the hydrogen-ion concentration $[H^+]$ *of a solution with the given pH.*

Example $pH = 7.4$

Solution   Substitute 7.4 for pH in the relationship $pH = \log_{10} \dfrac{1}{[H^+]}$.

$$\log_{10} \frac{1}{[H^+]} = 7.4$$

$$\frac{1}{[H^+]} = \text{antilog}_{10}\ 7.4 = 2.5 \times 10^7$$

$$[H^+] = \frac{1}{2.5 \times 10^7} = 0.4 \times 10^{-7} = 4 \times 10^{-8}$$

59. $pH = 3.0$ 60. $pH = 4.2$ 61. $pH = 5.6$

62. $pH = 8.3$ 63. $pH = 7.2$ 64. $pH = 6.9$

65. The period T of a simple pendulum is given by the formula $T = 2\pi\sqrt{L/g}$, where T is in seconds, L is the length of the pendulum in feet, and $g \approx 32$ ft/sec². Find the period of a pendulum 1 foot long.

66. The area A of a triangle in terms of the sides is given by the formula

$$A = \sqrt{s(s-a)(s-b)(s-c)},$$

where a, b, and c are the lengths of the sides of the triangle and s equals one-half of the perimeter. Find the area of a triangle in which the lengths of the three sides are 2.314 inches, 4.217 inches, and 5.618 inches.

67. The amount of a radioactive element remaining at any time t is given by $y = y_0\,e^{-0.4t}$, where t is in seconds and y_0 is the amount present initially. How much of the element would remain after three seconds if 40 grams were present initially? Use: $e \approx 2.718$.

68. The number of bacteria present in a culture is related to time by the formula $N = N_0\,e^{0.04t}$, where N_0 is the amount of bacteria present at time $t = 0$, and t is time in hours. If 10,000 bacteria are present 10 hours after the beginning of an experiment, how many were present when $t = 0$? Use $e \approx 2.718$.

69. The atmospheric pressure p, in inches of mercury, is given approximately by $p = 30.0(10)^{-0.09a}$, where a is the altitude in miles above sea level. What is the atmospheric pressure at sea level? At 3 miles above sea level?

70. The intensity I (in lumens) of a light beam after passing through a thickness t (in centimeters) of a medium having an absorption coefficient of 0.1 is given by $I = 1000e^{-0.1t}$. How many centimeters of the material would reduce the illumination to 800 lumens? Use $e \approx 2.718$.

6.5

Logarithms to bases other than 10

The number e $(e \approx 2.718218)$ mentioned in Section 6.3 is of great mathematical interest and importance. It is possible to determine $\log_e x$ provided we have a table of $\log_{10} x$. Indeed, given a table of $\log_b x$, we can always find $\log_a x$ for any $a > 0$, $a \neq 1$. We can do this as follows.

Since $x > 0$ and $a \neq 1$, we have

$$x = a^{\log_a x}. \tag{1}$$

By applying Property L-1 of Section 6.3, we can equate the logarithms to the base b $(b > 0, b \neq 1)$ of each member of (1) to obtain

$$\log_b x = \log_b a^{\log_a x}.$$

By Theorem 6.3-III, this yields

$$\log_b x = \log_a x \cdot \log_b a,$$

or

$$\log_a x = \frac{\log_b x}{\log_b a}. \tag{2}$$

Conversion to base e Equation (2) gives us a means of finding $\log_a x$ when we have a table of logarithms to the base b. In particular, if $a = e$ and $b = 10$, we have

$$\log_e x = \frac{\log_{10} x}{\log_{10} e}. \tag{3}$$

Since $\log_{10} e \approx 0.4343$, (3) can be written

$$\log_e x = \frac{\log_{10} x}{0.4343}, \tag{4}$$

or

$$\log_e x = \frac{1}{0.4343} \log_{10} x = 2.3026 \log_{10} x, \tag{5}$$

which gives us a direct means of approximating $\log_e x$ when we have a table of logarithms to the base 10.

A table of $\log_e x$ will be found on page 419. We cannot, however, do as we did with a table of $\log_{10} x$—that is, simply list mantissas for logarithms over a certain interval and then manipulate characteristics to take care of all other intervals. Our numeration system is based on 10 and not on e. With the help of logarithms, though, we can use formula (5), above, and the table of $\log_{10} x$ on pages 416 and 417 to find $\log_e x$.

Example   Find $\log_e 278$.

Solution   We shall do this in two ways.

1. Since

$$\log_e x = 2.3026 \log_{10} x,$$

we have

$$\log_e 278 = 2.3026 \log_{10} 278$$
$$= 2.3026 \ (2.4440)$$
$$\approx 5.63.$$

2. Alternatively, we observe that $278 \approx 2.8 \times 10^2$. Hence

$$\log_e 278 \approx \log_e (2.8 \times 10^2) = \log_e 2.8 + 2 \log_e 10$$
$$= 1.0296 + 2(2.3026)$$
$$= 1.0296 + 4.6052$$
$$\approx 5.63.$$

Example   Find $\log_e 0.278$.

Solution   We observe that $0.278 \approx 2.8 \times 10^{-1}$. Hence

$$\log_e 0.278 \approx \log_e (2.8 \times 10^{-1}) = \log_e 2.8 + \log_e 10^{-1}$$
$$= \log_e 2.8 - \log_e 10$$
$$= 1.0296 - 2.3026$$
$$= -1.2730.$$

As one would expect, the logarithm is negative.

The process used in the preceding example can be reversed to find antilog$_e$ x. If, however, a table for the exponential function $\{(x, y) | y = e^x\}$ is available (see Table III on page 418), the antilog$_e$ x can be read directly from this, since antilog$_e$ $x = e^x$.

Examples   a. antilog$_e$ $3.2 = e^{3.2}$ b. antilog$_e$ $(-3.2) = e^{-3.2}$

$$= 24.533. \qquad\qquad\qquad = 0.0408.$$

Exercise 6.5

Find each logarithm.

Example $\log_3 7$

Solution   Use logarithms to base 10 to evaluate.

$$\log_3 7 = \frac{\log_{10} 7}{\log_{10} 3} = \frac{0.8451}{0.4771} \approx 1.77$$

1. $\log_2 10$ 2. $\log_2 5$ 3. $\log_5 240$ 4. $\log_3 18$

5. $\log_7 8.1$ 6. $\log_5 60$ 7. $\log_{100} 38$ 8. $\log_{100} 240$

Find each logarithm directly from Table IV on page 419.

9. $\log_e 3$ 10. $\log_e 8$ 11. $\log_e 17$ 12. $\log_e 98$

13. $\log_e 327$ 14. $\log_e 107$ 15. $\log_e 450$ 16. $\log_e 605$

Find each antilogarithm directly from Table III on page 418.

17. antilog$_e$ 0.50   18. antilog$_e$ 1.5   19. antilog$_e$ 3.4

20. antilog$_e$ 4.5   21. antilog$_e$ 0.231   22. antilog$_e$ 1.43

23. antilog$_e$ (-0.15) 24. antilog$_e$ (-0.95) 25. antilog$_e$ (-2.5)

26. antilog$_e$ (-4.2) 27. antilog$_e$ (-0.255) 28. antilog$_e$ (-3.65)

Solve without using the tables of logarithms.

29. If $\log_2 8 = 3$, find $\log_8 2$.   30. If $\log_4 16 = 2$, find $\log_{16} 4$.

31. If $\log_{10} 3 = 0.4771$, find $\log_3 10$.   32. If $\log_{10} e = 0.4343$, find $\log_e 10$.

33. If $\log_{10} 5 = 0.6990$, find $\log_5 100$.   34. If $\log_{10} 3 = 0.4771$, find $\log_3 100$.

35. Show that $(\log_{10} 4 - \log_{10} 2) \log_2 10 = 1$.

36. Show that $(2 \log_2 3)(\log_9 2 + \log_9 4) = 3$.

7

Circular functions

7.1

The circular functions cosine and sine

Notion of periodicity The graphs of some functions display interesting cyclical characteristics. For example, the x-axis shown in Figure 7.1 can clearly be divided into equal successive subintervals in such a way that the graph over each subinterval is a repetition of the graph over every other equal subinterval. Functions that display this property are said to be "periodic," and the length of each of the equal subintervals is called a **period** of the function.

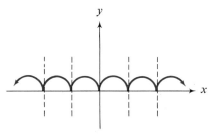

Figure 7.1

Definition 7.1 *If f is a function with domain $D \subset R$, such that for some $a \in R$, $a \neq 0$, the values $x + a$ and $x - a$ are in D for each $x \in D$, and*

$$f(x + a) = f(x),$$

then f is periodic with period a.

If there is a least positive number a for which the function is periodic, then a is called the **fundamental period** of the function. Of course, if a function is periodic with period a, it is also periodic with period $2a$, $3a$, $-a$, and, in general, ka, $k \in J$, $k \neq 0$; but we are primarily concerned only with the fundamental period, and when we refer to the period of a function we shall ordinarily mean its fundamental period. The most common periodic functions are the circular functions (defined using arc lengths on a circle) and trigonometric functions (defined using angles). In this chapter, we shall study circular functions.

Cosine and sine functions Periodic functions can be defined using the **unit circle** with equation $x^2 + y^2 = 1$. Consider, intuitively, a point moving steadily in a counterclockwise direction around the circle. At any given time, the moving point occupies a position on the circle; the point, in turn, is associated with an ordered pair (x, y). If the distance along the circle from the point $(1, 0)$ to the point (x, y) is designated by s, then we can associate the real number s with the ordered pair (x, y) (Figure 7.2). We shall

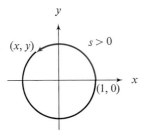

Figure 7.2

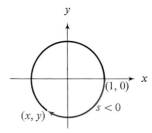

Figure 7.3

assume that every arc of a circle has a length and that there is a one-to-one correspondence between the members of the set R_+ of positive real numbers and the lengths of all arcs of the unit circle measured in a *counterclockwise* direction from the point $(1, 0)$ to points (x, y) on the circle. Since the circumference of a circle is $2\pi r$ and the radius of the unit circle is 1, the distance once around the unit circle counterclockwise is 2π, twice around is 4π, and so on.

Let us agree that values of $s < 0$ denote lengths of arc measured from $(1, 0)$ in a *clockwise* direction to points (x, y) on the circle (Figure 7.3). This extends the one-to-one correspondence of the set of lengths of arcs on the unit circle to the entire set R of real numbers.

Especially useful are the associations of the members of R with the members of the set of all first components of the ordered pairs (x, y), and of the members of R with the members of the set of all second components of the ordered pairs. Before making this association, let us first rename these components.

Definition 7.2 *If (x, y) is the point at arc length s from $(1, 0)$ on the unit circle $x^2 + y^2 = 1$, then x is the cosine of s, and y is the sine of s. We denote this by writing*

$$x = \cos s \quad and \quad y = \sin s.$$

In other words, the first component of the point (x, y) located at arc length s from $(1, 0)$ on the unit circle is called the cosine of s, and the second component is called the sine of s; and we denote these by cos s and sin s, respectively. Thus, $(x, y) = (\cos s, \sin s)$. Having given meaning to the symbolism cos s and sin s, we can now define two new functions.

Definition 7.3 *If $s \in R$, then*

$$\text{I} \quad \text{cosine} = \{(s, x) \mid x = \cos s\},$$

$$\text{II} \quad \text{sine} = \{(s, y) \mid y = \sin s\}.$$

Domain and range of sine and cosine   As you can see, the domain of each of these functions is the set R. The range of the cosine function is the set of all first components of the ordered pairs corresponding to points on the unit circle, and hence is the set $\{x \mid |x| \le 1\}$. Similarly, the range of the sine function is $\{y \mid |y| \le 1\}$.

Since, for $k \in J$,

$$\cos (s + 2k\pi) = \cos s \tag{1}$$

and

$$\sin (s + 2k\pi) = \sin s, \tag{2}$$

the functions cosine and sine are periodic, with fundamental period 2π (see Exercises 29 and 30, page 193).

Signs of cosine and sine   Because cos s and sin s are simply the coordinates of points on the unit circle, they are positive or negative in accord with the values of x and y in the various quadrants. Table 7.1 summarizes in a convenient way the sign associated with $x = \cos s$ and $y = \sin s$ in each quadrant.

Table 7.1

Quadrant II	Quadrant I
x or cos $s < 0$	x or cos $s > 0$
y or sin $s > 0$	y or sin $s > 0$
Quadrant III	Quadrant IV
x or cos $s < 0$	x or cos $s > 0$
y or sin $s < 0$	y or sin $s < 0$

cos s and sin s in terms of each other   By Definition 7.2, $x = \cos s$ and $y = \sin s$ are subject to the condition that $x^2 + y^2 = 1$. Hence we have the following basic identity relating cos s and sin s.

Theorem 7.1 *For every $s \in R$,*

$$\cos^2 s + \sin^2 s = 1. \qquad (3)$$

Note that for convenience we write $\cos^2 s$ for $(\cos s)^2$ and $\sin^2 s$ for $(\sin s)^2$. Now Theorem 7.1 can be used to write

$$\sin s = \begin{cases} \sqrt{1 - \cos^2 s} & \text{in Quadrants I and II,} & (4a) \\ -\sqrt{1 - \cos^2 s} & \text{in Quadrants III and IV,} & (4b) \end{cases}$$

and

$$\cos s = \begin{cases} \sqrt{1 - \sin^2 s} & \text{in Quadrants I and IV,} & (5a) \\ -\sqrt{1 - \sin^2 s} & \text{in Quadrants II and III.} & (5b) \end{cases}$$

Therefore, if either $\sin s$ or $\cos s$ is known and the quadrant in which the terminal point of the arc of length s lies can be determined, then we can find the value for the other function.

Example   Given that $\cos s = -3/5$ and $\pi < s < 3\pi/2$, find $\sin s$.

Solution   Using (4b), we have

$$\sin s = -\sqrt{1 - \cos^2 s} = -\sqrt{1 - \left(-\frac{3}{5}\right)^2} = -\sqrt{\frac{16}{25}} = -\frac{4}{5}.$$

$\cos(-s)$ and   Because the defining unit
$\sin(-s)$ circle is symmetric to the
                   horizontal axis, it follows
that if $(\cos s, \sin s) = (x, y)$, then

$$(\cos(-s), \sin(-s)) = (x, -y).$$

We thus have the following theorem, which is illustrated in Figure 7.4 for the case $0 < s < \pi/2$.

Theorem 7.2 *For every $s \in R$*

I $\cos(-s) = \cos s,$ (6)

II $\sin(-s) = -\sin s.$ (7)

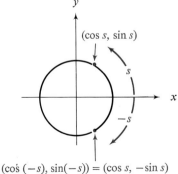

$(\cos(-s), \sin(-s)) = (\cos s, -\sin s)$

Figure 7.4

Even and odd Any function f with domain $D \subset R$, such that
functions
$$f(-s) = f(s)$$

for every $s \in D$, is called an **even function**. On the other hand, if

$$f(-s) = -f(s)$$

for every $s \in D$, then f is called an **odd function**. Hence, cosine is an even function and sine is an odd function, because

$$\cos(-s) = \cos s \quad \text{and} \quad \sin(-s) = -\sin x.$$

Exercise 7.1

Use relationship (1) or (2) on page 184 to determine a value of s, $0 \leq s < 2\pi$, for which the given equation is true. Then use Table 7.1 to state whether the given function value is positive or negative.

Example $\sin s = \sin \dfrac{27\pi}{4}$

Solution Since $\dfrac{27\pi}{4}$ lies between 6π and 8π, we can write

$$\sin \frac{27\pi}{4} = \sin \left(\frac{3\pi}{4} + 6\pi \right).$$

By relationship (2),

$$\sin \left(\frac{3\pi}{4} + 6\pi \right) = \sin \frac{3\pi}{4},$$

so that the desired value of s is $\dfrac{3\pi}{4}$. Since $\dfrac{3\pi}{4}$ lies between $\dfrac{\pi}{2}$ and π, we know from Table 7.1 that $\sin \dfrac{27\pi}{4} > 0$.

1. $\cos s = \cos \dfrac{13\pi}{3}$ 2. $\cos s = \cos \dfrac{27\pi}{5}$

3. $\sin s = \sin \dfrac{17\pi}{7}$ 4. $\sin s = \sin \dfrac{29\pi}{4}$

5. $\cos s = \cos \dfrac{41\pi}{5}$ 6. $\sin s = \sin \dfrac{36\pi}{7}$

7. $\sin s = \sin\left(-\dfrac{4\pi}{3}\right)$

8. $\cos s = \cos\left(-\dfrac{7\pi}{5}\right)$

9. $\cos s = \cos\left(-\dfrac{31\pi}{4}\right)$

10. $\sin s = \sin\left(-\dfrac{17\pi}{3}\right)$

In Problems 11–20, find the required function value and state the quadrant in which s terminates.

11. Given that $\sin s = 3/5$ and $\cos s > 0$,   find $\cos s$.

12. Given that $\cos s = 4/5$ and $\sin s < 0$,   find $\sin s$.

13. Given that $\cos s = 5/13$ and $\sin s < 0$,   find $\sin s$.

14. Given that $\sin s = 12/13$ and $\cos s < 0$,   find $\cos s$.

15. Given that $\sin s = -2/3$ and $\cos s > 0$,   find $\cos s$.

16. Given that $\cos s = -1/4$ and $\sin s > 0$,   find $\sin s$.

17. Given that $\cos s = -\dfrac{\sqrt{3}}{2}$ and $\sin s < 0$,   find $\sin s$.

18. Given that $\sin s = -\dfrac{\sqrt{3}}{2}$ and $\cos s > 0$,   find $\cos s$.

19. Given that $\sin s = -3/5$ and $\cos s > 0$,   find $\cos s$.

20. Given that $\cos s = -4/5$ and $\sin s > 0$,   find $\sin s$.

In Problems 21–24, use the symmetry of the circle and other geometric considerations to make a conjecture about the relationship between the given pairs for $0 < s < \pi/2$.

21. $\cos(\pi - s)$ and $\cos s$

22. $\sin(\pi + s)$ and $\sin s$

23. $\sin\left(\dfrac{\pi}{2} + s\right)$ and $\cos s$

24. $\cos\left(\dfrac{3\pi}{2} - s\right)$ and $\sin s$

25. Use a figure similar to Figure 7.4 to argue that Theorem 7.2 is valid for $\pi/2 < s < \pi$.

7.2

Values of cosine and sine for special arc lengths

In general, finding $\cos s$ and $\sin s$ for a given real number s is a difficult matter. However, some values of $\cos s$ and $\sin s$, corresponding to special values of s, are readily available.

Quadrantal
function values

Figure 7.5 shows the coordinates of four points on the unit circle with which we can associate specific values of s. For $s = 0$,

$$\cos 0 = 1 \quad \text{and} \quad \sin 0 = 0. \qquad (1)$$

Furthermore, since the arc of a circle included in a quadrant is one-fourth of the circumference of the circle, that is, $2\pi/4$, we know immediately that

$$\cos \frac{\pi}{2} = 0 \quad \text{and} \quad \sin \frac{\pi}{2} = 1, \qquad (2)$$

$$\cos \pi = -1 \quad \text{and} \quad \sin \pi = 0, \qquad (3)$$

$$\cos \frac{3\pi}{2} = 0 \quad \text{and} \quad \sin \frac{3\pi}{2} = -1. \qquad (4)$$

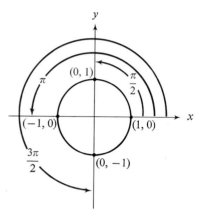

Figure 7.5

Equations (1) and (2) on page 184 can be used to find values of $\cos s$ and $\sin s$ for values of s differing by $2k\pi$, $k \in J$, from those obtained above.

Example

Find $\cos \dfrac{7\pi}{2}$.

Solution

Expressing $7\pi/2$ as the sum of an integral multiple of 2π and a number between 0 and 2π, we have $7\pi/2 = 3\pi/2 + 2\pi$. Using the fact that $\cos(s + 2k\pi) = \cos s$, we obtain

$$\cos \left(\frac{3\pi}{2} + 2\pi \right) = \cos \frac{3\pi}{2},$$

and from (4),

$$\cos \frac{7\pi}{2} = \cos \frac{3\pi}{2} = 0.$$

Notice that in finding values for $\cos s$ and $\sin s$ we are finding *coordinates of points* on the unit circle. The function values obtained in (1), (2), (3), and (4) above are sometimes called **quadrantal values** because the terminal point of each arc lies on an axis.

We can find other special values for $(\cos s, \sin s)$ by using the geometry of the unit circle and the distance formula,

$$d^2 = (x_2 - x_1)^2 + (y_2 - y_1)^2.$$

Values for
multiples
of $\pi/4$

First consider $s = \pi/4$. Figure 7.6 shows the unit circle and the designated value for s. Since (x, y) bisects the arc from $(1, 0)$ to $(0, 1)$, it follows from geometric considerations that $x = y$. Because $x^2 + y^2 = 1$, we have

$$x^2 + x^2 = 1,$$

$$2x^2 = 1,$$

$$x^2 = \frac{1}{2},$$

and

$$x = \frac{1}{\sqrt{2}} \quad \text{or} \quad x = -\frac{1}{\sqrt{2}}.$$

Now, both x and y are positive in the first quadrant, so the desired value for x is $1/\sqrt{2}$. Since $x = y$, y is also equal to $1/\sqrt{2}$, and hence we have

$$\cos \frac{\pi}{4} = \frac{1}{\sqrt{2}} \quad \text{and} \quad \sin \frac{\pi}{4} = \frac{1}{\sqrt{2}}.$$

Values for $\cos s$ and $\sin s$ for s equal to $3\pi/4$, $5\pi/4$, and $7\pi/4$ can be found using geometric symmetry as shown in Figure 7.7. These are listed in Table 7.2. on page 192.

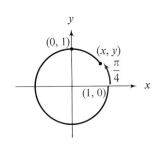

Figure 7.6

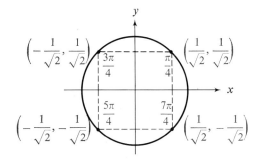

Figure 7.7

Values for
multiples
of $\pi/6$

Next, consider $s = \pi/6$, as pictured in Figure 7.8 on page 190. If the ordered pair corresponding to $s = \pi/6$ is (x, y), then the ordered pair corresponding to $s = -\pi/6$ is $(x, -y)$. Now the arc from $(x, -y)$ to (x, y) is of length $\pi/6 + \pi/6 = \pi/3$, and so is the length of the arc from (x, y) to $(0, 1)$. Because equal arcs of a circle subtend equal chords, the distance from $(0, 1)$ to (x, y) is the same as the distance from (x, y) to $(x, -y)$. Using the distance formula, we therefore have

$$(x - 0)^2 + (y - 1)^2 = (x - x)^2 + (-y - y)^2,$$

or

$$x^2 + y^2 - 2y + 1 = 4y^2.$$

Since $x^2 + y^2 = 1$, we substitute 1 for $x^2 + y^2$ in this equation and obtain

$$1 - 2y + 1 = 4y^2,$$

$$4y^2 + 2y - 2 = 0,$$

$$2(2y - 1)(y + 1) = 0,$$

so that

$$y = \frac{1}{2} \quad \text{or} \quad y = -1.$$

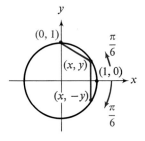

Figure 7.8

Because (x, y) is in the first quadrant, we must select the value $1/2$ as the y-coordinate. Next, we use $x^2 + y^2 = 1$ to find a value for x. Thus

$$x^2 + \left(\frac{1}{2}\right)^2 = 1,$$

from which

$$x = \pm\frac{\sqrt{3}}{2}.$$

Because (x, y) is in the first quadrant, we have $x = \sqrt{3}/2$, and it follows that

$$\cos\frac{\pi}{6} = \frac{\sqrt{3}}{2} \quad \text{and} \quad \sin\frac{\pi}{6} = \frac{1}{2}. \quad (5)$$

With this information, values for $\cos s$ and $\sin s$ for s equal to $5\pi/6$, $7\pi/6$, and $11\pi/6$ can be obtained using geometric symmetry, as shown in Figure 7.9. These values are also listed on page 192 in Table 7.2.

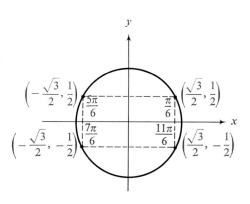

Figure 7.9

From Equation (5), page 190, we can quickly find values for $\cos(\pi/3)$ and $\sin(\pi/3)$ by symmetry. Figure 7.10 shows the point (x, y) associated with s equal to $\pi/3$; clearly, the abscissa of (x, y) is the ordinate of $(\sqrt{3}/2, 1/2)$, and the ordinate of (x, y) is the abscissa of $(\sqrt{3}/2, 1/2)$. Hence

$$\cos\frac{\pi}{3} = \frac{1}{2} \quad \text{and} \quad \sin\frac{\pi}{3} = \frac{\sqrt{3}}{2}.$$

Again, values for $\cos s$ and $\sin s$ for s equal to $2\pi/3$, $4\pi/3$, and $5\pi/3$ can be obtained from geometric considerations, as shown in Figure 7.11. These values are also listed in Table 7.2.

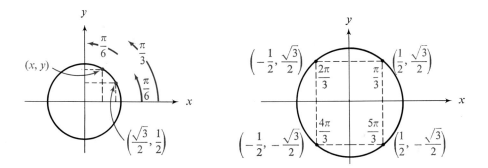

| Figure 7.10 | Figure 7.11 |

Table 7.2, in conjunction with Equations (1) and (2), page 184, can be used to find values of $\cos s$ and $\sin s$ for values of s differing by $2k\pi$, $k \in J$, from those listed in the table.

Example   Find $\cos\dfrac{14\pi}{3}$ and $\sin\dfrac{14\pi}{3}$.

Solution   We first observe that $\dfrac{14\pi}{3} = \dfrac{2\pi}{3} + \dfrac{12\pi}{3} = \dfrac{2\pi}{3} + 4\pi$. Then

$$\cos\frac{14\pi}{3} = \cos\left(\frac{2\pi}{3} + 4\pi\right) = \cos\frac{2\pi}{3} = -\frac{1}{2}$$

and

$$\sin\frac{14\pi}{3} = \sin\left(\frac{2\pi}{3} + 4\pi\right) = \sin\frac{2\pi}{3} = \frac{\sqrt{3}}{2}.$$

Table 7.2

s	$\cos s$	$\sin s$	s	$\cos s$	$\sin s$
0	1	0	π	-1	0
$\dfrac{\pi}{6}$	$\dfrac{\sqrt{3}}{2}$	$\dfrac{1}{2}$	$\dfrac{7\pi}{6}$	$-\dfrac{\sqrt{3}}{2}$	$-\dfrac{1}{2}$
$\dfrac{\pi}{4}$	$\dfrac{1}{\sqrt{2}}$	$\dfrac{1}{\sqrt{2}}$	$\dfrac{5\pi}{4}$	$-\dfrac{1}{\sqrt{2}}$	$-\dfrac{1}{\sqrt{2}}$
$\dfrac{\pi}{3}$	$\dfrac{1}{2}$	$\dfrac{\sqrt{3}}{2}$	$\dfrac{4\pi}{3}$	$-\dfrac{1}{2}$	$-\dfrac{\sqrt{3}}{2}$
$\dfrac{\pi}{2}$	0	1	$\dfrac{3\pi}{2}$	0	-1
$\dfrac{2\pi}{3}$	$-\dfrac{1}{2}$	$\dfrac{\sqrt{3}}{2}$	$\dfrac{5\pi}{3}$	$\dfrac{1}{2}$	$-\dfrac{\sqrt{3}}{2}$
$\dfrac{3\pi}{4}$	$-\dfrac{1}{\sqrt{2}}$	$\dfrac{1}{\sqrt{2}}$	$\dfrac{7\pi}{4}$	$\dfrac{1}{\sqrt{2}}$	$-\dfrac{1}{\sqrt{2}}$
$\dfrac{5\pi}{6}$	$-\dfrac{\sqrt{3}}{2}$	$\dfrac{1}{2}$	$\dfrac{11\pi}{6}$	$\dfrac{\sqrt{3}}{2}$	$-\dfrac{1}{2}$
π	-1	0	2π	1	0

Exercise 7.2

In Problems 1–16, use relationships 1 and 2 from page 184, together with Table 7.2, to find each given function value.

Example $\sin \dfrac{11\pi}{4}$

Solution $\sin \left(\dfrac{3\pi}{4} + 2\pi\right) = \sin \dfrac{3\pi}{4} = \dfrac{1}{\sqrt{2}}$

1. $\cos \dfrac{9\pi}{4}$

2. $\sin \dfrac{9\pi}{4}$

3. $\cos \dfrac{8\pi}{3}$

4. $\sin \dfrac{5\pi}{3}$

5. $\sin \dfrac{15\pi}{6}$

6. $\cos \dfrac{15\pi}{6}$

7. $\sin \dfrac{11\pi}{2}$

8. $\cos \dfrac{11\pi}{2}$

9. $\sin 8\pi$

10. $\cos 12\pi$

11. $\sin(-5\pi)$

12. $\cos(-7\pi)$

13. $\cos\left(-\dfrac{7\pi}{4}\right)$

14. $\sin\left(-\dfrac{9\pi}{4}\right)$

15. $\cos\left(-\dfrac{8\pi}{3}\right)$

16. $\sin\left(-\dfrac{11\pi}{2}\right)$

In Problems 17–24, assume $0 \le s < 2\pi$ and give the value of s for which both conditions are true. Use Table 7.2 as necessary.

17. $\sin s = 1/\sqrt{2}$ and $\cos s > 0$

18. $\cos s = 1/\sqrt{2}$ and $\sin s < 0$

19. $\cos s = -\sqrt{3}/2$ and $\sin s > 0$

20. $\sin s = -1/2$ and $\cos s < 0$

21. $\sin s = \sqrt{3}/2$ and $\cos s < 0$

22. $\cos s = -1/\sqrt{2}$ and $\sin s > 0$

23. $\cos s = -1/2$ and $\sin s < 0$

24. $\sin s = -\sqrt{3}/2$ and $\cos s > 0$

25. Over what subinterval of $0 \le s \le \pi$ is it true that $\sin s > \cos s$? *Hint:* Use Table 7.2.

26. Over what subinterval of $0 \le s \le \pi/2$ is it true that $\sin s < \cos s$?

27. On a pair of Cartesian axes, let the horizontal axis be the s-axis and the vertical axis be the x-axis. Use values from Table 7.2 to locate corresponding points of the cosine function for the interval $0 \le s \le 2\pi$, and connect these points with a smooth curve going from left to right.

28. In Problem 27, let the vertical axis be the y-axis, and similarly graph the sine function over the interval $0 \le s \le 2\pi$.

29. Use the fact that $(1, 0)$ is the only point on the unit circle with first coordinate 1 to show that there is no value a, $0 < a < 2\pi$, that is a period of the cosine function.

30. Use the fact that $(0, 1)$ is the only point on the unit circle with second coordinate 1 to show that there is no value a, $0 < a < 2\pi$, that is a period of the sine function.

31. Prove that $f = \{(s, y) | y = \cos s + \sin s\}$ is periodic with period 2π.

7.3

Properties of the cosine function

Figure 7.12 shows selected points on the unit circle together with their coordinates. Since the arc length from the point $(1, 0)$ to $(\cos (s_1 + s_2),\ \sin (s_1 + s_2))$ is $s_1 + s_2$, and from $(\cos s_1,\ -\sin s_1)$ to $(\cos s_2,\ \sin s_2)$ the arc length is also $s_1 + s_2$, the respective chords have equal lengths. Using the distance formula to express this fact, and substituting from Theorem 7.2, we have

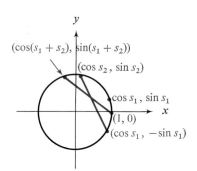

Figure 7.12

$$\sqrt{[\cos (s_1 + s_2) - 1]^2 + [\sin (s_1 + s_2) - 0]^2}$$
$$= \sqrt{(\cos s_2 - \cos s_1)^2 + [\sin s_2 - (-\sin s_1)]^2},$$

or, squaring both members,

$$[\cos (s_1 + s_2) - 1]^2 + [\sin (s_1 + s_2)]^2 = (\cos s_2 - \cos s_1)^2 + (\sin s_2 + \sin s_1)^2.$$

If we now perform the operations indicated, we have

$$\cos^2 (s_1 + s_2) - 2 \cos (s_1 + s_2) + 1 + \sin^2 (s_1 + s_2)$$
$$= \cos^2 s_2 - 2 \cos s_1 \cos s_2 + \cos^2 s_1 + \sin^2 s_2 + 2 \sin s_1 \sin s_2 + \sin^2 s_1.$$

Regrouping terms, we obtain

$$[\cos^2 (s_1 + s_2) + \sin^2 (s_1 + s_2)] - 2 \cos (s_1 + s_2) + 1$$
$$= (\cos^2 s_2 + \sin^2 s_2) + (\cos^2 s_1 + \sin^2 s_1) - 2 \cos s_1 \cos s_2 + 2 \sin s_1 \sin s_2,$$

and since $\cos^2 s + \sin^2 s = 1$, it follows that

$$1 - 2 \cos (s_1 + s_2) + 1 = 1 + 1 - 2 \cos s_1 \cos s_2 + 2 \sin s_1 \sin s_2.$$

This simplifies to the following result.

Theorem 7.3   *For each* $s_1, s_2 \in R$,

$$\cos (s_1 + s_2) = \cos s_1 \cos s_2 - \sin s_1 \sin s_2. \tag{1}$$

By replacing s_2 with $-s_2$ in the sum formula (1) and using formulas (6) and (7) in Theorem 7.2, we obtain the following companion result.

Theorem 7.4   *For each $s_1, s_2 \in R$,*

$$\cos(s_1 - s_2) = \cos s_1 \cos s_2 + \sin s_1 \sin s_2. \tag{2}$$

Reduction
formulas for
cosine   The extremely important relationships (1) and (2) are called, respectively, the **sum formula** and the **difference formula** for the cosine function. These are used in proving the following results.

Theorem 7.5   *For each $s \in R$,*

$$\text{I} \quad \cos(\pi - s) = -\cos s,$$
$$\text{II} \quad \cos(\pi + s) = -\cos s,$$
$$\text{III} \quad \cos(2\pi - s) = \cos s.$$

We shall prove only Part I here and leave Parts II and III as exercises.

Proof of 7.5-I   In relationship (2), replace s_1 with π and s_2 with s to obtain

$$\cos(\pi - s) = \cos \pi \cos s + \sin \pi \sin s.$$

From Table 7.2, $\cos \pi = -1$ and $\sin \pi = 0$, so that this equation becomes

$$\cos(\pi - s) = (-1)(\cos s) + 0 \sin s = -\cos s,$$

and the proof is complete.

The formulas in Theorem 7.5 are called **reduction formulas** for the cosine function. Geometric interpretations of I, II, and III are shown in Figure 7.13. In each case, it is apparent that if s is the length of an arc terminating in Quadrants II, III, or IV, then the arc can be visualized either as the sum or as the difference of two arcs, one of measure $n\pi$, $n \in J$, and the other measuring $\bar{s}$, where $0 < \bar{s} < \pi/2$. We refer to the arc measuring $\bar{s}$ as the **reference arc** for s. For example, the reference arc of an arc measuring $5\pi/4$ is one measuring $\pi/4$, because $5\pi/4 = \pi + \pi/4$, while

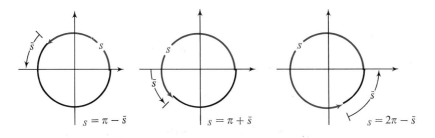

Figure 7.13

the reference arc for an arc with measure $-8\pi/3$ measures $\pi/3$ because $-8\pi/3 = -3\pi + \pi/3$.

Use of tables for cosine In Section 7.2, we found function values (elements in the range) for the cosine function for selected elements in the domain by using the distance formula and certain geometric considerations. It is necessary, however, that we be able to find—or, more precisely, to approximate—a function value for any given real number. To do this, we use tables that provide such approximations to the desired number of decimal places. The means by which the tables are constructed will not concern us at present.

Because the circular functions are periodic and reduction formulas are available, we need tables only over the interval $0 \leq s \leq \pi/2$. Table V in the Appendix gives approximate function values at intervals of 0.01 for the cosine function and other circular functions to be discussed in the following sections.

Examples   a. $\cos 0.73 \approx 0.7452$ b. $\cos 1.29 \approx 0.2771$

Notice that in Table V the letter x is used to represent an element in the domain. This is customary in many tables. In this usage, of course, x can be thought of as representing an arc length along the unit circle, just as s did. In fact, because x is ordinarily the variable used to represent an element in the domain of a function, *we shall hereafter be using x where we heretofore used s.*

Use of reduction formulas An appropriate reduction formula enables us to approximate function values of x outside the interval $0 \leq x \leq \pi/2$. In such cases, note that

$$\pi/2 \approx 1.57, \quad \pi \approx 3.14, \quad 3\pi/2 \approx 4.71, \quad \text{and} \quad 2\pi \approx 6.28.$$

Also note that $\bar{x}$ is used in the same way as $\bar{s}$.

Example   Find $\cos 3.03$.

Solution   Since $1.57 < 3.03 < 3.14$, we can use the reduction formula

$$\cos x = -\cos (\pi - x).$$

Thus,

$$\cos 3.03 \approx -\cos (3.14 - 3.03)$$
$$= -\cos 0.11$$
$$\approx -0.9940.$$

A sketch of the unit circle, including the reference arc, helps interpret the effect of the reduction formula.

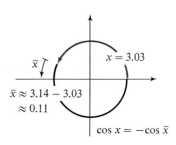

$\bar{x} \approx 3.14 - 3.03$
≈ 0.11

$x = 3.03$

$\cos x = -\cos \bar{x}$

To find approximations to function values for numbers with four-digit numerals, we can use the method of linear interpolation explained in Section 6.3 for tables of logarithms.

Finding elements
in the domain

Table V can also be used to find elements in the domain of the cosine function for specified elements in the range. For the time being, we shall restrict our attention to finding such elements only in the interval $\{x \mid 0 \le x \le \pi/2\}$.

Example

Approximate the member of $\{x \mid 0 < x < 1.57\}$ for which

$$\cos x = 0.9664.$$

Solution

From Table V, we see that x is approximately equal to 0.26.

Exercise 7.3

Reduce each of the following expressions to an equivalent one either of the form $\cos \bar{x}$ *or of the form* $-\cos \bar{x}$, *where* $0 \le \bar{x} \le \pi/2$.

Example

$\cos \dfrac{7\pi}{5}$

Solution

We first note that $7\pi/5$ lies in Quadrant III, since $7\pi/5 = \pi + 2/5\pi$. Thus the reference arc is $2\pi/5$. Using Theorem 7.5, Part II, we have

$$\cos \frac{7\pi}{5} = \cos\left(\pi + \frac{2}{5}\pi\right) = -\cos\frac{2}{5}\pi.$$

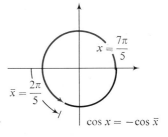

$x = \dfrac{7\pi}{5}$

$\bar{x} = \dfrac{2\pi}{5}$

$\cos x = -\cos \bar{x}$

1. $\cos \dfrac{3\pi}{5}$

2. $\cos \dfrac{7\pi}{8}$

3. $\cos \dfrac{9\pi}{8}$

4. $\cos \dfrac{11\pi}{9}$

5. $\cos \dfrac{13\pi}{7}$

6. $\cos \dfrac{13\pi}{16}$

7. $\cos \left(-\dfrac{7\pi}{6}\right)$

8. $\cos \left(-\dfrac{11\pi}{6}\right)$

Use the fact that cosine is periodic, along with an appropriate reduction formula, to write an equivalent expression either of the form $\cos \bar{x}$ *or of the form* $-\cos \bar{x}$, *where* $0 \le \bar{x} \le \pi/2$.

Example $\cos \dfrac{24\pi}{5}$

Solution   We first write $\dfrac{24\pi}{5} = \dfrac{4\pi}{5} + \dfrac{20\pi}{5} = \dfrac{4\pi}{5} + 4\pi$. Then

$$\cos \frac{24\pi}{5} = \cos \left(\frac{4\pi}{5} + 4\pi\right) = \cos \frac{4\pi}{5}.$$

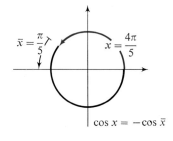

Next, because $4\pi/5$ is in Quadrant II, we note that $4\pi/5 = \pi - (\pi/5)$. Thus, the reference arc is $\pi/5$. Using Part I of Theorem 7.5, we have

$$\cos \frac{4\pi}{5} = \cos \left(\pi - \frac{\pi}{5}\right) = -\cos \frac{\pi}{5}.$$

9. $\cos \dfrac{18\pi}{7}$ 10. $\cos \dfrac{23\pi}{8}$ 11. $\cos \dfrac{13\pi}{5}$ 12. $\cos \dfrac{31\pi}{16}$

13. $\cos \dfrac{35\pi}{11}$ 14. $\cos \dfrac{53\pi}{10}$ 15. $\cos \left(-\dfrac{38\pi}{9}\right)$ 16. $\cos \left(-\dfrac{27\pi}{5}\right)$

Use Table V in the Appendix to find approximations for each of the following expressions. Use linear interpolation as required.

17. $\cos 0.59$ 18. $\cos 0.47$ 19. $\cos 1.32$ 20. $\cos 0.39$

21. $\cos 1.43$ 22. $\cos 0.235$ 23. $\cos 1.216$ 24. $\cos 1.042$

Use the fact that the cosine function is periodic, along with the reduction formulas, to find an approximate value for each of the following expressions.

Example $\cos 2$

Solution   Since $\pi \approx 3.14$ and $\pi/2 \approx 1.57$, an arc of length 2 terminates in the second quadrant. We have $\pi - 2 \approx 1.14$, so that $\bar{x} \approx 1.14$ is the length of the reference arc. Using Theorem 7.5-I, then, we obtain

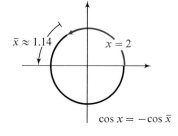

$$\begin{aligned} \cos 2 &= -\cos(\pi - 2) \\ &\approx -\cos(3.14 - 2) \\ &= -\cos 1.14 \\ &\approx -0.4176, \end{aligned}$$

which is the function value we seek.

25. cos 2.01   26. cos 4.52   27. cos 5.21   28. cos 2.24

29. cos 8.41   30. cos 10.32   31. cos (−9.52)   32. cos (−12.61)

Use Table V to approximate the element x in the domain of the function, $0 \le x \le 1.57$, for the specified element in the range.

33. cos $x = 0.9664$ 34. cos $x = 0.2579$ 35. cos $x = 0.7038$

36. cos $x = 0.1304$ 37. cos $x = 0.8600$ 38. cos $x = 0.2921$

Prove each of the following identities.

39. $\cos (x_1 - x_2) = \cos x_1 \cos x_2 + \sin x_1 \sin x_2$

40. $\cos (\pi + x) = -\cos x$

41. $\cos (2\pi - x) = \cos x$

42. $\cos \left(\dfrac{\pi}{2} - x \right) = \sin x$ *Hint:* Substitute $\dfrac{\pi}{2}$ for s_1 and x for s_2 in Theorem 7.4.

43. $\cos \left(\dfrac{\pi}{2} + x \right) = -\sin x$

44. $\cos \left(\dfrac{3\pi}{2} - x \right) = -\sin x$

45. $\cos \left(\dfrac{3\pi}{2} + x \right) = \sin x$

46. $\sin \left(\dfrac{\pi}{2} - x \right) = \cos x$ *Hint:* Substitute $\dfrac{\pi}{2} - x$ for x in Problem 42.

7.4
Properties of the sine function

From Problem 42 in Exercise 7.3 we have

$$\cos \left(\frac{\pi}{2} - x \right) = \sin x$$

If now we set $x = x_1 + x_2$, then we obtain

$$\cos \left[\frac{\pi}{2} - (x_1 + x_2) \right] = \sin (x_1 + x_2).$$

This can be rewritten as

$$\cos\left[\left(\frac{\pi}{2} - x_1\right) - x_2\right] = \sin(x_1 + x_2),$$

and if the left-hand member is expanded by means of Theorem 7.4, we find that

$$\cos\left(\frac{\pi}{2} - x_1\right)\cos x_2 + \sin\left(\frac{\pi}{2} - x_1\right)\sin x_2 = \sin(x_1 + x_2).$$

By the identity $\cos(\pi/2 - x_1) = \sin x_1$, and by Problem 46 in Exercise 7.3, we have

$$\cos\left(\frac{\pi}{2} - x_1\right)\cos x_2 + \sin\left(\frac{\pi}{2} - x_1\right)\sin x_2 = \sin x_1 \cos x_2 + \cos x_1 \sin x_2,$$

which establishes the following result.

Theorem 7.6 *If* $x_1, x_2 \in R$, *then*

$$\sin(x_1 + x_2) = \sin x_1 \cos x_2 + \cos x_1 \sin x_2. \tag{1}$$

By replacing x_2 with $-x_2$ in Theorem 7.6, we obtain a formula for $\sin(x_1 - x_2)$.

Theorem 7.7 *If* $x_1, x_2 \in R$, *then*

$$\sin(x_1 - x_2) = \sin x_1 \cos x_2 - \cos x_1 \sin x_2. \tag{2}$$

Reduction
formulas for sine

Relationships (1) and (2), respectively called the **sum formula** and the **difference formula** for the sine function, can be used in the same way as the sum and difference formulas for the cosine function to establish the following reduction formulas.

Theorem 7.8 *For each* $x \in R$,

$$\text{I} \quad \sin(\pi - x) = \sin x,$$

$$\text{II} \quad \sin(\pi + x) = -\sin x,$$

$$\text{III} \quad \sin(2\pi - x) = -\sin x.$$

Use of tables
for sine

Table V in the Appendix also gives approximate function values for the sine function in the interval $0 \le x \le \pi/2$. An appropriate reduction formula can be used to approximate function values of x outside this interval.

Example   Find sin 4.24.

Solution   Since we have $\pi \approx 3.14$ and $3\pi/2 \approx 4.71$, an arc of length 4.24 terminates in the third quadrant. The length of the reference arc, $\bar{x}$, therefore satisfies

$$\bar{x} \approx 4.24 - 3.14 = 1.10.$$

Using Part II of Theorem 7.8, we obtain

$$\sin 4.24 \approx \sin (3.14 + 1.10)$$
$$= -\sin 1.10$$
$$\approx -0.8912,$$

which is the function value we seek.

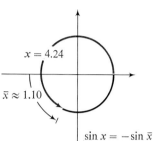

Exercise 7.4

Reduce each of the following to an equivalent expression either of the form $\sin \bar{x}$ *or of the form* $-\sin \bar{x}$, *where* $0 \leq \bar{x} \leq \pi/2$.

Example $\sin \dfrac{5\pi}{7}$

Solution   We first observe that $5\pi/7$ lies in Quadrant II. Thus $5\pi/7 = \pi - (2\pi/7)$. Thus the reference arc is $2\pi/7$. Then, using Part I of Theorem 7.8, we have

$$\sin \frac{5\pi}{7} = \sin \left(\pi - \frac{2\pi}{7} \right) = \sin \frac{2\pi}{7}.$$

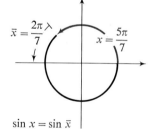

1. $\sin \dfrac{7\pi}{5}$ 2. $\sin \dfrac{7\pi}{8}$ 3. $\sin \dfrac{15\pi}{8}$ 4. $\sin \dfrac{12\pi}{13}$

5. $\sin \dfrac{15\pi}{11}$ 6. $\sin \dfrac{19\pi}{10}$ 7. $\sin \left(-\dfrac{8\pi}{7} \right)$ 8. $\sin \left(-\dfrac{7\pi}{8} \right)$

Use the fact that sine is periodic, along with an appropriate reduction formula, to write an equivalent expression of the form $\sin \bar{x}$ *or* $-\sin \bar{x}$, *where* $0 \leq \bar{x} \leq \pi/2$.

Example $\sin \dfrac{38\pi}{5}$

(Solution overleaf.)

Solution   We first write $\dfrac{38\pi}{5} = \dfrac{8\pi}{5} + \dfrac{30\pi}{5} = \dfrac{8\pi}{5} + 6\pi.$ Then

$$\sin \dfrac{38\pi}{5} = \sin\left(\dfrac{8\pi}{5} + 6\pi\right) = \sin \dfrac{8\pi}{5}.$$

Next, because $8\pi/5$ is in Quadrant IV, we have $8\pi/5 = 2\pi - 2\pi/5.$ Thus, the reference arc is $2\pi/5.$ Using Part III of Theorem 7.8, we have

$$\sin \dfrac{8\pi}{5} = -\sin\left(2\pi - \dfrac{2\pi}{5}\right) = -\sin \dfrac{2\pi}{5}.$$

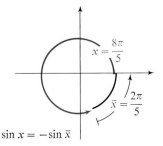

$\sin x = -\sin \bar{x}$

9. $\sin \dfrac{25\pi}{6}$ 10. $\sin \dfrac{27\pi}{7}$ 11. $\sin \dfrac{38\pi}{11}$ 12. $\sin \dfrac{35\pi}{17}$

13. $\sin \dfrac{43\pi}{5}$ 14. $\sin \dfrac{58\pi}{11}$ 15. $\sin\left(-\dfrac{24\pi}{5}\right)$ 16. $\sin\left(-\dfrac{38\pi}{7}\right)$

Use Table V in the Appendix to find approximations for each expression. Use linear interpolation as required.

17. $\sin 0.31$ 18. $\sin 1.36$ 19. $\sin 1.21$ 20. $\sin 0.86$

21. $\sin 1.50$ 22. $\sin 1.362$ 23. $\sin 1.042$ 24. $\sin 0.754$

Use the fact that the sine function is periodic, along with the reduction formulas, to find an approximate value for each expression.

25. $\sin 2.21$ 26. $\sin 3.68$ 27. $\sin 5.47$ 28. $\sin (-0.31)$

29. $\sin (-1.92)$ 30. $\sin 12.24$ 31. $\sin (-9.61)$ 32. $\sin (-15.32)$

Use Table V to approximate the element x $(0 \le \bar{x} \le 1.57)$ in the domain of the function for the specified element in the range.

33. $\sin x = 0.9975$ 34. $\sin x = 0.5396$ 35. $\sin x = 0.8573$

36. $\sin x = 0.2280$ 37. $\sin x = 0.4000$ 38. $\sin x = 0.9230$

Prove each of the following identities.

39. $\sin (x_1 - x_2) = \sin x_1 \cos x_2 - \cos x_1 \sin x_2$

40. $\sin (\pi - x) = \sin x$ 41. $\sin (\pi + x) = -\sin x$

42. $\sin (2\pi - x) = -\sin x$

43. $\sin \left(\dfrac{\pi}{2} + x\right) = \cos x$

44. $\sin \left(\dfrac{3\pi}{2} - x\right) = -\cos x$

45. $\sin \left(\dfrac{3\pi}{2} + x\right) = -\cos x$

7.5

Graphs of sine and cosine functions

We have been using the variables s and x as real numbers associated with arc length on a unit circle. Figure 7.14 illustrates how such arc lengths, in turn, can be

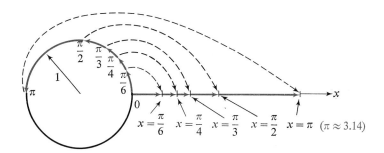

Figure 7.14

associated with line segments on the x-axis of a rectangular coordinate system. Here x is the measure of the line segment on the x-axis that can be thought of as obtained by "unwinding" the arc that has the same measure.

The periodic nature of the cosine and sine functions can be seen clearly by graphing these functions on a rectangular coordinate system in R^2, where values for s or x (elements in the domain) are associated with points on the horizontal axis.

Graph of sine   Although we have Table V available to find values for cos x and sin x for many values of x, the special values for 0, $\pi/6$, $\pi/4$, etc., in Table 7.2, page 192, will suffice for our purposes. Because the graphs of the cosine function and the sine function are called *sine waves*, we shall look first at the graph of

$$\{(x, f(x))\,|\,f(x) = \sin x\}, \quad \text{or simply} \quad \{(x, y)\,|\,y = \sin x\}.$$

Graphing some familiar ordered pairs in this function over the interval $0 \leq x \leq 2\pi$, gives us the graph in Figure 7.15. Assuming that sine is a continuous function— that is, that its graph contains no breaks or gaps (you will prove in calculus that

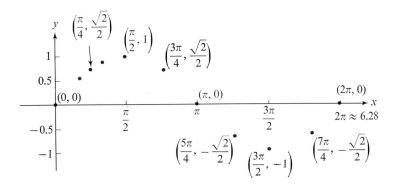

Figure 7.15

this is true)—and that it increases as x increases from 0 to $\pi/2$, then decreases from $\pi/2$ to π, and so forth, we can join these points with a smooth curve to produce Figure 7.16. Then, because $\sin(x + 2\pi) = \sin x$, this pattern repeats itself over intervals of length 2π in both directions. Thus we have the pattern shown in Figure 7.17 for the graph of the sine function. Units on the horizontal axis

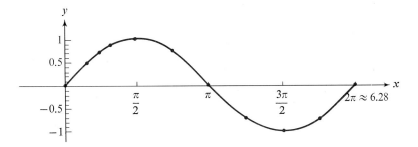

Figure 7.16

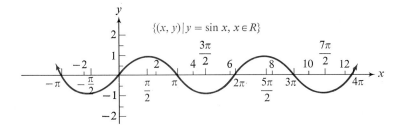

Figure 7.17

representing integers (short marks) and integral multiples of $\pi/2$ (long marks) are shown. The range of the function,

$$\{y \mid -1 \le y \le 1\},$$

is clearly evident from the graph. The zeros of the function, πn, where $n = 0, \pm 1, \pm 2, \pm 3, \ldots$, namely the values of x associated with the points at which the curve crosses the x-axis, also appear on the graph. Graphs with this characteristic form are called **sine waves** (as remarked earlier), or **sinusoids**. The portion of the graph over any fundamental period of the function is called a **cycle** of the sine wave. Half the difference of the maximum and minimum ordinates on such a curve is called the **amplitude** of the wave. Thus, for the graph of $y = \sin x$, the amplitude of the wave is $\frac{1}{2}[1 - (-1)] = 1$.

Graph of cosine The graph of the cosine function can be obtained in the same manner as the graph of the sine function. From Table 7.2 on page 192, or from memory, we obtain the coordinates of several points on the graph, as shown in Figure 7.18. By connecting these points with a smooth curve, we obtain the graph of $\{(x, y) \mid y = \cos x\}$ over one period of the function. Duplicating this pattern over several more periods, we have a representative portion of the entire graph of the cosine function (Figure 7.19). Notice that

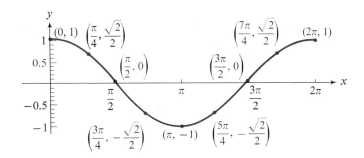

Figure 7.18

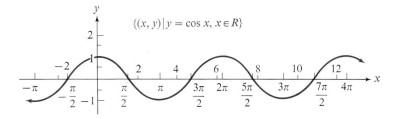

Figure 7.19

the graph of the cosine function is a sinusoid, with amplitude 1. Furthermore, the zeros of the function, $\pi/2 + n\pi$, where $n = 0, \pm 1, \pm 2, \pm 3, \ldots$, and the range $\{y| -1 \leq y \leq 1\}$ are also evident.

Functions defined by equations of the form

$$y = A \sin (Bx + C) \quad \text{and} \quad y = A \cos (Bx + C),$$

where A, B, and C are constants, and A and B are not 0, always have sine waves for graphs. With variations in the numbers A, B, and C, the graphs are variously situated with respect to the origin and have a variety of amplitudes and periods. To analyze such graphs, let us consider the effect of A on the graphs of the functions defined by $y = A \sin x$ and $y = A \cos x$.

Graphs of $y = A \sin x$ **and** $y = A \cos x$ For each value of x, each ordinate to the graph of $y = A \sin x$ is A times the ordinate to the graph of $y = \sin x$. Therefore, the amplitude of the graph of $y = A \sin x$ is $|A|$ times the amplitude of the graph of $y = \sin x$. Of course, the graph of $y = A \cos x$ is a similar modification of the graph of $y = \cos x$.

Example   Graph $y = 3 \sin x$, $-\pi \leq x \leq 4\pi$.

Solution   It may be helpful first to sketch $y = \sin x$, $0 \leq x \leq 2\pi$, as a reference. Then, since the amplitude of $y = 3 \sin x$ is 3, we can sketch the desired graph on the same coordinate system over the interval $0 \leq x \leq 2\pi$ by making each ordinate 3 times the corresponding ordinate of the graph of $y = \sin x$. We can then extend this cycle to include the entire interval $-\pi \leq x \leq 4\pi$, as shown in the figure. In this figure, the first cycle is sketched with a heavier line for emphasis. Note also that, in this and some of the succeeding figures, different unit lengths are used on the x- and y-axes. For $y = 3 \sin x$, the fundamental period p is the same as for $y = \sin x$, namely $p = 2\pi$.

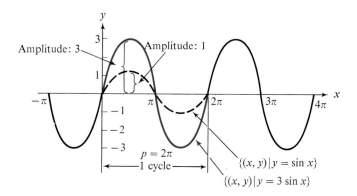

Graphs of
$y = \sin Bx$ and
$y = \cos Bx$

Next let us examine the way the graph of

$$y = \sin Bx, \quad B \neq 0,$$

differs from the graph of $y = \sin x$. Note first that $\sin Bx$ has values between -1 and 1 inclusive, as has $\sin x$. Also,

$$\sin (Bx + 2\pi) = \sin Bx,$$

just as $\sin (x + 2\pi) = \sin x$. If we factor B from $Bx + 2\pi$, however, we have $B(x + 2\pi/B)$, and hence the function defined by $y = \sin Bx$ is periodic with period $2\pi/|B|$. We use $|B|$ instead of B to ensure a positive number for the period. It can be shown, although it is not done here, that $2\pi/|B|$ is the fundamental period of the function. Hence the graph of $y = \sin Bx$ is a sine wave with amplitude 1; it completes one cycle over the interval $0 \leq x \leq 2\pi/|B|$.

Example   Graph $y = \cos 2x$, $\dfrac{-3\pi}{2} \leq x \leq 2\pi$.

Solution   Let us first sketch a cycle of $y = \cos x$, $0 \leq x \leq 2\pi$, as a reference. Since

$$p = \frac{2\pi}{|B|} = \frac{2\pi}{2} = \pi,$$

we next sketch a cycle of the graph of $y = \cos 2x$ over the interval $0 \leq x \leq \pi$ on the same coordinate system, and extend the cycle obtained over the interval $-3\pi/2 \leq x \leq 2\pi$.

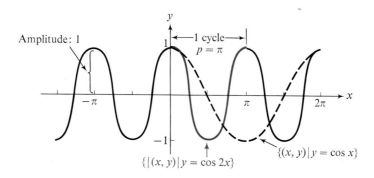

Example   Graph $y = -4 \sin \dfrac{1}{2}x$, $-2\pi \leq x \leq 4\pi$.

Solution   We first sketch the graph of $y = \sin x$ as a reference. Since $A = -4$, each ordinate of the graph of $y = -4 \sin x/2$ is the negative of the

(Solution continued.)

ordinate of the graph of $y = 4 \sin (x/2)$. Since

$$p = \frac{2\pi}{|B|} = \frac{2\pi}{1/2} = 4\pi,$$

there is one cycle in the interval $0 \le x \le 4\pi$. Hence, we sketch one cycle, and extend it over the interval $-2\pi \le x \le 4\pi$.

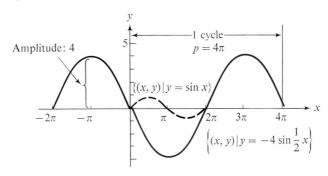

<table>
<tr><td>Graphs of
$y = \sin (x + C)$
and
$y = \cos (x + C)$</td><td>Finally, let us look at the difference between the graphs of

$$y = \sin (x + C) \quad \text{and} \quad y = \sin x,$$

where $C > 0$. For $x = -C$, we have

$$\sin (x + C) = \sin 0 = 0.$$</td></tr>
</table>

Similarly, for any real number x_1, the ordinate of $\sin (x + C)$ at $x_1 - C$ will be the same as the ordinate of $\sin x$ at x_1. Hence, the graph of $y = \sin (x + C)$ is said to **lead** the graph of $y = \sin x$ by C. The number C, itself, is called the **phase shift** of the wave. If $C < 0$, then the graph of $y = \sin (x + C)$ is shifted $|C|$ units to the right of the graph of $y = \sin x$ and is said to **lag** the graph of $y = \sin x$.

<table>
<tr><td>Graphs of
$y = A \sin (Bx + C)$
and
$y = A \cos (Bx + C)$</td><td>We use all of the foregoing information about the effect of A, B, and C on the graph of $y = A \sin (Bx + C)$, or of $y = A \cos (Bx + C)$, to help sketch the graph of such an equation.</td></tr>
</table>

Example   Sketch the graph of $y = 3 \sin \left(2x + \frac{\pi}{3}\right)$.

Solution   First, let us rewrite the equation by factoring 2 from the expression in parentheses:

$$y = 3 \sin 2\left(x + \frac{\pi}{6}\right).$$

By inspecting this equation, we note the following things about the graph:

1. It is a sine wave.
2. It has amplitude 3.
3. It has period $2\pi/2 = \pi$.
4. It leads the graph of $y = 3 \sin 2x$ by $\pi/6$.

With these facts, we can quickly sketch the graph shown.

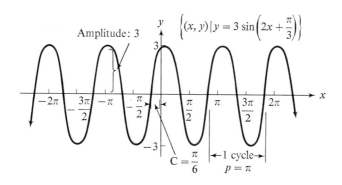

Example   Sketch the graph of $y = \frac{1}{2} \sin \frac{\pi x}{2}$.

Solution   By inspection, we note concerning the graph:

1. It is a sine wave.
2. It has amplitude $1/2$.
3. It has period $\dfrac{2\pi}{\pi/2} = 4$.

Since the period is 4, we use integers as elements of the domain. Scaling the x-axis in integral units facilitates sketching the graph.

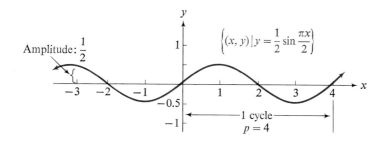

Exercise 7.5

Sketch the graphs of the following equations over the interval $-2\pi \leq x \leq 2\pi$.

1. $y = 2 \sin x$

2. $y = 3 \cos x$

3. $y = \dfrac{1}{2} \cos x$

4. $y = \dfrac{1}{3} \cos x$

5. $y = -4 \sin x$

6. $y = -\dfrac{1}{2} \cos x$

7. $y = \sin 2x$

8. $y = \cos 3x$

9. $y = \cos \dfrac{1}{3} x$

10. $y = \cos \dfrac{1}{2} x$

11. $y = -3 \sin 2x$

12. $y = -\dfrac{1}{2} \sin 3x$

13. $y = \sin (x + \pi)$

14. $y = \cos \left(x - \dfrac{\pi}{2} \right)$

15. $y = 2 \cos \left(x - \dfrac{\pi}{4} \right)$

16. $y = 3 \sin \left(x + \dfrac{\pi}{6} \right)$

17. $y = 3 \sin 2 \left(x - \dfrac{\pi}{3} \right)$

18. $y = 2 \cos 3 \left(x + \dfrac{\pi}{4} \right)$

19. $y = 2 \sin \pi x$

20. $y = -3 \cos \dfrac{\pi}{2} x$

21. $y = -\dfrac{1}{2} \cos \dfrac{\pi}{3} x$

22. $y = \dfrac{1}{4} \sin \dfrac{\pi}{4} x$

From the respective graph, determine the zeros (over the specified domain) of the function defined by the equation in the given problem.

23. Problem 7   24. Problem 8   25. Problem 9

26. Problem 10   27. Problem 19   28. Problem 20

Example

Sketch the graph of $y = \sin x + 2 \cos x$ over the interval $0 \leq x \leq 2\pi$.

Solution

First, sketch the graphs of $y = \sin x$ and $y = 2 \cos x$ on the same co-ordinate system over the given interval. The ordinate of the graph of $y = \sin x + 2 \cos x$ at each point x on the x-axis is the *algebraic* sum of the corresponding ordinates of $y = \sin x$ and $y = 2 \cos x$. Thus for $x = \pi/6$,

$$\sin x = 0.5, \quad 2 \cos x = 2\sqrt{3}/2 \approx 1.7, \quad \text{and} \quad \sin x + 2 \cos x \approx 2.2.$$

Now the ordinate can be approximated graphically by adding the directed line segments from the x-axis to the curves at $x = \pi/6$:

$$\overline{AB} + \overline{AC} = \overline{AB} + \overline{BD} = \overline{AD}.$$

If this is done for a few selected values of x, we obtain a good approximation for the curve.

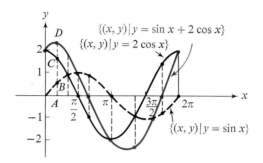

Sketch the graph of each equation over the interval $-2\pi \le x \le 2\pi$.

29. $y = \sin x + \cos x$

30. $y = 3 \sin x + \cos x$

31. $y = \sin 2x + \dfrac{1}{2} \cos x$

32. $y = \sin 3x + 2 \cos \dfrac{1}{2} x$

33. $y = \sin x - 2 \cos x$

34. $y = 3 \cos x - \sin 2x$

35. $y = 3 \sin x + 1$

36. $y = 4 \sin 2x - 3$

37. $y = x + \cos x$

38. $y = 2x - \cos x$

7.6

Additional properties
of the cosine and sine functions

Formulas for
cos 2x and sin 2x
The sum and difference formulas for cosine and sine that we developed in Sections 7.3 and 7.4 can be used to generate still other relationships between values of these functions.

Theorem 7.9 *If $x \in R$, then*

$$\text{I} \quad \cos 2x = \cos^2 x - \sin^2 x, \tag{1}$$

$$\text{II} \quad \sin 2x = 2 \sin x \cos x. \tag{2}$$

Proof Let $s_1 = s_2 = x$ in Equation (1) on page 194. Then

$$\cos (x + x) = \cos x \cos x - \sin x \sin x,$$
$$\cos 2x = \cos^2 x - \sin^2 x.$$

Let $x_1 = x_2 = x$ in Equation (1) on page 200. Then

$$\sin (x + x) = \sin x \cos x + \cos x \sin x,$$

$$\sin 2x = 2 \sin x \cos x,$$

as was to be shown.

Formulas for cos $x/2$ and sin $x/2$ Alternative forms of the equation for cos $2x$ can be obtained by using the relationship $\cos^2 x + \sin^2 x = 1$ to obtain

$$\cos^2 x = 1 - \sin^2 x \quad \text{and} \quad \sin^2 x = 1 - \cos^2 x,$$

and then replacing the appropriate term in the right-hand member of (1). The results are the formulas:

$$\cos 2x = 1 - 2 \sin^2 x, \tag{3}$$

$$\cos 2x = 2 \cos^2 x - 1. \tag{4}$$

Relationships (3) and (4) lead directly to the following formulas:

Theorem 7.10 *If $x \in R$, then*

I $\cos \dfrac{x}{2} = \begin{cases} \sqrt{\dfrac{1 + \cos x}{2}} & \text{when } \dfrac{x}{2} \text{ terminates in Quadrant I or IV,} \quad\quad (5) \\[4mm] -\sqrt{\dfrac{1 + \cos x}{2}} & \text{when } \dfrac{x}{2} \text{ terminates in Quadrant II or III;} \quad (5') \end{cases}$

II $\sin \dfrac{x}{2} = \begin{cases} \sqrt{\dfrac{1 - \cos x}{2}} & \text{when } \dfrac{x}{2} \text{ terminates in Quadrant I or II,} \quad\quad (6) \\[4mm] -\sqrt{\dfrac{1 - \cos x}{2}} & \text{when } \dfrac{x}{2} \text{ terminates in Quadrant III or IV.} \quad (6') \end{cases}$

Proof We shall prove only the formula for sin $(x/2)$ here and leave the proof of the formula for cos $(x/2)$ as an exercise. If we replace x in (3) with $x/2$, we have

$$\cos 2\left(\frac{x}{2}\right) = 1 - 2 \sin^2 \frac{x}{2},$$

from which

$$\cos x = 1 - 2 \sin^2 \frac{x}{2},$$

$$\sin^2 \frac{x}{2} = \frac{1 - \cos x}{2},$$

and

$$\sin \frac{x}{2} = \pm \sqrt{\frac{1 - \cos x}{2}}.$$

Now, because $\sin (x/2) > 0$ for $x/2$ in Quadrant I or II, we take the positive square root in these quadrants; and for a similar reason in the remaining quadrants we take the negative square root. Thus our proof is complete.

Example   Find $\sin \dfrac{\pi}{12}$.

Solution   Since $\dfrac{\pi}{12} = \dfrac{1}{2}\left(\dfrac{\pi}{6}\right)$, from (6) we have

$$\sin \frac{\pi}{12} = \sin \frac{1}{2}\left(\frac{\pi}{6}\right)$$

$$= \sqrt{\frac{1 - \cos (\pi/6)}{2}}$$

$$= \sqrt{\frac{1 - \sqrt{3}/2}{2}} = \sqrt{\frac{2 - \sqrt{3}}{4}} = \frac{1}{2}\sqrt{2 - \sqrt{3}}.$$

Exercise 7.6

Use Formulas (1)–(6) and Table 7.2 to find numerical values for each expression.

1. $\sin \dfrac{\pi}{8}$ 2. $\cos \dfrac{\pi}{8}$ 3. $\cos \left(-\dfrac{\pi}{8}\right)$ 4. $\sin \left(-\dfrac{\pi}{8}\right)$

5. $\cos \dfrac{3\pi}{8}$ 6. $\sin \dfrac{3\pi}{8}$ 7. $\sin \dfrac{5\pi}{24}$ 8. $\cos \dfrac{5\pi}{24}$

Example   Given that $\sin x = 4/5$, find $\sin 2x$, $\pi/2 \le x \le \pi$.

Solution   Since $\sin x = 4/5$ and x terminates in Quadrant II,

$$\cos x = -\sqrt{1 - \sin^2 x} = -\sqrt{1 - \left(\frac{4}{5}\right)^2} = -\sqrt{\frac{9}{25}} = -\frac{3}{5}.$$

From Formula (2) on page 211, we have

$$\sin 2x = 2 \sin x \cos x = 2 \left(\frac{4}{5}\right)\left(-\frac{3}{5}\right) = -\frac{24}{25}.$$

9. Given that $\cos x = -\dfrac{3}{5}$, find:

   a. $\cos 2x, \quad \dfrac{\pi}{2} \le x \le \pi$ b. $\sin 2x, \quad \pi \le x \le \dfrac{3\pi}{2}$

   c. $\sin \dfrac{1}{2}x, \quad 3\pi \le x \le \dfrac{7\pi}{2}$ d. $\cos \dfrac{1}{2}x, \quad \dfrac{\pi}{2} \le x \le \pi$

10. Given that $\sin x = \dfrac{\sqrt{5}}{3}$, find:

   a. $\cos 2x, \quad \dfrac{\pi}{2} \le x \le \pi$ b. $\sin 2x, \quad \dfrac{\pi}{2} \le x \le \pi$

   c. $\sin \dfrac{1}{2}x, \quad 0 \le x \le \dfrac{\pi}{2}$ d. $\cos \dfrac{1}{2}x, \quad 2\pi \le x \le \dfrac{5\pi}{2}$

Example   Given that $\sin x = 0.2$, find an approximation for $\sin 2x$, $\pi/2 \le x \le \pi$.

Solution   Since $\sin x = 0.2$ and x terminates in Quadrant II,

$$\cos x = -\sqrt{1 - \sin^2 x} = -\sqrt{1 - 0.04} = -\sqrt{0.96} \approx -0.98.$$

From Formula (2) on page 211, we have

$$\sin 2x = 2 \sin x \cos x \approx 2(0.2)(-0.98) \approx -0.39.$$

11. Given that $\cos x = 0.6$, find an approximation for:

   a. $\sin 2x, \quad 0 \le x \le \dfrac{\pi}{2}$ b. $\cos 2x, \quad \dfrac{3\pi}{2} \le x \le 2\pi$

   c. $\sin \dfrac{1}{2}x, \quad 0 \le x \le \dfrac{\pi}{2}$ d. $\cos \dfrac{1}{2}x, \quad \dfrac{3\pi}{2} \le x \le 2\pi$

12. Given that $\sin x = -0.3$, find an approximation for:

   a. $\sin 2x, \quad \pi \le x \le \dfrac{3\pi}{2}$ b. $\cos 2x, \quad \dfrac{3\pi}{2} \le x \le 2\pi$

   c. $\cos \dfrac{1}{2}x, \quad \pi \le x \le \dfrac{3\pi}{2}$ d. $\sin \dfrac{1}{2}x, \quad \dfrac{3\pi}{2} \le x \le 2\pi$

By means of the formulas in this section, evaluate each of the expressions in Problems 13–16 mentally.

13. $2 \sin \dfrac{\pi}{12} \cos \dfrac{\pi}{12}$ 14. $\cos^2 \dfrac{\pi}{12} - \sin^2 \dfrac{\pi}{12}$

15. $2 \cos^2 \dfrac{5\pi}{12} - 1$ 16. $1 - 2 \sin^2 \dfrac{5\pi}{12}$

17. Express $\sin^2 x$ in terms of $\cos 2x$. *Hint:* Use Formula (3) on page 212.

18. Express $\cos^2 x$ in terms of $\cos 2x$.

19. Show that for all $x \in R$, $\cos 3x = 4 \cos^3 x - 3 \cos x$. *Hint:* $\cos 3x = \cos (2x + x)$.

20. Show that for all $x \in R$, $\sin 3x = 3 \sin x - 4 \sin^3 x$.

21. Show that $\sin 2x$ is periodic with fundamental period π.

22. Show that $\cos 2x$ is periodic with fundamental period π.

23. Show that $\sin (x/2)$ is periodic with fundamental period 4π.

24. Show that $\cos (x/2)$ is periodic with fundamental period 4π.

25. Prove Theorem 7.10-I.

7.7
Other circular functions

The functions sine and cosine can be used to define other periodic functions. First, however, let us give names to some ratios of sine and cosine function values.

Definition 7.4   *Let $x \in R$.*

  I   *The tangent of x is denoted by* tan x, *and*

$$\tan x = \frac{\sin x}{\cos x} \qquad \left(x \neq \frac{\pi}{2} + k\pi, \, k \in J \right).$$

  II   *The cotangent of x is denoted by* cot x, *and*

$$\cot x = \frac{\cos x}{\sin x} \qquad (x \neq k\pi, \, k \in J).$$

  III   *The secant of x is denoted by* sec x, *and*

$$\sec x = \frac{1}{\cos x} \qquad \left(x \neq \frac{\pi}{2} + k\pi, \, k \in J \right).$$

  IV   *The cosecant of x is denoted by* csc x, *and*

$$\csc x = \frac{1}{\sin x} \qquad (x \neq k\pi, \, k \in J).$$

Example   If $\sin x = 3/5$ and $\pi/2 \leq x \leq \pi$, find $\cos x$, $\tan x$, $\sec x$, $\csc x$, and $\cot x$.

Solution   By Equation (5) on page 185, for $\pi/2 \leq x \leq \pi$,

$$\cos x = -\sqrt{1 - \sin^2 x}.$$

Then, since $\sin x = 3/5$,

$$\cos x = -\sqrt{1 - \left(\frac{3}{5}\right)^2} = -\frac{4}{5}.$$

By Definition 7.4,

$$\tan x = \frac{\sin x}{\cos x} = \frac{3/5}{-4/5} = -\frac{3}{4}, \quad \sec x = \frac{1}{\cos x} = \frac{1}{-4/5} = -\frac{5}{4},$$

$$\cot x = \frac{\cos x}{\sin x} = \frac{-4/5}{3/5} = -\frac{4}{3}, \quad \csc x = \frac{1}{\sin x} = \frac{1}{3/5} = \frac{5}{3}.$$

Now, let us take the ratios in Definition 7.4 one at a time and examine the function associated with each.

Tangent   We can use Definition 7.4 together with Table 7.2, page 192, to find $\tan x$ for all values of x included in the table. For example,

$$\tan 0 = \frac{\sin 0}{\cos 0} = \frac{0}{1} = 0,$$

$$\tan \frac{\pi}{6} = \frac{\sin (\pi/6)}{\cos (\pi/6)} = \frac{1/2}{\sqrt{3/2}} = \frac{1}{\sqrt{3}},$$

and so forth. The values of $\tan x$ obtained in this way are tabulated in Table 7.3. Just as we do for $\sin x$ and $\cos x$, we can use Table V, page 420, to obtain additional values for $\tan x$.

Definition 7.5   *If $x \in R$, and $x \neq \pi/2 + k\pi$, $k \in J$, then*

$$\text{tangent} = \{(x, y) \mid y = \tan x\}.$$

Because $\tan x = \sin x/\cos x$, the domain of the tangent function is R, with the exception of the real numbers x for which $\cos x = 0$. In other words, we except real numbers of the form $\pi/2 + k\pi$, $k \in J$. The range of the tangent function is R. Since the sine and cosine functions have period 2π, so has the tangent function. We note, however, that

$$\tan x = \frac{\sin x}{\cos x} = \frac{-\sin (x + \pi)}{-\cos (x + \pi)} = \tan (x + \pi),$$

Table 7.3

x	$\tan x$	x	$\tan x$
0	0	π	0
$\dfrac{\pi}{6}$	$\dfrac{1}{\sqrt{3}}$	$\dfrac{7\pi}{6}$	$\dfrac{1}{\sqrt{3}}$
$\dfrac{\pi}{4}$	1	$\dfrac{5\pi}{4}$	1
$\dfrac{\pi}{3}$	$\sqrt{3}$	$\dfrac{4\pi}{3}$	$\sqrt{3}$
$\dfrac{\pi}{2}$	not defined	$\dfrac{3\pi}{2}$	not defined
$\dfrac{2\pi}{3}$	$-\sqrt{3}$	$\dfrac{5\pi}{3}$	$-\sqrt{3}$
$\dfrac{3\pi}{4}$	-1	$\dfrac{7\pi}{4}$	-1
$\dfrac{5\pi}{6}$	$-\dfrac{1}{\sqrt{3}}$	$\dfrac{11\pi}{6}$	$-\dfrac{1}{\sqrt{3}}$
π	0	2π	0

so that this function actually has period π also. Clearly, $\tan x$ alternates in sign from quadrant to quadrant, being positive in Quadrants I and III and negative in Quadrants II and IV.

From Definition 7.4, we have

$$\tan(-x) = \frac{\sin(-x)}{\cos(-x)}.$$

From Theorem 7.2, $\sin(-x) = -\sin x$ and $\cos(-x) = \cos x$. Hence,

$$\tan(-x) = \frac{-\sin x}{\cos x} = -\tan x.$$

Thus, like the sine function, the tangent function is an odd function (see page 186).

Definition 7.6   *If $x \in R$, and $x \neq k\pi$, $k \in J$, then*

$$\text{cotangent} = \{(x, y) \mid y = \cot x\}.$$

Cotangent   Since $\cot x = \cos x/\sin x$, the domain of the cotangent function is R, excepting those real numbers for which $\sin x = 0$, that is, all real numbers except those of the form $k\pi$, where

$k \in J$. The range of cotangent is R, and, like the tangent function, contangent has fundamental period π. Note that because $\cot x = \cos x/\sin x$ and $\tan x = \sin x/\cos x$,

$$\cot x = \frac{1}{\tan x} \quad \text{and} \quad \tan x = \frac{1}{\cot x}$$

for those values of x for which each expression is defined.

Definition 7.7 *If $x \in R$, and $x \neq \pi/2 + k\pi$, $k \in J$, then*

$$\text{secant} = \{(x, y) \mid y = \sec x\}.$$

Secant   Since $\sec x$ is the reciprocal of $\cos x$, the domain of the secant function is R, except for those values of x for which $\cos x = 0$, namely, $x = \pi/2 + k\pi$, $k \in J$. Because $|\cos x| \leq 1$ for all $x \in R$, $\sec x$ is never less than 1 in absolute value; the range of the secant function is $\{y \mid |y| \geq 1\}$. Since the cosine function has fundamental period 2π, the secant function also has fundamental period 2π.

Definition 7.8 *If $x \in R$, and $x \neq k\pi$, $k \in J$, then*

$$\text{cosecant} = \{(x, y) \mid y = \csc x).$$

Cosecant   Because $\csc x$ is the reciprocal of $\sin x$, the domain of the cosecant function contains all real numbers except those for which $\sin x = 0$; that is, the domain is all of R except the numbers $k\pi$, $k \in J$. The cosecant function has range $\{y \mid |y| \geq 1\}$ and fundamental period 2π.

Since $\cot x$, $\sec x$, and $\csc x$ are the reciprocals of $\tan x$, $\cos x$, and $\sin x$, respectively, we can find these function values for commonly used values of x by using Table 7.2 or 7.3. Furthermore, we can use Table V in the Appendix for additional values.

Examples   a. $\sec \dfrac{7\pi}{6}$ b. $\csc 1.32$

Solutions   a. By using Definition 7.4-II and Table 7.2, we have

$$\sec \frac{7\pi}{6} = \frac{1}{\cos \left(\dfrac{7\pi}{6}\right)} = \frac{1}{-\sqrt{3}/2} = -\frac{2}{\sqrt{3}}.$$

b. By using Table V, we obtain

$$\csc 1.32 = 1.032.$$

Other formulas
for tangent

Each of the functions discussed in this section has its associated sum, difference, and reduction formulas, comparable to and derived from those of sine and cosine.

Theorem 7.11   *If $x_1, x_2 \in R$, and x_1, x_2, and $x_1 + x_2 \neq \pi/2 + k\pi$, $k \in J$, then*

$$\tan (x_1 + x_2) = \frac{\tan x_1 + \tan x_2}{1 - \tan x_1 \tan x_2}.$$

Proof   By definition, for $x_1 + x_2 \in R$, $x_1 + x_2 \neq \pi/2 + k\pi$, $k \in J$,

$$\tan (x_1 + x_2) = \frac{\sin (x_1 + x_2)}{\cos (x_1 + x_2)}.$$

By Theorems 7.6 and 7.3, we have

$$\tan (x_1 + x_2) = \frac{\sin x_1 \cos x_2 + \cos x_1 \sin x_2}{\cos x_1 \cos x_2 - \sin x_1 \sin x_2}.$$

Dividing numerator and denominator of the right-hand member by $\cos x_1 \cos x_2$ for $x_1, x_2 \in R$, and $x_1, x_2 \neq (\pi/2) + k\pi$, $k \in J$, we find

$$\tan (x_1 + x_2) = \frac{\dfrac{\sin x_1 \cos x_2}{\cos x_1 \cos x_2} + \dfrac{\cos x_1 \sin x_2}{\cos x_1 \cos x_2}}{\dfrac{\cos x_1 \cos x_2}{\cos x_1 \cos x_2} - \dfrac{\sin x_1 \sin x_2}{\cos x_1 \cos x_2}}.$$

After simplifying each term in the right-hand member, from Definition 7.4-I we obtain

$$\tan (x_1 + x_2) = \frac{\tan x_1 + \tan x_2}{1 - \tan x_1 \tan x_2},$$

as was to be shown.

Also, by substituting $x_1 - x_2$ for x in the definition of $\tan x$, we can establish a difference formula for tangents in a similar way.

Theorem 7.12   *If $x_1, x_2 \in R$ and x_1, x_2 and $x_1 - x_2 \neq \pi/2 + k\pi$, $k \in J$, then*

$$\tan (x_1 - x_2) = \frac{\tan x_1 - \tan x_2}{1 + \tan x_1 \tan x_2}.$$

The following reduction formulas follow directly from Theorems 7.11 and 7.12.

Theorem 7.13 *For each $x \in R$,*

$$\text{I} \quad \tan (\pi - x) = -\tan x,$$
$$\text{II} \quad \tan (\pi + x) = \tan x,$$
$$\text{III} \quad \tan (2\pi - x) = -\tan x.$$

Their proofs are left as exercises. Formulas for $\tan 2x$ and $\tan \dfrac{x}{2}$ also follow immediately from formulas we have already developed.

Theorem 7.14 *If $x \in R$, $x \neq \pi/4 + k\pi/2$, $k \in J$, then*

$$\tan 2x = \frac{2 \tan x}{1 - \tan^2 x}.$$

Proof Replacing x_1 and x_2 by x in Theorem 7.11, we obtain

$$\tan (x + x) = \frac{\tan x + \tan x}{1 - \tan x \tan x},$$

$$\tan 2x = \frac{2 \tan x}{1 - \tan^2 x}.$$

Theorem 7.15 *If $x \in R$ and $x \neq k\pi$, $k \in J$, then*

$$\text{I} \quad \tan \frac{x}{2} = \frac{1 - \cos x}{\sin x}, \qquad \text{II} \quad \tan \frac{x}{2} = \frac{\sin x}{1 + \cos x}.$$

Proof We shall prove only Part I here. The proof of Part II, which is similar, is left as an exercise. By Definition 7.4,

$$\tan \frac{x}{2} = \frac{\sin \dfrac{x}{2}}{\cos \dfrac{x}{2}},$$

from which, by multiplying numerator and denominator by $2 \sin \dfrac{x}{2}$, we have

$$\tan \frac{x}{2} = \frac{2 \sin^2 \dfrac{x}{2}}{2 \sin \dfrac{x}{2} \cos \dfrac{x}{2}}.$$

Therefore, by Theorems 7.9 and 7.10,

$$\tan \frac{x}{2} = \frac{1 - \cos x}{\sin x},$$

as desired.

Relationships for cotangent, secant, and cosecant

Relationships similar to those developed for tangent in Theorems 7.11 through 7.15 can be derived for contangent, secant, and cosecant. For practical purposes, however, the reduction formulas for cosine, sine, and tangent are the only ones necessary in view of the fact that

$$\sec x = \frac{1}{\cos x}, \quad \csc x = \frac{1}{\sin x}, \quad \text{and} \quad \cot x = \frac{1}{\tan x}.$$

Exercise 7.7

Using the definitions and theorems stated in this section, find values for the other five circular functions from the one function value and quadrant given.

Example

$\tan x = \dfrac{4}{3};$ x terminates in Quadrant III.

Solution

From Definition 7.4, $\tan x = \dfrac{\sin x}{\cos x}$, so $\dfrac{\sin x}{\cos x} = \dfrac{4}{3}$. Since, in Quadrant

III, $\cos x = -\sqrt{1 - \sin^2 x}$, we have

$$\frac{\sin x}{-\sqrt{1 - \sin^2 x}} = \frac{4}{3},$$

or

$$3 \sin x = -4\sqrt{1 - \sin^2 x}.$$

Squaring each member and simplifying, we find that

$$9 \sin^2 x = 16(1 - \sin^2 x),$$
$$25 \sin^2 x = 16,$$
$$\sin^2 x = \frac{16}{25}.$$

(Solution continued.)

It follows that either $\sin x = 4/5$ or $\sin x = -4/5$. Since x terminates in Quadrant III, we have $\sin x = -4/5$. Using $\sin^2 x + \cos^2 x = 1$, we find that

$$\frac{16}{25} + \cos^2 x = 1,$$

from which either

$$\cos x = \frac{3}{5} \quad \text{or} \quad \cos x = -\frac{3}{5}.$$

In Quadrant III, $\cos x < 0$, and hence $\cos x = -3/5$. From these results, and using Definition 7.4, we find

$$\sin x = -\frac{4}{5}, \quad \cos x = -\frac{3}{5}, \quad \sec x = -\frac{5}{3}, \quad \csc x = -\frac{5}{4}, \quad \cot x = \frac{3}{4}.$$

1. $\sin x = -8/17$; x terminates in Quadrant III.

2. $\sec x = -5/3$; x terminates in Quadrant II.

3. $\tan x = 5/12$; x terminates in Quadrant I.

4. $\tan x = 8/15$; x terminates in Quadrant III.

5. $\cot x = -3$; x terminates in Quadrant IV.

6. $\csc x = -17/15$; x terminates in Quadrant IV.

Prove the following formulas. In each case, state restrictions on x or function values of x.

7. $\tan(\pi - x) = -\tan x$ 8. $\tan(2\pi - x) = -\tan x$

9. $\tan\left(\dfrac{\pi}{2} - x\right) = \cot x$ 10. $\tan\left(\dfrac{\pi}{2} + x\right) = -\cot x$

Use the fact that the period of tangent is π to express each function value in the form $\tan \bar{x}$ or $-\tan \bar{x}$, where $0 \le \bar{x} \le \dfrac{\pi}{2}$.

Example $\tan \dfrac{19\pi}{5}$

Solution   Since the period of tangent is π, we express $\dfrac{19\pi}{5}$ as

$$\frac{15\pi}{5} + \frac{4\pi}{5}, \quad \text{or} \quad 3\pi + \frac{4\pi}{5}.$$

Then we have

$$\tan\left(\frac{19\pi}{5}\right) = \tan\left(3\pi + \frac{4\pi}{5}\right)$$

$$= \tan\left(\frac{4\pi}{5}\right).$$

Since $\dfrac{\pi}{2} < \dfrac{4\pi}{5} < \pi$, we can use the reduction formula in Theorem 7.13-I on page 220 to obtain

$$\tan\frac{4\pi}{5} = -\tan\frac{\pi}{5}.$$

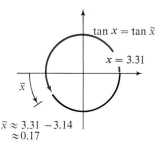

11. $\tan\dfrac{8\pi}{7}$ 12. $\tan\dfrac{13\pi}{5}$ 13. $\tan\dfrac{21\pi}{11}$ 14. $\tan\dfrac{23\pi}{12}$

15. $\tan\dfrac{47\pi}{13}$ 16. $\tan\dfrac{39\pi}{5}$ 17. $\tan\left(\dfrac{-11\pi}{7}\right)$ 18. $\tan\left(\dfrac{-15\pi}{7}\right)$

Use the definitions and periodicity of the circular functions, together with Table V in the Appendix, to obtain an approximation for each function value.

Example $\tan 3.31$

Solution $\tan 3.31 \approx \tan(3.31 - 3.14)$

$$= \tan 0.17$$

$$\approx 0.1717$$

19. $\tan 4.52$ 20. $\tan 5.61$ 21. $\tan 7.45$

22. $\tan 8.30$ 23. $\tan 12.50$ 24. $\tan 27.30$

25. $\tan(-9.32)$ 26. $\tan(-6.41)$ 27. $\sec 6.92$

28. $\cot 8.31$ 29. $\csc(-7.09)$ 30. $\sec(-9.42)$

Use Table V to approximate the element x, $0 \le x \le 1.57$, in the domain of the function, for the specified element in the range. Interpolate as required.

31. $\tan x = 1.459$ 32. $\cot x = 0.6563$ 33. $\sec x = 2.448$

34. $\csc x = 1.010$ 35. $\cot x = 0.0208$ 36. $\tan x = 0.5000$

7.8

Graphs of other circular functions

The circular functions defined by $y = \tan x$, $y = \cot x$, $y = \sec x$, and $y = \csc x$ do not have sine waves for graphs in R^2, although each of these graphs does display periodic properties and all have certain other distinctive features.

Tangent   To graph $y = \tan x$, we first recall that its fundamental period is π rather than 2π. Next, from Table 7.3 we can locate the points shown in Figure 7.20 over the interval $0 \le x \le \pi$.

Because tan x is undefined for $x = \pi/2$, and $|\tan x|$ increases indefinitely as x approaches $\pi/2$, we expect no point on the graph corresponding to $x = \pi/2$. The line $x = \pi/2$ is an asymptote to the curve.

Assuming, as in the case of sine and cosine, that tangent is a continuous function wherever it is defined, and that tan x increases as x increases from 0 to $\pi/2$, and from $\pi/2$ to π, we can connect the points in Figure 7.20 with a smooth curve to obtain the graph over one period of the function. Repeating this pattern for a number of periods gives us the characteristic shape of the graph of the tangent function over its entire domain. Observe that the asymptotes to the curve in Figure 7.21 are the graphs of $x = \pi/2 + k\pi$, the zeros of the function defined by $y = \tan x$ are the real numbers $k\pi$, $k \in J$, and the range of tangent is R.

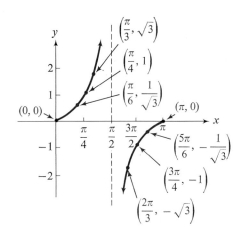

Figure 7.20

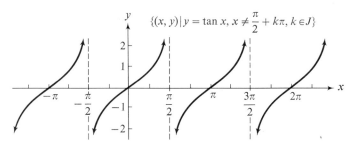

Figure 7.21

Reciprocal circular
functions

We can sketch the graphs of the other circular functions in a similar way, that is, by plotting some points over $0 < x < 2\pi$, and joining the points with a curve. These are shown in Figure 7.22 along with the graphs of their respective reciprocal functions.

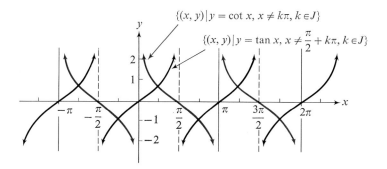

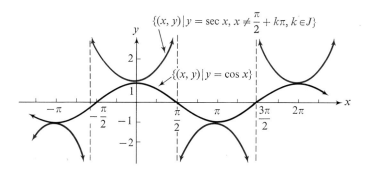

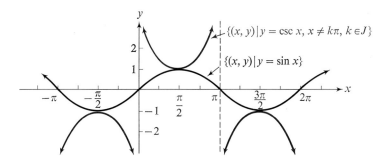

Figure 7.22

Exercise 7.8

Graph each of the following equations over the interval $-2\pi \le x \le 2\pi$.

1. $y = \csc 2x$

2. $y = \sec 2x$

3. $y = \cot 2x$

4. $y = \tan 2x$

5. $y = \tan \frac{1}{2}x$

6. $y = \cot \frac{1}{2}x$

7. $y = 2 \sec \frac{1}{2}x$

8. $y = \frac{1}{2}\cot 2x$

9. $y = -\tan x$

10. $y = -\cot x$

11. $y = -\sec x$

12. $y = -\csc x$

13. $y = -\frac{1}{2}\cot x$

14. $y = -2 \tan x$

15. $y = \tan\left(x + \frac{\pi}{3}\right)$

16. $y = \cot\left(x - \frac{\pi}{3}\right)$

17. $y = \sec\left(x - \frac{\pi}{4}\right)$

18. $y = \csc\left(x + \frac{\pi}{4}\right)$

7.9
Identities

The fact that all of the circular functions are defined either in terms of the unit circle or in terms of other circular functions suggests that these functions are interrelated in a number of ways. These relationships are called **identities**, since they are true for every permissible replacement of any variables involved. Thus, sine and cosine are related by the identity

$$\sin^2 x + \cos^2 x = 1,$$

as we observed earlier. Also, the definitions of tan x, cot x, sec x, and csc x are identities. Thus, the defining relations,

$$\tan x = \frac{\sin x}{\cos x}, \quad \sec x = \frac{1}{\cos x}, \quad \cot x = \frac{\cos x}{\sin x}, \quad \text{and} \quad \csc x = \frac{1}{\sin x},$$

are true for all real numbers x for which the denominators in the right-hand members are not zero.

Proofs of   These four identities can be used to prove still other identities.
identities

Example   Show that

$$\tan^2 x + 1 = \sec^2 x$$

is an identity.

Solution   We can "prove an identity" by showing that the given equation is equivalent to a true statement or to another equation known to be an identity. Since, for every real number x such that $x \neq \pi/2 + k\pi$, $k \in J$, we have

$$\tan x = \frac{\sin x}{\cos x} \quad \text{and} \quad \sec x = \frac{1}{\cos x},$$

we can substitute appropriately in the given equation and obtain

$$\frac{\sin^2 x}{\cos^2 x} + 1 = \frac{1}{\cos^2 x}.$$

Because $\cos^2 x \neq 0$ for any x for which $\tan x$ and $\sec x$ are defined, we can multiply each member here by $\cos^2 x$ to obtain for all such x the equivalent equation

$$\sin^2 x + \cos^2 x = 1.$$

This is a known identity, and hence the given equation is also an identity.

When we prove an identity, we are actually proving a theorem. However, we shall present such relationships simply as exercises. We can prove similarly, for example, that

$$\cot^2 x + 1 = \csc^2 x$$

is also an identity.

In proving identities, it is sometimes more convenient to restrict manipulation to one member of the given equation, and sometimes more convenient to work with both members. In the foregoing example, both members were multiplied by $\cos^2 x$ to prove the identity. The following example restricts the transformations to the left-hand member.

Example   Show that

$$\frac{\sin x}{1 - \cos x} - \cot x = \frac{1}{\sin x} \tag{1}$$

is an identity.

Solution
In proving identities, it is often helpful to rewrite a given equation in terms of $\sin x$ and $\cos x$ only. Writing $\cos x/\sin x$ for $\cot x$, we have

$$\frac{\sin x}{1-\cos x} - \frac{\cos x}{\sin x} = \frac{1}{\sin x}.$$

Writing the left-hand member as a single fraction, we have

$$\frac{\sin^2 x - (1 - \cos x)\cos x}{(1 - \cos x)\sin x} = \frac{1}{\sin x},$$

from which

$$\frac{\sin^2 x - \cos x + \cos^2 x}{(1 - \cos x)\sin x} = \frac{1}{\sin x},$$

$$\frac{(\sin^2 x + \cos^2 x) - \cos x}{(1 - \cos x)\sin x} = \frac{1}{\sin x},$$

$$\frac{1 - \cos x}{(1 - \cos x)\sin x} = \frac{1}{\sin x}.$$

Since $1 - \cos x$ is restricted from 0 in (1) above, we can rewrite the left-hand member here and arrive at the equivalent equation

$$\frac{1}{\sin x} = \frac{1}{\sin x},$$

which is clearly an identity, and the demonstration is complete.

The sum and difference formulas, as well as the formulas derived from these, are often involved in identities.

Example
Show that

$$\frac{\sin 2x}{1 + \cos 2x} = \tan x$$

is an identity

Solution
Replacing $\sin 2x$ and $\cos 2x$ in the given equation with $2 \sin x \cos x$ and $2 \cos^2 x - 1$, respectively, we have

$$\frac{2 \sin x \cos x}{1 + (2 \cos^2 x - 1)} = \tan x,$$

$$\frac{2 \sin x \cos x}{2 \cos^2 x} = \tan x,$$

$$\frac{\sin x \cos x}{\cos^2 x} = \tan x.$$

With the restriction that $\cos x \neq 0$, the left-hand member of this equation can be written $\dfrac{\sin x}{\cos x}$, and we have

$$\frac{\sin x}{\cos x} = \tan x,$$

which is true by definition.

For convenience, let us list again some important identities we have encountered. While no restrictions are given for variables here, it is important to keep such restrictions in mind when using any identity.

Summary of identities

1. $\sin^2 x + \cos^2 x = 1$ 2. $\cos(-x) = \cos x$ 3. $\sin(-x) = -\sin x$

4. $\cos(x_1 + x_2) = \cos x_1 \cos x_2 - \sin x_1 \sin x_2$

5. $\cos(x_1 - x_2) = \cos x_1 \cos x_2 + \sin x_1 \sin x_2$

6. $\sin(x_1 + x_2) = \sin x_1 \cos x_2 + \cos x_1 \sin x_2$

7. $\sin(x_1 - x_2) = \sin x_1 \cos x_2 - \cos x_1 \sin x_2$

8. a. $\cos 2x = \cos^2 x - \sin^2 x$ b. $\cos 2x = 2\cos^2 x - 1$

   c. $\cos 2x = 1 - 2\sin^2 x$

9. $\sin 2x = 2 \sin x \cos x$

10. $\cos \dfrac{x}{2} = \pm \sqrt{\dfrac{1 + \cos x}{2}}$ 11. $\sin \dfrac{x}{2} = \pm \sqrt{\dfrac{1 - \cos x}{2}}$

12. $\tan x = \dfrac{\sin x}{\cos x}$ 13. $\tan(-x) = -\tan x$

14. $\tan^2 x + 1 = \sec^2 x$ 15. $\cot^2 x + 1 = \csc^2 x$

16. $\sec x = \dfrac{1}{\cos x}$ 17. $\csc x = \dfrac{1}{\sin x}$

18. $\cot x = \dfrac{\cos x}{\sin x}$ 19. $\cot x = \dfrac{1}{\tan x}$

20. $\tan(x_1 + x_2) = \dfrac{\tan x_1 + \tan x_2}{1 - \tan x_1 \tan x_2}$ 21. $\tan(x_1 - x_2) = \dfrac{\tan x_1 - \tan x_2}{1 + \tan x_1 \tan x_2}$

22. $\tan 2x = \dfrac{2 \tan x}{1 - \tan^2 x}$ 23. $\tan \dfrac{x}{2} = \dfrac{1 - \cos x}{\sin x}$ 24. $\tan \dfrac{x}{2} = \dfrac{\sin x}{1 + \cos x}$

Exercise 7.9

Each of the following expressions may be written as a circular function of kx, or $k(x_1 \pm x_2)$, where k is a positive integer, with one or at most two steps. Write the answer directly. Assume that x, x_1, and x_2 take on no value for which a denominator vanishes.

Examples	a. $\cos(-x)$	b. $2 \sin x \cos x$	c. $2 \cos^2 4x - 1$
Solutions	a. $\cos x$	b. $\sin 2x$	c. $\cos 2(4x)$
			$\cos 8x$

1. $\tan(-x)$ 2. $-\sin(-x)$

3. $\dfrac{1}{\cot x}$ 4. $1 - 2 \sin^2 x$

5. $\dfrac{\tan x_1 + \tan x_2}{1 - \tan x_1 \tan x_2}$ 6. $\dfrac{2 \tan x}{1 - \tan^2 x}$

7. $\cos^2 x - \sin^2 x$ 8. $\cos^2 3x - \sin^2 3x$

9. $\sin x_1 \cos x_2 - \cos x_1 \sin x_2$ 10. $\sin 5x \cos 3x + \cos 5x \sin 3x$

11. $\dfrac{2 \tan 3x}{1 - \tan^2 3x}$ 12. $\tan^2 x + 1$

13. $1 - \cos^2 x$ 14. $1 - \sec^2 x$

15. $\cos(x_1 + x_2) \cos(x_1 - x_2) - \sin(x_1 + x_2) \sin(x_1 - x_2)$

16. $\sin(x_1 - x_2) \cos(x_1 + x_2) + \cos(x_1 - x_2) \sin(x_1 + x_2)$

Transform each first expression and show that it is identical to the second expression. (Assume suitable restrictions on x in each case.)

17. $\cos x \tan x$; $\sin x$ 18. $\sin x \sec x$; $\tan x$

19. $\sin^2 x \cot^2 x$; $\cos^2 x$ 20. $\dfrac{\cos^2 x}{\cot^2 x}$; $\sin^2 x$

21. $\cos^2 x (1 + \tan^2 x)$; 1 22. $(\csc^2 x - 1)$; $\cot^2 x$

23. $\sec x \csc x - \cot x$; $\tan x$ 24. $(1 - \cos^2 x)(1 + \cot^2 x)$; 1

25. $\dfrac{\sin x \sec x}{\tan x}$; 1 26. $\cos x \tan x \csc x$; 1

27. $(\sec^2 x - 1)(\csc^2 x - 1)$; 1 28. $\dfrac{\cos x - \sin x}{\cos x}$; $1 - \tan x$

29. $\dfrac{1}{1+\sin x} + \dfrac{1}{1-\sin x}$; $\quad 2\sec^2 x$
30. $\dfrac{1+\tan^2 x}{\csc^2 x}$; $\quad \tan^2 x$

Verify that the formulas of Problems 31–46 are identities.

31. $\sin x \cot x = \cos x$
32. $\tan x \csc x = \sec x$

33. $\sec x - \cos x = \sin x \tan x$
34. $\sec x - \sin x \tan x = \cos x$

35. $\dfrac{1+\tan^2 x}{\tan^2 x} = \csc^2 x$
36. $\dfrac{\sin^2 x}{1-\cos x} = \dfrac{1+\sec x}{\sec x}$

37. $\tan^2 x - \sin^2 x = \sin^2 x \tan^2 x$
38. $\cot^2 x + \sec^2 x = \tan^2 x + \csc^2 x$

39. $\tan x + \sec x = \dfrac{1}{\sec x - \tan x}$
40. $\dfrac{\sec x + \csc x}{1 + \tan x} = \csc x$

41. $\sin 2x = \dfrac{2\tan x}{1+\tan^2 x}$
42. $\dfrac{2}{\sin 2x} = \tan x + \cot x$

43. $\cot x - \cot 2x = \csc 2x$
44. $\dfrac{2}{1+\cos 2x} = \sec^2 x$

45. $\dfrac{1+\cos 2x}{\sin 2x} = \cot x$
46. $\cos 2x = \dfrac{1-\tan^2 x}{1+\tan^2 x}$

47. Express $\sin^2 x$ in terms of $\cos 2x$.

48. Express $\cos^2 x$ in terms of $\cos 2x$.

8

Trigonometric functions

8.1

Angles and their measure

In studying geometry, we learn that an angle is the union of two rays with a common endpoint (Figure 8.1-a) and that an angle can be designated by naming a point on each ray together with the common endpoint of the rays, or else by simply assigning a single symbol, say α, to the angle. The common endpoint of the rays is called the **vertex** of the angle, and the rays are called the **sides** of the angle.

Now each angle in the plane is congruent ($\cong$) to an angle with one side along the positive x-axis and vertex at the origin (Figure 8.1-b). Such an angle is said to be in **standard position**. The side $\overrightarrow{OC'}$ of $\angle A'OC'$ in Figure 8.1-b is called the **initial side** of the angle, the side $\overrightarrow{OA'}$ is called the **terminal side**, and the angle can be visualized as being formed by a rotation from the initial side $\overrightarrow{OC'}$ into the terminal side $\overrightarrow{OA'}$. In trigonometry, the *amount of rotation* is considered along with the angle itself. If the terminal side of an angle in standard position lies in a given quadrant, we say that the angle is *in* that quadrant.

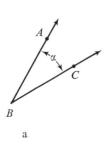

a

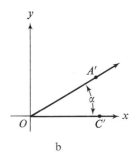

b

Figure 8.1

Angle measurement A measure is assigned to an angle by means of a circle with center at the vertex of the angle. For convenience, we shall restrict this discussion to the set of angles in standard position, although the process described is perfectly general. If the circumference of a circle of radius $r > 0$ is divided into p arcs of equal length, then each of the arcs will have length $2\pi r/p$. Starting at the point $(r, 0)$, we can scale the circumference of the circle in both the counterclockwise and clockwise directions from this point, in terms of this arc length as a unit. In doing this, it is customary to assign *negative* numbers in the *clockwise* direction, and *positive* numbers in the *counterclockwise* direction. As an example, consider the circle in Figure 8.2, where the circumference is divided into 16 equal parts and the arc length of each part is $2\pi r/16$. The terminal side of any angle α in standard position intercepts the circumference at a point, and one of the scale numbers (arc lengths) assigned to that point becomes the measure of the angle, depending on the amount of rotation involved.

Note that although angles such as α_1, α_2, and α_3 in Figure 8.3 have the same initial side and the same terminal side, their measures are different because the amounts of rotation involved are different. Such angles are called **coterminal**.

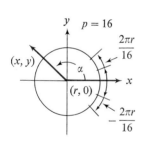

Figure 8.2   Figure 8.3

Recall from geometry that the lengths of arc intercepted by a given central angle on concentric circles are proportional to the circumferences of the circles. *It follows that the measure assigned to an angle in terms of a given unit is independent of the radius of the measuring circle.*

The two most commonly encountered units of angle measure are the **degree** and the **radian**. In degree measure, the circumference of the circle is divided into 360 arcs of equal length, and hence the unit arc is of length

$$\frac{2\pi r}{360} = \frac{\pi}{180}r,$$

where r is the radius of the circle. For this measure, we use the notation $m°(\alpha)$.

In radian measure, the circumference is divided into 2π arcs of equal length, and hence the length of the unit arc is

$$\frac{2\pi r}{2\pi} = r.$$

For this measure we use the notation $m^R(\alpha)$. Thus, in radian measure, the length of the unit arc is the length of the radius of the circle, and the circumference contains just 2π of these units. In particular, if the unit circle is used as the basis for $m^R(\alpha)$, then the length of the unit arc is 1. For this reason, the unit circle is customarily used as a reference for the measure of an angle in radians.

Comparison of degree and radian measure   Because the measures of a given angle in different units are proportional to the circumference of the measuring circle in these same units, we have

$$\frac{m°(\alpha)}{360} = \frac{m^R(\alpha)}{2\pi},$$

so that

$$m°(\alpha) = \frac{180}{\pi} m^R(\alpha) \tag{1}$$

and

$$m^R(\alpha) = \frac{\pi}{180} m°(\alpha). \tag{2}$$

Equations (1) and (2) are **conversion formulas** for expressing the relationships between degrees and radians.

Symbolism for angles and their measure   Note that $m°$ and m^R are functions, each with the set of angles as domain and R as range. Several conventions regarding the use of symbolism for angles and their measure are customarily made in mathematics. For one thing, to express a relation such as $m°(\alpha) = 40$ or $m^R(\beta) = \pi/4$, we usually write $\alpha = 40°$ or $\beta = (\pi/4)^R$. Thus, for example, if α, β, and γ are the angles of a triangle, then

$$m^R(\alpha) + m^R(\beta) + m^R(\gamma) = \pi,$$

and this equation is generally abbreviated

$$\alpha + \beta + \gamma = \pi^R.$$

This notation is consistent with the fact that if α and β are two angles, and $\alpha + \beta$ denotes their **sum**, then

$$m^R(\alpha + \beta) = m^R(\alpha) + m^R(\beta).$$

Here the sum $\alpha + \beta$ is construed to be the result of taking the terminal side of α as the initial side of β, with the same point for vertex, and then viewing $\alpha + \beta$ as the angle with initial side that of α and terminal side that of β, as in Figure 8.4.

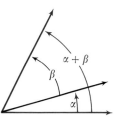

Following another convention, we sometimes write, for example,

$$m^\circ(\alpha) = 60^\circ, \quad \text{or} \quad m^R(\alpha) = \frac{\pi}{3}^R.$$

Figure 8.4

Here, the appearance of the degree or radian symbol in the right-hand member of the equation should be interpreted as a clarifying redundancy.

Examples
     Find the degree measure, to the nearest tenth of a degree, of the angle whose radian measure is given.

     a. $\dfrac{\pi}{3}^R$ b. 0.62^R

Solutions
     a. $\dfrac{\pi}{3}^R = \left(\dfrac{180}{\pi} \cdot \dfrac{\pi}{3}\right)^\circ = 60^\circ$

     b. $0.62^R = \left(\dfrac{180}{\pi} \cdot 0.62\right)^\circ = \dfrac{111.6^\circ}{\pi} \approx 35.5^\circ$

Examples
     Find the radian measure, to the nearest hundredth of a radian, of the angle whose degree measure is given.

     a. 30° b. 300°

Solutions
     a. $30^\circ = \left(\dfrac{\pi}{180} \cdot 30\right)^R = \dfrac{\pi}{6}^R \approx 0.52^R$

     b. $300^\circ = \left(\dfrac{\pi}{180} \cdot 300\right)^R = \dfrac{5\pi}{3}^R \approx 5.24^R$

Note that, in these examples, the number of units in one measure does not equal the number of units in another measure; for example, $30 \neq \pi/6$. But an angle whose measure is 30° is congruent to an angle whose measure is $(\pi/6)^R$, and we express this fact by writing $30° = (\pi/6)^R$. This is analogous to 36 inches = 3 feet, 6 feet = 2 yards, and so forth, when indicating lengths of line segments in different units.

If the measure of an angle is given in radians, then the length s of the intercepted arc of a circle with given radius r can be found directly. That is,

$$s = r \cdot m^R(\alpha). \tag{3}$$

If the measure of an angle is given in degrees, it can first be changed to radian measure and then the length of the intercepted arc can be found directly.

Exercise 8.1

Find the degree measure of the angle whose radian measure is as given.

1. a. 0^R b. $\dfrac{\pi^R}{2}$ c. π^R d. $\dfrac{3\pi^R}{2}$ e. $2\pi^R$

2. By filling in the blank spaces, complete the following table, which compares the radian measure and degree measure of angles that are of frequent occurrence.

$m°(\alpha)$	30°	45°		120°	135°	
$m^R(\alpha)$			$\dfrac{\pi^R}{3}$			$\dfrac{5\pi^R}{6}$

$m°(\alpha)$	210°				315°	330°
$m^R(\alpha)$		$\dfrac{5\pi^R}{4}$	$\dfrac{4\pi^R}{3}$	$\dfrac{5\pi^R}{3}$		

Find the degree measure, to the nearest tenth of a degree, of the angle whose radian measure is as given.

3. $\dfrac{2\pi^R}{9}$ 4. $\dfrac{3\pi^R}{5}$ 5. $\dfrac{7\pi^R}{5}$ 6. $\dfrac{5\pi^R}{8}$

7. 0.30^R 8. 1.25^R 9. 3.62^R 10. 9.14^R

Find the radian measure, to the nearest hundredth of a radian, of the angle whose degree measure is as given.

11. 20° 12. 50° 13. 130° 14. 310°

15 420° 16. 580° 17. 750° 18. 800°

19. What is the degree measure, to the nearest hundredth of a degree, of an angle whose radian measure is 1^R?

20. What is the radian measure, to the nearest thousandth of a radian, of an angle whose degree measure is 1°?

Find all angles α satisfying $-360° \leq \alpha \leq 720°$ that are coterminal with the angle whose measure is given. Write a representation for the measures of all *angles coterminal with the angle.*

Examples a. 24° b. $-184°$

Solutions a. $24° + 360° = 384°$ b. $-184° + 360° = 176°$

   $24° - 360° = -336°$ $-184° + 720° = 536°$

   $24° + 360°k, \ k \in J$ $-184° + 360°k, \ k \in J$

21. 30° 22. $-150°$ 23. $-240°$ 24. 190°

25. 420° 26. 683° 27. $-330°$ 28. $-271°$

On a circle with given radius, find the length of the arc intercepted by the angle whose measure is as given.

Example $r = 3''; \quad m°(\alpha) = 120$

Solution The measure of the angle is first changed to radian measure. Thus

$$m^R(\alpha) = \left(\frac{\pi}{180} \cdot 120\right)^R = \frac{2\pi^R}{3}.$$

Then, by Equation (3) on page 236,

$$s = r \cdot m^R(\alpha) = 3 \cdot \frac{2\pi}{3} = 2\pi \approx 6.28.$$

Therefore, the desired arc length is approximately 6.28″.

29. $r = 4''$; $(\pi/6)^R$ 30. $r = 5.2''$; π^R 31. $r = 1.2'$; 0.60^R

32. $r = 3.6'$; 1^R 33. $r = 2''$; $135°$ 34. $r = 4.3''$; $180°$

8.2

Functions of angles

Historically, interest in the circular functions arose from the study of angles and triangles. In defining the circular functions, we used the notion of arc length on a unit circle to associate a real number [the arc length from the point $(1, 0)$ to a point on the circle] with another real number, the first or second coordinate of the point. Thus we defined functions with real numbers for *both* domain and range.

Geometric ratios in the coordinate plane   Now, consider any angle α in standard position and let (x_1, y_1) and (x_2, y_2) be any two distinct points (except the origin) in its terminal side or ray (Figure 8.5). We can show that the following equalities of ratios hold:

$$\frac{x_1}{y_1} = \frac{x_2}{y_2} \quad (y_1, y_2 \neq 0),$$

$$\frac{x_1}{\sqrt{x_1^2 + y_1^2}} = \frac{x_2}{\sqrt{x_2^2 + y_2^2}},$$

and

$$\frac{y_1}{\sqrt{x_1^2 + y_1^2}} = \frac{y_2}{\sqrt{x_2^2 + y_2^2}}.$$

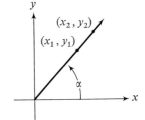

Figure 8.5

The proofs of these are left as exercises.
    Although the figure shows the terminal side of α in the first quadrant, the ratios are equal for coordinates of points on the terminal side of an angle in any quadrant. Because these ratios do not depend on the choice of a point in the terminal side of α, we can use them to define some new functions, each with *the set of angles in standard position* as domain, and with sets of real numbers as ranges. Traditionally, six such assignments are made, and the names given the resulting functions are those we have used to name the circular functions.

Definition 8.1 *If α is an angle in standard position and if $(x, y) \neq (0, 0)$ is any point on the terminal side of α, then*

I $\qquad$ cosine $= \left\{ (\alpha, \cos \alpha) \,\middle|\, \cos \alpha = \dfrac{x}{\sqrt{x^2 + y^2}} \right\}$,

II $\qquad$ sine $= \left\{ (\alpha, \sin \alpha) \,\middle|\, \sin \alpha = \dfrac{y}{\sqrt{x^2 + y^2}} \right\}$,

III $\qquad$ tangent $= \left\{ (\alpha, \tan \alpha) \,\middle|\, \tan \alpha = \dfrac{y}{x}, \quad x \neq 0 \right\}$,

IV $\quad$ cotangent $= \left\{ (\alpha, \cot \alpha) \,\middle|\, \cot \alpha = \dfrac{x}{y}, \quad y \neq 0 \right\}$,

V $\qquad$ secant $= \left\{ (\alpha, \sec \alpha) \,\middle|\, \sec \alpha = \dfrac{\sqrt{x^2 + y^2}}{x}, \quad x \neq 0 \right\}$,

VI $\qquad$ cosecant $= \left\{ (\alpha, \csc \alpha) \,\middle|\, \csc \alpha = \dfrac{\sqrt{x^2 + y^2}}{y}, \quad y \neq 0 \right\}$.

These functions are called **trigonometric**. Observe that in each place where the expression $\sqrt{x^2 + y^2}$ occurs, the *positive* root is used.

Example   Find the value of each of the six trigonometric functions of α, if the terminal side of α contains the point $(-3, 5)$.

Solution   By Definition 8.1,

$$\cos \alpha = \frac{x}{\sqrt{x^2 + y^2}} \qquad\qquad \sec \alpha = \frac{\sqrt{x^2 + y^2}}{x}$$

$$= \frac{-3}{\sqrt{9 + 25}} = -\frac{3}{\sqrt{34}}, \qquad = \frac{\sqrt{9 + 25}}{-3} = -\frac{\sqrt{34}}{3},$$

$$\sin \alpha = \frac{y}{\sqrt{x^2 + y^2}} \qquad\qquad \csc \alpha = \frac{\sqrt{x^2 + y^2}}{y}$$

$$= \frac{5}{\sqrt{9 + 25}} = \frac{5}{\sqrt{34}}, \qquad = \frac{\sqrt{9 + 25}}{5} = \frac{\sqrt{34}}{5},$$

$$\tan \alpha = \frac{y}{x} = \frac{5}{-3} = -\frac{5}{3}, \qquad \cot \alpha = \frac{x}{y} = \frac{-3}{5} = -\frac{3}{5}.$$

Because every angle in the plane is congruent to an angle in standard position, the definitions of the trigonometric functions can be extended to assign the same numbers to every angle congruent to a given angle α. Thus, while these functions are defined in terms of angles in standard position, they can be viewed as applying to the set of all angles in the plane. Moreover, since congruent angles have the same measure, we can identify angles in the domain of each function with a particular unit of measure. Thus, we write

$$\sin 30° \quad \text{and} \quad \sin \frac{\pi^R}{6}$$

as abbreviations for "the sine of an angle whose measure is 30 degrees" and "the sine of an angle whose measure is $\pi/6$ radians," respectively.

Relationship between circular and trigonometric functions   The fact that the circular functions are defined using the unit circle, together with the fact that the unit circle can be used to assign measures to angles, makes it reasonable to expect a very close relationship to exist between the circular and trigonometric functions. Such is indeed the case.

Since the trigonometric functions of an angle α have been defined in terms of the coordinates of *any* point other than the origin on the terminal side of α, we can arbitrarily choose the point $P(\cos x, \sin x)$ where the terminal side of the angle intersects the unit circle. Thus,

$$\cos \alpha = \frac{\cos x}{1} = \cos x,$$

and, similarly,

$$\sin \alpha = \frac{\sin x}{1} = \sin x$$

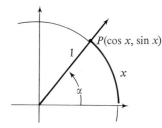

Figure 8.6

(Figure 8.6). From this, we see that the elements in the ranges of the trigonometric functions are equal to the corresponding elements in the ranges of the analogous circular functions. Thus, if we denote any of the six trigonometric functions by T, and the circular functions of the same name by C, then

$$T(\alpha) = C(x), \tag{1}$$

where x is the length of the arc intercepted by α on the unit circle. Notice that elements x in the domain of C are *real numbers,* while elements α in the domain of T are *angles.* The elements in the range of each function are real numbers.

In Section 8.1 we agreed to name an angle by its measure. In particular, if $m^R(\alpha) = x$, then we write $\alpha = x^R$, and if $m°(\alpha) = t$, then we write $\alpha = t°$. Thus by replacing α by x^R and $t°$ in (1), we obtain

$$T(x^R) = C(x) \quad \text{and} \quad T(t°) = C(x).$$

For example,

$$\cos \frac{\pi^R}{3} = \cos \frac{\pi}{3} \quad \text{and} \quad \cos 60° = \cos \frac{\pi}{3},$$

where in each equation the left-hand member is an element in the range of the *trigonometric* function and the right-hand member is an element in the range of the analogous *circular* function.

Using relationship (1), we obtain the entries of Table 8.1 directly from Table 7.2, page 192, Definition 7.4 and the relationship between degree and radian measures of angles. These are left for you to verify.

Table 8.1

$m°(\alpha)$	$m^R(\alpha)$	$\sin \alpha$	$\csc \alpha$	$\cos \alpha$	$\sec \alpha$	$\tan \alpha$	$\cot \alpha$
$0°$	0^R	0	not defined	1	1	0	not defined
$30°$	$\dfrac{\pi^R}{6}$	$\dfrac{1}{2}$	2	$\dfrac{\sqrt{3}}{2}$	$\dfrac{2}{\sqrt{3}}$	$\dfrac{1}{\sqrt{3}}$	$\sqrt{3}$
$45°$	$\dfrac{\pi^R}{4}$	$\dfrac{1}{\sqrt{2}}$	$\sqrt{2}$	$\dfrac{1}{\sqrt{2}}$	$\sqrt{2}$	1	1
$60°$	$\dfrac{\pi^R}{3}$	$\dfrac{\sqrt{3}}{2}$	$\dfrac{2}{\sqrt{3}}$	$\dfrac{1}{2}$	2	$\sqrt{3}$	$\dfrac{1}{\sqrt{3}}$
$90°$	$\dfrac{\pi^R}{2}$	1	1	0	not defined	not defined	0
$180°$	π^R	0	not defined	-1	-1	0	not defined
$270°$	$\dfrac{3\pi^R}{2}$	-1	-1	0	not defined	not defined	0

Examples   a. $\cos 45° = \dfrac{1}{\sqrt{2}}$ b. $\tan \pi^R = 0$

Identities for
trigonometric
functions
Since, as observed on page 235,

$$m^R(\alpha + \beta) = m^R(\alpha) + m^R(\beta)$$

for any angles α and β, all of the sum and difference formulas are immediately valid for sums and differences of angles. Thus, for any α and β,

$$\cos (\alpha + \beta) = \cos \alpha \cos \beta - \sin \alpha \sin \beta,$$
$$\cos (\alpha - \beta) = \cos \alpha \cos \beta + \sin \alpha \sin \beta,$$
$$\sin (\alpha + \beta) = \sin \alpha \cos \beta + \cos \alpha \sin \beta,$$
$$\sin (\alpha - \beta) = \sin \alpha \cos \beta - \cos \alpha \sin \beta.$$

If half of an angle is interpreted to mean an angle formed by the initial side of the angle together with the bisector of the angle (Figure 8.7), then it is also true that

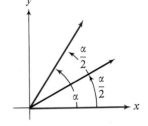

$$m^R(\tfrac{1}{2}\alpha) = \tfrac{1}{2}m^R(\alpha),$$

and the formulas developed in Sections 7.6 and 7.7 are also valid for the trigonometric functions.

Figure 8.7

In summary, the correspondence between the values of the circular functions and those of the trigonometric functions,

$$T(\alpha) = C(x), \tag{1}$$

renders all of the relationships thus far studied for circular functions equally applicable to the trigonometric functions. Because the elements in the domains of the trigonometric functions are angles, an identity such as

$$\sin 2\alpha = 2 \sin \alpha \cos \alpha$$

is sometimes called a **double-angle formula**, while an identity such as

$$\sin \frac{\alpha}{2} = \pm \sqrt{\frac{1 - \cos \alpha}{2}}$$

is called a **half-angle formula**.

Tables for
trigonometric
functions
Trigonometric function values that are not listed in Table 8.1 can be obtained from Table V in the Appendix if $m^R(\alpha)$ is known. Table VI in the Appendix gives trigonometric function values of angles with given degree measure. The table is graduated in intervals of 10 minutes (10′), one minute being equal to one-sixtieth of a degree. We can interpolate as necessary to find, to four significant figures, values between those listed in the table. Observe that the table reads from top to

bottom for $0° \leq \alpha \leq 45°$, where the function values are identified at the top of the page, and from bottom to top for $45° \leq \alpha \leq 90°$, where the function values are identified at the bottom of the page.

Example   Find tan 149°.

Solution   From the reduction formula

$$\tan \alpha = -\tan (180° - \alpha),$$

we have

$$\tan 149° = -\tan (180° - 149°)$$

$$= -\tan 31°.$$

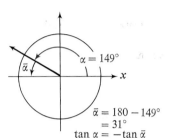

$\bar{\alpha} = 180 - 149°$
$\quad = 31°$
$\tan \alpha = -\tan \bar{\alpha}$

From Table VI in the Appendix, we find that $\tan 31° \approx 0.6009$. Thus

$$\tan 149° \approx -0.6009.$$

Observe that in the preceding example we again used the symbol $\approx$ because the function values obtained from the table are approximations to irrational numbers. Note also how a reference angle can be used in the same way that we used a reference arc in Chapter 7 to provide a geometric interpretation of a reduction formula.

Exercise 8.2

Find the value of each of the six trigonometric functions of α for the terminal side of α containing the given point.

1. $(3, 4)$ 2. $(5, -12)$ 3. $(\sqrt{2}, -\sqrt{2})$ 4. $(-1, \sqrt{3})$

5. $(-6, -8)$ 6. $(-4, -3)$ 7. $(0, 3)$ 8. $(-7, 0)$

Determine in which quadrant the terminal side of each angle lies.

Example $\sin \alpha > 0, \cos \alpha < 0$

Solution   Since $\sin \alpha > 0$, the terminal side of α lies in Quadrant I or II; since $\cos \alpha < 0$, the terminal side of α lies in Quadrant II or III. Therefore, the terminal side of α lies in Quadrant II, the intersection of {I, II} and {II, III}.

9. $\sin \alpha < 0$, $\cos \alpha > 0$ 10. $\cos \alpha > 0$, $\tan \alpha < 0$

11. $\sec \alpha > 0$, $\sin \alpha > 0$ 12. $\tan \alpha < 0$, $\sin \alpha > 0$

13. $\sec \alpha > 0$, $\csc \alpha < 0$ 14. $\cot \alpha > 0$, $\sin \alpha < 0$

Use Table 8.1 and reduction formulas to find the value of each expression.

Example $\cos 150°$

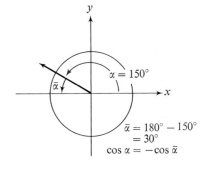

Solution   From the reduction formula,
$\cos \alpha = -\cos (180° - \alpha)$, of
Theorem 7.5-I, we have

$$\cos 150° = -\cos (180 - 150)°$$

$$= -\cos 30°$$

$$= -\frac{\sqrt{3}}{2}.$$

15. $\tan 330°$ 16. $\cot 120°$ 17. $\cos 225°$

18. $\tan 150°$ 19. $\sin 330°$ 20. $\cos 240°$

21. $\sin (-30°)$ 22. $\tan (-60°)$ 23. $\sin (-240°)$

24. $\cos (-135°)$ 25. $\tan (-300°)$ 26. $\tan (-750°)$

Use Table V or Table VI, as appropriate, to find function values to four significant figures for each expression.

27. $\sin 32°$ 28. $\cos 49°$ 29. $\tan 33°$ 30. $\cot 51°$

31. $\sec 17°$ 32. $\csc 84°$ 33. $\cos 0.63^R$ 34. $\sin 0.42^R$

35. $\cot 1.42^R$ 36. $\tan 1.03^R$ 37. $\csc \dfrac{5\pi^R}{9}$ 38. $\sec \dfrac{5\pi^R}{12}$

39. $\sin 132°$ 40. $\cos 153°$ 41. $\tan 320°$ 42. $\sin 312°$

43. $\cos (-130°)$ 44. $\tan (-605°)$ 45. $\sin 2.07^R$ 46. $\cos 3.51^R$

47. $\tan 4.21^R$ 48. $\tan 6.00^R$ 49. $\cos (-1.63^R)$ 50. $\sin (-12.32^R)$

51. Show (geometrically) that if (x_1, y_1) and (x_2, y_2) are the coordinates of any two points (except the origin) on the terminal side (ray) of an angle in the first quadrant, then the following equalities of ratios hold:

$$\frac{x_1}{y_1} = \frac{x_2}{y_2} (y_1, y_2 \neq 0), \quad \frac{x_1}{\sqrt{x_1^2 + y_1^2}} = \frac{x_2}{\sqrt{x_2^2 + y_2^2}}, \quad \text{and} \quad \frac{y_1}{\sqrt{x_1^2 + y_1^2}} = \frac{y_2}{\sqrt{x_2^2 + y_2^2}}.$$

52. Under the same conditions as in Problem 51, explain how the equalities of these ratios hold for the coordinates of two points on the terminal side of an angle whose terminal side is located in the second, third, and fourth quadrants.

8.3

Right triangles

An examination of Figure 8.8-a makes it evident that in the first quadrant, any point (x, y) on the terminal side of an angle α determines a right triangle with sides

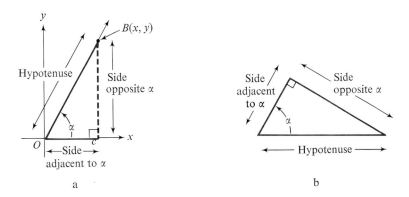

Figure 8.8

measuring x and y and with hypotenuse of length $\sqrt{x^2 + y^2}$. Therefore, as special cases of trigonometric function values, we can consider the following **trigonometric ratios** of the sides of a right triangle:

$$\sin \alpha = \frac{\text{length of side opposite } \alpha}{\text{length of hypotenuse}},$$

$$\cos \alpha = \frac{\text{length of side adjacent to } \alpha}{\text{length of hypotenuse}},$$

$$\tan \alpha = \frac{\text{length of side opposite } \alpha}{\text{length of side adjacent to } \alpha}.$$

Similar statements can be made about $\csc \alpha$, $\sec \alpha$, and $\cot \alpha$. It is not necessary that the right triangle be oriented in such a way that α is in standard position. The foregoing relations involving ratios of the lengths of the sides of a right triangle are equally applicable to a right triangle in any position, as suggested in Figure 8.8-b.

If we are given some parts of a triangle, that is, the measures of some angles and the lengths of some sides, and asked to find the remaining measures, we are asked to **solve** the triangle. In general, we shall give solutions to the nearest tenth of a unit for lengths and to the nearest 10′ for the degree measures of angles.

Example If one angle of a right triangle measures 47° and the side adjacent to this angle has length 24, solve the triangle and determine its area.

Solution We first make a sketch illustrating the situation. Generally the hypotenuse is labeled c, the legs a and b, and the angles opposite $a, b,$ and c are labeled α, β, and γ, respectively, or A, B, and C, respectively. In the present case, we note first that

$$\beta = 90° - \alpha = 90° - 47° = 43°.$$

Next we have

$$\tan 47° = \frac{a}{b} = \frac{a}{24}, \quad \text{or} \quad a = 24 \tan 47°,$$

and

$$\sec 47° = \frac{c}{b} = \frac{c}{24}, \quad \text{or} \quad c = 24 \sec 47°.$$

From Table VI in the Appendix, $\tan 47° \approx 1.072$ and $\sec 47° \approx 1.466$, so that

$$a \approx 24(1.072) \approx 25.7,$$

$$c \approx 24(1.466) \approx 35.2.$$

The area, which is equal to half the product of the length of the base by the length of the altitude, is given by

$$\mathcal{A} = \frac{1}{2} ab \approx \frac{1}{2}(25.7)(24) = 308.4.$$

Note that the ratio selected in each case in the foregoing example was one in which the length of the remaining side and the length of the hypotenuse appeared in the numerator. With such a selection of ratios, the operation in the computation is multiplication rather than division.

Logarithms can be used to perform the computations in the preceding example and in many of the problems that follow. Handbooks are available with the logarithms of the trigonometric ratios to various degrees of accuracy if you wish to use them, although devices such as the slide rule, desk calculator, and computer have largely replaced logarithmic tables for such computations.

Right triangles play a useful role in finding one trigonometric function value, given another.

Example   If $\tan \alpha = \dfrac{2}{3}$ and α is in Quadrant I, find $\sin \alpha$.

Solution   Sketch a right triangle and label the sides so that
$\tan \alpha$ is as given. By the Pythagorean theorem,
the hypotenuse of the triangle has length
$\sqrt{2^2 + 3^2} = \sqrt{13}$, so that $\sin \alpha = \dfrac{2}{\sqrt{13}}$.

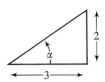

Note in this example that once the length of the hypotenuse is determined, *all* of the values for trigonometric functions of α are available by inspection.

Right triangles are helpful also in quadrants other than the first if care is taken to assign *directed distances* as lengths to the sides, while viewing the length of the hypotenuse as always positive. This is simply a variation on our procedures for finding function values using either reduction formulas or reference angles.

Example   If $\cot \alpha = \dfrac{3}{7}$, and $\sin \alpha < 0$, find $\cos \alpha$.

Solution   Since $\cot \alpha > 0$ while $\sin \alpha < 0$, α can lie only in Quadrant III. Sketch
a right triangle in Quadrant III and label the sides with directed
lengths so that $\cot \bar\alpha = \dfrac{3}{7}$. By the Pythagorean theorem,

length of $\overline{OA} = l(\overline{OA})$

$\qquad = \sqrt{(-3)^2 + (-7)^2}$

$\qquad = \sqrt{9 + 49} = \sqrt{58}.$

By inspection, then, $\cos \alpha = \dfrac{-3}{\sqrt{58}}.$

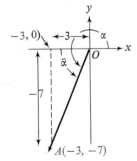

Exercise 8.3

In all problems in this exercise set, give lengths to the nearest tenth of a unit and angle measures to the nearest 10′.

Solve each right triangle ABC, and determine its area. Consider $C = 90°$.

1. $a = 6,\ b = 8$　　　　　　　　　　2. $c = 24,\ A = 32°$
3. $b = 120,\ B = 54°$　　　　　　　　4. $a = 3.5,\ B = 48°$

5. $c = 16$, $A = 22°$ 6. $a = 5$, $c = 16$

7. In the right triangle ABC, find a if $\sin A = 4/5$ and $c = 20$.

8. In the right triangle ABC, find b if $\tan A = 3/4$ and $a = 12$.

9. Find the other trigonometric function values of θ if $\sin \theta = -1/2$ and the terminal side of θ is in the third quadrant.

10. Find the other trigonometric function values of θ if $\tan \theta = -7/24$ and the terminal side of θ is in the fourth quadrant.

11. Find the other trigonometric function values of θ if $\cos \theta = -3/7$ and the terminal side of θ is in the second quadrant.

12. Find the other trigonometric function values of θ if $\sin \theta = 2/3$ and the terminal side of θ is in the second quadrant.

In each case, sketch the angle θ and an appropriate right triangle, and find the other trigonometric function values.

13. $\sin \theta = 3/5$, $\cos \theta < 0$ 14. $\cos \theta = -12/13$, $\sin \theta > 0$

15. $\tan \theta = 5/12$, $\sec \theta < 0$ 16. $\cot \theta = 4/3$, $\sin \theta < 0$

17. $\sec \theta = -2$, $\sin \theta < 0$ 18. $\csc \theta = 3$, $\sec \theta < 0$

19. Find the value of $\dfrac{\sin \theta + 2 \cos \theta - \tan \theta}{1 - \cot \theta + \sec \theta}$ if $\tan \theta = 1$ and the terminal side of θ is in the third quadrant.

20. Find the value of $\dfrac{\csc^2 \theta + \sin \theta - 1}{2 - \tan^2 \theta + \cot^2 \theta}$ if $\sin \theta = \dfrac{1}{2}$ and the terminal side of θ is in the second quadrant.

In Problems 21 and 22, given the information pertaining to the figure, find the length of the line segment denoted by x.

21. Given: $l(\overline{BA}) = 25$,

          $\alpha = 16°$,

          $\beta = 12°$,

          BC is a straight line.

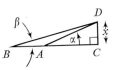

22. Given: $l(\overline{AB}) = 350$,

          $\alpha = 21°$,

          $\beta = 8°$,

          BD is a straight line.

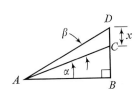

23. Find the area of a parallelogram if two of its adjacent sides are 22 and 28 inches in length and the measure of the included angle is 40°.

24. In a circle, the length of the radius is 10 inches and the length of a chord is 12 inches. Find the measure of the angle made by two lines tangent to the circle at the ends of the chord.

25. Find the length of the base of an isosceles triangle if the length of one of the equal sides is 12 inches and the measure of one of the equal angles is 40°.

26. Find the perimeter of a regular pentagon inscribed in a circle of radius 8 inches.

27. A rectangle is 80 feet long and 64 feet wide. Find the measures of the angles between a diagonal and the sides.

28. The sides of an isosceles triangle have lengths 6, 6, and 8. Find a measure for each angle in the triangle.

29. The angle of elevation (the angle between the line of sight and the horizontal) from a point 200 feet from the base of a building to the base of a flagpole on top of the building is 60°. The angle of elevation from the same spot to the top of the flagpole is 65°. How tall is the flagpole?

30. Each of two surveyors is located 200 feet from a flagpole. If the angle between the flagpole and one surveyor, when measured by the other surveyor, is 36°, how far apart are the two surveyors?

8.4

The law of sines

The fact that the area $\mathscr{A}$ of a triangle is equal to one-half the product of the length of its base and the length of its altitude gives us an immediate expression for its area in terms of the lengths of two sides of the triangle and the measure of the included angle. Thus, from Figure 8.9,

$$\mathscr{A} = \tfrac{1}{2}c(b \sin \alpha) = \tfrac{1}{2}bc \sin \alpha,$$

$$\mathscr{A} = \tfrac{1}{2}a(c \sin \beta) = \tfrac{1}{2}ac \sin \beta,$$

$$\mathscr{A} = \tfrac{1}{2}b(a \sin \gamma) = \tfrac{1}{2}ab \sin \gamma.$$

Because each triangle has only one area, we can equate the right-hand members of these equations to obtain

$$\frac{1}{2} bc \sin \alpha = \frac{1}{2} ac \sin \beta = \frac{1}{2} ab \sin \gamma.$$

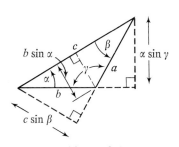

Figure 8.9

Multiplying each member here by $2/(abc)$, we obtain the following result, called the **law of sines**.

Theorem 8.1 *If α, β, and γ are the angles of a triangle and a is the length of the side opposite α, b is the length of the side opposite β, and c is the length of the side opposite γ, then*

$$\frac{\sin \alpha}{a} = \frac{\sin \beta}{b} = \frac{\sin \gamma}{c}.$$

Use of the law of sines in solving triangles The law of sines can be used to solve certain triangles. Let us first consider the case in which two angles and one side of a triangle are given.

Example Solve the triangle for which

$$\alpha = 45°, \quad \beta = 60°, \quad \text{and} \quad a = 10.$$

Solution It is helpful first to make a sketch. Then, from Theorem 8.1, we have

$$\frac{\sin 45°}{10} = \frac{\sin 60°}{b},$$

from which

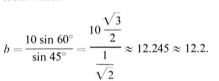

$$b = \frac{10 \sin 60°}{\sin 45°} = \frac{10 \dfrac{\sqrt{3}}{2}}{\dfrac{1}{\sqrt{2}}} \approx 12.245 \approx 12.2.$$

Next, we observe that

$$\gamma = 180° - \alpha - \beta = 180° - 45° - 60° = 75°.$$

From Theorem 8.1, we have

$$\frac{\sin 45°}{10} = \frac{\sin 75°}{c}.$$

Then, from Table VI,

$$c = \frac{10 \sin 75°}{\sin 45°} \approx 10 \left(\frac{0.9659}{0.7071}\right) \approx 10(1.36) = 13.6.$$

Thus, we have

$$b \approx 12.2, \quad c \approx 13.6, \quad \text{and} \quad \gamma = 75°.$$

Ambiguous case If the lengths of two sides of a triangle, say *a* and *b*, and the measure of an angle opposite one of them, say α, are given, we may encounter ambiguity, depending on the value of *a* in relation to those of b and α. In Figure 8.10, we hold b and α constant and observe the possible situations as a assumes different values.

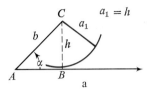

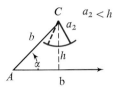

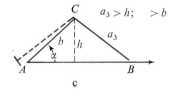

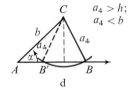

Figure 8.10

First we determine the length h of the altitude of a triangle with the given measures for the angle α and the adjacent side b. Since $\sin \alpha = h/b$,

$$h = b \sin \alpha.$$

Now consider the case (Figure 8.10-a) in which the length a of the given side satisfies the equation

$$a = b \sin \alpha.$$

We have a unique solution—a right triangle. Next consider the case (Figure 8.10-b) in which the length a of the given side satisfies the inequality

$$a < b \sin \alpha.$$

Since a is less than h, the given values are such that no triangle is possible. Next consider the case (Figure 8.10-c) in which the length a of the given side is such that

$$a > b \sin \alpha \quad \text{and} \quad a \geq b.$$

Here we have a unique solution. For the fourth and last possibility (Figure 8.10-d), in which the length a of the given side is such that

$$a > b \sin \alpha \quad \text{and} \quad a < b,$$

we have two triangles possible, $\triangle ABC$ and $\triangle AB'C$.

Of course, if the given angle α is obtuse (Figure 8.11), then any specified lengths of the sides and measures of the angles permit only two possibilities:

1. $a \leq b$, no triangle (as shown in Figure 8.11-a);
2. $a > b$, one oblique triangle (as shown in Figure 8.11-b).

a

b

Figure 8.11

Example

Solve the triangle for which $a = 4$, $b = 3$, and $\beta = 45°$.

Solution

We first sketch a figure and observe that this gives rise to the ambiguous case in which we are given the lengths of two sides and the measure of an angle opposite one of them. We first check for the number of possible solutions and observe that $b > a \sin \beta$ and $b < a$. Therefore, there are two triangles with the given measurements. Using Theorem 8.1, we next determine a value for $\sin \alpha$. We have

$$\frac{\sin \alpha}{4} = \frac{\sin 45°}{3},$$

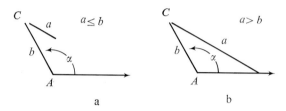

$$\sin \alpha = \frac{4}{3} \cdot \frac{\sqrt{2}}{2} = \frac{2}{3}\sqrt{2}$$

$$\approx \frac{2}{3}(1.414) \approx 0.9426.$$

Because $\sin \alpha > 0$ in Quadrants I and II, from Table VI we have either

$$\alpha \approx 70° \; 30' \quad \text{or} \quad \alpha' \approx 180° - 70° \; 30'$$
$$= 109° \; 30'.$$

If $\alpha \approx 70° \; 30'$, then $\gamma \approx 180° - 45° - 70° \; 30' = 64° \; 30'$, and we therefore have

$$\frac{\sin 45°}{3} \approx \frac{\sin 64° \; 30'}{c},$$

so that

$$c \approx 3\left(\frac{0.9026}{0.7071}\right) \approx 3.8.$$

If $\alpha' \approx 109° \; 30'$, then $\gamma' \approx 180° - 45° - 109° \; 30' = 25° \; 30'$, and we have

$$\frac{\sin 45°}{3} \approx \frac{\sin 25° \; 30'}{c},$$

so that

$$c' \approx 3\left(\frac{0.4305}{0.7071}\right) \approx 1.8.$$

Thus, the two solutions for the triangle are

$$\alpha \approx 70° \; 30', \qquad \gamma \approx 64° \; 30', \qquad c \approx 3.8$$

and

$$\alpha' \approx 109° \; 30', \qquad \gamma' \approx 25° \; 30', \qquad c' \approx 1.8.$$

Exercise 8.4

Solve the following triangles. In all problems in this exercise, give lengths to the nearest tenth and angle measures to the nearest 10′.

1. $b = 10, \; B = 30°, \; A = 80°$ 2. $a = 64, \; B = 36° \; 10', \; C = 82° \; 20'$

3. $c = 78.1, \; A = 58° \; 50', \; C = 63° \; 10'$ 4. $b = 1.02, \; B = 41° \; 10', \; C = 80° \; 20'$

5. $a = 84.2, \; A = 110°, \; C = 22° \; 20'$ 6. $c = 0.94, \; A = 41° \; 10', \; B = 96° \; 50'$

Determine the number of triangles that satisfy the conditions in each problem.

Example $b = 34, \quad c = 12, \quad C = 30°$

Solution   Sketch a figure and determine

$$h = 34 \; \sin 30° = 34 \cdot \frac{1}{2} = 17.$$

Since $12 < 17$, we have

$$c < b \sin C,$$

and no triangle exists for the given data.

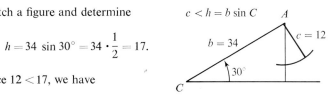

7. $a = 5.4$, $b = 7.0$, $B = 30°$ 8. $b = 4.9$, $c = 3.2$, $C = 30°$

9. $a = 31.1$, $c = 41.3$, $C = 30°$ 10. $a = 42.3$, $b = 20.7$, $B = 30°$

11. $b = 16.2$, $c = 14.3$, $C = 20°$ 12. $a = 141$, $b = 182$, $B = 20°$

13. $a = 4.6$, $b = 2.3$, $B = 30°$ 14. $b = 68.1$, $c = 41.3$, $C = 30°$

Solve the following triangles.

15. $a = 4.8$, $c = 3.9$, $A = 113°$ 16. $a = 3.2$, $b = 2.6$, $B = 54°$

17. $b = 6.21$, $c = 4.39$, $B = 42° \, 40'$ 18. $a = 179$, $b = 212$, $B = 114° \, 10'$

19. $b = 1.8$, $c = 1.5$, $C = 32° \, 30'$ 20. $a = 9.4$, $b = 8.6$, $B = 54° \, 20'$

21. $a = 8.4$, $b = 6.9$, $B = 62° \, 10'$ 22. $b = 0.42$, $c = 0.21$, $B = 31° \, 50'$

23. $a = 13.84$, $b = 6.92$, $A = 60°$ 24. $a = 4.72$, $c = 9.44$, $A = 30°$

25. $b = 420$, $c = 610$, $B = 33° \, 20'$ 26. $a = 5.42$, $b = 6.82$, $A = 43° \, 30'$

Find the area $\mathscr{A}$ of each of the following triangles to the nearest tenth of a unit.

Example $a = 23.1$, $A = 44°$, $C = 26°$

Solution   Make a sketch. Since we are given measures for two angles and one side, only one solution is possible. First we have

$$B = 180° - 44° - 26° = 110°.$$

From the law of sines,

$$\frac{b}{\sin 110°} = \frac{23.1}{\sin 44°},$$

and

$$b = \frac{23.1 \sin 110°}{\sin 44°}.$$

Since $h = 23.1 \sin 26°$,

$$\mathscr{A} = \frac{1}{2} hb = \frac{1}{2} (23.1 \sin 26°) \left(\frac{23.1 \sin 110°}{\sin 44°} \right) \approx 158.$$

27. $a = 6.4$, $b = 8.2$, $B = 30°$ 28. $c = 4.5$, $A = 24°$, $C = 61°$

29. $b = 4.95$, $A = 16° \, 10'$, $B = 60° \, 30'$ 30. $b = 6.3$, $c = 4.3$, $B = 42° \, 10'$

31. $a = 1.14$, $b = 8.31$, $B = 29°$ 32. $b = 2.16$, $c = 4.32$, $C = 60°$

33. From a window in a tower, 85 feet above the ground, the angle of elevation to the top of a nearby building measures 34° 30′. From a point on the ground directly below the window, the angle of elevation to the top of the same building measures 50° 20′. Find the height of the building.

34. Two men, 500 feet apart, observe a balloon between them that is in the same vertical plane with them. The respective angles of elevation of the balloon are observed by the men to measure 80° 10′ and 52° 50′. Find the height of the balloon above the ground.

Show that, in any triangle ABC, each of the following equalities is true.

35. $\dfrac{a+b}{b} = \dfrac{\sin \alpha + \sin \beta}{\sin \beta}$

36. $\dfrac{a-b}{b} = \dfrac{\sin \alpha - \sin \beta}{\sin \beta}$

37. $\dfrac{a-b}{a+b} = \dfrac{\tan \frac{1}{2}(\alpha - \beta)}{\tan \frac{1}{2}(\alpha + \beta)}$ (This is called the **law of tangents**.)

8.5

The law of cosines

The x- and y-coordinates of a point on the terminal side of an angle β in standard position, when the point is located b units from the origin, are

$$b \cos \beta \quad \text{and} \quad b \sin \beta,$$

respectively (Figure 8.12). We can use this fact to derive a very useful formula. Figure 8.13 shows points $A(x_1, y_1)$ and $B(x_2, y_2)$ lying on the terminal sides of

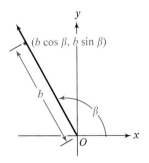

Figure 8.12

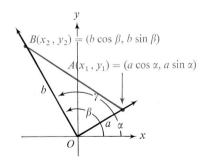

Figure 8.13

angles α and β, respectively. Now, if $B(x_2, y_2)$ is located b units from the origin, while $A(x_1, y_1)$ is located a units from the origin, then

$$[l(\overline{AB})]^2 = (x_2 - x_1)^2 + (y_2 - y_1)^2$$
$$= (b\cos\beta - a\cos\alpha)^2 + (b\sin\beta - a\sin\alpha)^2$$
$$= b^2\cos^2\beta - 2ab\cos\beta\cos\alpha + a^2\cos^2\alpha$$
$$+ b^2\sin^2\beta - 2ab\sin\alpha\sin\beta + a^2\sin^2\alpha$$
$$= b^2(\cos^2\beta + \sin^2\beta) + a^2(\cos^2\alpha + \sin^2\alpha)$$
$$- 2ab(\cos\beta\cos\alpha + \sin\beta\sin\alpha).$$

Thus, by Identities 1 and 5 on page 229, we have the following formula for the square of the distance between two points A and B in the plane:

$$[l(\overline{AB})]^2 = a^2 + b^2 - 2ab\cos(\beta - \alpha).$$

Note that $\beta - \alpha$, or γ, is just the angle between the terminal sides of the angles β and α.

In view of the fact that the location of the axes in the plane is purely a matter of convenience, we have established the following theorem, which is known as the **law of cosines**.

Theorem 8.2 *If α, β, and γ are the angles of a triangle, and a, b, and c are the lengths of the sides opposite α, β and γ, respectively, then*

$$c^2 = a^2 + b^2 - 2ab\cos\gamma \tag{1}$$
$$b^2 = a^2 + c^2 - 2ac\cos\beta \tag{2}$$
$$a^2 = b^2 + c^2 - 2bc\cos\alpha. \tag{3}$$

Solving triangles This theorem has many applications, among them the solution of certain triangles. If we are given the measure of an angle and the lengths of the adjacent sides, we can use the law of cosines to help us find the remaining parts of the triangle.

Example Solve the triangle for which

$$\alpha = 100°, \quad b = 10, \quad \text{and} \quad c = 12.$$

Solution Make a sketch of the triangle and label the sides and angles. We can first find a by using the law of cosines; thus,

$$a^2 = b^2 + c^2 - 2bc\cos\alpha,$$
$$a^2 = 10^2 + 12^2 - 2(10)(12)\cos 100°,$$

$$a^2 \approx 100 + 144 - 240(-0.1736)$$

$$= 244 + 41.664 \approx 285.7.$$

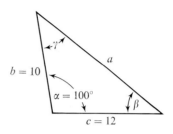

From the table of squares in the Appendix, we find that $16^2 = 256$, while $17^2 = 289$; therefore, a is a little less than 17. We shall use 17. Next, to find β, we again use the law of cosines (we could use the law of sines because now we have the length of a side opposite a given angle) in the form

$$b^2 = a^2 + c^2 - 2ac \cos \beta,$$

or

$$\cos \beta = \frac{b^2 - a^2 - c^2}{-2ac}.$$

With $a^2 = 285.7$, $b^2 = 100$, and $c^2 = 144$, we obtain

$$\cos \beta \approx \frac{100 - 285.7 - 144}{-2(17)(12)} \approx 0.8.$$

From Table VI in the Appendix, $\beta \approx 36° \, 50'$. Since the sum of the angles of a triangle is $180°$, we can find an approximation for γ by noting that

$$\gamma = 180° - \alpha - \beta \approx 180° - 100° - 36° \, 50' = 43° \, 10'.$$

We have, then, for the remaining parts of the triangle,

$$a \approx 17, \quad \beta \approx 36° \, 50', \quad \text{and} \quad \gamma \approx 43° \, 10'.$$

Of course, if greater precision were desired in the preceding example, we could resort to various other means of approximating $\sqrt{285.7}$, and dispense with rounding measures of angles to the nearest $10'$.

Since Equations (1), (2), and (3) in Theorem 8.2 are relationships between three sides and one angle if a triangle, we can, as we did in the preceding example, use *any* of these forms to solve for the measure of an angle, given the lengths of the three sides.

Exercise 8.5

Find the remaining parts of the triangle. In all problems in this exercise, give lengths and areas to the nearest tenth and angle measurements to the nearest $10'$.

1. $a = 10$, $b = 4$, $\gamma = 30°$ 2. $b = 14.2$, $c = 7.9$, $\alpha = 64° \, 10'$

3. $a = 4.9$, $c = 6.8$, $\beta = 122°\ 20'$ 4. $a = 241$, $c = 104$, $\beta = 148°\ 10'$

5. $b = 14.6$, $c = 6.21$, $\alpha = 80°\ 40'$ 6. $b = 9.4$, $c = 10.2$, $\alpha = 100°\ 50'$

7. $a = 5$, $b = 8$, $c = 7$ 8. $a = 4.5$, $b = 5.3$, $c = 2.8$

9. Find the greatest angle of the triangle whose sides are 5.1, 4.2, and 4.5.

10. Find the least angle of the triangle whose sides are 29.5, 33.2, and 41.4.

11. Find the area of the triangle in Problem 1, page 257.

12. Find the area of the triangle in Problem 2, page 257.

13. Show that $1 + \cos \alpha = \dfrac{(b + c + a)(b + c - a)}{2bc}$.

14. Show that $1 - \cos \alpha = \dfrac{(a - b + c)(a + b - c)}{2bc}$.

15. Show that if $s = \dfrac{a + b + c}{2}$, then $\cos \dfrac{1}{2} \alpha = \sqrt{\dfrac{s(s - a)}{bc}}$. *Hint:* Recall (page 229) that

$\cos \dfrac{1}{2} \alpha = \sqrt{\dfrac{1 + \cos \alpha}{2}}$, and use this information in conjunction with the result from

Problem 13, above.

16. Show that if $s = \dfrac{a + b + c}{2}$, then $\sin \dfrac{1}{2} \alpha = \sqrt{\dfrac{(s - b)(s - c)}{bc}}$. *Hint:* See suggestion in

Problem 15.

In Problems 17 and 18, use the results of either Problem 15 or Problem 16.

17. Find the greatest angle of the triangle whose sides are 6.3, 4.8, and 4.3.

18. Find the least angle of the triangle whose sides are 16.4, 23.4, and 20.1.

19. Use the results of Problems 13 and 14 to show that the area $\mathcal{A}$ of a triangle is given by

$$\mathcal{A} = \sqrt{s(s - a)(s - b)(s - c)},$$

where $s = \dfrac{a + b + c}{2}$. This formula is known as **Heron's formula**.

20. Use the results of Problem 19 to find the area of the triangle in

     (a) Problem 7;   (b) Problem 8.

21. Show that the Pythagorean theorem is a special case of the law of cosines.

22. Find the least angle of the triangle with vertices at the points $(0, 0)$, $(5, -2)$, and $(-7, -3)$.

23. Find the area of the triangle with vertices at the points $(1, 1)$, $(5, 5)$, and $(-2, 6)$. (See Problem 19.)

8.6

Polar coordinates

The Cartesian coordinates that we have been using specify the location of a point in the plane by giving the directed distances of the point from a pair of fixed perpendicular lines, the axes. There is an alternative coordinate system that is frequently used in the plane, in which the location of a point is specified in a different manner.

In the plane, consider a fixed ray $\overrightarrow{PB}$ and any point A. We can describe the location of A by giving the distance r from P to A and specifying the angle BPA (Figure 8.14), which is customarily designated by θ. By stating the ordered pair $(r, m°(\theta))$ or $(r, m^R(\theta))$, we clearly identify the location of A. We ordinarily write (r, θ) for either of these ordered pairs, where the meaning should be clear from the context. The components of such an ordered pair are called **polar coordinates** of A. The fixed ray $\overrightarrow{PB}$ is called the **polar axis**, and the initial point P of the polar axis is called the **pole** of the system.

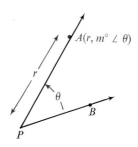

Figure 8.14

Notice that while there is a one-to-one correspondence between the set of ordered pairs in a Cartesian coordinate system and the points in the geometric plane, each point in the plane has infinitely many pairs of polar coordinates. In the first place, if (r, θ) are polar coordinates of A, then so are

$$(r, \theta + k360°), \quad k \in J$$

(Figure 8.15-a). In the second place, if we let $-r < 0$ denote the directed distance from P to A along the negative extension of the ray $\overrightarrow{PA'}$ in a direction opposite that

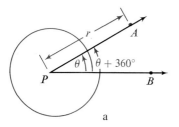

a

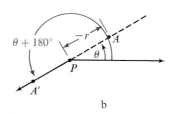

b

Figure 8.15

of $\overrightarrow{PA}$ (Figure 8.15-b), then we see that also $(-r, \theta + 180°)$, and more generally

$$(-r, \theta + 180° + k360°), \quad k \in J,$$

are polar coordinates of A. The pole P itself is represented by $(0, \theta)$ for any θ.

Example   Write four additional sets of polar coordinates for the point having polar coordinates $(3, 30°)$.

Solution   With positive values for r, two more pairs of polar coordinates for $(3, 30°)$ are $(3, 390°)$ and $(3, -330°)$. Using negative values for r, we have $(-3, 210°)$ and $(-3, -150°)$. The figures show these cases. Of course there are infinitely many other possible polar coordinates for the same point.

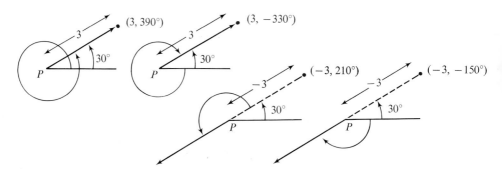

Relationships between polar and rectangular coordinates   Cartesian and polar coordinates of a point can be related by means of the trigonometric functions. If the pole P in a polar coordinate system is also the origin O in a Cartesian coordinate system, and if the polar axis coincides with the positive x-axis of the Cartesian system, as shown in Figure 8.16, then the coordinates (x, y) can be expressed in terms of the polar coordinates (r, θ) by the following equations:

$$x = r \cos \theta, \quad y = r \sin \theta. \qquad (1)$$

Conversely, we have

$$r = \pm\sqrt{x^2 + y^2}, \quad \cos \theta = \frac{x}{\pm\sqrt{x^2 + y^2}},$$

$$\sin \theta = \frac{y}{\pm\sqrt{x^2 + y^2}} \quad [(x, y) \neq (0, 0)]. \qquad (2)$$

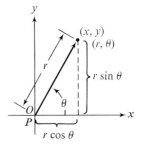

Figure 8.16

The sets of equations (1) and (2) enable us to find rectangular coordinates for a point with a given pair of polar coordinates, and vice versa. Often, in determining θ, it is simplest first to determine the quadrant from the signs of x and y, and then to use the relation

$$\tan \theta = \frac{y}{x}.$$

Example   Find the rectangular coordinates of the point with polar coordinates $(4, 30°)$. Show the graph of the point on a combined polar and rectangular coordinate system.

Solution   Using (1), we obtain

$$x = 4 \cos 30° = 4 \cdot \frac{\sqrt{3}}{2} = 2\sqrt{3},$$

$$y = 4 \sin 30° = 4 \cdot \frac{1}{2} = 2.$$

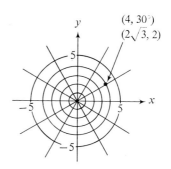

The rectangular coordinates are $(2\sqrt{3}, 2)$.

Example   Find a pair of polar coordinates for the point with Cartesian coordinates $(7, -2)$. Show the graph of the point on a combined polar and rectangular coordinate system.

Solution   By (2), $r = \pm\sqrt{x^2 + y^2} = \pm\sqrt{49 + 4} = \pm\sqrt{53}$. Choosing the positive sign, and noting that the point $(7, -2)$ is in the fourth quadrant, we have

$$\tan \theta = \frac{-2}{7},$$

from which

$$\theta \approx -16° \, 00'.$$

A pair of polar coordinates is therefore

$$(\sqrt{53}, -16° \, 00'),$$

where the given angle measure is an approximation.

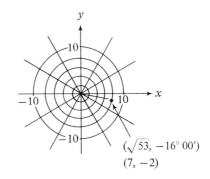

Equations in polar form and in rectangular form Equations (1) can be used to transform Cartesian equations to polar form, and Equations (2) can be used to transform polar equations to Cartesian form.

Example Transform $x^2 + y^2 - 2x + 3 = 0$ to polar form.

Solution From (1), we find that $x^2 = r^2 \cos^2 \theta$ and $y^2 = r^2 \sin^2 \theta$. Thus we have

$$r^2 \cos^2 \theta + r^2 \sin^2 \theta - 2r \cos \theta + 3 = 0,$$

from which

$$r^2(\cos^2 \theta + \sin^2 \theta) - 2r \cos \theta + 3 = 0,$$

$$r^2 - 2r \cos \theta + 3 = 0.$$

Example Transform $r(1 - 2 \cos \theta) = 3$ to Cartesian form.

Solution Using (2), we have

$$\pm\sqrt{x^2 + y^2} \left[1 - 2\left(\frac{x}{\pm\sqrt{x^2 + y^2}} \right) \right] = 3,$$

from which

$$\pm\sqrt{x^2 + y^2} - 2x = 3,$$

$$\pm\sqrt{x^2 + y^2} = 2x + 3.$$

Upon squaring each member, we have the equation

$$x^2 + y^2 = 4x^2 + 12x + 9,$$

from which

$$3x^2 - y^2 + 12x + 9 = 0.$$

Exercise 8.6

Find four sets of polar coordinates $(-360° < \theta \le 360°)$ *for the point with polar coordinates as given.*

1. $(6, 485°)$ 2. $(-3, 518°)$ 3. $(-2, -450°)$

4. $(5, 720°)$ 5. $(6, -600°)$ 6. $(3, -395°)$

Find the Cartesian coordinates of a point with polar coordinates as given.

7. $(5, 45°)$ 8. $(-3, 30°)$ 9. $\left(\frac{1}{2}, 330°\right)$

10. $\left(\frac{3}{4}, 225°\right)$ 11. $(10, -135°)$ 12. $(-6, -240°)$

Find two sets of polar coordinates, one involving an angle of positive measure and one an angle of negative measure, for the point with Cartesian coordinates as given.

13. $(3\sqrt{2}, 3\sqrt{2})$ 14. $\left(-\frac{\sqrt{3}}{2}, \frac{1}{2}\right)$ 15. $(-1, -\sqrt{3})$

16. $(0, -4)$ 17. $(0, 0)$ 18. $(-6, 0)$

Transform the given equation to an equation in polar form.

19. $x^2 + y^2 = 25$ 20. $x = 3$ 21. $y = -4$

22. $x^2 + y^2 - 4y = 0$ 23. $x^2 + 9y^2 = 9$ 24. $x^2 - 4y^2 = 4$

Transform the given equation to an equation in Cartesian form.

25. $r = 5$ 26. $r = 4 \sin \theta$ 27. $r = 9 \cos \theta$

28. $r \cos \theta = 3$ 29. $r(1 - \cos \theta) = 2$ 30. $r(1 + \sin \theta) = 2$

31. Show by transformation of coordinates that the graph of $r = \sec^2 (\theta/2)$ is a parabola.

32. Show by transformation of coordinates that the graph of $r = \csc^2 (\theta/2)$ is a parabola.

33. Graph $\{(r, \theta) \mid r = 4 \sin \theta, 0 \leq \theta < 360°\}$.

34. Graph $\{(r, \theta) \mid r = 9 \cos \theta, 0 \leq \theta < 360°\}$.

Inverse functions and conditional equations

9.1
Inverses of circular functions

Inverse relations   Each of the circular functions has an inverse relation, but none of these inverses is a function. We recall from Sections 4.1 and 4.2 that the inverse of a function is the relation obtained by interchanging the components x and y of each ordered pair in the function and that the graph of the inverse of a function can be obtained by reflecting the graph of the function in the line with equation $y = x$. Furthermore, the resulting inverse is a function if and only if the function is one-to-one. Now consider the sine function

$$\{(x, y) | y = \sin x\}, \tag{1}$$

and the inverse of this function,

$$\{(x, y) | x = \sin y\}. \tag{2}$$

Both graphs are shown in Figure 9.1, where it is evident that each is the reflection of the other in the line $y = x$. Equation (2), $x = \sin y$, does not define a function, because for each element x_i in its domain, $\{x | -1 \le x \le 1, x \in R\}$, there are an unlimited number of elements in its range.

The inverse relation of the sine function is called the **arcsine relation**:

$$\text{arcsine} = \{(x, y) | x = \sin y\}.$$

Each circular and trigonometric function has an inverse relation. Thus we have

$$\text{arccosine} = \{(x, y) | x = \cos y\},$$

and

$$\text{arctangent} = \{(x, y) | x = \tan y\},$$

whose graphs are shown in Figure 9.2. The arccosecant, arcsecant, and arccotangent are less frequently used and their graphs are not shown.

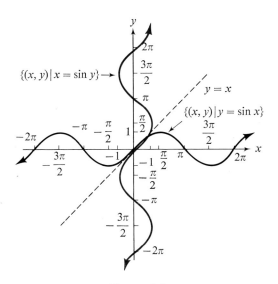

Figure 9.1

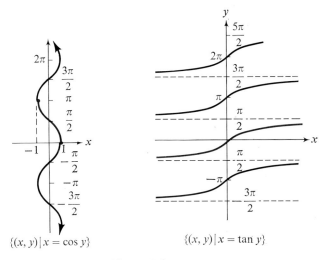

Figure 9.2

Inverse functions   By suitably restricting the *domains* of the circular functions—that is, by suitably restricting the *ranges* of the respective inverse relations—we can define an **inverse function** for each circular and each trigonometric function. This is called the **principal-valued** inverse function. To distinguish it, a capital initial letter is often used. Thus for the sine function, we have the principal-valued inverse function

$$\text{Arcsine} = \left\{ (x, y) \,\middle|\, x = \sin y, \ -\frac{\pi}{2} \leq y \leq \frac{\pi}{2} \right\}.$$

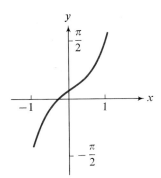

The graph of Arcsine is shown in Figure 9.3.

Because there is now a unique element y in the range $\{y \mid -\pi/2 \leq y \leq \pi/2\}$ corresponding to each x in the domain $\{x \mid -1 \leq x \leq 1\}$, we can use inverse-function notation and write

$$y = \text{Arcsin } x$$

to mean

$$x = \sin y \quad \text{and} \quad -\frac{\pi}{2} \leq y \leq \frac{\pi}{2}.$$

Figure 9.3

Alternatively, we write

$$y = \text{Sin}^{-1} x$$

with this same meaning (but with the understanding, of course, that $\text{Sin}^{-1} x$ does *not* denote $1/\text{Sin } x$). Similar notation is used for the other inverse circular functions.

The restricted ranges for the inverse functions for the cosine and tangent functions are indicated in color in Figure 9.2. These ranges and those of the other inverse circular functions are specified in the following definition.

Definition 9.1   *The inverse circular functions corresponding to the six circular functions are*

I $\text{Arcsine} = \left\{ (x, y) \,\middle|\, y = \text{Sin}^{-1} x, \ -\frac{\pi}{2} \leq y \leq \frac{\pi}{2} \right\},$

II $\text{Arccosine} = \{ (x, y) \mid y = \text{Cos}^{-1} x, \ 0 \leq y \leq \pi \},$

III $\text{Arctangent} = \left\{ (x, y) \,\middle|\, y = \text{Tan}^{-1} x, \ -\frac{\pi}{2} < y < \frac{\pi}{2} \right\},$

IV $\text{Arccotangent} = \{ (x, y) \mid y = \text{Cot}^{-1} x, \ 0 < y < \pi \},$

V $\text{Arcsecant} = \left\{(x, y)\,|\, y = \text{Sec}^{-1}\, x, \quad 0 \le y \le \pi, y \neq \dfrac{\pi}{2}\right\},$

VI $\text{Arccosecant} = \left\{(x, y)\,|\, y = \text{Csc}^{-1}\, x, \quad -\dfrac{\pi}{2} \le y \le \dfrac{\pi}{2}, y \neq 0\right\}.$

Choice of ranges   An analogous definition can be made for the inverse *trigonometric* functions provided that members of the ranges shown are interpreted as radian measures for angles in standard position. Although these particular ranges are the ones customarily chosen, the choices actually are quite arbitrary. They have generally been selected to involve small values of y, to have relatively simple graphs, and of course to yield a one-to-one correspondence between domain and range.

Example   Find Arctan 0.2236.

Solution   It may be helpful first to let $y = \text{Arctan}\ 0.2236$ and then rewrite the equation equivalently as

$$\tan y = 0.2236, \quad -\frac{\pi}{2} < y < \frac{\pi}{2}.$$

From Table V in the Appendix, we find 0.2236 in the column headed tan x. Since this corresponds to tan 0.22 and since $-\pi/2 < 0.22 < \pi/2$, we have

$$\text{Arctan}\ 0.2236 \approx 0.22.$$

Triangle interpretation   It is helpful to interpret symbolism such as $\cos(\text{Arcsin}\ \sqrt{3}/4)$ by using triangles. If we let

$$\alpha = \text{Arcsin}\ \sqrt{3}/4,$$

then equivalently

$$\sin \alpha = \frac{\sqrt{3}}{4}, \quad -\frac{\pi}{2} < \alpha < \frac{\pi}{2},$$

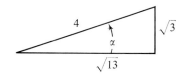

Figure 9.4

and $\sqrt{3}/4$ can be interpreted as the ratio of the length of the side of a triangle opposite an angle α to the length of the hypotenuse of the triangle, as shown in Figure 9.4. From the Pythagorean relationship, the remaining side of the triangle has length

$$\sqrt{4^2 - (\sqrt{3})^2} = \sqrt{16 - 3} = \sqrt{13},$$

which is also labeled. Then, since Arcsin $\sqrt{3}/4 = \alpha$, by inspecting the triangle sketched we can see that

$$\cos\left(\text{Arcsin } \frac{\sqrt{3}}{4}\right) = \cos \alpha = \frac{\sqrt{13}}{4}.$$

Consider now a case in which the element in the domain is not specified.

Example   Write $\sin(\text{Cos}^{-1} x)$ as an equivalent expression without inverse notation.

Solution   Let $\text{Cos}^{-1} x = \alpha$. Then, $\cos \alpha = x$. Notice also that $0 \leq \text{Cos}^{-1} x \leq \pi$ for each x for which it is defined, and hence $\sin(\text{Cos}^{-1} x) \geq 0$. Therefor, using $\sin^2 \alpha + \cos^2 \alpha = 1$, we have

$$\sin(\text{Cos}^{-1} x) = \sin \alpha$$

$$= \sqrt{1 - \cos^2 \alpha} = \sqrt{1 - x^2}.$$

Using a sketch, we can visualize the relationship as shown at the right, where, by inspection,

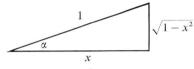

$$\sin(\cos^{-1} x) = \sqrt{1 - x^2}.$$

Observe that it follows from the definitions of the inverse functions that $\sin(\text{Sin}^{-1} x) = x$, $\cos(\text{Cos}^{-1} x) = x$, etc. For example, if $\text{Sin}^{-1} x$ is a real number y, then

$$x = \sin y = \sin(\text{Sin}^{-1} x).$$

Exercise 9.1

Find the value of each expression if it exists. Express results in terms of rational multiples of π. You may be able to recall these values from memory, or you may use Table 8.1, page 241.

Examples   a. $\text{Arccos } \dfrac{1}{2}$ b. $\text{Tan}^{-1}(-1)$

Solutions   a. We seek a number y such that   b. We seek a number y such that

$$\cos y = \frac{1}{2}, \quad 0 \leq y \leq \frac{\pi}{2}. \qquad\qquad \tan y = -1, -\frac{\pi}{2} < y < \frac{\pi}{2}.$$

The number is $\dfrac{\pi}{3}$.   The number is $-\dfrac{\pi}{4}$.

1. $\text{Arcsin } \dfrac{1}{2}$ 2. $\text{Arctan } \sqrt{3}$ 3. $\text{Cot}^{-1} 1$ 4. $\text{Cos}^{-1} \dfrac{1}{\sqrt{2}}$

5. $\text{Cos}^{-1} 2$ 6. $\text{Sec}^{-1} 2$ 7. $\text{Arctan}\left(-\dfrac{1}{\sqrt{3}}\right)$ 8. $\text{Arcsin } \dfrac{\sqrt{3}}{2}$

Use Table V in the Appendix as necessary to find each expression.

9. Arctan 0.1003   10. Arcsec 1.053   11. Sin^{-1} 0.3802

12. Cos^{-1} 0.6675   13. Arcsin (-0.8624) 14. Arccos (-0.3902)

15. Tan^{-1} 3.467   16. Csc^{-1} 1.422

Find the value for each expression if it exists.

Examples   a. $\text{Cos}^{-1}(\tan \pi)$ b. $\sin \dfrac{1}{2}\left(\text{Cos}^{-1}\dfrac{1}{2}\right)$

Solutions   a. Since $\tan \pi = 0,$ b. Since $\text{Cos}^{-1}\dfrac{1}{2}=\dfrac{\pi}{3},$

$$\text{Cos}^{-1}(\tan \pi) = \text{Cos}^{-1} 0 = \dfrac{\pi}{2}. \qquad \sin \dfrac{1}{2}\left(\text{Cos}^{-1}\dfrac{1}{2}\right) = \sin \dfrac{1}{2}\left(\dfrac{\pi}{3}\right) = \dfrac{1}{2}.$$

17. $\text{Sin}^{-1}\left(\cos \dfrac{\pi}{4}\right)$ 18. $\text{Cos}^{-1}\left(\sin \dfrac{\pi}{2}\right)$

19. $\text{Tan}^{-1}\left(\tan \dfrac{\pi}{3}\right)$ 20. $\text{Sin}^{-1}\left(\sin \dfrac{3\pi}{2}\right)$

21. $\sin\left(\text{Arccos }\dfrac{1}{2}\right)$ 22. $\tan\left(\text{Arcsin }\dfrac{\sqrt{3}}{2}\right)$

23. $\cos\left(\text{Cot}^{-1}(-\sqrt{3})\right)$ 24. $\sin\left(\text{Tan}^{-1}(-1)\right)$

25. $\sin\left(2 \text{ Arcsin }\dfrac{1}{2}\right)$ 26. $\sin\left(2 \text{ Cos}^{-1}\dfrac{3}{5}\right)$

27. $\tan \dfrac{1}{2}\left(\text{Arcsin }\dfrac{12}{13}\right)$ 28. $\cos \dfrac{1}{2}(\text{Tan}^{-1} 0)$

29. $\sin\left(\text{Sin}^{-1}\dfrac{1}{2} + \text{Cos}^{-1}\dfrac{3}{5}\right)$ 30. $\cos\left(\text{Sin}^{-1}\dfrac{1}{\sqrt{2}} + \text{Cos}^{-1}\dfrac{4}{5}\right)$

31. $\text{Arccos}(\sin(\text{Arctan}(-1)))$ 32. $\sin(\text{Cos}^{-1}(\tan 0))$

Write each expression without inverse notation.

Example $\sin (\text{Cos}^{-1} x + \text{Sin}^{-1} y)$

Solution   From the formula for $\sin (x_1 + x_2)$, page 229, we have

$\sin (\text{Cos}^{-1} x + \text{Sin}^{-1} y)$
$$= \sin (\text{Cos}^{-1} x) \cos (\text{Sin}^{-1} y) + \cos (\text{Cos}^{-1} x) \sin (\text{Sin}^{-1} y).$$

Using sketches, we can visualize the relationships in each factor of each term in the right-hand member. By inspection, we obtain

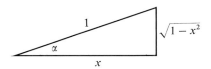

$\sin (\text{Cos}^{-1} x + \text{Sin}^{-1} y)$
$$= (\sqrt{1 - x^2})(\sqrt{1 - y^2}) + xy$$
$$= \sqrt{(1 - x^2)(1 - y^2)} + xy.$$

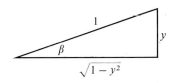

33. $\cos (\text{Arcsin } x)$ 34. $\cot (\text{Arctan } x)$

35. $\tan (\text{Sin}^{-1} y)$ 36. $\sin (\text{Tan}^{-1} y)$

37. $\cos \left(\dfrac{1}{2} \text{Arccos } x\right)$ 38. $\sin (\text{Sin}^{-1} x - \text{Sin}^{-1} y)$

Solve the following equations for x in terms of y.

Example $y = 2 \text{ Arcsin } 3x.$

Solution   We have

$$\frac{y}{2} = \text{Arcsin } 3x,$$

and from the definition of Arcsine, we write the expression as

$$3x = \sin \frac{y}{2}, \quad \text{and then as} \quad x = \frac{1}{3} \sin \frac{y}{2},$$

with $-\dfrac{\pi}{2} \leq \dfrac{y}{2} \leq \dfrac{\pi}{2}, \quad \text{or} \quad -\pi \leq y \leq \pi.$

39. $y = 3 \text{ Arccos } 2x$

40. $y = 2 \text{ Sin}^{-1} \dfrac{x}{5}$

41. $y = \dfrac{1}{2} \text{ Tan}^{-1}(x + \pi)$

42. $y = \dfrac{2}{3} \text{ Cot}^{-1}(\pi x)$

43. Show that $\text{Arcsin } \dfrac{2}{5} = \text{Arctan } \dfrac{2}{\sqrt{21}}$.   *Hint*: Sketch a triangle.

44. Show that $\text{Arccos } \dfrac{1}{3} = \text{Arccot } \dfrac{1}{2\sqrt{2}}$.

45. Is $\text{Arccos }(\cos x) = x$? Why or why not?

46. Is $\text{Arcsin }(\sin x) = x$? Why or why not?

9.2
Conditional equations

In Sections 7.9 and 8.2, we observed that certain equations involving circular or trigonometric function values are *identities*; that is, they are true for all permissible values of any variables involved. In this section, we shall be concerned with *conditional* equations of a similar kind, namely, equations that involve circular or trigonometric function values but that are not satisfied by all permissible values of any variables involved.

Solution of conditional equations   Various procedures exist for solving conditional equations that involve circular or trigonometric function values. Of course, the methods applicable to algebraic equations are also valid for such equations, as you will see in the following examples.

Because the circular and trigonometric functions are periodic, we should expect equations involving $\sin x$, $\tan x$, or other circular or trigonometric function values to have infinite solution sets.

Example   Solve $\sin x = -1$ for:

a. $x \in R$,

b. $x \in$ {angles with measures given in radians},

c. $x \in$ {angles with measures given in degrees}.

(Solution overleaf.)

Solution

From the entries in Table 7.2, or using a unit circle where the point $(0, -1)$ is associated with an arc of length x, we observe that the solutions are:

a. $\left\{ x \mid x = \dfrac{3\pi}{2} + 2k\pi, \quad k \in J \right\}$,

b. $\left\{ x \mid x = \left(\dfrac{3\pi}{2} + 2k\pi \right)^{R}, \quad k \in J \right\}$,

c. $\{ x \mid x = (270 + k \cdot 360)°, \quad k \in J \}$.

If, in the foregoing example, the replacement set for x were restricted to the interval $-\pi/2 \leq x \leq \pi/2$, then, of course, the solution could be written

$$x = \text{Arcsin}(-1) = -\frac{\pi}{2},$$

$$x = \text{Arcsin}(-1) = -\frac{\pi}{2}^{R},$$

and

$$x = \text{Arcsin}(-1) = -90°.$$

Example

Specify $\{ w \mid \sqrt{3}/2 = \sin w, \ w \in R \}$ in terms of rational multiples of π.

Solution

As in the preceding example, we can use the unit circle to help us visualize the situation here. Since we know (from Table 7.2 or from memory) that $\sqrt{3}/2 = \sin (\pi/3)$, we then know that w_1, the first-quadrant value for w, must be $\pi/3$. Using $\pi/3$ as a reference arc, and the fact

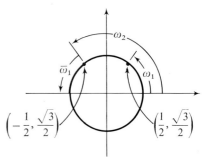

$$\left(-\frac{1}{2}, \frac{\sqrt{3}}{2} \right) \qquad \left(\frac{1}{2}, \frac{\sqrt{3}}{2} \right)$$

that $\sin w$ is positive only in the first and second quadrants, we observe that the only other value for w in our set over the interval $0 \leq w < 2\pi$ is $w_2 = 2\pi/3$. Hence,

$$\left\{ w \mid \frac{\sqrt{3}}{2} = \sin w \right\} = \left\{ w \mid w = \frac{\pi}{3} + 2k\pi, k \in J \right\} \cup \left\{ w \mid w = \frac{2\pi}{3} + 2k\pi, k \in J \right\}.$$

The identities we have developed for the circular and trigonometric functions are useful in solving some equations. We can use these identities to rewrite equations involving more than one function value as equivalent equations involving values of only one function.

Example   Solve $\sin \alpha = \cos \alpha$ over the interval $0 \leq \alpha \leq \pi/2$.

Solution   Since $\sin \alpha \neq 0$ when $\cos \alpha = 0$, the equation is not satisfied if $\cos \alpha = 0$. We can accordingly assume that $\cos \alpha \neq 0$, and hence can multiply each member by $1/\cos \alpha$ to obtain

$$\frac{\sin \alpha}{\cos \alpha} = 1,$$

or, equivalently,

$$\tan \alpha = 1.$$

Therefore we have as our solution set $\{\pi/4\}$.

Some equations that involve circular or trigonometric function values are not of the first degree. In such cases, an algebraic factorization will sometimes lead to a solution.

Example   Solve $2 \sin u \cos u - \sin u - 2 \cos u + 1 = 0$ over $0 \leq u < 2\pi$.

Solution   If $\sin u$ is factored from the first two terms of the left-hand member of the given equation and -1 is factored from the second two terms, the result is

$$(\sin u)(2 \cos u - 1) - 1(2 \cos u - 1) = 0.$$

Then, factoring $2 \cos u - 1$ from each term in the left-hand member, we have

$$(2 \cos u - 1)(\sin u - 1) = 0.$$

Therefore, we seek

$$\{u \mid 2 \cos u - 1 = 0\} \cup \{u \mid \sin u - 1 = 0\}.$$

If $2 \cos u - 1 = 0$, then $\cos u = 1/2$, and if $\sin u - 1 = 0$, then $\sin u = 1$. Thus, the values of u satisfying these equations are of the form

$$\frac{\pi}{3} + 2k\pi, \qquad \frac{5\pi}{3} + 2k\pi, \qquad \text{or} \qquad \frac{\pi}{2} + 2k\pi, \quad k \in J.$$

Over the interval $0 \leq u \leq 2\pi$, the solution set is $\{\pi/3, 5\pi/3, \pi/2\}$.

Exercise 9.2

Specify the given set in terms of rational multiples of π.

Example $\{w \mid \tan w = -1\}$.

Solution   From memory or the entries in Table 7.2, we obtain

$$\{w \mid \tan w = -1\} = \left\{ w \mid w = \frac{3\pi}{4} + k\pi \right\}, \quad \text{where } k \in J.$$

1. $\left\{ p \mid \sin p = \dfrac{1}{2} \right\}$ 2. $\{q \mid \cot q = 1\}$ 3. $\{y \mid \cos y = 2\}$

4. $\{y \mid \tan y = \sqrt{3}\}$ 5. $\left\{ r \mid \cos r = \dfrac{1}{\sqrt{2}} \right\}$ 6. $\{s \mid \sec s = 2\}$

7. $\left\{ x \mid \sin x = -\dfrac{\sqrt{3}}{2} \right\}$ 8. $\left\{ x \mid \cos x = -\dfrac{\sqrt{3}}{2} \right\}$

In all the following problems in the exercise, use Table V or VI as necessary, and give results to the nearest reading in the table.

In Problems 9–20, find the solution set of each equation for
a. $x \in R$,   *b. x an angle whose measure is given in (1) radians, and (2) degrees.*

9. $\cos x = \dfrac{1}{2}$ 10. $\sin x = \dfrac{\sqrt{2}}{2}$ 11. $\tan x = \sqrt{3}$

12. $\cot x = -1$ 13. $\sec x - \sqrt{2} = 0$ 14. $\csc x + 1 = 0$

15. $4 \sin x - 1 = 0$ 16. $2 \tan x + 3 = 0$ 17. $3 \cot x - 1 = 0$

18. $3 \cos x + 1 = 0$ 19. $2 \sec x - 5 = 0$ 20. $3 \csc x + 8 = 0$

In Problems 21–26, find the solution set of each equation over R.

Example $\sqrt{3} \sin x + \cos x = 0$

Solution   By first noting that $\cos x \neq 0$ in any solution, we can divide each member by $\cos x$ to produce

$$\sqrt{3} \, \frac{\sin x}{\cos x} + 1 = 0$$

from which

$$\sqrt{3} \tan x + 1 = 0,$$

$$\tan x = -\frac{1}{\sqrt{3}}.$$

Therefore, the solution set is $\left\{ x \,\middle|\, x = \dfrac{5\pi}{6} + k\pi, \quad k \in J \right\}.$

21. $\sin x - \sqrt{3} \cos x = 0$ 22. $3 \sin x + \cos x = 0$

23. $\tan^2 x - 1 = 0$ 24. $2 \sin^2 x - 1 = 0$

25. $\sin^2 x - \cos^2 x = 1$ 26. $\sin x + \cos x \tan x = 3$

In Problems 27–32, find the solution set of each equation for θ a member of the set of angles such that $0° \le \theta < 360°$.

Example $2 \cos^2 \theta \tan \theta - \tan \theta = 0$

Solution   Factoring $\tan \theta$ from the terms in the left-hand member, we have

$$\tan \theta \, (2 \cos^2 \theta - 1) = 0,$$

from which either

$$\tan \theta = 0 \quad \text{or} \quad 2 \cos^2 \theta - 1 = 0.$$

If $\tan \theta = 0$, then either $\theta = 0°$ or $\theta = 180°$. If $2 \cos^2 \theta - 1 = 0$, then $\cos^2 \theta = 1/2$, $\cos \theta = \pm 1/\sqrt{2}$, and θ is an odd multiple of $45°$. Therefore, the solution set over $0° \le \theta < 360°$ is

$$\{0°, 45°, 135°, 180°, 225°, 315°\}.$$

27. $(2 \sin \theta - 1)(2 \sin^2 \theta - 1) = 0$

28. $(\tan \theta - 1)(2 \cos \theta + 1) = 0$

29. $2 \sin \theta \cos \theta + \sin \theta = 0$

30. $\tan \theta \sin \theta - \tan \theta = 0$

31. $2 \sin \theta \cos \theta + 2 \sin \theta - \cos \theta - 1 = 0$ *Hint:*   First factor by grouping.

32. $2 \sin \theta \tan \theta + \tan \theta - 2 \sin \theta - 1 = 0$

In Problems 33–40, find the solution set of each equation over the interval $0^R \leq \alpha < 2\pi^R$.

Example $\tan^2 \alpha + \sec \alpha - 1 = 0.$

Solution   Since $\tan^2 \alpha = \sec^2 \alpha - 1$, the given equation can be written as

$$\sec^2 \alpha - 1 + \sec \alpha - 1 = 0,$$

$$\sec^2 \alpha + \sec \alpha - 2 = 0,$$

$$(\sec \alpha + 2)(\sec \alpha - 1) = 0.$$

Now observe that the solution set is

$$\{\alpha \,|\, \sec \alpha + 2 = 0\} \cup \{\alpha \,|\, \sec \alpha - 1 = 0\}$$

over the interval $0^R \leq \alpha < 2\pi^R$. If $\sec \alpha + 2 = 0$, then $\sec \alpha = -2$, and if $\sec \alpha - 1 = 0$, then $\sec \alpha = 1$. Over the interval $0^R \leq \alpha < 2\pi^R$ the required solution set is

$$\left\{ \frac{2\pi^R}{3}, \frac{4\pi^R}{3} \right\} \cup \{0^R\} = \left\{ 0^R, \frac{2\pi^R}{3}, \frac{4\pi^R}{3} \right\}.$$

33. $\tan^2 \alpha - 2 \tan \alpha + 1 = 0$ 34. $4 \sin^2 \alpha - 4 \sin \alpha + 1 = 0$

35. $\cos^2 \alpha + \cos \alpha = 2$ 36. $\cot^2 \alpha = 5 \cot \alpha - 4$

37. $\sec^2 \alpha + 3 \tan \alpha - 11 = 0$ 38. $\tan^2 \alpha + 4 = 2 \sec^2 \alpha$

39. $\sin^2 \alpha + \sin \alpha - 1 = 0$ *Hint*: Use the quadratic formula.

40. $\tan^2 \alpha = \tan \alpha + 3$

41. Approximate a solution of $x/2 - \sin x = 0$, $x \in R$, $x \neq 0$, by graphical methods. *Hint*: Graph $y_1 = x/2$ and $y_2 = \sin x$ on the same coordinate system and determine a value of x for which $y_1 = y_2$.

42. Approximate a solution of $\cos x = x^2$, $x \in R$, by graphical methods.

9.3

Conditional equations for multiples

Equations that contain circular or trigonometric function values such as $\sin 2x$, $\cos 3\alpha$, etc., need further consideration.

Example   Solve $\sqrt{2} \cos 3\alpha = 1$ over the interval $0° \leq \alpha < 360°$.

Solution   The equation can be written equivalently as $\cos 3\alpha = \dfrac{1}{\sqrt{2}}$.

Therefore, we have

$$3\alpha = 45° + k \cdot 360° \qquad \text{or} \qquad 3\alpha = 315° + k \cdot 360°,$$

from which, by dividing each member by 3, we obtain

$$\alpha = 15° + k \cdot 120° \qquad \text{or} \qquad \alpha = 105° + k \cdot 120°, \quad k \in J.$$

Because we want solutions over the interval $0° \leq \alpha < 360°$, we consider 0, 1, 2 as replacements for k and obtain the solution set

$$\{15°, 135°, 255°, 105°, 225°, 345°\}.$$

A judicious selection of one or more of the identities encountered earlier can frequently help you solve certain kinds of equations. The following examples illustrate two such cases.

Example   Solve $\sin \alpha \cos \alpha = 1/4$ over the interval $0^R \leq \alpha < 2\pi^R$.

Solution   Multiplying each member of the given equation by 2, we obtain the equivalent equation

$$2 \sin \alpha \cos \alpha = \frac{1}{2}.$$

Since $2 \sin \alpha \cos \alpha = \sin 2\alpha$ for every α in the desired interval, this latter equation is equivalent to

$$\sin 2\alpha = \frac{1}{2}.$$

Therefore, we have

$$2\alpha = \left(\frac{\pi}{6} + 2k\pi\right)^R \qquad \text{or} \qquad 2\alpha = \left(\frac{5\pi}{6} + 2k\pi\right)^R, \quad k \in J,$$

so that

$$\alpha = \left(\frac{\pi}{12} + k\pi\right)^R \qquad \text{or} \qquad \alpha = \left(\frac{5\pi}{12} + k\pi\right)^R, \quad k \in J.$$

Because we want the solutions over the interval $0^R \leq \alpha < 2\pi^R$, we consider 0 and 1 as replacements for k and obtain the solution set

$$\left\{\frac{\pi^R}{12}, \frac{13\pi^R}{12}, \frac{5\pi^R}{12}, \frac{17\pi^R}{12}\right\}.$$

Example   Solve $\cos 2x = \sin x, \; x \in R.$

Solution   Since $\cos 2x = 1 - 2\sin^2 x,$ the equation $\cos 2x = \sin x$ can be
written equivalently as

$$1 - 2\sin^2 x = \sin x,$$
$$2\sin^2 x + \sin x - 1 = 0,$$
$$(2\sin x - 1)(\sin x + 1) = 0,$$

from which

$$\sin x = \frac{1}{2} \quad \text{or} \quad \sin x = -1.$$

Then as solution set we have

$$\left\{ x \mid x = \frac{\pi}{6} + 2\pi k \right\} \cup \left\{ x \mid x = \frac{5\pi}{6} + 2\pi k \right\} \cup \left\{ x \mid x = \frac{3\pi}{2} + 2\pi k \right\}, \quad k \in J.$$

Exercise 9.3

In all problems in this exercise, use Table V or VI as necessary.

In Problems 1–12, solve each equation over the interval $0° \le \theta < 360°$.

1. $\cos 2\theta = \dfrac{\sqrt{2}}{2}$

2. $\tan 2\theta = \sqrt{3}$

3. $\sin \dfrac{1}{2}\theta = \dfrac{1}{2}$

4. $\cot \dfrac{1}{3}\theta = -1$

5. $\tan 3\theta = 0$

6. $\sin 4\theta = 1$

7. $\sin \theta \cos \theta = \dfrac{1}{2}$

8. $\cos^2 \theta - \sin^2 \theta = -1$

9. $\cos 2\theta + \sin 2\theta = 0$

10. $\sin \theta \cos \theta = \dfrac{\cos 2\theta}{2}$

11. $2\cos^2 2\theta + \cos 2\theta - 1 = 0$

12. $\tan^2 2\theta + 2\tan 2\theta + 1 = 0$

In Problems 13–22, solve each equation over R.

13. $\sin 2x - \cos x = 0$

14. $\cos 2x = \cos^2 x - 1$

15. $\cos 2x = \cos x - 1$

16. $\sin x = \sin 2x$

17. $\cos 2x \sin x + \sin x = 0$ 18. $\sin 2x \cos x - \sin x = 0$

19. $\sin 4x - 2 \sin 2x = 0$ *Hint:* $\sin 4x = \sin 2(2x)$.

20. $\sin 3x + 4 \sin^2 x = 0$ *Hint:* $\sin 3x = \sin (x + 2x)$.

21. $\sin 2x + 2 \sin x - \cos x - 1 = 0$

22. $\sin 2x + \sin x + 2 \cos x + 1 = 0$

10

Matrices
and determinants

10.1

Definitions; matrix addition

Matrices are today much used in mathematics and engineering, and also in the physical, social, and life sciences.

A **matrix** is a rectangular array of real numbers (or other suitable entities), which are called **entries**, or **elements**, of the matrix. In this book, we shall consider only real numbers as entries. A matrix is customarily displayed in a pair of brackets or parentheses (we shall use brackets). Thus

$$\begin{bmatrix} 1 & 2 & 3 \\ 4 & 5 & 6 \end{bmatrix} \quad \text{and} \quad \begin{bmatrix} 2 \\ 1 \end{bmatrix}$$

are matrices. The **order**, or **dimension**, of a matrix is the ordered pair having as first component the number of (horizontal) **rows** and as second component the number of (vertical) **columns** in the matrix. Thus,

$$\begin{bmatrix} 1 & 2 & 3 \\ 4 & 5 & 6 \end{bmatrix}, \quad \begin{bmatrix} 1 \\ 2 \\ 3 \end{bmatrix}, \quad \text{and} \quad \begin{bmatrix} a_1 & a_2 & a_3 & a_4 \\ b_1 & b_2 & b_3 & b_4 \\ c_1 & c_2 & c_3 & c_4 \\ d_1 & d_2 & d_3 & d_4 \end{bmatrix}$$

are 2×3 (read "two by three"), 3×1 (read "three by one"), and 4×4 (read "four by four") matrices, respectively. Note that the number of *rows* is given first, and then the number of *columns*. A matrix consisting of a single row is called a **row matrix**, or a **row vector**, whereas a matrix consisting of a single column is called a **column matrix**, or a **column vector**.

Matrices are frequently denoted by capital letters. Thus, we might want to talk about the matrices A and B, where

$$A = \begin{bmatrix} a_1 & a_2 \\ b_1 & b_2 \end{bmatrix} \quad \text{and} \quad B = [b_1 \quad b_2].$$

To show that A is a 2×2 matrix, we can write $A_{2 \times 2}$. Similarly, $B_{1 \times 2}$ is a matrix with one row and two columns.

To represent the entries of a matrix, either single or double subscript notation is employed. Consider any 3×3 matrix, A. We can represent A by

$$A = \begin{bmatrix} a_1 & a_2 & a_3 \\ b_1 & b_2 & b_3 \\ c_1 & c_2 & c_3 \end{bmatrix},$$

where a different letter is used for each row, and a single subscript denotes the column in which each particular entry is located. Alternatively, we can use a different letter for each column, and let the subscript denote the row. Thus, we can also use the notation

$$A = \begin{bmatrix} a_1 & b_1 & c_1 \\ a_2 & b_2 & c_2 \\ a_3 & b_3 & c_3 \end{bmatrix}.$$

A much more useful convention involves double subscripts, where a single letter, say a, is used to denote an entry in a matrix, and then *two* subscripts are appended, the first subscript telling in which *row* the entry occurs, and the second telling which *column*. Thus, we write

$$A = \begin{bmatrix} a_{11} & a_{12} & a_{13} \\ a_{21} & a_{22} & a_{23} \\ a_{31} & a_{32} & a_{33} \end{bmatrix},$$

where a_{21} is the element in the second *row* and first *column*, a_{33} is the element in the third *row* and third *column*; if we wish to generalize, a_{ij} is the element in the *ith row* and *jth column*.

Definition 10.1   *Two matrices, A and B, are equal if and only if both matrices are of the same order and $a_{ij} = b_{ij}$ for each i, j.*

Thus,

$$\begin{bmatrix} 2 & 1 \\ 3 & 0 \end{bmatrix} = \begin{bmatrix} \frac{4}{2} & 2-1 \\ \sqrt{9} & 0 \end{bmatrix}, \quad \text{but} \quad \begin{bmatrix} 2 & 1 \\ 3 & 0 \end{bmatrix} \neq \begin{bmatrix} 2 & 3 \\ 1 & 0 \end{bmatrix}.$$

Definition 10.2   *The transpose of a matrix A, denoted by A^t, is the matrix in which the rows are the columns of A and the columns are the rows of A.*

Examples   a. $\begin{bmatrix} 2 & 1 \\ 3 & 0 \end{bmatrix}^t = \begin{bmatrix} 2 & 3 \\ 1 & 0 \end{bmatrix}$ b. $\begin{bmatrix} 1 & 2 & 3 \\ 4 & 5 & 6 \end{bmatrix}^t = \begin{bmatrix} 1 & 4 \\ 2 & 5 \\ 3 & 6 \end{bmatrix}$

Definition 10.3 *The sum of two matrices of the same order, $A_{m \times n}$ and $B_{m \times n}$, is the matrix $(A + B)_{m \times n}$ in which the entry in the ith row and jth column is $a_{ij} + b_{ij}$, for $i = 1, 2, 3, \ldots, m$ and $j = 1, 2, 3, \ldots, n$.*

Example
$$\begin{bmatrix} 3 & 1 & 2 \\ 2 & 1 & 4 \end{bmatrix} + \begin{bmatrix} 1 & 0 & 2 \\ -1 & 3 & 0 \end{bmatrix} = \begin{bmatrix} 3+1 & 1+0 & 2+2 \\ 2+(-1) & 1+3 & 4+0 \end{bmatrix}$$
$$= \begin{bmatrix} 4 & 1 & 4 \\ 1 & 4 & 4 \end{bmatrix}.$$

The sum of two matrices of different order is not defined.

Definition 10.4 *A matrix with each entry equal to 0 is a zero matrix.*

Zero matrices are generally denoted by the symbol **0**. This distinguishes a zero matrix from the real number 0. For example,

$$\mathbf{0}_{2 \times 4} = \begin{bmatrix} 0 & 0 & 0 & 0 \\ 0 & 0 & 0 & 0 \end{bmatrix}$$

is the 2×4 zero matrix.

Definition 10.5 *The negative of the matrix $A_{m \times n}$, denoted by $-A_{m \times n}$, is the matrix formed by replacing each entry in $A_{m \times n}$ with its additive inverse.*

For example, if

$$A_{3 \times 2} = \begin{bmatrix} 3 & -1 \\ 2 & -2 \\ -4 & 5 \end{bmatrix}, \quad \text{then } -A_{3 \times 2} = \begin{bmatrix} -3 & 1 \\ -2 & 2 \\ 4 & -5 \end{bmatrix}.$$

The sum $B_{m \times n} + (-A_{m \times n})$ is called the **difference** of $B_{m \times n}$ and $A_{m \times n}$ and is denoted $B_{m \times n} - A_{m \times n}$.

Properties
of sums
At this point, we are able to establish the following facts concerning the set of all $m \times n$ matrices with real-number entries, for any given m and n.

Theorem 10.1 *If A, B, and C are $m \times n$ matrices with real-number entries:*

I *$(A + B)_{m \times n}$ is a matrix with real-number*   *Closure law for addition.*
  entries.

II $(A + B) + C = A + (B + C)$.   *Associative law for addition.*

III *The matrix* $\mathbf{0}_{m \times n}$ *has the property that for*   *Additive-identity law.*
     every matrix $A_{m \times n}$,

$$A + \mathbf{0} = A \quad and \quad \mathbf{0} + A = A.$$

IV *For every matrix* $A_{m \times n}$, *the matrix* $-A_{m \times n}$ *Additive-inverse law.*
    has the property that

$$A + (-A) = \mathbf{0} \quad and \quad (-A) + A = \mathbf{0}.$$

V $A + B = B + A$.   *Commutative law for addition.*

Proof of 10.1-III Since each entry of the zero matrix is 0, it follows that the entries of $A_{m \times n} + \mathbf{0}_{m \times n}$ are $a_{ij} + 0 = a_{ij}$ and the entries of $\mathbf{0}_{m \times n} + A_{m \times n}$ are $0 + a_{ij} = a_{ij}$, and the theorem is proved.

For example,

$$\begin{bmatrix} a_{11} & a_{12} \\ a_{21} & a_{22} \end{bmatrix} + \begin{bmatrix} 0 & 0 \\ 0 & 0 \end{bmatrix} = \begin{bmatrix} a_{11} & a_{12} \\ a_{21} & a_{22} \end{bmatrix}.$$

Proof of 10.1-IV Let the entries of $A_{m \times n}$ and $-A_{m \times n}$ be a_{ij} and $-a_{ij}$, respectively. Since each entry of $A + (-A)$ is $a_{ij} - a_{ij}$, or 0, we have $A + (-A) = \mathbf{0}$. Similarly, $(-A) + A = \mathbf{0}$, and the theorem is proved.

For example, if

$$A = \begin{bmatrix} 1 & -1 & 2 \\ 3 & -1 & 1 \end{bmatrix},$$

then

$$A + (-A) = \begin{bmatrix} 1 & -1 & 2 \\ 3 & -1 & 1 \end{bmatrix} + \begin{bmatrix} -1 & 1 & -2 \\ -3 & 1 & -1 \end{bmatrix} = \begin{bmatrix} 0 & 0 & 0 \\ 0 & 0 & 0 \end{bmatrix} = \mathbf{0}.$$

The proof of Parts I, II, and V of Theorem 10.1 for the case of 2×2 matrices are left as exercises.

Exercise 10.1

State the order and find the transpose of each matrix.

Example $\begin{bmatrix} 2 & 4 \\ 1 & -3 \\ 6 & 0 \end{bmatrix}$

(Solution overleaf.)

Solution 3×2 matrix; $\begin{bmatrix} 2 & 4 \\ 1 & -3 \\ 6 & 0 \end{bmatrix}^t = \begin{bmatrix} 2 & 1 & 6 \\ 4 & -3 & 0 \end{bmatrix}$.

1. $\begin{bmatrix} 6 & -1 \\ 2 & 3 \end{bmatrix}$ 2. $\begin{bmatrix} 4 & 1 \\ 0 & -2 \end{bmatrix}$ 3. $\begin{bmatrix} 2 & -7 & 3 \\ 1 & 4 & 0 \end{bmatrix}$

4. $\begin{bmatrix} -3 & 1 \\ 6 & 0 \\ 0 & 2 \end{bmatrix}$ 5. $\begin{bmatrix} 2 & 3 & -1 \\ 4 & 0 & 1 \\ -2 & 3 & 1 \end{bmatrix}$ 6. $\begin{bmatrix} 4 & -1 & -2 \\ 3 & 0 & 0 \\ 2 & 1 & 1 \end{bmatrix}$

7. $\begin{bmatrix} 4 & -3 & -1 & 0 \\ 2 & 1 & 1 & 6 \end{bmatrix}$ 8. $\begin{bmatrix} -2 & 1 & 3 & 2 \\ 4 & 0 & 0 & -2 \\ -1 & 3 & 2 & 4 \end{bmatrix}$ 9. $[0 \ \ 0 \ \ 0 \ \ 0 \ \ 0]$

Write each sum as a single matrix.

Example $\begin{bmatrix} 2 & 1 & 4 \\ 3 & -1 & 0 \end{bmatrix} + \begin{bmatrix} 6 & 3 & 0 \\ -2 & 1 & 0 \end{bmatrix}$

Solution $\begin{bmatrix} 2+6 & 1+3 & 4+0 \\ 3-2 & -1+1 & 0+0 \end{bmatrix} = \begin{bmatrix} 8 & 4 & 4 \\ 1 & 0 & 0 \end{bmatrix}$

10. $\begin{bmatrix} 2 & 3 \\ 1 & 6 \end{bmatrix} + \begin{bmatrix} 1 & -2 \\ 2 & 3 \end{bmatrix}$ 11. $\begin{bmatrix} 4 & -1 & 3 \\ 2 & 1 & 0 \end{bmatrix} + \begin{bmatrix} 3 & -1 & 0 \\ 4 & 0 & -2 \end{bmatrix}$

12. $\begin{bmatrix} 3 & 0 & -1 \\ 2 & 1 & 2 \end{bmatrix} + \begin{bmatrix} 6 & -1 & 0 \\ 0 & 2 & 4 \end{bmatrix}$ 13. $[1 \ \ 3 \ \ 5 \ \ 7] + [0 \ \ -2 \ \ 1 \ \ 3]$

14. $\begin{bmatrix} 4 \\ 3 \\ -1 \end{bmatrix} + \begin{bmatrix} 6 \\ 0 \\ -2 \end{bmatrix}$ 15. $\begin{bmatrix} 2 & 3 \\ 1 & 0 \\ -1 & 2 \end{bmatrix} + \begin{bmatrix} -2 & 0 \\ -3 & 0 \\ 4 & -1 \end{bmatrix}$

16. $\begin{bmatrix} 2 & 3 & 4 \\ -1 & 6 & 2 \\ 1 & 0 & 3 \end{bmatrix} + \begin{bmatrix} 0 & 0 & 0 \\ 0 & 0 & 0 \\ 0 & 0 & 0 \end{bmatrix}$ 17. $\begin{bmatrix} 2 & -3 \\ 4 & -1 \\ -2 & 1 \end{bmatrix} + \begin{bmatrix} -2 & 3 \\ -4 & 1 \\ 2 & -1 \end{bmatrix}$

18. Show that $\begin{bmatrix} b_{11} - a_{11} & b_{12} - a_{12} \\ b_{21} - a_{21} & b_{22} - a_{22} \end{bmatrix}$ is a solution of the matrix equation

$X + A = B,$ where $A = \begin{bmatrix} a_{11} & a_{12} \\ a_{21} & a_{22} \end{bmatrix}$ and $B = \begin{bmatrix} b_{11} & b_{12} \\ b_{21} & b_{22} \end{bmatrix}.$

19. Use the results of Problem 18 to argue that $X + A = B$ and $X = B - A$ are equivalent matrix equations in the system of 2×2 matrices.

Solve each of the following matrix equations.

20. $X + \begin{bmatrix} 3 & -1 \\ 2 & 1 \end{bmatrix} = \begin{bmatrix} 5 & 1 \\ -3 & 5 \end{bmatrix}$

21. $X - \begin{bmatrix} -1 & 0 \\ 0 & 0 \end{bmatrix} = \begin{bmatrix} 3 & -1 \\ 2 & 1 \end{bmatrix}^{t}$

22. $X + \begin{bmatrix} 3 & 2 \\ -1 & 4 \end{bmatrix} = \begin{bmatrix} -2 & -2 \\ -1 & 5 \end{bmatrix}$

23. $\begin{bmatrix} 1 & 3 \\ 1 & 0 \end{bmatrix}^{t} - \begin{bmatrix} 0 & 1 \\ 1 & 0 \end{bmatrix} = \begin{bmatrix} 2 & 2 \\ 1 & 3 \end{bmatrix}^{t} - X$

24. Prove Theorem 10.1-I.   25. Prove Theorem 10.1-II.   26. Prove Theorem 10.1-V.

10.2
Matrix multiplication

We shall be interested in two kinds of products involving matrices: (1) the product of a matrix and a real number, and (2) the product of two matrices. Let us consider them one at a time.

Definition 10.6 *The product of a real number c and an $m \times n$ matrix A with entries a_{ij} is the matrix cA with corresponding entries ca_{ij}, for $i = 1, 2, 3, \ldots, m$ and $j = 1, 2, 3, \ldots, n$.*

Example $3\begin{bmatrix} 2 & 1 \\ 0 & 5 \end{bmatrix} = \begin{bmatrix} 3 \times 2 & 3 \times 1 \\ 3 \times 0 & 3 \times 5 \end{bmatrix} = \begin{bmatrix} 6 & 3 \\ 0 & 15 \end{bmatrix}.$

Properties of products of matrices and real numbers   The following theorem asserts some simple algebraic laws for the multiplication of matrices by real numbers.

Theorem 10.2 *If A and B are $m \times n$ matrices and c and d are real numbers:*

I	*cA is an $m \times n$ matrix,*	V	*$1A = A$,*
II	*$c(dA) = (cd)A$,*	VI	*$(-1)A = -A$,*
III	*$(c + d)A = cA + dA$,*	VII	*$0A = \mathbf{0}$,*
IV	*$c(A + B) = cA + cB$,*	VIII	*$c\mathbf{0} = \mathbf{0}$.*

Proof We shall prove only Part IV, leaving the remaining parts as exercises. Since the elements of $A + B$ are of the form $a_{ij} + b_{ij}$, it follows, by definition, that the elements of $c(A + B)$ are of the form $c(a_{ij} + b_{ij})$. But, since a_{ij}, b_{ij}, and c denote real numbers, $c(a_{ij} + b_{ij}) = ca_{ij} + cb_{ij}$. Now, the elements of cA are of the form ca_{ij}, and those of cB are of the form cb_{ij}, so that the elements of $cA + cB$ are also of the form $ca_{ij} + cb_{ij}$, and the theorem is proved.

Mathematical systems having the properties listed in Theorems 10.1 and 10.2 are called **vector spaces** over the field R of real numbers. For example, the set $S_{2 \times 3}$ of 2×3 matrices—or more generally the set $S_{m \times n}$ of $m \times n$ matrices, with m and n fixed—is a vector space over R. Actually, properties VI, VII, and VIII of Theorem 10.2 are logical consequences of the other ten properties listed in Theorems 10.1 and 10.2 and could therefore be omitted as fundamental properties of a vector space. We have included them here for ready reference, since they are often used in computations.

Turning now to the product of two matrices, we have the following definition.

Definition 10.7 *The product of the matrices $A_{m \times p}$ and $B_{p \times n}$ is the matrix $(AB)_{m \times n}$ with entries determined as follows: The entry c_{ij} in the ith row and jth column of $(AB)_{m \times n}$ is found by multiplying the first element in the ith row of A by the first element of the jth column of B, to this product adding the product of the second element in the ith row of A with the second element in the jth column of B, to this sum adding the product of the third element in the ith row of A with the third element in the jth column of B, and so on.*

For example, the product of the matrices

$$A_{2 \times 2} = \begin{bmatrix} a_{11} & a_{12} \\ a_{21} & a_{22} \end{bmatrix} \quad \text{and} \quad B_{2 \times 2} = \begin{bmatrix} b_{11} & b_{12} \\ b_{21} & b_{22} \end{bmatrix}$$

is the matrix

$$(AB)_{2 \times 2} = \begin{bmatrix} a_{11}b_{11} + a_{12}b_{21} & a_{11}b_{12} + a_{12}b_{22} \\ a_{21}b_{11} + a_{22}b_{21} & a_{21}b_{12} + a_{22}b_{22} \end{bmatrix}.$$

The following schematic shows how to find the entry in the first row and first column of AB.

$$\begin{bmatrix} a_{11} & a_{12} \\ a_{21} & a_{22} \end{bmatrix} \begin{bmatrix} b_{11} & b_{12} \\ b_{21} & b_{22} \end{bmatrix} = \begin{bmatrix} a_{11}b_{11} + a_{12}b_{21} & \end{bmatrix}.$$

Example   If $A = \begin{bmatrix} 1 & 2 \\ -1 & 3 \end{bmatrix}$ and $B = \begin{bmatrix} 2 & 1 \\ 1 & 1 \end{bmatrix}$, find AB and BA.

Solution
$$AB = \begin{bmatrix} 1 & 2 \\ -1 & 3 \end{bmatrix}\begin{bmatrix} 2 & 1 \\ 1 & 1 \end{bmatrix} = \begin{bmatrix} 2+2 & 1+2 \\ -2+3 & -1+3 \end{bmatrix} = \begin{bmatrix} 4 & 3 \\ 1 & 2 \end{bmatrix},$$

$$BA = \begin{bmatrix} 2 & 1 \\ 1 & 1 \end{bmatrix}\begin{bmatrix} 1 & 2 \\ -1 & 3 \end{bmatrix} = \begin{bmatrix} 2-1 & 4+3 \\ 1-1 & 2+3 \end{bmatrix} = \begin{bmatrix} 1 & 7 \\ 0 & 5 \end{bmatrix}.$$

Properties of products of matrices   The foregoing example shows very clearly that the multiplication of matrices, in general, is *not commutative*. Thus, when discussing products of matrices, we must specify the *order* in which the matrices are to be considered as factors. For the product AB, we say that A is right-multiplied by B, and that B is left-multiplied by A.

Note that the definition of the product of two matrices A and B requires that the matrix A have the same number of *columns* as B has *rows*; the result, AB, then has the same number of rows as A and the same number of columns as B. Such matrices A and B are said to be **conformable** for multiplication. The fact that two matrices are conformable in the order AB, however, does not mean that they necessarily are conformable in the order BA.

Example   If $A = \begin{bmatrix} 3 & 1 & 2 \\ 1 & 0 & 1 \end{bmatrix}$ and $B = \begin{bmatrix} 1 & -1 \\ 2 & 1 \\ 3 & 1 \end{bmatrix}$, find AB.

Solution   Since A is a 2×3 matrix, and B is a 3×2 matrix, they are conformable for multiplication. We have

$$AB = \begin{bmatrix} 3 & 1 & 2 \\ 1 & 0 & 1 \end{bmatrix}\begin{bmatrix} 1 & -1 \\ 2 & 1 \\ 3 & 1 \end{bmatrix} = \begin{bmatrix} 3+2+6 & -3+1+2 \\ 1+0+3 & -1+0+1 \end{bmatrix} = \begin{bmatrix} 11 & 0 \\ 4 & 0 \end{bmatrix}.$$

In much of the matrix work in this book, we shall focus our attention on matrices having the same number of rows as columns. For brevity, a matrix of order $n \times n$ is often called a **square matrix of order** n. Although many of the ideas we shall discuss are applicable to matrices of any order, we shall apply the notions only to square matrices. If A is a square matrix, then A^2, A^3, etc., denote AA, $(AA)A$, etc.

Theorem 10.3   *If A, B, and C are $n \times n$ square matrices, then*
$$(AB)C = A(BC).$$

Theorem 10.4   *If A, B, and C are $n \times n$ square matrices, then*

     I $A(B + C) = AB + AC,$

     II $(B + C)A = BA + CA.$

The proofs of these theorems involve some complicated symbolism and are omitted here, but you will be asked to show their validity for the case of 2×2 matrices in the exercises. Observe that, because matrix multiplication is not, in general, commutative, we must establish both the left-hand and the right-hand distributive property.

Definition 10.8 *The principal diagonal of a square matrix is the ordered set of entries a_{ij}, where $i = j$, extending from the upper left-hand corner to the lower right-hand corner of the matrix.*

For example, the principal diagonal of

$$\begin{bmatrix} 1 & 3 & -1 \\ 5 & 2 & 3 \\ 6 & 4 & 0 \end{bmatrix}$$

consists of 1, 2, and 0, in that order.

Definition 10.9 *A diagonal matrix is a square matrix in which all entries not in the principal diagonal are 0.*

Thus,

$$\begin{bmatrix} 4 & 0 \\ 0 & 2 \end{bmatrix} \quad \text{and} \quad \begin{bmatrix} 1 & 0 & 0 \\ 0 & 1 & 0 \\ 0 & 0 & 0 \end{bmatrix}$$

are diagonal matrices.

Definition 10.10 *$I_{n \times n}$ denotes the diagonal matrix having 1's for entries on the principal diagonal.*

For example,

$$I_{2 \times 2} = \begin{bmatrix} 1 & 0 \\ 0 & 1 \end{bmatrix} \quad \text{and} \quad I_{4 \times 4} = \begin{bmatrix} 1 & 0 & 0 & 0 \\ 0 & 1 & 0 & 0 \\ 0 & 0 & 1 & 0 \\ 0 & 0 & 0 & 1 \end{bmatrix}.$$

The importance of $I_{n \times n}$ to the operation of multiplication of $n \times n$ matrices is apparent in the following result.

Theorem 10.5 *For each matrix $A_{n \times n}$ we have*

$$A_{n \times n} I_{n \times n} = I_{n \times n} A_{n \times n} = A_{n \times n}.$$

Further, if for the matrix $B_{n \times n}$ we have

$$A_{n \times n} B_{n \times n} = B_{n \times n} A_{n \times n} = A_{n \times n}$$

for all matrices $A_{n \times n}$, then

$$B_{n \times n} = I_{n \times n}.$$

Accordingly, $I_{n \times n}$ is the **identity element** for multiplication in the set of $n \times n$ square matrices, and $I_{n \times n}$ is unique. The proof of this theorem, for the illustrative case $n = 2$, is left as an exercise.

We can summarize the properties of multiplication of $n \times n$ matrices by listing them as we did the analogous properties for addition. That is, corresponding to Theorem 10.1, we can state the following theorem.

Theorem 10.6 *If $S_{n \times n}$ is the set of $n \times n$ square matrices, for n a fixed positive integer, and A, B, $C \in S_{n \times n}$, then:*

I	$AB \in S_{n \times n}$.	*Closure law for multiplication.*
II	$(AB)C = A(BC)$.	*Associative law for multiplication.*
III	$A(B + C) = AB + AC$ *and*	*Distributive laws.*
	$(B + C)A = BA + CA$.	
IV	$AI_{n \times n} = A$ *and* $I_{n \times n} A = A$.	*Multiplicative-identity law.*

Proof Property I is an immediate consequence of Definition 10.7. Properties II, III, and IV are repetitions of the properties stated in Theorems 10.3, 10.4, and 10.5, respectively.

The set of matrices not a field   If you compare the properties for sums and products of $n \times n$ matrices as listed in Theorems 10.1 and 10.6, with those that characterize a field on page 11, you will see that the set is not a field, because two of the field properties do *not* hold in $S_{n \times n}$. As we have already pointed out, the commutative law does not hold for matrix multiplication, and in Section 10.5 we shall see that there are square $n \times n$ matrices other than the zero $n \times n$ matrix that do not have multiplicative inverses.

Order of multiplication   The following result states the order in which matrices can be multiplied by real numbers and by other matrices.

Theorem 10.7 *If A and B are $n \times n$ square matrices, and a is a real number, then*

$$a(AB) = (aA)B = A(aB).$$

The proof for the illustrative case $n = 2$ is left as an exercise.

Exercise 10.2

Write each product as a single matrix.

Examples
a. $3\begin{bmatrix} 2 & 1 \\ -1 & 3 \\ 2 & 0 \end{bmatrix}$

b. $\begin{bmatrix} 3 & 1 & -1 \\ 0 & -1 & 2 \end{bmatrix} \cdot \begin{bmatrix} 1 & -1 \\ 0 & 2 \\ 1 & 0 \end{bmatrix}$

Solutions
a. $\begin{bmatrix} 6 & 3 \\ -3 & 9 \\ 6 & 0 \end{bmatrix}$

b. $\begin{bmatrix} 3+0-1 & -3+2+0 \\ 0+0+2 & 0-2+0 \end{bmatrix} = \begin{bmatrix} 2 & -1 \\ 2 & -2 \end{bmatrix}$

1. $-5\begin{bmatrix} 0 & 1 & -1 \\ 3 & -1 & 2 \end{bmatrix}$

2. $2\begin{bmatrix} 2 & 1 & 3 & -2 \\ 4 & 2 & 0 & -1 \\ 0 & 0 & -1 & 2 \end{bmatrix}$

3. $\begin{bmatrix} 1 & -2 \end{bmatrix} \cdot \begin{bmatrix} 3 \\ 2 \end{bmatrix}$

4. $\begin{bmatrix} 3 & -2 & 2 \end{bmatrix} \cdot \begin{bmatrix} 1 \\ 0 \\ -2 \end{bmatrix}$

5. $\begin{bmatrix} 3 & -1 \\ 2 & 1 \end{bmatrix} \cdot \begin{bmatrix} 1 & -4 \\ 2 & 1 \end{bmatrix}$

6. $\begin{bmatrix} 1 & -5 \\ 0 & 2 \end{bmatrix} \cdot \begin{bmatrix} 3 & 1 \\ -1 & 2 \end{bmatrix}$

7. $\begin{bmatrix} 4 & -5 \\ 7 & 3 \end{bmatrix} \cdot \begin{bmatrix} 5 & -1 \\ -2 & 7 \end{bmatrix}$

8. $\begin{bmatrix} 1 & -2 \\ -3 & 1 \end{bmatrix} \cdot \begin{bmatrix} 5 & 1 \\ 0 & 2 \end{bmatrix}$

9. $\begin{bmatrix} -3 & 1 & 0 \\ 2 & 1 & 1 \end{bmatrix} \cdot \begin{bmatrix} 2 & 0 \\ 1 & -1 \\ 3 & 0 \end{bmatrix}$

10. $\begin{bmatrix} 1 & -1 & 0 \\ 2 & 1 & 3 \end{bmatrix} \cdot \begin{bmatrix} 4 & -1 \\ 2 & 0 \\ 1 & 1 \end{bmatrix}$

11. $\begin{bmatrix} -1 & 0 & 1 \\ 2 & 1 & 0 \\ 1 & 0 & 0 \end{bmatrix} \cdot \begin{bmatrix} 0 & 1 & 3 \\ 1 & 0 & 2 \\ -1 & 1 & 1 \end{bmatrix}$

12. $\begin{bmatrix} 2 & -3 & 1 \\ 0 & 1 & -1 \\ 2 & 0 & 0 \end{bmatrix} \cdot \begin{bmatrix} 1 & 0 & 0 \\ 0 & 1 & 0 \\ 0 & 0 & 1 \end{bmatrix}$

13. $\begin{bmatrix} 2 & -2 & -1 \\ 1 & 1 & -2 \\ 1 & 0 & -1 \end{bmatrix} \cdot \begin{bmatrix} -1 & -2 & 5 \\ -1 & -1 & 3 \\ -1 & -2 & 4 \end{bmatrix}$

14. $\begin{bmatrix} -1 & -2 & 5 \\ -1 & -1 & 3 \\ -1 & -2 & 4 \end{bmatrix} \cdot \begin{bmatrix} 2 & -2 & -1 \\ 1 & 1 & -2 \\ 1 & 0 & -1 \end{bmatrix}$

Let $A = \begin{bmatrix} 1 & -2 \\ 1 & 0 \end{bmatrix}$ *and* $B = \begin{bmatrix} -1 & 2 \\ -1 & 1 \end{bmatrix}$. *Compute each of the following products.*

15. AB

16. BA

17. $(AB)A$

18. $(BA)B$

19. A^tB

20. AB^t

Find a matrix X satisfying each matrix equation.

21. $3X + \begin{bmatrix} 1 & 0 \\ 2 & 1 \end{bmatrix} = \begin{bmatrix} -2 & 3 \\ -1 & -2 \end{bmatrix}$ 22. $2X + 3 \begin{bmatrix} 1 & 1 \\ 0 & 1 \end{bmatrix} = \begin{bmatrix} 7 & -1 \\ 3 & -5 \end{bmatrix}$

23. $X + 2I = \begin{bmatrix} 3 & -1 \\ 1 & 2 \end{bmatrix}$ *Hint*: $2I = \begin{bmatrix} 2 & 0 \\ 0 & 2 \end{bmatrix}.$

24. $3X - 2I = \begin{bmatrix} 7 & 3 \\ 6 & 4 \end{bmatrix}$

25. Show that if $A = \begin{bmatrix} -1 & 2 \\ 0 & 1 \end{bmatrix}$ and $B = \begin{bmatrix} 1 & 0 \\ -1 & 2 \end{bmatrix}$, then

   a. $(A + B)(A + B) \neq A^2 + 2AB + B^2$,
   b. $(A + B)(A - B) \neq A^2 - B^2$.

26. a. Show that for each matrix $A_{2 \times 2}$,

   $$A_{2 \times 2} \cdot I_{2 \times 2} = I_{2 \times 2} \cdot A_{2 \times 2} = A_{2 \times 2}.$$

   b. Show that if, for a given matrix $B_{2 \times 2}$ and for *all* $A_{2 \times 2}$,

   $$A_{2 \times 2} \cdot B_{2 \times 2} = B_{2 \times 2} \cdot A_{2 \times 2} = A_{2 \times 2},$$

   then $B_{2 \times 2} = I_{2 \times 2}$.

27. Show that

   $$\left(\begin{bmatrix} a_{11} & a_{12} \\ a_{21} & a_{22} \end{bmatrix} \cdot \begin{bmatrix} b_{11} & b_{12} \\ b_{21} & b_{22} \end{bmatrix} \right) \cdot \begin{bmatrix} c_{11} & c_{12} \\ c_{21} & c_{22} \end{bmatrix} = \begin{bmatrix} a_{11} & a_{12} \\ a_{21} & a_{22} \end{bmatrix} \cdot \left(\begin{bmatrix} b_{11} & b_{12} \\ b_{21} & b_{22} \end{bmatrix} \cdot \begin{bmatrix} c_{11} & c_{12} \\ c_{21} & c_{22} \end{bmatrix} \right).$$

28. Show that

   $$\begin{bmatrix} a_{11} & a_{12} \\ a_{21} & a_{22} \end{bmatrix} \cdot \left(\begin{bmatrix} b_{11} & b_{12} \\ b_{21} & b_{22} \end{bmatrix} + \begin{bmatrix} c_{11} & c_{12} \\ c_{21} & c_{22} \end{bmatrix} \right)$$

   $$= \begin{bmatrix} a_{11} & a_{12} \\ a_{21} & a_{22} \end{bmatrix} \cdot \begin{bmatrix} b_{11} & b_{12} \\ b_{21} & b_{22} \end{bmatrix} + \begin{bmatrix} a_{11} & a_{12} \\ a_{21} & a_{22} \end{bmatrix} \cdot \begin{bmatrix} c_{11} & c_{12} \\ c_{21} & c_{22} \end{bmatrix}.$$

29. Show that

   $$\left(\begin{bmatrix} b_{11} & b_{12} \\ b_{21} & b_{22} \end{bmatrix} + \begin{bmatrix} c_{11} & c_{12} \\ c_{21} & c_{22} \end{bmatrix} \right) \cdot \begin{bmatrix} a_{11} & a_{12} \\ a_{21} & a_{22} \end{bmatrix}$$

   $$= \begin{bmatrix} b_{11} & b_{12} \\ b_{21} & b_{22} \end{bmatrix} \cdot \begin{bmatrix} a_{11} & a_{12} \\ a_{21} & a_{22} \end{bmatrix} + \begin{bmatrix} c_{11} & c_{12} \\ c_{21} & c_{22} \end{bmatrix} \cdot \begin{bmatrix} a_{11} & a_{12} \\ a_{21} & a_{22} \end{bmatrix},$$

30. Show that $\begin{bmatrix} 0 & a \\ a & 0 \end{bmatrix}^2 = a^2 I.$

31. Show that $(A_{2 \times 2} \cdot B_{2 \times 2})^t = B_{2 \times 2}^t \cdot A_{2 \times 2}^t.$

32. Show that $A_{2 \times 2}^2 = (-A_{2 \times 2})^2.$

In Problems 33–39, prove the specified part of Theorem 10.2 for 2 × 2 matrices.

33. Part I   34. Part II   35. Part III   36. Part V

37. Part VI   38. Part VII   39. Part VIII

40. Prove that if A and B are 2×2 matrices and a is a real number, then $a(AB) = (aA)B = A(aB)$.

10.3

The determinant function

Associated with each square matrix A having real-number entries is a real number called the **determinant** of A and denoted by δA or $\delta(A)$ (read "the determinant of A"). Thus δ (delta) is a function with domain the set of all square matrices having real-number entries and with range the set of all real numbers. We say that $\delta(A_{n \times n})$ is a determinant of **order** n. Let us begin by examining δ over the set $S_{2 \times 2}$ of 2×2 matrices.

Definition 10.11   *The determinant of the matrix*

$$\begin{bmatrix} a_{11} & a_{12} \\ a_{21} & a_{22} \end{bmatrix}$$

is the number $a_{11}a_{22} - a_{12}a_{21}$.

The determinant of a matrix is customarily displayed in the same form as a matrix, but with vertical bars in lieu of brackets. Thus,

$$\delta \begin{bmatrix} a_{11} & a_{12} \\ a_{21} & a_{22} \end{bmatrix} = \begin{vmatrix} a_{11} & a_{12} \\ a_{21} & a_{22} \end{vmatrix} = a_{11}a_{22} - a_{12}a_{21}.$$

Example   Given $A = \begin{bmatrix} 3 & 1 \\ -2 & 3 \end{bmatrix}$, find δA.

Solution $\delta A = \delta \begin{bmatrix} 3 & 1 \\ -2 & 3 \end{bmatrix} = \begin{vmatrix} 3 & 1 \\ -2 & 3 \end{vmatrix} = 3 \cdot 3 - (1)(-2) = 9 + 2 = 11.$

Turning next to 3×3 matrices, we have the following definition.

Definition 10.12   *The determinant of the matrix*

$$\begin{bmatrix} a_{11} & a_{12} & a_{13} \\ a_{21} & a_{22} & a_{23} \\ a_{31} & a_{32} & a_{33} \end{bmatrix}, \quad \textit{denoted by} \quad \begin{vmatrix} a_{11} & a_{12} & a_{13} \\ a_{21} & a_{22} & a_{23} \\ a_{31} & a_{32} & a_{33} \end{vmatrix},$$

is given by the expression

$$a_{11}a_{22}a_{33} - a_{11}a_{23}a_{32} + a_{12}a_{23}a_{31} - a_{12}a_{21}a_{33} + a_{13}a_{21}a_{32} - a_{13}a_{22}a_{31}.$$

Sign of a term in the expression of a determinant   An inspection of the subscripts of the factors of the products involved in this determinant (as well as those of the determinant of a 2×2 matrix) will show that each product is formed by taking one entry from each row and one entry from each column, with the restriction that no two factors be entries in the same row or column. The determinant consists of the sum of all such $\pm$ products as are possible. Whether a product or its negative is used in computing the sum of the terms in a determinant depends on the number of **inversions** in the second subscripts in the product when the first subscripts are in natural order, 1 2 3 .... An inversion occurs in a sequence of natural numbers each time a natural number is preceded by a greater natural number. For example, in the sequence 1 4 3 2, there are three inversions, because 4 precedes 3, 4 precedes 2, and 3 precedes 2. Now, if there is an odd number of inversions in the sequence formed by the *second subscripts of the factors in a product* when the first subscripts are in natural order, then the negative of the product is used; otherwise, the product itself is used.

With this means of distinguishing the signs associated with products, we can generalize our definition of the determinant of a matrix.

Definition 10.13   *The determinant of the square matrix*

$$\begin{bmatrix} a_{11} & a_{12} & \cdots & a_{1n} \\ a_{21} & a_{22} & \cdots & a_{2n} \\ \vdots & \vdots & & \vdots \\ a_{n1} & a_{n2} & \cdots & a_{nn} \end{bmatrix}, \quad \textit{denoted by} \quad \begin{vmatrix} a_{11} & a_{12} & \cdots & a_{1n} \\ a_{21} & a_{22} & \cdots & a_{2n} \\ \vdots & \vdots & & \vdots \\ a_{n1} & a_{n2} & \cdots & a_{nn} \end{vmatrix},$$

is equal to the sum of all products, $\pm a_{1j_1}a_{2j_2}a_{3j_3}\cdots a_{nj_n}$, where each j takes on all values from 1 to n, and no j's in the same product are equal. For each term in the sum, the negative sign is used if the number of inversions in the sequence formed by the j's is odd; otherwise, the positive sign is used.

Value of a determinant   It is evident from this definition that the determinant of a matrix with real entries is a real number. We shall refer to this real number as the **value** of the determinant; the process of computing the number is called **expanding** the determinant. In particular, the value of the determinant of the matrix $[a_{11}]$ is a_{11}.

Example In expanding the determinant of a 4×4 square matrix, one of the products is $a_{11} a_{23} a_{32} a_{44}$. Is this product a term in the expansion or is its negative a term in the expansion?

Solution The first subscripts are in natural order, 1 2 3 4, and the second subscripts are in the order 1 3 2 4. Since the only inversion is that 3 appears before 2, the negative of $a_{11} a_{23} a_{32} a_{44}$ is in the expansion of the determinant.

Definition 10.14 *The minor M_{ij} of the element a_{ij} in a given determinant is the determinant that remains after the ith row and jth column in the given determinant have been deleted.*

For example, in the determinant

$$\begin{vmatrix} a_{11} & a_{12} & a_{13} \\ a_{21} & a_{22} & a_{23} \\ a_{31} & a_{32} & a_{33} \end{vmatrix}, \tag{1}$$

the minor of the element a_{11} is $M_{11} = \begin{vmatrix} a_{22} & a_{23} \\ a_{32} & a_{33} \end{vmatrix}$,

the minor of the element a_{23} is $M_{23} = \begin{vmatrix} a_{11} & a_{12} \\ a_{31} & a_{32} \end{vmatrix}$,

the minor of the element a_{31} is $M_{31} = \begin{vmatrix} a_{12} & a_{13} \\ a_{22} & a_{23} \end{vmatrix}$.

Definition 10.15 *The cofactor A_{ij} of the element a_{ij} is the minor of a_{ij} if $i + j$ is an even integer, and the negative of the minor of a_{ij} if $i + j$ is an odd integer.*

For example, in the determinant (1),

$$\text{the cofactor of } a_{11} \text{ is } \begin{vmatrix} a_{22} & a_{23} \\ a_{32} & a_{33} \end{vmatrix},$$

because $1 + 1$ is 2, an even integer;

$$\text{the cofactor of } a_{23} \text{ is } -\begin{vmatrix} a_{11} & a_{12} \\ a_{31} & a_{32} \end{vmatrix},$$

because $2 + 3$ is 5, an odd integer;

$$\text{the cofactor of } a_{31} \text{ is } \begin{vmatrix} a_{12} & a_{13} \\ a_{22} & a_{23} \end{vmatrix},$$

because $3 + 1$ is 4, an even integer.

The following sign array is a convenient means of determining whether the cofactor of a given element equals the minor or whether it equals the negative of the minor.

$$
\begin{array}{ccccccc}
+ & - & + & \cdot & \cdot & \cdot & (-)^{n+1} \\
- & + & - & \cdot & \cdot & \cdot & \cdot \\
+ & - & + & \cdot & \cdot & \cdot & \cdot \\
\cdot & \cdot & \cdot & \cdot & \cdot & \cdot & \cdot \\
\cdot & \cdot & \cdot & \cdot & \cdot & \cdot & \cdot \\
\cdot & \cdot & \cdot & \cdot & \cdot & \cdot & \cdot \\
(-)^{n+1} & & & \cdot & \cdot & \cdot & +
\end{array}
$$

Expansion by cofactors   With the definition of a cofactor in mind, let us look again at the definition on page 293 for the determinant of the matrix

$$
A = \begin{bmatrix} a_{11} & a_{12} & a_{13} \\ a_{21} & a_{22} & a_{23} \\ a_{31} & a_{32} & a_{33} \end{bmatrix}.
$$

By Definition 10.12, the value of this determinant is

$$
\delta(A) = a_{11}a_{22}a_{33} - a_{11}a_{23}a_{32} + a_{12}a_{23}a_{31} - a_{12}a_{21}a_{33} \\
+ a_{13}a_{21}a_{32} - a_{13}a_{22}a_{31}.
$$

By suitably factoring pairs of terms in the right-hand member, we obtain

$$
\delta(A) = a_{11}(a_{22}a_{33} - a_{23}a_{32}) + a_{12}(a_{23}a_{31} - a_{21}a_{33}) + a_{13}(a_{21}a_{32} - a_{22}a_{31}).
$$

If the binomial factor in the middle term is rewritten $-(a_{21}a_{33} - a_{23}a_{31})$, we have

$$
\delta(A) = a_{11}(a_{22}a_{33} - a_{23}a_{32}) + a_{12}[-(a_{21}a_{33} - a_{23}a_{31})] + a_{13}(a_{21}a_{32} - a_{22}a_{31}),
$$

which is equal to

$$
a_{11}\begin{vmatrix} a_{22} & a_{23} \\ a_{32} & a_{33} \end{vmatrix} - a_{12}\begin{vmatrix} a_{21} & a_{23} \\ a_{31} & a_{33} \end{vmatrix} + a_{13}\begin{vmatrix} a_{21} & a_{22} \\ a_{31} & a_{32} \end{vmatrix}.
$$

Accordingly, we have

$$
\delta(A) = a_{11}A_{11} + a_{12}A_{12} + a_{13}A_{13}.
$$

Thus, the determinant $\delta(A)$ is equal to the sum formed by multiplying each entry in the first row by its cofactor and then adding these products.

Similar methods can be used to show that this determinant is also equal to the sum formed by multiplying each entry in *any* row (or column) by its cofactor and then adding the products.

Although we shall not show it here, Definition 10.12 is logically equivalent to the following one.

Definition 10.16 *The determinant of the square matrix*

$$\begin{bmatrix} a_{11} & a_{12} & \cdots & a_{1n} \\ a_{21} & a_{22} & \cdots & a_{2n} \\ \vdots & & & \vdots \\ a_{n1} & a_{n2} & \cdots & a_{nn} \end{bmatrix}$$

is the sum of the n products formed by multiplying each entry in any single row (or any single column) by its cofactor.

When this latter definition is used to rewrite a determinant, the determinant is said to be expanded about whatever row (or column) is chosen.

Example   If $A = \begin{bmatrix} 3 & 2 & 1 \\ 0 & 1 & -2 \\ 1 & 3 & 4 \end{bmatrix}$, find $\delta(A)$ by expansion about the first

column.

Solution   Noting that $a_{11} = 3$, $a_{21} = 0$, and $a_{31} = 1$, we have

$$\delta(A) = 3 \begin{vmatrix} 1 & -2 \\ 3 & 4 \end{vmatrix} - 0 \begin{vmatrix} 2 & 1 \\ 3 & 4 \end{vmatrix} + 1 \begin{vmatrix} 2 & 1 \\ 1 & -2 \end{vmatrix}$$

$$= 3(10) + 0 + (-5) = 25.$$

Exercise 10.3

$$\text{Let } A = \begin{bmatrix} 2 & 1 & -2 & 0 \\ 1 & 0 & 3 & -1 \\ -2 & 1 & 2 & 2 \\ 1 & -1 & 3 & 1 \end{bmatrix}.$$

Each of the following products is a term in an expansion of $\delta(A)$. Determine the product, and write the product or its negative in accordance with the number of inversions in the second subscripts.

1. $a_{11}a_{22}a_{33}a_{44}$ 2. $a_{11}a_{23}a_{34}a_{42}$ 3. $a_{11}a_{24}a_{32}a_{43}$

4. $a_{12}a_{21}a_{33}a_{44}$ 5. $a_{13}a_{24}a_{32}a_{41}$ 6. $a_{14}a_{23}a_{32}a_{41}$

Let A be the matrix given above. Determine the minor M_{ij} and cofactor A_{ij} (in determinant form) of each of the following entries.

7. a_{11} 8. a_{13} 9. a_{23} 10. a_{41}

11. a_{31} 12. a_{33} 13. a_{44} 14. a_{14}

Evaluate each determinant.

Examples   a. $\begin{vmatrix} 2 & -3 \\ 1 & 4 \end{vmatrix}$ b. $\begin{vmatrix} 1 & 2 & 0 \\ 3 & -1 & 4 \\ -2 & 1 & 3 \end{vmatrix}$

Solutions   a. $\begin{vmatrix} 2 & -3 \\ 1 & 4 \end{vmatrix} = (2)(4) - (-3)(1) = 11.$

        b. Expand about any row or column; the first row is used here:

$$\begin{vmatrix} 1 & 2 & 0 \\ 3 & -1 & 4 \\ -2 & 1 & 3 \end{vmatrix} = 1 \begin{vmatrix} -1 & 4 \\ 1 & 3 \end{vmatrix} - 2 \begin{vmatrix} 3 & 4 \\ -2 & 3 \end{vmatrix} + 0 \begin{vmatrix} 3 & -1 \\ -2 & 1 \end{vmatrix}$$

$$= 1[(-1)(3) - (4)(1)] - 2[(3)(3) - (4)(-2)] + 0$$

$$= -41.$$

15. $\begin{vmatrix} 1 & 0 \\ 2 & 1 \end{vmatrix}$ 16. $\begin{vmatrix} 3 & -2 \\ 4 & 1 \end{vmatrix}$ 17. $\begin{vmatrix} -3 & -1 \\ 3 & 1 \end{vmatrix}$ 18. $\begin{vmatrix} -1 & 6 \\ 0 & -2 \end{vmatrix}$

19. $\begin{vmatrix} -4 & 2 \\ 1 & 7 \end{vmatrix}$ 20. $\begin{vmatrix} 8 & -1 \\ 3 & 2 \end{vmatrix}$ 21. $\begin{vmatrix} x & -1 \\ 1 & 3 \end{vmatrix}$ 22. $\begin{vmatrix} -3 & x \\ -1 & 3 \end{vmatrix}$

23. $\begin{vmatrix} 2 & 0 & 1 \\ 1 & 1 & 2 \\ -1 & 0 & 1 \end{vmatrix}$ 24. $\begin{vmatrix} 1 & 3 & 1 \\ -1 & 2 & 1 \\ 0 & 2 & 0 \end{vmatrix}$ 25. $\begin{vmatrix} 1 & 2 & 3 \\ 3 & -1 & 2 \\ 2 & 0 & 2 \end{vmatrix}$

26. $\begin{vmatrix} 1 & 0 & 0 \\ 0 & 1 & 2 \\ 0 & 3 & 4 \end{vmatrix}$ 27. $\begin{vmatrix} -1 & 0 & 2 \\ -2 & 1 & 0 \\ 0 & 1 & -3 \end{vmatrix}$ 28. $\begin{vmatrix} 2 & 1 & 4 \\ 3 & 2 & 6 \\ 5 & -3 & 10 \end{vmatrix}$

29. $\begin{vmatrix} a & b & 1 \\ a & b & 1 \\ 1 & 1 & 1 \end{vmatrix}$ 30. $\begin{vmatrix} a & a & a \\ 1 & 2 & 3 \\ 4 & 5 & 6 \end{vmatrix}$ 31. $\begin{vmatrix} x & 0 & 0 \\ 0 & x & 0 \\ 0 & 0 & x \end{vmatrix}$ 32. $\begin{vmatrix} 0 & 0 & x \\ 0 & x & 0 \\ x & 0 & 0 \end{vmatrix}$

Solve for x.

33. $\begin{vmatrix} x & -1 \\ 2 & 3 \end{vmatrix} = 17$ 34. $\begin{vmatrix} 1 & -5 \\ x & 3 \end{vmatrix} = -7$

35. $\begin{vmatrix} x & 0 & 0 \\ 2 & 1 & 3 \\ 0 & 1 & 4 \end{vmatrix} = 3$ 36. $\begin{vmatrix} x^2 & x & 1 \\ 0 & 2 & 1 \\ 3 & 1 & 4 \end{vmatrix} = 28$

Expand by cofactors and verify.

37. $\begin{vmatrix} 0 & 1 & 0 & 0 \\ 1 & 0 & 3 & 2 \\ 5 & -1 & 2 & 1 \\ 1 & 0 & 1 & 1 \end{vmatrix} = 5$ 38. $\begin{vmatrix} 1 & 2 & 0 & -1 \\ 1 & 0 & -1 & 2 \\ 0 & 1 & 1 & 1 \\ 2 & -1 & 0 & 1 \end{vmatrix} = 17$

39. In accordance with Definition 10.13, the determinant of an $n \times n$ matrix is the sum of a certain number of products. What is this number for $n = 2$? For $n = 3$? For $n = 4$?

40. Generalize the results of Problem 39, making a conjecture about the number of such products in the determinant of an $n \times n$ matrix.

41. Show that for any 2×2 matrix A, $\delta(aA) = a^2\delta(A)$.

42. Show that for any 2×2 matrix A, $\delta(A') = \delta(A)$.

43. Show that for any 2×2 matrices A and B, $\delta(AB) = \delta(A) \cdot \delta(B)$.

10.4

Properties of determinants

Determinants have some properties that are useful by virtue of the fact that they permit us to generate equal determinants with different and simpler configurations of entries. This, in turn, helps us find values for determinants. We shall list these properties in the form of theorems and leave the proofs for the case of 2×2 determinants as exercises.

Theorem 10.8 *If each entry in any row or each entry in any column of a determinant is 0, then the determinant is equal to 0.*

Examples   a. $\begin{vmatrix} 0 & 0 \\ 1 & 2 \end{vmatrix} = 0$ b. $\begin{vmatrix} 1 & 1 & 0 \\ 3 & 5 & 0 \\ 2 & 7 & 0 \end{vmatrix} = 0$ c. $\begin{vmatrix} 0 & 1 & 0 & 0 \\ 1 & 0 & 0 & 0 \\ 0 & 0 & 0 & 1 \end{vmatrix} = 0.$

Theorem 10.9 *If any two rows (or columns) of a determinant are interchanged, the resulting determinant is the negative of the original determinant.*

Examples   a. $\begin{vmatrix} 1 & 2 \\ 3 & 4 \end{vmatrix} = - \begin{vmatrix} 3 & 4 \\ 1 & 2 \end{vmatrix}$ b. $\begin{vmatrix} 1 & 2 & 3 \\ 4 & 5 & 6 \\ 7 & 8 & 9 \end{vmatrix} = - \begin{vmatrix} 3 & 2 & 1 \\ 6 & 5 & 4 \\ 9 & 8 & 7 \end{vmatrix}$

In the first example, rows 1 and 2 were interchanged. In the second example, columns 1 and 3 were interchanged.

Theorem 10.10 *If two rows (or two columns) in a determinant have corresponding entries that are equal, the determinant is equal to 0.*

Examples   a. $\begin{vmatrix} 1 & 1 \\ 3 & 3 \end{vmatrix} = 0$ b. $\begin{vmatrix} 1 & 2 & 1 \\ 3 & 1 & 0 \\ 1 & 2 & 1 \end{vmatrix} = 0$ c. $\begin{vmatrix} 1 & 2 & 3 & 4 \\ 5 & 6 & 7 & 8 \\ 0 & 0 & 1 & 0 \\ 1 & 2 & 3 & 4 \end{vmatrix} = 0$

Theorem 10.11 *If each of the entries of one row (or column) of a determinant is multiplied by k, then the determinant is multiplied by k.*

Examples   a. $\begin{vmatrix} 1 & 0 & 0 \\ 2 & 1 & 3 \\ 1 \times 2 & 3 \times 2 & 4 \times 2 \end{vmatrix} = 2 \begin{vmatrix} 1 & 0 & 0 \\ 2 & 1 & 3 \\ 1 & 3 & 4 \end{vmatrix}$

           b. $\begin{vmatrix} 4 & 5 & 8 \\ 1 & 1 & 2 \\ 3 & 1 & 6 \end{vmatrix} = 2 \begin{vmatrix} 4 & 5 & 4 \\ 1 & 1 & 1 \\ 3 & 1 & 3 \end{vmatrix}$

Note that this process is different from that of the multiplication of a matrix by a real number. In the latter, each entry in the matrix is multiplied by the real number, rather than, as here, only the entries in a single row or column being so multiplied.

Theorem 10.12 *If each entry in a row (or column) of a determinant is written as the sum of two terms, then the determinant can be written as the sum of two determinants as follows: If*

$$D = \begin{vmatrix} a_{11} & a_{12} & \cdots & a_{1n} \\ \vdots & \vdots & & \vdots \\ b_{i1} + c_{i1} & b_{i2} + c_{i2} & \cdots & b_{in} + c_{in} \\ \vdots & \vdots & & \vdots \\ a_{n1} & a_{n2} & \cdots & a_{nn} \end{vmatrix},$$

(Theorem continued overleaf.)

then

$$D = \begin{vmatrix} a_{11} & a_{12} & \cdots & a_{1n} \\ \vdots & \vdots & & \vdots \\ b_{i1} & b_{i2} & \cdots & b_{in} \\ \vdots & \vdots & & \vdots \\ a_{n1} & a_{n2} & \cdots & a_{nn} \end{vmatrix} + \begin{vmatrix} a_{11} & a_{12} & \cdots & a_{1n} \\ \vdots & \vdots & & \vdots \\ c_{i1} & c_{i2} & \cdots & c_{in} \\ \vdots & \vdots & & \vdots \\ a_{n1} & a_{n2} & \cdots & a_{nn} \end{vmatrix};$$

and if

$$D = \begin{vmatrix} a_{11} & \cdots & b_{1j} + c_{1j} & \cdots & a_{1n} \\ a_{21} & \cdots & b_{2j} + c_{2j} & \cdots & a_{2n} \\ \vdots & & \vdots & & \vdots \\ a_{n1} & \cdots & b_{nj} + c_{nj} & \cdots & a_{nn} \end{vmatrix},$$

then

$$D = \begin{vmatrix} a_{11} & \cdots & b_{1j} & \cdots & a_{1n} \\ a_{21} & \cdots & b_{2j} & \cdots & a_{2n} \\ \vdots & & \vdots & & \vdots \\ a_{n1} & \cdots & b_{nj} & \cdots & a_{nn} \end{vmatrix} + \begin{vmatrix} a_{11} & \cdots & c_{1j} & \cdots & a_{1n} \\ a_{21} & \cdots & c_{2j} & \cdots & a_{2n} \\ \vdots & & \vdots & & \vdots \\ a_{n1} & \cdots & c_{nj} & \cdots & a_{nn} \end{vmatrix}.$$

Examples   a. $\begin{vmatrix} 1 & 3 \\ 2 & 5 \end{vmatrix} = \begin{vmatrix} 1 & 1 \\ 2 & 4 \end{vmatrix} + \begin{vmatrix} 1 & 2 \\ 2 & 1 \end{vmatrix}$

b. $\begin{vmatrix} 4 & 0 & 0 \\ 0 & 4 & 0 \\ 0 & 0 & 4 \end{vmatrix} = \begin{vmatrix} 2 & 0 & 0 \\ 0 & 4 & 0 \\ 0 & 0 & 4 \end{vmatrix} + \begin{vmatrix} 2 & 0 & 0 \\ 0 & 4 & 0 \\ 0 & 0 & 4 \end{vmatrix}$

Theorem 10.13 *If each entry of one row (or column) of a determinant is multiplied by a real number k and the resulting product is added to the corresponding entry in another row (or column, respectively) in the determinant, then the resulting determinant is equal to the original determinant.*

Examples   a. $\begin{vmatrix} 1 & 1 \\ 2 & 1 \end{vmatrix} = \begin{vmatrix} 1 & 1 \\ 2 + 3(1) & 1 + 3(1) \end{vmatrix} = \begin{vmatrix} 1 & 1 \\ 5 & 4 \end{vmatrix}$

b. $\begin{vmatrix} 1 & 2 & 3 \\ 4 & 5 & 6 \\ 7 & 8 & 9 \end{vmatrix} = \begin{vmatrix} 1 + 2(3) & 2 & 3 \\ 4 + 2(6) & 5 & 6 \\ 7 + 2(9) & 8 & 9 \end{vmatrix} = \begin{vmatrix} 7 & 2 & 3 \\ 16 & 5 & 6 \\ 25 & 8 & 9 \end{vmatrix}$

The preceding theorems can be used to write sequences of equal determinants, leading from one form of a determinant to another and more useful form.

Example   Expand

$$D = \begin{vmatrix} 2 & -1 & 1 & -3 \\ 1 & 3 & -4 & 2 \\ 1 & 0 & -2 & 1 \\ 3 & -1 & 5 & 2 \end{vmatrix}.$$

Solution   As a step toward expanding the determinant, we shall use Theorem 10.13 to produce an equal determinant with a row or a column containing zero entries in all but one place. Let us arbitrarily select the second column for this role, because one entry is already zero. Multiplying a_{1j} by 3 and adding the result to a_{2j}, we obtain

$$D = \begin{vmatrix} 2 & -1 & 1 & -3 \\ 1+3(2) & 3+3(-1) & -4+3(1) & 2+3(-3) \\ 1 & 0 & -2 & 1 \\ 3 & -1 & 5 & 2 \end{vmatrix} = \begin{vmatrix} 2 & -1 & 1 & -3 \\ 7 & 0 & -1 & -7 \\ 1 & 0 & -2 & 1 \\ 3 & -1 & 5 & 2 \end{vmatrix}.$$

Next, multiplying a_{1j} by -1 and adding the result to a_{4j}, we find that

$$D = \begin{vmatrix} 2 & -1 & 1 & -3 \\ 7 & 0 & -1 & -7 \\ 1 & 0 & -2 & 1 \\ 3-1(2) & -1-1(-1) & 5-1(1) & 2-1(-3) \end{vmatrix} = \begin{vmatrix} 2 & -1 & 1 & -3 \\ 7 & 0 & -1 & -7 \\ 1 & 0 & -2 & 1 \\ 1 & 0 & 4 & 5 \end{vmatrix}.$$

If we now expand the determinant about the second column, we have

$$D = \begin{vmatrix} 2 & -1 & 1 & -3 \\ 7 & 0 & -1 & -7 \\ 1 & 0 & -2 & 1 \\ 1 & 0 & 4 & 5 \end{vmatrix} = -(-1) \begin{vmatrix} 7 & -1 & -7 \\ 1 & -2 & 1 \\ 1 & 4 & 5 \end{vmatrix} + 0A_{22} - 0A_{32} + 0A_{42}.$$

From this point, we can reduce the third-order determinant to a second-order determinant by a similar procedure or, alternatively, expand directly about the elements in any row or column. Expanding about the elements of the first row, we obtain

$$\begin{vmatrix} 7 & -1 & -7 \\ 1 & -2 & 1 \\ 1 & 4 & 5 \end{vmatrix} = 7 \begin{vmatrix} -2 & 1 \\ 4 & 5 \end{vmatrix} - (-1) \begin{vmatrix} 1 & 1 \\ 1 & 5 \end{vmatrix} + (-7) \begin{vmatrix} 1 & -2 \\ 1 & 4 \end{vmatrix},$$

from which

$$D = 7(-14) + (4) - 7(6) = -98 + 4 - 42 = -136.$$

Exercise 10.4

Without evaluating, state why each statement is true.

1. $\begin{vmatrix} 2 & 3 & 1 \\ 0 & 0 & 0 \\ -1 & 2 & 0 \end{vmatrix} = 0$

2. $\begin{vmatrix} 3 & 1 & 3 \\ 0 & 1 & 0 \\ 1 & 2 & 1 \end{vmatrix} = 0$

3. $\begin{vmatrix} 2 & 3 & 1 & 1 \\ 2 & 0 & 1 & 2 \\ 2 & 3 & 1 & 1 \\ 0 & 1 & 2 & 0 \end{vmatrix} = 0$

4. $\begin{vmatrix} 7 & 3 & 2 & 0 \\ 2 & 1 & 2 & 0 \\ 4 & 1 & 1 & 0 \\ 0 & 2 & 1 & 0 \end{vmatrix} = 0$

5. $\begin{vmatrix} 4 & 2 & 1 \\ 0 & -1 & -2 \\ 1 & 0 & 2 \end{vmatrix} = - \begin{vmatrix} 4 & 2 & 1 \\ 0 & 1 & 2 \\ 1 & 0 & 2 \end{vmatrix}$

6. $\begin{vmatrix} -2 & 3 & 1 \\ -1 & 0 & 1 \\ -2 & 1 & 0 \end{vmatrix} = - \begin{vmatrix} 2 & 3 & 1 \\ 1 & 0 & 1 \\ 2 & 1 & 0 \end{vmatrix}$

7. $2 \begin{vmatrix} 1 & 0 & 2 \\ -1 & 2 & 0 \\ 1 & 1 & 1 \end{vmatrix} = \begin{vmatrix} 1 & 0 & 2 \\ -1 & 2 & 0 \\ 2 & 2 & 2 \end{vmatrix}$

8. $\begin{vmatrix} 3 & -4 & 2 \\ 1 & -2 & 0 \\ 0 & 8 & 1 \end{vmatrix} = -2 \begin{vmatrix} 3 & 2 & 2 \\ 1 & 1 & 0 \\ 0 & -4 & 1 \end{vmatrix}$

9. $\begin{vmatrix} 1 & 2 \\ 3 & 4 \end{vmatrix} = \begin{vmatrix} 1+2 & 2 \\ 3+4 & 4 \end{vmatrix}$

10. $\begin{vmatrix} 1 & 2 \\ 3 & 4 \end{vmatrix} = \begin{vmatrix} 1+4 & 2 \\ 3+8 & 4 \end{vmatrix}$

11. $\begin{vmatrix} 1 & 2 & 1 \\ 0 & 2 & 3 \\ 2 & -1 & 2 \end{vmatrix} = \begin{vmatrix} 1 & 2 & 1 \\ 0 & 2 & 3 \\ 0 & -5 & 0 \end{vmatrix}$

12. $\begin{vmatrix} -1 & 1 & 0 \\ 2 & 3 & -1 \\ 2 & 1 & 2 \end{vmatrix} = \begin{vmatrix} 0 & 1 & 0 \\ 5 & 3 & -1 \\ 3 & 1 & 2 \end{vmatrix}$

Theorem 10.13 was used on the left-hand member of each of the following equalities to produce the elements in the right-hand member. Complete the entries.

13. $\begin{vmatrix} 1 & 3 \\ 2 & 2 \end{vmatrix} = \begin{vmatrix} 1 & 3 \\ 0 & \end{vmatrix}$

14. $\begin{vmatrix} 2 & -1 \\ 3 & 1 \end{vmatrix} = \begin{vmatrix} & 0 \\ 3 & 1 \end{vmatrix}$

15. $\begin{vmatrix} 1 & -2 & 1 \\ 3 & 1 & 4 \\ 0 & 2 & 1 \end{vmatrix} = \begin{vmatrix} 1 & -2 & 1 \\ 0 & 7 & \\ 0 & 2 & 1 \end{vmatrix}$

16. $\begin{vmatrix} 3 & -1 & 0 \\ 1 & 2 & 1 \\ 2 & 3 & 1 \end{vmatrix} = \begin{vmatrix} 3 & -1 & 0 \\ 1 & 2 & 1 \\ 1 & & 0 \end{vmatrix}$

17. $\begin{vmatrix} 2 & 3 & 1 & 4 \\ 0 & 2 & 1 & 2 \\ 1 & 1 & 2 & 3 \\ 0 & 1 & 1 & 1 \end{vmatrix} = \begin{vmatrix} 0 & 1 & & -2 \\ 0 & 2 & 1 & 2 \\ 1 & 1 & 2 & 3 \\ 0 & 1 & 1 & 1 \end{vmatrix}$

18. $\begin{vmatrix} 2 & 1 & 1 & 0 \\ 1 & 2 & 0 & 2 \\ 3 & 1 & 0 & 3 \\ 2 & 1 & 4 & 2 \end{vmatrix} = \begin{vmatrix} 2 & 1 & 1 & 0 \\ 1 & 2 & 0 & 2 \\ 3 & 1 & 0 & 3 \\ -3 & 0 & 0 & 2 \end{vmatrix}$

First reduce each determinant to an equal 2×2 *determinant and then evaluate.*

19. $\begin{vmatrix} 2 & 1 & 0 \\ 3 & 2 & 1 \\ -1 & 2 & 0 \end{vmatrix}$
20. $\begin{vmatrix} 1 & 2 & 1 \\ 2 & -1 & 2 \\ 0 & 1 & 0 \end{vmatrix}$
21. $\begin{vmatrix} 1 & 0 & 3 \\ 2 & -1 & 1 \\ 1 & 2 & 1 \end{vmatrix}$

22. $\begin{vmatrix} 1 & 2 & -1 \\ 2 & 1 & 3 \\ 0 & 1 & 2 \end{vmatrix}$
23. $\begin{vmatrix} 1 & 2 & 1 \\ -1 & 2 & 3 \\ 2 & -1 & 1 \end{vmatrix}$
24. $\begin{vmatrix} 3 & -1 & 2 \\ 1 & 2 & 1 \\ -2 & 1 & 3 \end{vmatrix}$

25. $\begin{vmatrix} 11 & -10 & 32 \\ 12 & -11 & 35 \\ 12 & -6 & 33 \end{vmatrix}$
26. $\begin{vmatrix} 9 & 31 & 16 \\ 11 & 38 & 19 \\ 10 & 34 & 21 \end{vmatrix}$
27. $\begin{vmatrix} 28 & 27 & 25 \\ 31 & 30 & 26 \\ 36 & 35 & 30 \end{vmatrix}$

28. $\begin{vmatrix} 26 & 29 & 29 \\ 25 & 27 & 30 \\ 25 & 26 & 28 \end{vmatrix}$
29. $\begin{vmatrix} 13 & 16 & 19 \\ 28 & 34 & 40 \\ 27 & 33 & 39 \end{vmatrix}$
30. $\begin{vmatrix} 19 & 54 & 17 \\ 20 & 57 & 18 \\ 19 & 55 & 19 \end{vmatrix}$

31. $\begin{vmatrix} 0 & 0 & 1 & 2 \\ 6 & 0 & 0 & 1 \\ 6 & 1 & 0 & -1 \\ 6 & 1 & 0 & 2 \end{vmatrix}$
32. $\begin{vmatrix} 4 & 2 & 0 & 2 \\ -1 & 0 & 2 & 1 \\ 3 & 0 & -1 & 1 \\ 0 & 0 & 2 & 1 \end{vmatrix}$
33. $\begin{vmatrix} 0 & 1 & 0 & 2 \\ 0 & 2 & 0 & 3 \\ 2 & -1 & 1 & 0 \\ 0 & 0 & 8 & 8 \end{vmatrix}$

34. $\begin{vmatrix} 0 & 2 & -1 & 3 \\ 0 & 0 & 2 & 1 \\ 3 & 0 & 1 & 0 \\ -6 & 6 & 0 & 0 \end{vmatrix}$
35. $\begin{vmatrix} 1 & 2 & 3 & -1 \\ 0 & 4 & 8 & 4 \\ -2 & 0 & 1 & 1 \\ 2 & 1 & 0 & 1 \end{vmatrix}$
36. $\begin{vmatrix} 1 & 2 & 1 & 1 \\ 2 & -1 & 0 & 1 \\ 0 & 6 & 3 & 9 \\ 2 & 0 & -1 & 1 \end{vmatrix}$

37. Show that

$$\begin{vmatrix} x & y & 1 \\ x_1 & y_1 & 1 \\ x_2 & y_2 & 1 \end{vmatrix} = 0$$

represents an equation of the straight line through the points (x_1, y_1) and (x_2, y_2).

38. Use the results in Problem 37 to find an equation of the line through the points $(3, -1)$ and $(-2, 5)$.

39. Show that $\begin{vmatrix} 1 & a & a^2 \\ 1 & b & b^2 \\ 1 & c & c^2 \end{vmatrix} = (b - c)(c - a)(a - b).$

40. Show that $\begin{vmatrix} a_{11} & a_{12} & a_{13} & a_{14} \\ a_{21} & a_{22} & a_{23} & a_{24} \\ 0 & 0 & a_{33} & a_{34} \\ 0 & 0 & a_{43} & a_{44} \end{vmatrix} = \begin{vmatrix} a_{11} & a_{12} \\ a_{21} & a_{22} \end{vmatrix} \cdot \begin{vmatrix} a_{33} & a_{34} \\ a_{43} & a_{44} \end{vmatrix}.$

Prove each theorem in the case of 2×2 matrices.

41. Theorem 10.7   42. Theorem 10.8   43. Theorem 10.9

44. Theorem 10.10   45. Theorem 10.11   46. Theorem 10.12

10.5
The inverse of a square matrix

In the field of real numbers, every element a except 0 has a multiplicative inverse $1/a$, with the property that $a \cdot (1/a) = 1$. The question should (and does) arise, " Does every square matrix A have a multiplicative inverse A^{-1}?"

Definition 10.17   *For a given square matrix A of order n, if there is a square matrix A^{-1} of order n such that*

$$AA^{-1} = I \quad and \quad A^{-1}A = I,$$

where I is the multiplicative identity of order n, then A^{-1} is the multiplicative inverse of A.

To answer the question about the existence of a multiplicative inverse for a matrix, we shall begin by considering the simple case of 2×2 matrices. If we let

$$A = \begin{bmatrix} a_{11} & a_{12} \\ a_{21} & a_{22} \end{bmatrix},$$

then we must see whether there exists a 2×2 matrix A^{-1} such that $AA^{-1} = I$. If so, let $A^{-1} = \begin{bmatrix} b & c \\ d & e \end{bmatrix}$. We wish to have

$$\begin{bmatrix} a_{11} & a_{12} \\ a_{21} & a_{22} \end{bmatrix} \begin{bmatrix} b & c \\ d & e \end{bmatrix} = \begin{bmatrix} 1 & 0 \\ 0 & 1 \end{bmatrix}.$$

This leads to

$$\begin{bmatrix} a_{11}b + a_{12}d & a_{11}c + a_{12}e \\ a_{21}b + a_{22}d & a_{21}c + a_{22}e \end{bmatrix} = \begin{bmatrix} 1 & 0 \\ 0 & 1 \end{bmatrix},$$

which is true if and only if

$$a_{11}b + a_{12}d = 1, \qquad a_{11}c + a_{12}e = 0,$$
$$a_{21}b + a_{22}d = 0, \qquad a_{21}c + a_{22}e = 1.$$

Solving these equations for b, c, d, and e, we have

$$(a_{11}a_{22} - a_{12}a_{21})b = a_{22}, \quad (a_{11}a_{22} - a_{12}a_{21})c = -a_{12},$$

$$(a_{11}a_{22} - a_{12}a_{21})d = -a_{21}, \quad (a_{11}a_{22} - a_{12}a_{21})e = a_{11}, \tag{1}$$

from which

$$b = \frac{a_{22}}{a_{11}a_{22} - a_{12}a_{21}}, \quad c = \frac{-a_{12}}{a_{11}a_{22} - a_{12}a_{21}},$$

$$d = \frac{-a_{21}}{a_{11}a_{22} - a_{12}a_{21}}, \quad e = \frac{a_{11}}{a_{11}a_{22} - a_{12}a_{21}},$$

provided that $a_{11}a_{22} - a_{12}a_{21} \neq 0$. Now the denominator of each of these fractions is just $\delta(A)$, so that

$$A^{-1} = \begin{bmatrix} \dfrac{a_{22}}{\delta(A)} & \dfrac{-a_{12}}{\delta(A)} \\[2mm] \dfrac{-a_{21}}{\delta(A)} & \dfrac{a_{11}}{\delta(A)} \end{bmatrix} = \frac{1}{\delta(A)}\begin{bmatrix} a_{22} & -a_{12} \\ -a_{21} & a_{11} \end{bmatrix}.$$

By direct multiplication, it can be verified not only that

$$AA^{-1} = I,$$

but also (surprisingly, since matrix multiplication is not always commutative) that

$$A^{-1}A = I.$$

The inverse of a 2×2 matrix   Thus, to write the inverse of a 2 × 2 square matrix A for which $\delta(A) \neq 0$, we interchange the entries on the principal diagonal, replace each of the other two entries with its negative, and multiply the result by $1/\delta(A)$.

Example   If $A = \begin{bmatrix} 1 & 3 \\ 2 & -1 \end{bmatrix}$, find A^{-1}.

Solution   We first observe that $\delta(A) = -7$. Hence,

$$A^{-1} = -\frac{1}{7}\begin{bmatrix} -1 & -3 \\ -2 & 1 \end{bmatrix} = \begin{bmatrix} \dfrac{1}{7} & \dfrac{3}{7} \\[2mm] \dfrac{2}{7} & -\dfrac{1}{7} \end{bmatrix}.$$

It is a good idea always to check the result when finding A^{-1}, because there is much room for blundering in the process of determining the inverse. In the above example, we have

$$A^{-1}A = -\frac{1}{7}\begin{bmatrix} -1 & -3 \\ -2 & 1 \end{bmatrix}\begin{bmatrix} 1 & 3 \\ 2 & -1 \end{bmatrix} = -\frac{1}{7}\begin{bmatrix} -7 & 0 \\ 0 & -7 \end{bmatrix} = \begin{bmatrix} 1 & 0 \\ 0 & 1 \end{bmatrix}.$$

Matrices that have no inverse   Moreover, from the above analysis we can now answer the question, "Does every 2×2 square matrix A have an inverse?" The answer is "No," for if $\delta(A)$ is 0, then the foregoing equations for b, c, d, e would have no solution. Square matrices A for which $\delta(A) = 0$ are called **singular matrices**.

Example   Show that $\begin{bmatrix} 3 & 5 \\ 6 & 10 \end{bmatrix}$ is singular, and hence has no inverse.

Solution   Since $\delta(A) = 3(10) - 6(5) = 0$, no inverse exists.

The inverse of an $n \times n$ matrix   More generally, and without proving it, we have the following result for the inverse of a square matrix.

Theorem 10.14   *If*

$$A = \begin{bmatrix} a_{11} & a_{12} & \cdots & a_{1n} \\ a_{21} & a_{22} & \cdots & a_{2n} \\ \vdots & \vdots & & \vdots \\ a_{n1} & a_{n2} & \cdots & a_{nn} \end{bmatrix}$$

and if $\delta(A) \neq 0$, then

$$A^{-1} = \frac{1}{\delta(A)}\begin{bmatrix} A_{11} & A_{21} & \cdots & A_{n1} \\ A_{12} & A_{22} & \cdots & A_{n2} \\ \vdots & \vdots & & \vdots \\ A_{1n} & A_{2n} & \cdots & A_{nn} \end{bmatrix},$$

where A_{ij} is the cofactor of a_{ij} in A. If $\delta(A) = 0$, then A has no inverse.

Observe that A^{-1} is the matrix having as its entries the cofactors of the entries in A multiplied by $1/\delta(A)$, but the cofactors of the *row* entries in A are the *column* entries in A^{-1}. One way to obtain A^{-1} is to replace each entry in A with its cofactor and multiply the *transpose* of the resulting matrix by $1/\delta(A)$.

Example

If $A = \begin{bmatrix} 1 & 0 & 1 \\ 2 & 1 & 0 \\ 1 & -1 & 1 \end{bmatrix}$, find A^{-1}.

Solution

We first observe that $\delta(A) = -2$. Therefore, since $\delta(A)$ is not zero, A has an inverse. Next, replacing each entry in A with its cofactor, we obtain the matrix

$$\begin{bmatrix} 1 & -2 & -3 \\ -1 & 0 & 1 \\ -1 & 2 & 1 \end{bmatrix}, \quad \text{whose transpose is} \quad \begin{bmatrix} 1 & -1 & -1 \\ -2 & 0 & 2 \\ -3 & 1 & 1 \end{bmatrix},$$

so that

$$A^{-1} = -\frac{1}{2}\begin{bmatrix} 1 & -1 & -1 \\ -2 & 0 & 2 \\ -3 & 1 & 1 \end{bmatrix}.$$

As a check, we have

$$A^{-1}A = -\frac{1}{2}\begin{bmatrix} 1 & -1 & -1 \\ -2 & 0 & 2 \\ -3 & 1 & 1 \end{bmatrix}\begin{bmatrix} 1 & 0 & 1 \\ 2 & 1 & 0 \\ 1 & -1 & 1 \end{bmatrix}$$

$$= -\frac{1}{2}\begin{bmatrix} -2 & 0 & 0 \\ 0 & -2 & 0 \\ 0 & 0 & -2 \end{bmatrix} = \begin{bmatrix} 1 & 0 & 0 \\ 0 & 1 & 0 \\ 0 & 0 & 1 \end{bmatrix}.$$

Theorem 10.14 is applicable to $n \times n$ square matrices, although, clearly, the process of actually determining A^{-1} becomes very laborious for matrices much larger than 3×3.

Properties of matrices and their inverses   There are a number of useful properties associated with matrices and their inverses. The following theorem gives one example.

Theorem 10.15 *If A and B are $n \times n$ nonsingular square matrices, then AB has an inverse, namely*

$$(AB)^{-1} = B^{-1}A^{-1}.$$

Proof If we right-multiply AB by $B^{-1}A^{-1}$ and apply the associative law for the multiplication of matrices, we have

$$AB \cdot B^{-1}A^{-1} = A \cdot I \cdot A^{-1} = A \cdot A^{-1} = I.$$

(*Proof continued overleaf.*)

Moreover, if we left-multiply AB by $B^{-1}A^{-1}$, we have

$$B^{-1}A^{-1} \cdot AB = B^{-1} \cdot I \cdot B = B^{-1} \cdot B = I.$$

Thus, since $(AB)(B^{-1}A^{-1}) = (B^{-1}A^{-1})(AB) = I$, by the definition of the inverse of a matrix we have

$$(AB)^{-1} = B^{-1}A^{-1}.$$

This theorem can be used to find the inverse of products of any number of nonsingular matrices. For example, if there are three factors A, B, and C in such a product, then

$$(ABC)^{-1} = [(AB)C]^{-1} = C^{-1}(AB)^{-1} = C^{-1}B^{-1}A^{-1}.$$

Exercise 10.5

Find the inverse of each matrix if one exists.

Example $B = \begin{bmatrix} 1 & 0 & -1 \\ 1 & 3 & 1 \\ 0 & 1 & 2 \end{bmatrix}$

Solution The determinant $\delta(B)$ is given by

$$\delta \begin{bmatrix} 1 & 0 & -1 \\ 1 & 3 & 1 \\ 0 & 1 & 2 \end{bmatrix} = 1(5) - 0 - 1(1) = 4.$$

Replacing each entry of B with its cofactor gives

$$\begin{bmatrix} 5 & -2 & 1 \\ -1 & 2 & -1 \\ 3 & -2 & 3 \end{bmatrix}; \quad \begin{bmatrix} 5 & -2 & 1 \\ -1 & 2 & -1 \\ 3 & -2 & 3 \end{bmatrix}^t = \begin{bmatrix} 5 & -1 & 3 \\ -2 & 2 & -2 \\ 1 & -1 & 3 \end{bmatrix};$$

$$B^{-1} = \frac{1}{\delta(B)} \begin{bmatrix} \text{each } b_{ij} \text{ of} \\ B \text{ replaced} \\ \text{by } B_{ij} \end{bmatrix}^t = \frac{1}{4} \begin{bmatrix} 5 & -1 & 3 \\ -2 & 2 & -2 \\ 1 & -1 & 3 \end{bmatrix} = \begin{bmatrix} \dfrac{5}{4} & -\dfrac{1}{4} & \dfrac{3}{4} \\ -\dfrac{2}{4} & \dfrac{2}{4} & -\dfrac{2}{4} \\ \dfrac{1}{4} & -\dfrac{1}{4} & \dfrac{3}{4} \end{bmatrix}.$$

1. $\begin{bmatrix} 1 & 2 \\ 1 & 3 \end{bmatrix}$

2. $\begin{bmatrix} 3 & 1 \\ 2 & -1 \end{bmatrix}$

3. $\begin{bmatrix} 2 & -3 \\ 1 & 1 \end{bmatrix}$

4. $\begin{bmatrix} 3 & -2 \\ 2 & 1 \end{bmatrix}$ 5. $\begin{bmatrix} -2 & -1 \\ 4 & 2 \end{bmatrix}$ 6. $\begin{bmatrix} 3 & 1 \\ 9 & 3 \end{bmatrix}$

7. $\begin{bmatrix} 5 & 7 \\ 3 & 4 \end{bmatrix}$ 8. $\begin{bmatrix} 5 & -4 \\ 4 & -3 \end{bmatrix}$ 9. $\begin{bmatrix} 7 & 4 \\ -4 & -2 \end{bmatrix}$

10. $\begin{bmatrix} -9 & 5 \\ -4 & 2 \end{bmatrix}$ 11. $\begin{bmatrix} -2 & -6 \\ -3 & -9 \end{bmatrix}$ 12. $\begin{bmatrix} 21 & 7 \\ 9 & 3 \end{bmatrix}$

13. $\begin{bmatrix} 1 & -1 & 2 \\ 2 & 1 & 3 \\ 0 & 0 & 2 \end{bmatrix}$ 14. $\begin{bmatrix} 0 & 4 & 2 \\ 1 & 0 & 2 \\ 0 & -1 & 1 \end{bmatrix}$ 15. $\begin{bmatrix} 2 & -1 & 1 \\ 3 & 0 & 1 \\ 2 & 2 & 1 \end{bmatrix}$

16. $\begin{bmatrix} 1 & 2 & 1 \\ 0 & 2 & 1 \\ -2 & 2 & 3 \end{bmatrix}$ 17. $\begin{bmatrix} 2 & 1 & 1 \\ 1 & 0 & 2 \\ 4 & 2 & 2 \end{bmatrix}$ 18. $\begin{bmatrix} -3 & 1 & -6 \\ 2 & 1 & 4 \\ 2 & 0 & 4 \end{bmatrix}$

19. $\begin{bmatrix} 1 & 2 & -3 \\ 3 & -1 & 0 \\ 5 & 3 & -6 \end{bmatrix}$ 20. $\begin{bmatrix} 2 & 4 & -1 \\ 1 & 6 & 2 \\ 5 & 14 & 0 \end{bmatrix}$ 21. $\begin{bmatrix} 2 & -1 & -5 \\ 1 & 3 & 4 \\ 0 & 1 & 2 \end{bmatrix}$

22. $\begin{bmatrix} 2 & 1 & -8 \\ 1 & 1 & -2 \\ 1 & 2 & 3 \end{bmatrix}$ 23. $\begin{bmatrix} 0 & 0 & 1 \\ 0 & 1 & 0 \\ 1 & 0 & 0 \end{bmatrix}$ 24. $\begin{bmatrix} 1 & 0 & 1 \\ 0 & 1 & 0 \\ 1 & 0 & 0 \end{bmatrix}$

25. Verify that

$$\left(\begin{bmatrix} 2 & 3 \\ 1 & -1 \end{bmatrix} \cdot \begin{bmatrix} 0 & 1 \\ 3 & 1 \end{bmatrix} \right)^{-1} = \begin{bmatrix} 0 & 1 \\ 3 & 1 \end{bmatrix}^{-1} \cdot \begin{bmatrix} 2 & 3 \\ 1 & -1 \end{bmatrix}^{-1}.$$

26. Verify that

$$\left(\begin{bmatrix} 1 & 2 \\ -1 & 0 \end{bmatrix} \cdot \begin{bmatrix} 1 & 1 \\ 2 & 0 \end{bmatrix} \cdot \begin{bmatrix} 2 & -1 \\ 0 & 1 \end{bmatrix} \right)^{-1} = \begin{bmatrix} 2 & -1 \\ 0 & 1 \end{bmatrix}^{-1} \cdot \begin{bmatrix} 1 & 1 \\ 2 & 0 \end{bmatrix}^{-1} \cdot \begin{bmatrix} 1 & 2 \\ -1 & 0 \end{bmatrix}^{-1}.$$

27. Verify that

$$\left(\begin{bmatrix} 3 & 0 & 1 \\ 2 & 1 & 0 \\ 0 & 1 & 2 \end{bmatrix} \cdot \begin{bmatrix} 2 & 1 & 0 \\ 1 & 1 & 2 \\ 0 & 1 & 0 \end{bmatrix} \right)^{-1} = \begin{bmatrix} 2 & 1 & 0 \\ 1 & 1 & 2 \\ 0 & 1 & 0 \end{bmatrix}^{-1} \cdot \begin{bmatrix} 3 & 0 & 1 \\ 2 & 1 & 0 \\ 0 & 1 & 2 \end{bmatrix}^{-1}.$$

28. Show that $[A^t]^{-1} = [A^{-1}]^t$ for each nonsingular 2×2 matrix.

29. Show that $\delta(A^{-1}) = 1/\delta(A)$ for each nonsingular 2×2 matrix.

30. Prove that if a and b are any real numbers, then $\delta(aA^2 + bA) = \delta(aA + bI)\delta(A)$ for all 2×2 matrices A.

31. Prove that $\delta(B^{-1}AB) = \delta(A)$ for all nonsingular 2×2 matrices A and B.

32. Prove that if A is a 2×2 matrix and a, b, and c are real numbers, with $c \neq 0$, and if $aA^2 + bA + cI = 0$, then A has an inverse.

10.6

Solution of linear systems

Solution by
substitution

You should recall from earlier studies of algebra that systems of linear equations can be solved in a variety of ways. One elementary way to solve a system is by *substitution*.

Example

Solve

$$2x + 3y = 1 \tag{1}$$

$$3x - y = 7 \tag{2}$$

in R^2, that is, in $R \times R$.

Solution

Solving Equation (2) for y in terms of x produces $y = 3x - 7$. Then, replacing y in (1) with $3x - 7$, we have

$$2x + 3(3x - 7) = 1,$$
$$2x + 9x - 21 = 1,$$
$$11x = 22,$$
$$x = 2.$$

Upon replacing x in $y = 3x - 7$ with 2, we find that

$$y = 3(2) - 7 = -1.$$

Therefore, the solution set in R^2 of the given system is $\{(2, -1)\}$.

Solution by linear
combinations

A second method commonly used to solve systems of equations is that of eliminating a variable by forming one or more *linear combinations* of the left-hand members of the equations in a system in which the right-hand members are zero. This technique is sometimes referred to as solution by *addition* (or *subtraction*).

Example

Solve

$$x + 2y - 3z + 4 = 0 \tag{3}$$

$$2x - y + z - 3 = 0 \tag{4}$$

$$3x + 2y + z - 10 = 0 \tag{5}$$

in R^3, that is, in $R \times R \times R$.

Solution

To eliminate x between Equations (3) and (4), we can multiply each member of Equation (3) by -2 and add the results to the corresponding members of Equation (4) to obtain

$$-2(x + 2y - 3z + 4) + (2x - y + z - 3) = 0,$$
$$-5y + 7z - 11 = 0. \qquad (6)$$

We can next eliminate x between Equation (3) and Equation (5) by multiplying each member of (3) by -3 and adding the result to the corresponding member of (5) to produce

$$-3(x + 2y - 3z + 4) + (3x + 2y + z - 10) = 0,$$
$$-4y + 10z - 22 = 0. \qquad (7)$$

Next, to solve the system consisting of (6) and (7), we add -4 times each member of (6) to 5 times the corresponding member of (7), obtaining

$$-4(-5y + 7z - 11) + 5(-4y + 10z - 22) = 0,$$
$$22z = 66,$$
$$z = 3.$$

By successive substitutions in, say, (6) and (3), we obtain $y = 2$ and $x = 1$, so that the solution set of the given system in R^3 is $\{(1, 2, 3)\}$.

Solution by matrices

Matrices also offer a means of finding solutions of systems of linear equations. We first verify the matrix-product equation

$$\begin{bmatrix} a_{11} & a_{12} & \cdots & a_{1n} \\ \vdots & \vdots & & \vdots \\ a_{n1} & a_{n2} & \cdots & a_{nn} \end{bmatrix} \begin{bmatrix} x_1 \\ \vdots \\ x_n \end{bmatrix} = \begin{bmatrix} a_{11}x_1 + a_{12}x_2 + \cdots + a_{1n}x_n \\ \vdots & & \vdots \\ a_{n1}x_1 + a_{n2}x_n + \cdots + a_{nn}x_n \end{bmatrix},$$

and hence note that the linear system

$$a_{11}x_1 + a_{12}x_2 + \cdots + a_{1n}x_n = c_1$$
$$a_{21}x_1 + a_{22}x_2 + \cdots + a_{2n}x_n = c_2$$
$$\vdots \quad \vdots \quad \vdots \quad \vdots$$
$$a_{n1}x_1 + a_{n2}x_2 + \cdots + a_{nn}x_n = c_n$$

can be written as the matrix equation

$$\begin{bmatrix} a_{11} & a_{12} & \cdots & a_{1n} \\ \vdots & \vdots & & \vdots \\ a_{n1} & a_{n2} & \cdots & a_{nn} \end{bmatrix} \begin{bmatrix} x_1 \\ \vdots \\ x_n \end{bmatrix} = \begin{bmatrix} c_1 \\ \vdots \\ c_n \end{bmatrix},$$

where the first factor in the left-hand member is called the **coefficient matrix** for the system. In more concise notation, this latter equation can be written

$$AX = B, \qquad (8)$$

where A is an $n \times n$ square matrix, and X and B are $n \times 1$ column matrices. If now A is nonsingular, we can left-multiply both members of this equation by A^{-1} to obtain

$$A^{-1}AX = A^{-1}B,$$
$$IX = A^{-1}B,$$
$$X = A^{-1}B, \tag{9}$$

where $A^{-1}B$ is an $n \times 1$ column matrix. Since X and $A^{-1}B$ are equal, each entry in X is equal to the corresponding entry in $A^{-1}B$, and hence these latter entries constitute the components of the solution of the given linear system. Conversely, left-multiplying (9) by A shows that the X given by (9) satisfies (8). If A is a singular matrix, then of course it has no inverse, and either the system has no solution or the solution is not unique.

Example

Use matrices to find the solution set of

$$2x + y + z = 1$$
$$x - 2y - 3z = 1$$
$$3x + 2y + 4z = 5$$

in R^3.

Solution

We first write the system as a matrix equation of the form $AX = B$, thus:

$$\begin{bmatrix} 2 & 1 & 1 \\ 1 & -2 & -3 \\ 3 & 2 & 4 \end{bmatrix} \begin{bmatrix} x \\ y \\ z \end{bmatrix} = \begin{bmatrix} 1 \\ 1 \\ 5 \end{bmatrix}.$$

We next expand $\delta(A)$ about its first row, obtaining

$$\delta(A) = \delta \begin{bmatrix} 2 & 1 & 1 \\ 1 & -2 & -3 \\ 3 & 2 & 4 \end{bmatrix} = 2(-2) - 1(13) + 1(8) = -9,$$

and observe that A is nonsingular. Then

$$A^{-1} = \begin{bmatrix} 2 & 1 & 1 \\ 1 & -2 & -3 \\ 3 & 2 & 4 \end{bmatrix}^{-1} = -\frac{1}{9} \begin{bmatrix} -2 & -2 & -1 \\ -13 & 5 & 7 \\ 8 & -1 & -5 \end{bmatrix}.$$

Now, because $X = A^{-1}B$, we have

$$\begin{bmatrix} x \\ y \\ z \end{bmatrix} = -\frac{1}{9} \begin{bmatrix} -2 & -2 & -1 \\ -13 & 5 & 7 \\ 8 & -1 & -5 \end{bmatrix} \begin{bmatrix} 1 \\ 1 \\ 5 \end{bmatrix} = -\frac{1}{9} \begin{bmatrix} -9 \\ 27 \\ -18 \end{bmatrix} = \begin{bmatrix} 1 \\ -3 \\ 2 \end{bmatrix}.$$

Hence $x = 1$, $y = -3$, and $z = 2$, and the solution set of the system in R^3 is $\{(1, -3, 2)\}$.

The computation of A^{-1} is laborious when A is a square matrix containing many rows and columns. The foregoing method is not always the easiest to use in determining an inverse, particularly when an electronic digital computer is available, but it is most valuable for theoretical developments.

Exercise 10.6

Solve each system by (a) *substitution and/or linear combinations and* (b) *matrices. If the system has no solution, so state.*

1. $2x - 3y = -1$
   $x + 4y = 5$

2. $3x - 4y = -2$
   $x - 2y = 0$

3. $3x - 4y = -2$
   $6x + 12y = 36$

4. $2x - 4y = 7$
   $x - 2y = 1$

5. $2x - 3y = 0$
   $2x + y = 16$

6. $2x + 3y = 3$
   $3x - 4y = 0$

7. $x + y = 2$
   $2x - z = 1$
   $2y - 3z = -1$

8. $2x - 6y + 3z = -12$
   $3x - 2y + 5z = -4$
   $4x + 5y - 2z = 10$

9. $x - 2y + z = -1$
   $3x + y - 2z = 4$
   $y - z = 1$

10. $2x + 5z = 9$
    $4x + 3y = -1$
    $3y - 4z = -13$

11. $2x + 2y + z = 1$
    $x - y + 6z = 21$
    $3x + 2y - z = -4$

12. $4x + 8y + z = -6$
    $2x - 3y + 2z = 0$
    $x + 7y - 3z = -8$

13. $x + y + z = 0$
    $2x - y - 4z = 15$
    $x - 2y - z = 7$

14. $x + y - 2z = 3$
    $3x - y + z = 5$
    $3x + 3y - 6z = 9$

10.7
Cramer's rule

If the matrix technique for solving linear systems discussed in the preceding section is viewed in terms of determinants, we arrive at a general solution for such systems. This also, however, is more of theoretical than computational value.

Consider the system

$$a_{11}x_1 + a_{12}x_2 + \cdots + a_{1n}x_n = c_1$$
$$a_{21}x_1 + a_{22}x_2 + \cdots + a_{2n}x_n = c_2$$
$$\vdots \qquad \vdots \qquad \qquad \vdots \qquad \vdots$$
$$a_{n1}x_1 + a_{n2}x_2 + \cdots + a_{nn}x_n = c_n$$

or simply $Ax = B$. If the coefficient matrix A in the matrix equation $AX = B$ is nonsingular, then its inverse, A^{-1}, is

$$A^{-1} = \frac{1}{\delta(A)} \begin{bmatrix} A_{11} & A_{21} & \cdots & A_{n1} \\ \vdots & \vdots & & \vdots \\ A_{1n} & A_{2n} & \cdots & A_{nn} \end{bmatrix}.$$

Now, $B = \begin{bmatrix} c_1 \\ c_2 \\ \vdots \\ c_n \end{bmatrix}$, so that

$$A^{-1}B = \frac{1}{\delta(A)} \begin{bmatrix} c_1 A_{11} + c_2 A_{21} + \cdots + c_n A_{n1} \\ c_1 A_{12} + c_2 A_{22} + \cdots + c_n A_{n2} \\ \vdots & \vdots & \vdots \\ c_1 A_{1n} + c_2 A_{2n} + \cdots + c_n A_{nn} \end{bmatrix}.$$

Each entry in $A^{-1}B$ can be seen to be of the form

$$\frac{c_1 A_{1j} + c_2 A_{2j} + \cdots + c_n A_{nj}}{\delta(A)}.$$

But $c_1 A_{1j} + c_2 A_{2j} + \cdots + c_n A_{nj}$ is just the expansion of the determinant

$$\begin{array}{c} j\text{th} \\ \text{column} \\ \downarrow \end{array}$$

$$\begin{vmatrix} a_{11} & a_{12} & \cdots & c_1 & \cdots & a_{1n} \\ a_{21} & a_{22} & \cdots & c_2 & \cdots & a_{2n} \\ \vdots & \vdots & & \vdots & & \vdots \\ a_{n1} & a_{n2} & \cdots & c_n & \cdots & a_{nn} \end{vmatrix}$$

about the jth column, which has entries $c_1, c_2, c_3, \ldots c_n$. Thus, if the variables in a linear system are denoted by $x_1, x_2, \ldots x_n$, then each entry x_j in $A^{-1}B$ is given by

$$\begin{array}{c} j\text{th} \\ \text{column} \\ \downarrow \end{array}$$

$$x_j = \frac{\delta(A_j)}{\delta(A)} = \frac{\begin{vmatrix} a_{11} & a_{12} & \cdots & c_1 & \cdots & a_{1n} \\ a_{21} & a_{22} & \cdots & c_2 & \cdots & a_{2n} \\ \vdots & \vdots & & \vdots & & \vdots \\ a_{n1} & a_{n2} & \cdots & c_n & \cdots & a_{nn} \end{vmatrix}}{\begin{vmatrix} a_{11} & a_{12} & \cdots & & a_{1n} \\ a_{21} & a_{22} & \cdots & & a_{2n} \\ \vdots & \vdots & & & \vdots \\ a_{n1} & a_{n2} & \cdots & & a_{nn} \end{vmatrix}}.$$

Application of
Cramer's rule

This relationship expresses **Cramer's rule**. Cramer's rule is the assertion that if the determinant of the coefficient matrix of an $n \times n$ linear system *is not* 0, then the equations are consistent (the system has a solution), and the unique solution in R^n can be found for each variable in the system as follows:

1. Find the determinant of the coefficient matrix for the system.
2. Replace each entry in the jth column of the coefficient matrix A with the corresponding entry from the column matrix B (the constant terms in the equations), and find the determinant of the resulting matrix.
3. Divide the result in Step 2 by the result in Step 1. The quotient is x_j.

Example

Use Cramer's rule to solve the system

$$-4x + 2y - 9z = 2$$
$$3x + 4y + z = 5$$
$$x - 3y + 2z = 8$$

in R^3.

Solution

By inspection,

$$\delta(A) = \begin{vmatrix} -4 & 2 & -9 \\ 3 & 4 & 1 \\ 1 & -3 & 2 \end{vmatrix}$$

$$= -4(11) - 2(5) - 9(-13)$$

$$= -44 - 10 + 117 = 63.$$

Replacing the entries in the first column of A with corresponding constants 2 5, and 8 we have

$$\delta(A_x) = \begin{vmatrix} 2 & 2 & -9 \\ 5 & 4 & 1 \\ 8 & -3 & 2 \end{vmatrix}$$

$$= 2(11) - 2(2) - 9(-47)$$

$$= 22 - 4 + 423 = 441.$$

Hence

$$x = \frac{\delta(A_x)}{\delta(A)} = \frac{441}{63} = 7.$$

Similarly, by replacing, in turn, the entries of the second and third columns of A with the corresponding constants 2, 5 and 8, we have

$$\delta(A_y) = \begin{vmatrix} -4 & 2 & -9 \\ 3 & 5 & 1 \\ 1 & 8 & 2 \end{vmatrix} \quad \text{and} \quad \delta(A_z) = \begin{vmatrix} -4 & 2 & 2 \\ 3 & 4 & 5 \\ 1 & -3 & 8 \end{vmatrix}.$$

(*Solution continued overleaf.*)

Now,

$$\delta(A_y) = -4(2) - 2(5) - 9(19)$$

$$= -8 - 10 - 171 = -189,$$

$$\delta(A_z) = -4(47) - 2(19) + 2(-13)$$

$$= -188 - 38 - 26 = -252,$$

so that

$$y = \frac{\delta(A_y)}{\delta(A)} = \frac{-189}{63} = -3,$$

$$z = \frac{\delta(A_z)}{\delta(A)} = \frac{-252}{63} = -4.$$

Hence the solution set in R^3 of the system is $\{(7, -3, -4)\}$.

Test for consistency If $\delta(A) = 0$ for a linear system, then the system either has infinitely many members in its solution set (the equations are consistent, and one of them can be obtained from the others by linear combinations) or has an empty solution set (the equations are inconsistent). The distinction can be determined as follows: Consider the matrix of coefficients

$$\begin{bmatrix} a_{11} & \cdots & a_{1n} \\ \vdots & & \vdots \\ a_{n1} & \cdots & a_{nn} \end{bmatrix}$$

of a linear system and the **augmented matrix**

$$\begin{bmatrix} a_{11} & \cdots & a_{1n} & c_1 \\ \vdots & & \vdots & \vdots \\ a_{n1} & \cdots & a_{nn} & c_n \end{bmatrix};$$

in each find a determinant (obtained by striking out certain rows and columns) of order as great as possible with value not 0. The order of such a nonvanishing determinant is called the **rank** of the matrix. The rank of the augmented matrix is either the same as, or 1 greater than, that of the matrix of coefficients. It can be shown that the equations are consistent if and only if the two ranks are the same.

For example, the coefficient matrix C and the augmented matrix C_A of the system

$$x + 2y + 3z = 2$$

$$2x + 4y + 2z = -1$$

$$x + 2y - 2z = 5$$

are given by

$$C = \begin{bmatrix} 1 & 2 & 3 \\ 2 & 4 & 2 \\ 1 & 2 & -2 \end{bmatrix} \quad \text{and} \quad C_A = \begin{bmatrix} 1 & 2 & 3 & 2 \\ 2 & 4 & 2 & -1 \\ 1 & 2 & -2 & 5 \end{bmatrix}.$$

Since $\delta(C) = 0$ (check this), the system does not have a unique solution. If the first column and third row are deleted, the remaining determinant,

$$\begin{vmatrix} 2 & 3 \\ 4 & 2 \end{vmatrix},$$

is not zero, so C has rank 2 (check this). Now, if the first column of C_A is deleted, the remaining entries form the determinant

$$\delta \begin{bmatrix} 2 & 3 & 2 \\ 4 & 2 & -1 \\ 2 & -2 & 5 \end{bmatrix},$$

which is also not zero (check this), so C_A has rank 3. Therefore, the system of equations is inconsistent.

Exercise 10.7

Find the solution set of each of the following systems by Cramer's rule. If $\delta(A) = 0$ in any of the systems, use the ranks of the coefficient matrix and the augmented matrix to determine whether or not the equations in the system are consistent.

1. $x - y = 2$
   $x + 4y = 5$

2. $x + y = 4$
   $x - 2y = 0$

3. $3x - 4y = -2$
   $x + y = 6$

4. $\frac{2}{3}x + y = 1$
   $x - \frac{4}{3}y = 0$

5. $x - 2y = 5$
   $\frac{2}{3}x - \frac{4}{3}y = 6$

6. $2x - 3y = 12$
   $x = 4$

7. $ax + by = 1$
   $bx + ay = 1$

8. $x + y = a$
   $x - y = b$

9. $x - 2y + z = -1$
   $3x + y - 2z = 4$
   $y - z = 1$

10. $2x + 5z = 9$
    $4x + 3y = -1$
    $3y - 4z = -13$

11. $2x + 2y + z = 1$
    $x - y + 6z = 21$
    $3x + 2y - z = -4$

12. $4x + 8y + z = -6$
    $2x - 3y + 2z = 0$
    $x + 7y - 3z = -8$

13. $x + y + z = 0$
    $2x - y - 4z = 15$
    $x - 2y - z = 7$

14. $x + y - 2z = 2$
    $3x - y + z = 5$
    $3x + 3y - 6z = 6$

15. $x - 2y - 2z = 3$
    $2x - 4y + 4z = 1$
    $3x - 3y - 3z = 4$

16. $3x - 2y + 5z = 6$
    $4x - 4y + 3z = 0$
    $5x - 4y + z = -5$

17. $x + y + z = 0$
    $w + 2y - z = 4$
    $2w - y + 2z = 3$
    $-2w + 2y - z = -2$

18. $x + y + z = 0$
    $x + z + w = 0$
    $x + y + w = 0$
    $y + z + w = 0$

19. Show that if both $\delta(A_y) = 0$ and $\delta(A_x) = 0$, and if c_1 and c_2 are not both 0, then $\delta(A) = 0$ and the equations in the linear system (a_1 and b_1 not both 0, a_2 and b_2 not both 0)

$$a_1 x + b_1 y + c_1 = 0$$
$$a_2 x + b_2 y + c_2 = 0$$

are consistent. *Hint*: Show that the first two determinant equations imply that $a_1 c_2 = a_2 c_1$ and $b_1 c_2 = b_2 c_1$ and that the rest follows from the formation of a proportion with these equations.

20. Show that if $\delta(A) = 0$ and $\delta(A_x) = 0$ and if a_1 and a_2 are not both 0, then $\delta(A_y) = 0$, where $\delta(A)$ is the determinant of the coefficient matrix of the system in Problem 19.

10.8

Solution of linear systems
by using row-equivalent matrices

Elementary matrices The results of left-multiplying an arbitrary 2×2 matrix

$$M = \begin{bmatrix} a & b \\ c & d \end{bmatrix}$$

by special matrices, called **elementary matrices**, are shown in color in Table 10.1. The verifications of these facts are left as exercises.

Row-equivalent matrices The result of the left-multiplication of a 2×2 matrix A by an elementary matrix is a new matrix B which differs from A in one or more rows and is said to result from an **elementary transformation** of A. We say that the matrix B is *row-equivalent* to A. More generally, we have the following definition.

Definition 10.18 *If B is a matrix resulting from a finite number of elementary transformations on a matrix A, then A and B are* **row-equivalent** *matrices. This is expressed by writing $A \sim B$ or $B \sim A$.*

Row equivalence of a nonsingular matrix and I The effects of an elementary transformation on a 2×2 matrix, as detailed in the right-hand column of Table 10.1, can be extended to $n \times n$ square matrices. Row-equivalent matrices can be obtained in a routine way by observing the results in

Table 10.1

Left multiplication by elementary matrices

Multiplication	Result
$\begin{bmatrix} k & 0 \\ 0 & 1 \end{bmatrix} \begin{bmatrix} a & b \\ c & d \end{bmatrix} = \begin{bmatrix} ka & kb \\ c & d \end{bmatrix}$	Entries of first row multiplied by $k \neq 0$
$\begin{bmatrix} 1 & 0 \\ 0 & k \end{bmatrix} \begin{bmatrix} a & b \\ c & d \end{bmatrix} = \begin{bmatrix} a & b \\ kc & kd \end{bmatrix}$	Entries of second row multiplied by $k \neq 0$
$\begin{bmatrix} 0 & 1 \\ 1 & 0 \end{bmatrix} \begin{bmatrix} a & b \\ c & d \end{bmatrix} = \begin{bmatrix} c & d \\ a & b \end{bmatrix}$	First and second rows interchanged
$\begin{bmatrix} 1 & k \\ 0 & 1 \end{bmatrix} \begin{bmatrix} a & b \\ c & d \end{bmatrix} = \begin{bmatrix} a+kc & b+kd \\ c & d \end{bmatrix}$	Entries of second row multiplied by k and added to corresponding entries of first row
$\begin{bmatrix} 1 & 0 \\ k & 1 \end{bmatrix} \begin{bmatrix} a & b \\ c & d \end{bmatrix} = \begin{bmatrix} a & b \\ ka+c & kb+d \end{bmatrix}$	Entries of first row multiplied by k and added to corresponding entries of second row

this table. Some of these results are analogous to statements for determinants given in Section 10.4.

Example   Show that $\begin{bmatrix} 1 & 2 & 1 \\ -1 & 1 & 0 \\ 1 & 0 & 1 \end{bmatrix} \sim \begin{bmatrix} 1 & 0 & 0 \\ 0 & 1 & 0 \\ 0 & 0 & 1 \end{bmatrix}$.

Solution   We first make appropriate transformations to obtain "0" elements in each entry (except for the principal diagonal) in columns 1, 2, and 3. Each reference to a row indicates the row of the preceding matrix.

$$\begin{bmatrix} 1 & 2 & 1 \\ -1 & 1 & 0 \\ 1 & 0 & 1 \end{bmatrix} \sim \begin{bmatrix} 1 & 2 & 1 \\ 0 & 3 & 1 \\ 0 & -2 & 0 \end{bmatrix} \begin{array}{l} \\ \text{Row 2 + Row 1} \\ \text{Row 3 + } [(-1) \times \text{Row 1}] \end{array}$$

$$\sim \begin{bmatrix} 1 & 0 & 1 \\ 0 & 3 & 1 \\ 0 & 0 & \dfrac{2}{3} \end{bmatrix} \begin{array}{l} \text{Row 1 + Row 3} \\ \\ \text{Row 3 + } \left[\left(\dfrac{2}{3}\right) \times \text{Row 2} \right] \end{array}$$

(Solution continued overleaf.)

$$\sim \begin{bmatrix} 1 & 0 & 0 \\ 0 & 3 & 0 \\ 0 & 0 & \dfrac{2}{3} \end{bmatrix} \begin{array}{l} \text{Row } 1 + \left[\left(-\dfrac{3}{2}\right) \times \text{Row } 3\right] \\[1.5em] \text{Row } 2 + \left[\left(-\dfrac{3}{2}\right) \times \text{Row } 3\right] \\[1.5em] \end{array}$$

$$\sim \begin{bmatrix} 1 & 0 & 0 \\ 0 & 1 & 0 \\ 0 & 0 & 1 \end{bmatrix} \begin{array}{l} \\ \left(\dfrac{1}{3}\right) \times \text{Row } 2 \\[1em] \left(\dfrac{3}{2}\right) \times \text{Row } 3 \end{array}$$

The foregoing example is a special case of a more general result, which we state here without proof.

Theorem 10.16 *If A is a nonsingular n × n square matrix, then A is row-equivalent to $I_{n \times n}$.*

Starting with the augmented matrix of a linear system, and generating a sequence of row-equivalent matrices, we can obtain a matrix from which the solution set of the system is evident simply by inspection. The validity of the method, which is illustrated by example below, follows from the fact that performing elementary row transformations on the augmented matrix of a system corresponds in this context to forming equivalent systems of equations.

Example   Solve $x + 2y - 3z = -4$

$$2x - y + z = 3$$

$$3x + 2y + z = 10$$

Solution   The augmented matrix of the system is

$$\begin{bmatrix} 1 & 2 & -3 & \vdots & -4 \\ 2 & -1 & 1 & \vdots & 3 \\ 3 & 2 & 1 & \vdots & 10 \end{bmatrix}.$$

The system of equations corresponding to each successive matrix is shown on the right of the matrix in the following solution.

$$\begin{array}{r} \\ \text{Row } 2 + [-2 \times \text{Row } 1] \\ \text{Row } 3 + [-3 \times \text{Row } 1] \end{array} \begin{bmatrix} 1 & 2 & -3 & \vdots & -4 \\ 0 & -5 & 7 & \vdots & 11 \\ 0 & -4 & 10 & \vdots & 22 \end{bmatrix} \begin{array}{l} x + 2y - 3z = -4 \\ 0x - 5y + 7z = 11 \\ 0x - 4y + 10z = 22 \end{array}$$

$$\text{Row } 1 + \left[\frac{2}{5} \times \text{Row } 2\right]\begin{bmatrix} 1 & 0 & -\frac{1}{5} & \vdots & \frac{2}{5} \\ 0 & -5 & 7 & \vdots & 11 \\ \text{Row } 3 + \left[-\frac{4}{5}\times\text{Row }2\right]\ 0 & 0 & \frac{22}{5} & \vdots & \frac{66}{5} \end{bmatrix} \begin{array}{l} x + 0y - \frac{1}{5}z = \frac{2}{5} \\ 0x - 5y + 7z = 11 \\ 0x + 0y + \frac{22}{5}z = \frac{66}{5} \end{array}$$

$$\begin{array}{l} 5\times\text{Row }1 \\ \\ \frac{5}{22}\times\text{Row }3 \end{array}\begin{bmatrix} 5 & 0 & -1 & \vdots & 2 \\ 0 & -5 & 7 & \vdots & 11 \\ 0 & 0 & 1 & \vdots & 3 \end{bmatrix} \begin{array}{l} 5x + 0y - z = 2 \\ 0x - 5y + 7z = 11 \\ 0x + 0y + z = 3 \end{array}$$

$$\begin{array}{l}\text{Row }1+\text{Row }3 \\ \text{Row }2 + [-7\times\text{Row }3] \\ \\ \end{array}\begin{bmatrix} 5 & 0 & 0 & \vdots & 5 \\ 0 & -5 & 0 & \vdots & -10 \\ 0 & 0 & 1 & \vdots & 3 \end{bmatrix} \begin{array}{l} 5x + 0y + 0z = 5 \\ 0x - 5y + 0z = -10 \\ 0x + 0y + z = 3 \end{array}$$

$$\begin{array}{l}\frac{1}{5}\times\text{Row }1 \\ \\ -\frac{1}{5}\times\text{Row }2 \\ \\ \end{array}\begin{bmatrix} 1 & 0 & 0 & \vdots & 1 \\ 0 & 1 & 0 & \vdots & 2 \\ 0 & 0 & 1 & \vdots & 3 \end{bmatrix} \begin{array}{l} x + 0y + 0z = 1 \\ 0x + y + 0z = 2 \\ 0x + 0y + z = 3 \end{array}$$

The last system is equivalent to

$$x = 1$$
$$y = 2$$
$$z = 3.$$

From this, the solution set, $\{(1, 2, 3)\}$, for the given system is evident by inspection.

For any given $n \times n$ linear system, there are many sequences of row operations which will transform the augmented matrix of a system equivalently to one of the form

$$\begin{bmatrix} 1 & 0 & 0 & \cdots & 0 & \vdots & x_1 \\ 0 & 1 & 0 & & 0 & \vdots & \cdot \\ 0 & 0 & 1 & & 0 & \vdots & \cdot \\ \vdots & & & & \vdots & \vdots & \cdot \\ 0 & 0 & 0 & \cdots & 1 & \vdots & x_r \end{bmatrix},$$

from which the solution set, $\{(x_1, \cdots x_n)\}$, of the original system is evident by inspection. Finding the most efficient sequence depends on experience and insight.

Exercise 10.8

Use row transformations on an augmented matrix to solve each system of equations.

1. $x - 2y = 4$
   $x + 3y = -1$

2. $x + y = -1$
   $x - 4y = -14$

3. $3x - 2y = 13$
   $4x - y = 19$

4. $4x - 3y = 16$
   $2x + y = 8$

5. $x - 2y = 6$
   $3x + y = 25$

6. $x - y = -8$
   $x + 2y = 9$

7. $x + y - z = 2$
   $2x - y + z = 4$
   $x + 2y - 2z = 2$

8. $2x - y + 3z = 1$
   $x + 2y - z = -1$
   $3x + y + z = 2$

9. $2x - y = 0$
   $3y + z = 7$
   $2x + 3z = 1$

10. $3x - z = 7$
    $2x + y = 6$
    $3y - z = 7$

11. $2x - 5y + 3z = -1$
    $-3x - y + 2z = 11$
    $-2x + 7y + 5z = 9$

12. $2x + y + z = 4$
    $3x - z = 3$
    $2x + 3z = 13$

By completing Exercises 13–17, prove each of the results summarized in Table 10.1.

13. $\begin{bmatrix} k & 0 \\ 0 & 1 \end{bmatrix} \begin{bmatrix} a & b \\ c & d \end{bmatrix} = \begin{bmatrix} ka & kb \\ c & d \end{bmatrix}$

14. $\begin{bmatrix} 1 & 0 \\ 0 & k \end{bmatrix} \begin{bmatrix} a & b \\ c & d \end{bmatrix} = \begin{bmatrix} a & b \\ kc & kd \end{bmatrix}$

15. $\begin{bmatrix} 0 & 1 \\ 1 & 0 \end{bmatrix} \begin{bmatrix} a & b \\ c & d \end{bmatrix} = \begin{bmatrix} c & d \\ a & b \end{bmatrix}$

16. $\begin{bmatrix} 1 & k \\ 0 & 1 \end{bmatrix} \begin{bmatrix} a & b \\ c & d \end{bmatrix} = \begin{bmatrix} a + kc & b + kd \\ c & d \end{bmatrix}$

17. $\begin{bmatrix} 1 & 0 \\ k & 1 \end{bmatrix} \begin{bmatrix} a & b \\ c & d \end{bmatrix} = \begin{bmatrix} a & b \\ ka + c & kb + d \end{bmatrix}$

11

Complex numbers and vectors

11.1

Definitions and their consequences

Some equations do not have solutions in the set of real numbers. For example, if $b > 0$ then $x^2 = -b$ has no real-number solution, because there is no real number whose square is negative. For this reason, the expression $\sqrt{-b}$, for $b \in R$, $b > 0$, is undefined in the set of real numbers. In this chapter, we wish to consider a set of numbers containing members whose squares are negative real numbers and also containing a subset of members that can be identified with the set of real numbers. We shall see that this new set of numbers, called the set C of **complex numbers**, provides solutions for all polynomial equations with real coefficients in one variable. Furthermore, we shall see that the field postulates F-1 through F-11 are satisfied by the complex numbers and thus that C constitutes a field.

Complex numbers as ordered pairs
To begin, consider the set of all ordered pairs of real numbers. This set is, of course, $R \times R$, but for our purposes let us use C to denote it, and let us also use z to represent an unspecified element of C. That is, $z \in C$ is a complex number.

Definition 11.1 $C = R \times R = R^2 = \{z \mid z = (a, b), a \in R \text{ and } b \in R\}$.

Definition 11.2 *Let* $z_1 = (a_1, b_1) \in C$ *and* $z_2 = (a_2, b_2) \in C$. *Then* $z_1 = z_2$ *if and only if* $a_1 = a_2$ *and* $b_1 = b_2$.

Equality properties in the set C
Definition 11.2 gives meaning to **equality** in C. Since the equality postulates E-1 through E-4 hold in the set R of real numbers, it follows from Definition 11.2 that they hold also in the set C. Formal verification is left as an exercise.

Next, let us define sums and products of complex numbers. Although these definitions may appear arbitrary and unusual to start with (particularly in the case of products), we shall see in Section 11.3 that they relate to the corresponding operations in R in a direct and useful way.

Definition 11.3   *If* $z_1 = (a_1, b_1) \in C$, *and* $z_2 = (a_2, b_2) \in C$, *then*

$$\text{I} \quad z_1 + z_2 = (a_1, b_1) + (a_2, b_2) = (a_1 + a_2, b_1 + b_2),$$

$$\text{II} \quad z_1 z_2 = (a_1, b_1) \cdot (a_2, b_2) = (a_1 a_2 - b_1 b_2, a_1 b_2 + a_2 b_1).$$

Examples

a. $(3, 5) + (7, 9) = (3 + 7, 5 + 9) = (10, 14)$

b. $(3, 5) \cdot (7, 9) = (3 \cdot 7 - 5 \cdot 9, 3 \cdot 9 + 5 \cdot 7)$

$$= (21 - 45, 27 + 35)$$

$$= (-24, 62)$$

Addition properties in *C*   With the operations of addition and multiplication thus defined, we can now prove that the set C constitutes a field. The proof is divided into two parts, the first dealing with properties of addition and the second with those of multiplication.

Theorem 11.1   *If* $z_1, z_2, z_3 \in C$, *then:*

F-1 $z_1 + z_2 \in C$.   *Closure law for addition.*

F-2 $(z_1 + z_2) + z_3 = z_1 + (z_2 + z_3)$.   *Associative law for addition.*

F-3   *There exists an element*   *Additive-identity law.*
$z_0 = (0, 0) \in C$ *such that*

$z + z_0 = z \quad and \quad z_0 + z = z$

for all z in C.

F-4   *For each* $z \in C$ *there exists an*   *Additive-inverse law.*
element $-z \in C$ *such that*

$z + (-z) = z_0 \quad and \quad (-z) + z = z_0.$

F-5 $z_1 + z_2 = z_2 + z_1$.   *Commutative law for addition.*

Proof of F-1   Let $z_1 = (a_1, b_1)$ and $z_2 = (a_2, b_2)$. By Definition 11.3-I,

$$z_1 + z_2 = (a_1, b_1) + (a_2, b_2) = (a_1 + a_2, b_1 + b_2),$$

and since the operation of addition is closed in the field R of real numbers, we have $(a_1 + a_2)$ and $(b_1 + b_2) \in R$. Hence, by Definition 11.1, $(a_1 + a_2, b_1 + b_2) \in C$; that is, $z_1 + z_2 \in C$.

The proof of F-2 is left as an exercise. It depends on Definition 11.3-I and the fact that addition is associative in the field R of real numbers.

To give meaning to the expressions $z_1 + z_2 + z_3$ and $z_1 z_2 z_3$, let us agree to the following statement.

Definition 11.4   *If* $z_1, z_2, z_3 \in C$, *then*

$$z_1 + z_2 + z_3 = (z_1 + z_2) + z_3 \quad and \quad z_1 z_2 z_3 = (z_1 z_2)z_3 .$$

This definition can be extended to cover $z_1 + z_2 + z_3 + z_4$, etc. According to Property F-2, it is of course immaterial whether $z_1 + z_2 + z_3$ is considered in accordance with Definition 11.4 or as $z_1 + (z_2 + z_3)$.

Proof of F-3   Let $z = (a, b) \in C$, and consider the element $z_0 = (0, 0)$. Then $z_0 \in C$. By Definition 11.3-I,

$$z + z_0 = (a, b) + (0, 0) = (a + 0, b + 0) = (a, b) = z$$

and

$$z_0 + z = (0, 0) + (a, b) = (0 + a, 0 + b) = (a, b) = z.$$

Here, of course, we have used the fact that 0 is the identity element for addition in the field R of real numbers. We say that $z_0 = (0, 0)$ is the **identity element for addition**, or the **zero element**, in the set C.

Proof of F-4   Let $z = (a, b) \in C$, and consider the element $-z = (-a, -b)$. Then $-z \in C$. We have

$$z + (-z) = (a, b) + (-a, -b) = (a + (-a), b + (-b)) = (0, 0).$$

and similarly

$$(-z) + z = (-a, -b) + (a, b) = ((-a) + a, (-b) + b) = (0, 0).$$

We say that $-z = (-a, -b)$ is the **additive inverse**, or the **negative**, of $z = (a, b)$.

Multiplication   Turning now to properties of multiplication, we have the
properties in C following result.

Theorem 11.2 *If $z_1, z_2, z_3 \in C$, then:*

F-6 $z_1 z_2 \in C.$ *Closure law for multiplication.*

F-7 $(z_1 z_2)z_3 = z_1(z_2 z_3).$ *Associative law for multiplication.*

F-8 $z_1(z_2 + z_3) = z_1 z_2 + z_1 z_3$ *and*   *Distributive law.*

      $(z_2 + z_3)z_1 = z_2 z_1 + z_3 z_1.$

F-9   *There exists an element*   *Multiplicative-identity law.*
      $z_I = (1, 0) \in C$ *such that for all* $z \in C$
      $zz_I = z$ *and* $z_I z = z.$

F-10 $z_1 z_2 = z_2 z_1.$ *Commutative law for multiplication.*

F-11 *For each* $z \in C$ *other*   *Multiplicative-inverse law.*
      than the zero element z_0, *there*
      exists an element $z^{-1} \in C$ *such that*
      $zz^{-1} = z_I$ *and* $z^{-1}z = z_I.$

The proof of F-6 is analogous to that of F-1, with multiplication in place of addition. It is left as an exercise, as are the proofs of F-7 and F-8. For efficiency, let us prove F-10 before turning to F-9 and F-11.

Proof of F-10   Let $z_1 = (a_1, b_1)$ and $z_2 = (a_2, b_2)$; then by Definition 11.3-II,

$$z_1 z_2 = (a_1, b_1)(a_2, b_2) = (a_1 a_2 - b_1 b_2, a_1 b_2 + a_2 b_1)$$

and

$$z_2 z_1 = (a_2, b_2)(a_1, b_1) = (a_2 a_1 - b_2 b_1, a_2 b_1 + a_1 b_2).$$

But since multiplication and addition are commutative in the field R of real numbers, we have

$$a_1 a_2 - b_1 b_2 = a_2 a_1 - b_2 b_1 \quad \text{and} \quad a_1 b_2 + a_2 b_1 = a_2 b_1 + a_1 b_2,$$

from which

$$z_1 z_2 = z_2 z_1.$$

Proof of F-9   Let $z = (a, b)$, and consider the element $z_I = (1, 0) \in C$. We have

$$zz_I = (a, b)(1, 0) = (a \cdot 1 - b \cdot 0, a \cdot 0 + 1 \cdot b) = (a, b) = z.$$

Similarly,

$$z_I z = (1, 0)(a, b) = (1 \cdot a - 0 \cdot b, 1 \cdot b + a \cdot 0) = (a, b) = z,$$

though this follows equally well from the former result together with F-10.

We say that $(1, 0)$ is the **identity element for multiplication** in the set C.

Proof of F-11   Let $z = (a, b)$. Since $a \in R$ and $b \in R$ and $(a, b) \neq (0, 0)$, we have $a^2 + b^2 \neq 0$. Now consider the element

$$z^{-1} = \left(\frac{a}{a^2 + b^2}, \frac{-b}{a^2 + b^2} \right) \in C. \tag{1}$$

By direct computation, we obtain

$$zz^{-1} = (a, b) \left(\frac{a}{a^2 + b^2}, \frac{-b}{a^2 + b^2} \right) = \left(\frac{a^2 + b^2}{a^2 + b^2}, \frac{-ab + ab}{a^2 + b^2} \right) = (1, 0),$$

as desired. By F-10, we observe also that

$$z^{-1}z = (1, 0),$$

thus completing the proof of F-11.

We say that z^{-1}, given by (1), is the **multiplicative inverse** of $z = (a, b)$. You might wonder how the expression (1) entered the picture. Actually, it can be found by solving the equation

$$(a, b)(x, y) = (1, 0)$$

for (x, y), a task that is left as an exercise.

The field   Let us recapitulate. We started with the set C of ordered pairs
properties for C of real numbers, defined equality and the operations of addition
and multiplication on the elements of this set, and then,
one-by-one, established as theorems the properties F-1 through
F-11 that characterize a field. Thus we have shown that the elements of R^2, when viewed as the set of complex numbers, are the elements of a field.

Exercise 11.1

Write each sum as an ordered pair (a, b).

1. $(3, 6) + (2, 1)$ 2. $(7, 1) + (3, -5)$ 3. $(-6, -2) + (0, 1)$

4. $(3, -2) + (-2, 0)$ 5. $(0, 7) + (3, 0)$ 6. $(-2, -1) + (2, 1)$

7. $(2, 3) + (1, 1)$ 8. $(4, 5) + (0, 0)$

Write each product as an ordered pair (a, b).

9. $(1, 1) \cdot (1, 1)$ 10. $(1, 0) \cdot (2, 1)$ 11. $(2, 3) \cdot (4, 1)$

12. $(3, 1) \cdot (0, 2)$ 13. $(2, 2) \cdot (3, 4)$ 14. $(-2, 1) \cdot (1, 3)$

15. $(3, -2) \cdot (1, -1)$ 16. $(0, 1) \cdot (1, 0)$ 17. $(3, 4) \cdot (1, 1)$

18. $(0, 1) \cdot (0, 2)$

19. Write the sum $(a_1, 0) + (a_2, 0)$ as an ordered pair. Write the product $(a_1, 0) \cdot (a_2, 0)$ as an ordered pair.

20. Use Definition 11.1 to show that since the equality postulates E-1 through E-4 of Chapter 1 hold in the set R of real numbers, they hold also in the set C.

21. Make a chart showing F-1 through F-11 for the set R of real numbers compared with the corresponding properties of the set C of ordered pairs (a, b). Start the chart thus:

If $a, b \in R$, then: If $(a, b), (c, d) \in C$, then:

F-1 $a + b \in R$ $(a, b) + (c, d) \in C$ Closure law for addition.

Let $z_1, z_2, z_3 \in C$.

22. Show that $(z_1 + z_2) + z_3 = z_1 + (z_2 + z_3)$.

23. Show that $z_1 + z_2 = z_2 + z_1$.

24. Show that $z_1 z_2 \in C$.

25. Show that $(z_1 \cdot z_2) \cdot z_3 = z_1 \cdot (z_2 \cdot z_3)$.

26. Show that $z_1 \cdot (z_2 + z_3) = z_1 \cdot z_2 + z_1 \cdot z_3$ and $(z_2 + z_3)z_1 = z_2 z_1 + z_3 z_1$.

27. Solve the equation $(a, b)(x, y) = (1, 0)$ for (x, y), given that $(a, b) \neq (0, 0)$.

28. Show that if $z_1 \cdot z_2 = (0, 0)$, then either $z_1 = (0, 0)$, $z_2 = (0, 0)$, or both.

11.2

Properties of complex numbers

Subtraction and division in C We can extend the work of the preceding section to explore a few of the properties of complex numbers. First, let us define two more operations, subtraction and division, for the field C, in terms of addition and multiplication, respectively.

Definition 11.5 *If $z_1, z_2 \in C$, then*

$$z_1 - z_2 = z_1 + (-z_2).$$

You will recall that if $z_2 = (a, b)$, then $-z_2 = (-a, -b)$. The number $z_1 - z_2$ is called the **difference** of z_1 and z_2 and is viewed as the result of **subtracting** z_2 from z_1.

Definition 11.6 *If $z_1, z_2 \in C$ and $z_2 \neq (0, 0)$, then*

$$\frac{z_1}{z_2} = z_1 \cdot z_2^{-1}$$

You will recall from Equation (1) in Section 11.1 that if $z_2 = (a, b)$, then

$$z_2^{-1} = \left(\frac{a}{a^2 + b^2}, \frac{-b}{a^2 + b^2} \right).$$

The number z_1/z_2 is called the **quotient** of z_1 and z_2, and is viewed as the result of **dividing** z_1 by z_2.

Examples
a. $(2, 3) - (5, 6) = (2, 3) + (-5, -6) = (-3, -3)$

b. $\dfrac{(2, 3)}{(5, 6)} = (2, 3) \left(\dfrac{5}{25 + 36}, \dfrac{-6}{25 + 36} \right)$

$= \left(\dfrac{2 \cdot 5 + 3 \cdot 6}{61}, \dfrac{-2 \cdot 6 + 5 \cdot 3}{61} \right) = \left(\dfrac{28}{61}, \dfrac{3}{61} \right)$

It might be noted that the foregoing definitions of subtraction and division in the field C are analogous to the respective definitions given in Chapter 1 for subtraction and division in the field R of real numbers. Indeed, these definitions might be extended to *any* field, since in any field each element has an additive inverse, and each element other than the zero element has a multiplicative inverse.

The structure of algebraic systems We can now obtain a considerable advantage from our structural study of algebraic systems. In Section 1.4, many properties of the field R of real numbers were established. Since, however, the theorems stated there were derived exclusively from the field postulates along with the equality postulates, and did not otherwise depend on the fact that we were dealing with real numbers, it follows that the results are valid in any field—in particular, in the field C. Thus from Theorem 1.3 and Problem 22 of Exercise 1.4, we conclude that for $z \in C$, the additive inverse and the multiplicative inverse (except for the zero element z_0) are unique.

The remaining results of Section 1.4 also extend, of course, to the field C. As an example, the analogue of Theorem 1.10, the *fundamental principle of fractions*, is stated as follows for elements of C.

Theorem 11.3 *If $z_1, z_2,$ and z_3 are elements of C, and z_2 and z_3 are not the zero element z_0, then*

$$\frac{z_1}{z_2} = \frac{z_1 z_3}{z_2 z_3}.$$

In applying this theorem to quotients, it is helpful to have the following result available, as well as the succeeding definition.

Theorem 11.4 *If* $(a, b) \in C$ *and* $c \neq 0$, *then*

$$\frac{(a, b)}{(c, 0)} = \left(\frac{a}{c}, \frac{b}{c}\right).$$

Proof From Definition 11.6 and Equation (1) in Section 11.1,

$$\frac{(a, b)}{(c, 0)} = (a, b)(c, 0)^{-1} = (a, b)\left(\frac{c}{c^2 + 0^2}, \frac{0}{c^2 + 0^2}\right) = (a, b)\left(\frac{1}{c}, 0\right),$$

from which, by Definition 11.3, we obtain

$$\frac{(a, b)}{(c, 0)} = \left(\frac{a}{c}, \frac{b}{c}\right).$$

For example,

$$\frac{(13, -13)}{(26, 0)} = \left(\frac{13}{26}, \frac{-13}{26}\right) = \left(\frac{1}{2}, -\frac{1}{2}\right).$$

Definition 11.7 *The conjugate of* $z = (a, b) \in C$, *denoted by* $\bar{z}$, *is*

$$\bar{z} = (a, -b).$$

Thus, the conjugate of $(5, 1)$ is $(5, -1)$; and if $z = (-2, -7)$, then $\bar{z} = (-2, 7)$.

Now, for $(c, d) \neq (0, 0)$, we already know that the quotient $(a, b)/(c, d)$ is an element of C. This quotient can be written as an ordered pair by using Theorem 11.3 to multiply the numerator and the denominator of $(a, b)/(c, d)$ by the conjugate of the denominator $(c, -d)$, and then applying Theorem 11.4.

Example Write the quotient $\dfrac{(3, -2)}{(5, 1)}$ as an ordered pair by first using Theorem 11.3 and then using Theorem 11.4.

Solution By Theorem 11.3 and Definition 11.3, we have

$$\frac{(3, -2)}{(5, 1)} = \frac{(3, -2)(5, -1)}{(5, 1)(5, -1)} = \frac{(15 - 2, -3 - 10)}{(25 + 1, 0)} = \frac{(13, -13)}{(26, 0)},$$

from which we obtain, by Theorem 11.4,

$$\frac{(13, -13)}{(26, 0)} = \left(\frac{13}{26}, \frac{-13}{26}\right) = \left(\frac{1}{2}, -\frac{1}{2}\right).$$

Exercise 11.2

Write each difference as an ordered pair.

1. $(4, 2) - (1, 1)$ 2. $(-3, 4) - (0, 5)$ 3. $(-6, 1) - (3, 0)$

4. $(0, 1) - (6, 6)$ 5. $(4, -4) - (-4, 4)$ 6. $(0, 0) - (2, -3)$

Write each quotient as an ordered pair. Use Definition 11.6, $\dfrac{(a, b)}{(c, d)} = (a, b) \cdot (c, d)^{-1}.$

7. $\dfrac{(4, 3)}{(2, 2)}$ 8. $\dfrac{(6, 1)}{(1, 3)}$ 9. $\dfrac{(2, 1)}{(-1, 3)}$

10. $\dfrac{(6, -3)}{(4, 1)}$ 11. $\dfrac{(-2, -2)}{(1, 1)}$ 12. $\dfrac{(0, 0)}{(3, 3)}$

13. Write the product $\dfrac{(4, 1)}{(1, 2)} \cdot \dfrac{(-1, 3)}{(6, 1)}$ as an ordered pair. *Hint*: Use any of the field properties.

14. Write the product $\dfrac{(0, 1)}{(3, -1)} \cdot \dfrac{(2, 0)}{(2, 1)}$ as an ordered pair.

Write each quotient as an ordered pair by first multiplying the numerator and denominator by the conjugate of the denominator.

15. $\dfrac{(4, 1)}{(-1, 2)}$ 16. $\dfrac{(6, -1)}{(0, 4)}$ 17. $\dfrac{(2, 3)}{(-1, -1)}$

18. $\dfrac{(0, 1)}{(2, -3)}$ 19. $\dfrac{(4, -4)}{(2, -2)}$ 20. $\dfrac{(0, 0)}{(-2, 3)}$

21. Show that $\dfrac{(a, b)}{(a, b)} = (1, 0)$ for every nonzero ordered pair (a, b).

22. Show that if z_1, z_2, and z_3 are elements of C, and z_2 and z_3 are not $(0, 0)$, then

$$\frac{z_1}{z_2} = \frac{z_1 z_3}{z_2 z_3}.$$

23. Show that $\overline{z_1 + z_2} = \overline{z_1} + \overline{z_2}$.   24. Show that $\overline{z_1 \cdot z_2} = \overline{z_1} \cdot \overline{z_2}$.

25. Show that the set of ordered pairs of the form $(a, 0)$, $a \in R$, with the operations of addition and multiplication in C, constitutes a field. (Show that the field postulates are all satisfied in this set.)

11.3

Subsets of complex numbers

The set R as a
subset of C Let us turn now to a consideration of the relationship between
complex numbers and real numbers. First, recall that, graphically, the set of ordered pairs of real numbers (a, b) is in one-to-
one correspondence with the points in the geometric (x, y)-plane,
just as the set R of real numbers a is in one-to-one correspondence with the points
on the x-axis. Thus, the subset $\{(a, 0)\,|\,a \in R\}$ of C can be considered as correspond-
ing in a one-to-one way to the set R. Moreover, from Definitions 11.3, 11.5, 11.6,
and Theorem 11.4, we have the following results:

$$\left.\begin{aligned}
(a_1, 0) + (a_2, 0) &= (a_1 + a_2, 0), \\
(a_1, 0) \cdot (a_2, 0) &= (a_1 a_2 \ 0), \\
(a_1, 0) - (a_2, 0) &= (a_1 - a_2, 0), \\
\frac{(a_1, 0)}{(a_2, 0)} &= \left(\frac{a_1}{a_2}, 0\right), \quad a_2 \neq 0.
\end{aligned}\right\} \quad (1)$$

The behavior exhibited under the four basic operations by the first components a
of the numbers, $(a, 0) \in C$, and by the numbers $a \in R$, is identical. Let us, then,
identify $(a, 0)$ and a by the following definition.

Definition 11.8   *The element $(a, 0)$ in the set C is identified with the element a
in the set R, and we write*

$$a = (a, 0).$$

It should be noted that a and $(a, 0)$ actually are *conceptually* different, but the
identification is useful and should not be confusing. Under this convention, we
consider that $R \subset C$.

The set I of elements (a, b) of C in which $b \neq 0$ is called the set of **imaginary
numbers**; thus $I \subset C$.

Now, consider the subset of I consisting of all ordered pairs of the form $(0, b)$. In
particular, observe that if we square $(0, b)$, we obtain

$$(0, b)^2 = (0, b) \cdot (0, b) = (-b^2, 0).$$

Since we have agreed to identify $(-b^2, 0)$ with the real number $-b^2$, and since
$b^2 > 0$ for every real number $b \neq 0$, we have $-b^2 < 0$ for every such number.
Thus, we see that the set C provides us with a square root, $(0, b)$, for each negative
real number $-b^2$! Moreover, since

$$(0, -b)^2 = (0, -b) \cdot (0, -b) = (-b^2, 0),$$

C provides us with *two* such square roots. That is, $(0, b)$ and $(0, -b)$ are square roots of $(-b^2, 0)$, or $-b^2$. The square roots of nonnegative real numbers are in $\{(a, 0) \mid a \in R\}$, and those of negative real numbers are in $\{(0, b) \mid b \in R, b \neq 0\}$. As a special case, the square roots of $(-1, 0)$, or -1, are $(0, 1)$ and $(0, -1)$.

The foregoing discussion implies that the set C of complex numbers can be partitioned as follows:

1. $\{(a, 0) \mid a \in R\}$ or $\{a \mid a \in R\}$, the set of real numbers R;

2. $\{(a, b) \mid a, b \in R, b \neq 0\}$, the set of imaginary numbers I.

The set $\{(0, b) \mid b \in R, b \neq 0\}$, a subset of I, is called the set of **pure imaginary numbers**.

Complex numbers in the form $a + bi$　Historically, and for practical reasons, complex numbers are not ordinarily represented in ordered-pair form. To obtain an alternate representation, let us begin by adopting the following convention.

Definition 11.9　*In the set* C, $i = (0, 1)$.

With this definition, and with the convention $b = (b, 0)$ of Definition 11.8,

$$bi = (b, 0)(0, 1) = (0, b).$$

In accordance with this definition, we can extend our symbolism $\sqrt{b^2} = |b|$, which was first introduced on page 59.

Definition 11.10　*In the set* C, *for* $b \in R$, $\sqrt{-b^2} = |b| i$.

In particular, for $b = 1$, we have

$$\sqrt{-1} = 1 \cdot i = i.$$

Any ordered pair (a, b) can now be represented by the sum $(a, 0) + (0, b)$, so that, in our new notation, for $a, b \in R$,

$$(a, b) = a + bi = a + b\sqrt{-1}.$$

The number i is called the **imaginary unit**, a is called the **real part** of $a + bi$, and b is called its **imaginary part**. Definitions 11.2, 11.3, 11.5, and 11.6 can now be expressed using $a + bi$.

Definition 11.11   *If a_1, b_1, a_2, $b_2 \in R$ and $i = (0, 1) = \sqrt{-1}$, then*

I $a_1 + b_1 i = a_2 + b_2 i$ *if and only if* $a_1 = a_2$ *and* $b_1 = b_2$,

II $(a_1 + b_1 i) + (a_2 + b_2 i) = (a_1 + a_2) + (b_1 + b_2)i$,

III $(a_1 + b_1 i) \cdot (a_2 + b_2 i) = (a_1 a_2 - b_1 b_2) + (a_1 b_2 + a_2 b_1)i$,

IV $(a_1 + b_1 i) - (a_2 + b_2 i) = (a_1 - a_2) + (b_1 - b_2)i$,

V $\dfrac{a_1 + b_1 i}{a_2 + b_2 i} = \dfrac{a_1 a_2 + b_1 b_2}{a_2^2 + b_2^2} + \dfrac{b_1 a_2 - a_1 b_2}{a_2^2 + b_2^2} i$ *(a_2, b_2 not both 0)*.

From the definitions we now have, we may rewrite expressions involving complex numbers in the form $a + bi$ in the same way that we rewrote real polynomial expressions, except that i^2 is replaced with -1.

Examples

a. $(2 + 3i) + (6 - 2i) = (2 + 6) + (3 - 2)i = 8 + i$

b. $(2 - i)(1 + 3i) = 2 + 6i - i - 3i^2 = 5 + 5i$

c. $i - (2 + 3i) = (0 - 2) + (1 - 3)i = -2 - 2i$

d. $\dfrac{4 - i}{1 + i} = \dfrac{(4 - i)(1 - i)}{(1 + i)(1 - i)} = \dfrac{4 - 5i + i^2}{1 - i^2} = \dfrac{3}{2} - \dfrac{5}{2}i$

Observe that, in example (d) above, Theorem 11.3 was applied to multiply both $4 - i$ and $1 + i$ by $1 - i$, the conjugate of $1 + i$.

The symbol $\sqrt{-b}$, $b > 0$ In accordance with Definition 11.10, $\sqrt{-b} = i\sqrt{b}$, $b > 0$. The symbol $\sqrt{-b}$, $b > 0$, should be used with care since certain relationships involving the square root symbol that are valid for real numbers are not valid when the symbol does not represent a real number. For instance,

$$\sqrt{-2}\,\sqrt{-3} = (i\sqrt{2})(i\sqrt{3}) = i^2\sqrt{6} = -\sqrt{6} \neq \sqrt{(-2)(-3)}.$$

To avoid difficulty with this point, you should rewrite all expressions of the form $\sqrt{-b}$, $b > 0$, in the form $i\sqrt{b}$ or $\sqrt{b}\,i$.

Examples

a. $3 + \sqrt{-5} = 3 + \sqrt{5}\,i$

b. $(2 + \sqrt{-3})(2 - \sqrt{-3}) = (2 + \sqrt{3}\,i)(2 - \sqrt{3}\,i) = 4 - 3i^2 = 7$

c. $\dfrac{1}{1 + \sqrt{-4}} = \dfrac{1(1 - 2i)}{(1 + 2i)(1 - 2i)} = \dfrac{1 - 2i}{1 - 4i^2} = \dfrac{1}{5} - \dfrac{2}{5}i$

Exercise 11.3

Write each ordered pair (complex number) in the form $a + bi$.

1. $(2, 6)$ 2. $(-3, 4)$ 3. $(5, -2)$ 4. $(0, 6)$

5. $(-7, -3)$ 6. $(-3, 2)$ 7. $(4, 0)$ 8. $(0, 0)$

Write each complex number as an ordered pair.

9. $2 + 3i$ 10. $4 - 2i$ 11. $-3 + i$ 12. $-6 - 3i$

13. $4i$ 14. 0 15. 7 16. $-i$

17-24. Write the conjugate of each complex number in Problems 9–16 in the form $a + bi$.

Find real numbers x and y for which the following statements are true.

Example $(x - 2i)^2 = yi$

Solution   Write each member in the form $a + bi$.

$$x^2 - 4xi + 4i^2 = yi$$

$$(x^2 - 4) - 4xi = yi$$

For equality,

$$x^2 - 4 = 0 \quad \text{and} \quad -4x = y.$$

Therefore, $x = 2$ or -2. If $x = 2$, then $y = -8$; and if $x = -2$, then $y = 8$. The desired real numbers are 2 and -8, and -2 and 8.

25. $2x - yi = 3 + 2i$ 26. $-2i = 3x + yi$ 27. $4 + xi = x^2 - yi$

28. $x + 9i = y + y^2i$ 29. $(x + 3i)^2 = 2yi$ 30. $(x - 2i)^2 = 3x + yi$

Write each of the following expressions in the form $a + bi$.

31. $(2 + 4i) + (3 + i)$ 32. $(4 - i) - (6 - 2i)$ 33. $(2 - i) + (3 - 2i)$

34. $(2 + i) - (4 - 2i)$ 35. $3 - (4 + 2i)$ 36. $(2 - 6i) - 3$

37. $\dfrac{2}{1 - i}$ 38. $\dfrac{6}{3 + 2i}$ 39. $\dfrac{2 + i}{1 - 3i}$

40. $\dfrac{3 - i}{2i}$ 41. $(1 - 3i)^2$ 42. $(2 + i)^2$

43. $(1 - i)^2(1 + i)$ 44. $(3 - 4i)^2(1 - i)^2$ 45. $4 - \sqrt{-7}$

46. $2 + \sqrt{-1}$ 47. $5 - \sqrt{-4}$ 48. $\sqrt{-9}$

49. $\sqrt{-7}\sqrt{-1}$ 50. $\sqrt{-4}\sqrt{-5}$ 51. $\dfrac{2}{1 + \sqrt{-1}}$

52. $\dfrac{3}{1 - \sqrt{-4}}$ 53. $\dfrac{2 + \sqrt{-1}}{3 - \sqrt{-4}}$ 54. $\dfrac{\sqrt{-3}}{1 - \sqrt{-7}}$

55. Show that $c(a + bi) = ca + cbi$. *Hint*: take $c = c + 0i$.

56. Show that $\dfrac{a + bi}{c} = \dfrac{a}{c} + \dfrac{b}{c}i$.

11.4

Complex zeros of polynomial functions

Zeros of a polynomial over C In Section 5.2, it was observed that every polynomial function of degree $n \geq 1$ over the field of complex numbers has exactly n zeros. This is a consequence of the following, which is called the **fundamental theorem of algebra**.

Theorem 11.5 *Every polynomial function of degree $n \geq 1$ over the field C of complex numbers has at least one complex zero.*

The proofs of this and the following theorems involve concepts beyond those available to us and are omitted. As a consequence of Theorem 11.5, we have the following result.

Theorem 11.6 *Every polynomial, of degree $n \geq 1$, over the field C of complex numbers can be expressed as a product of a constant and n linear factors.*

Thus if $P(x) = a_0 x^n + a_1 x^{n-1} + \cdots + a_n$, where $a_i \in C$, $a_0 \neq 0$, $x \in C$, and $n \in N, n \geq 1$, then

$$P(x) = a_0(x - x_1)(x - x_2) \cdots (x - x_n).$$

If a factor $(x - x_i)$ occurs k times in such a linear factorization, then x_i is said to be a zero of $P(x)$ of **multiplicity** k. With this agreement, Theorem 11.6 shows that every polynomial function defined by a polynomial $P(x)$ of degree n with complex coefficients has exactly n complex zeros.

Examples   a. $P(x) = 2x^5 + x^3 - x^2 + 3x + 1$ has exactly five zeros.

           b. $P(x) = (x + 3)(x + 3)(x - i)(x + i)$ has exactly four zeros; -3 is a zero of multiplicity 2.

Zeros and roots   Note that, as observed in Section 4.4, any theorem stated in terms of zeros of polynomial functions applies to roots (solutions) of polynomial equations, and vice versa; a *zero* of

$$\{(x, P(x)) \mid P(x) = a_0 x^n + a_1 x^{n-1} + \cdots + a_n\}$$

is a *solution*, or *root*, of $P(x) = 0$.

Imaginary zeros of   Another important property of a polynomial with real coeffi-
real polynomials   cients establishes the fact that the imaginary zeros of such a polynomial occur in conjugate pairs.

Theorem 11.7   *If $P(z)$ is a polynomial over the field R of real numbers, and $P(z) = 0$ for some $z \in C$, then $P(\bar{z}) = 0$.*

Example   Given that $-1 + i$ is a zero of

$$P(x) = x^2 + 2x + 2,$$

then, by Theorem 11.7, $-1 - i$ is also a zero.

We can use Theorems 11.6 and 11.7 and the theorems in Chapter 5 applicable to real zeros to help us study zeros of polynomial functions with real coefficients.

Example   Find the zeros of $\{(x, P(x)) \mid P(x) = x^3 - 2x^2 + 3x - 6\}$.

Solution   The only possible rational zeros of $P(x)$ are $1, -1, 2, -2, 3, -3, 6$, and -6. Using synthetic division, we obtain:

		1	-2	3	-6
1		1	-1	2	-4
-1		1	-3	6	-12
2		1	0	3	0

From the entries in the last row, we note that $x - 2$ is a factor of $P(x)$, and so is $x^2 + 3$. Hence we can conclude that

(Solution continued overleaf.)

$$P(x) = (x - 2)(x^2 + 3)$$

$$= (x - 2)(x - 3i)(x + 3i).$$

For zeros of P, then, we have solutions of $(x - 2)(x - 3i)(x + 3i) = 0$, which are 2, $3i$, and $-3i$.

In the foregoing example, note that the imaginary zeros of $P(x)$, a polynomial with real coefficients, occur as a conjugate pair, as required by Theorem 11.7.

Exercise 11.4

In Problems 1–8, find all complex zeros.

Example $\{(x, P(x)) \mid P(x) = x^2 - 2x + 4\}$

Solution   Using the quadratic formula on the defining equation for $P(x) = 0$, we have

$$x = \frac{-(-2) \pm \sqrt{(-2)^2 - 4(1)(4)}}{2}$$

$$= \frac{2 \pm 2i\sqrt{3}}{2} = 1 \pm i\sqrt{3}.$$

Thus, zeros of the function are $1 + i\sqrt{3}$ and $1 - i\sqrt{3}$.

1. $\{(x, P(x)) \mid P(x) = x^2 + 1\}$ 2. $\{(x, P(x)) \mid P(x) = x^2 + x + 1\}$

3. $\{(x, P(x)) \mid P(x) = 2x^2 - 3x + 1\}$ 4. $\{(x, P(x)) \mid P(x) = 3x^2 - 2x + 4\}$

5. $\{(x, P(x)) \mid P(x) = 3x^3 - 5x^2 - 14x - 4\}$ *Hint*: First find all rational zeros.

6. $\{(x, P(x)) \mid P(x) = x^3 - 4x^2 - 5x + 14\}$

7. $\{(x, P(x)) \mid P(x) = 2x^4 + 3x^3 + 2x^2 - 1\}$

8. $\{(x, P(x)) \mid P(x) = 8x^4 - 22x^3 + 29x^2 - 66x + 15\}$

In Problems 9–16, one or more zeros are given for each of the polynomial functions; find the other zeros.

Example $\{(x, P(x)) \mid P(x) = x^4 + 3x^3 + 4x^2 + 12x\}$; $2i$ is one zero.

Solution From Theorem 11.7, since $2i$ is a zero, $-2i$ is also a zero. Hence $x + 2i$, $x - 2i$, and their product $x^2 + 4$ are factors of

$$P(x) = x^4 + 3x^3 + 4x^2 + 12x.$$

Dividing $P(x)$ by $x^2 + 4$, we obtain $x^2 + 3x$. Hence,

$$P(x) = (x^2 + 4)(x^2 + 3x)$$
$$= (x + 2i)(x - 2i)x(x + 3).$$

Hence, the three other zeros in addition to $2i$ are $-2i$, 0, and -3.

9. $\{(x, P(x)) \mid P(x) = x^2 + 4\}$; $2i$ is one zero.

10. $\{(x, P(x)) \mid P(x) = 3x^2 + 27\}$; $-3i$ is one zero.

11. $\{(x, P(x)) \mid P(x) = x^3 - 3x^2 + x - 3\}$; 3 and i are zeros.

12. $\{(x, Q(x)) \mid Q(x) = x^3 - 5x^2 + 7x + 13\}$; -1 and $3 - 2i$ are zeros.

13. $\{(x, Q(x)) \mid Q(x) = x^4 + 5x^2 + 4\}$; $-i$ and $2i$ are zeros.

14. $\{(x, P(x)) \mid P(x) = x^4 + 11x^2 + 18\}$; $3i$ and $\sqrt{2}\,i$ are zeros.

15. $\{(x, Q(x)) \mid Q(x) = x^4 + 3x^3 + 4x^2 + 27x - 45\}$; $-3i$ is a zero.

16. $\{(x, Q(x)) \mid Q(x) = x^5 - 2x^4 + 8x^3 - 16x^2 + 16x - 32\}$; $2i$ (multiplicity 2) and 2 are zeros.

17. A cubic equation with real coefficients has roots -2 and $1 + i$. What is the third root? Write the equation in the form $P(x) = 0$, given that the **leading coefficient** (the coefficient of the highest power of x) is 1.

18. A cubic equation with real coefficients has roots 4 and $2 - i$. What is the third root? Write the equation in the form $P(x) = 0$, given that the leading coefficient is 1.

19. One zero of $P(x) = 2x^3 - 11x^2 + 28x - 24$ is $2 - 2i$. Factor $P(x)$ over the complex numbers.

20. One zero of $Q(x) = 3x^3 - 10x^2 + 7x + 10$ is $2 + i$. Factor $Q(x)$ over the complex numbers.

11.5

Trigonometric form of complex numbers

Graphs of complex numbers In Chapter 4, we used Cartesian coordinates to establish a one-to-one correspondence between the set of ordered pairs (a, b) in R^2 and the set of points P in the geometric plane. Since a complex number can also be represented by an ordered pair (a, b), each point in the plane can be viewed as the graph of a complex

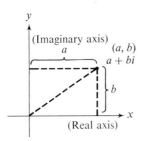

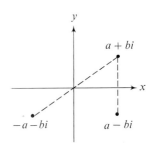

Figure 11.1   Figure 11.2

number (see Figure 11.1). Since the real part a of $a + bi$ is taken as the abscissa, or x-coordinate, of P, in this context the x-axis is called the **real axis**. Similarly, since the imaginary part b is taken as the ordinate or y-coordinate of P, the y-axis is called the **imaginary axis**. Just as we sometimes speak, for instance, of the point $(2, 3)$, meaning of course the point having coordinates 2 and 3, we likewise speak of the point $2 + 3i$, meaning the point representing the complex number $2 + 3i$.

A plane on which complex numbers are thus represented is often called a **complex plane**. It is also sometimes called an **Argand plane**, after the French mathematician Jean Robert Argand (1768–1822), who systematically used it, or a **Gauss plane**, after the great German mathematician Carl Friedrich Gauss (1777–1855).

A complex number $z = a + bi$, its conjugate $\bar{z} = a - bi$, and also its negative $-z = -a - bi$ are represented in Figure 11.2. It is evident that $\bar{z}$ is the reflection of z in the real axis, and $-z$ is the reflection of z in the origin as well as the reflection of $\bar{z}$ in the imaginary axis.

Modulus and argument   Since each nonzero complex number $z = a + bi$ lies on a ray with the origin as endpoint (Figure 11.3), we can associate with each z two useful concepts.

Definition 11.12   *The absolute value, or modulus, of the complex number* $z = a + bi$ *is denoted by* r, $|z|$, *or* $|a + bi|$, *and is given by*

$$r = |z| = |a + bi| = \sqrt{a^2 + b^2}.$$

Thus the modulus $|a + bi|$ is just the distance from the origin to the point $a + bi$.

Definition 11.13   *An argument, or an amplitude, of the complex number* $z = a + bi$, *denoted by* $\arg(a + bi)$, *is an angle* θ *with initial side the positive* x-*axis and terminal side the ray from the origin containing* $a + bi$.

Note that if θ is an argument of $a + bi$, then so is $\theta + 2k\pi^R$, or $\theta + k360°$, for each $k \in J$. For $a + bi = 0$, that is, for $a^2 + b^2 = 0$, any angle θ might be used as arg $(a + bi)$. Also if θ is an argument of $a + bi$, then $b/a = \tan \theta$ when $a \neq 0$.

Trigonometric form　　Because any ordered pair (a, b) can be written in the form $(r \cos \theta, r \sin \theta)$, where $r = \sqrt{a^2 + b^2}$ is the modulus of $a + bi$ and θ is an argument of $a + bi$, it follows that any complex number $a + bi$ can be written in the form

$$r \cos \theta + ir \sin \theta, \quad \text{or} \quad r(\cos \theta + i \sin \theta),$$

which is called the **trigonometric form**, or **polar form**, for a complex number (see Figure 11.4). More generally, using degree measure for angles, we have

$$a + bi = r[\cos (\theta + k360°) + i \sin (\theta + k360°)], \quad k \in J.$$

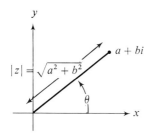

Figure 11.3

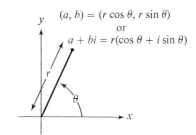

Figure 11.4

A convenient abbreviation that is used for the expression $\cos \theta + i \sin \theta$ is **cis θ** (read "cosine θ plus i sine θ"), so that we can write

$$a + bi = r \text{ cis } (\theta + k360°), \quad k \in J.$$

We ordinarily use for θ the angle of least nonnegative measure that is a solution of $a + bi = r \text{ cis } \theta$.

Example　　Represent $1 + \sqrt{3}\,i$ in trigonometric form.

Solution　　$r = |1 + \sqrt{3}\,i| = \sqrt{1 + 3} = 2$; $\tan \theta = \sqrt{3}/1$, and $\theta = 60°$.

Hence,

$$1 + \sqrt{3}\,i = 2(\cos 60° + i \sin 60°)$$
$$= 2 \text{ cis } 60°.$$

Example   Represent 4 cis 225° graphically, and write the number in rectangular
                  form.

Solution $a = r \cos \theta = 4 \left(-\dfrac{\sqrt{2}}{2} \right) = -2\sqrt{2},$

                  $b = r \sin \theta = 4 \left(-\dfrac{\sqrt{2}}{2} \right) = -2\sqrt{2},$

                  and

                  $a + bi = -2\sqrt{2} - 2\sqrt{2}\,i.$

Products and quotients of complex numbers can be found quite easily when the
complex numbers are in trigonometric form, as shown by the following results.

Theorem 11.8   *If $z_1, z_2 \in C$, with $z_1 = r_1$ cis θ_1 and $z_2 = r_2$ cis θ_2, then*

    I $z_1 \cdot z_2 = r_1 r_2 \text{ cis } (\theta_1 + \theta_2),$

    II $\dfrac{z_1}{z_2} = \dfrac{r_1}{r_2} \text{ cis } (\theta_1 - \theta_2) \quad (z_2 \neq 0 + 0i).$

We shall prove only Part I here and leave the proof of Part II as an exercise.

Proof of Theorem 11.8-I   We have

$$z_1 = r_1 (\cos \theta_1 + i \sin \theta_1) \quad \text{and} \quad z_2 = r_2 (\cos \theta_2 + i \sin \theta_2),$$

from which

$z_1 \cdot z_2 = r_1 (\cos \theta_1 + i \sin \theta_1) \cdot r_2 (\cos \theta_2 + i \sin \theta_2)$

$\qquad = r_1 \cdot r_2 \cdot [\cos \theta_1 \cos \theta_2 + i \cos \theta_1 \sin \theta_2 + i \sin \theta_1 \cos \theta_2 + i^2 \sin \theta_1 \sin \theta_2]$

$\qquad = r_1 \cdot r_2 \cdot [(\cos \theta_1 \cos \theta_2 - \sin \theta_1 \sin \theta_2) + i(\cos \theta_1 \sin \theta_2 + \sin \theta_1 \cos \theta_2)].$

By Theorems 7.3 and 7.6, the right-hand member can be written

$$r_1 r_2 [\cos (\theta_1 + \theta_2) + i \sin (\theta_1 + \theta_2)],$$

so that

$$z_1 \cdot z_2 = r_1 r_2 \text{ cis } (\theta_1 + \theta_2),$$

as was to be shown.

Example   Write the product $3 \text{ cis } 80° \cdot 5 \text{ cis } 40°$ in the form $a + bi$.

Solution   By Theorem 11.8-I,

$$3 \text{ cis } 80° \cdot 5 \text{ cis } 40° = 15 \text{ cis } 120° = 15(\cos 120° + i \sin 120°).$$

Since $\cos 120° = -\dfrac{1}{2}$ and $\sin 120° = \dfrac{\sqrt{3}}{2}$, we have

$$15(\cos 120° + i \sin 120°) = 15\left(-\frac{1}{2} + \frac{\sqrt{3}}{2}i\right) = -\frac{15}{2} + \frac{15\sqrt{3}}{2}i.$$

Example   Write the quotient $\dfrac{8 \text{ cis } 540°}{2 \text{ cis } 225°}$ as a complex number in the form $a + bi$.

Solution   By Theorem 11.8-II.

$$\frac{8 \text{ cis } 540°}{2 \text{ cis } 225°} = \frac{8}{2} \text{ cis } (540° - 225°) = 4 \text{ cis } 315° = 4(\cos 315° + i \sin 315°).$$

Since $\cos 315° = \dfrac{1}{\sqrt{2}}$ and $\sin 315° = \dfrac{-1}{\sqrt{2}}$, we have

$$4(\cos 315° + i \sin 315°) = 4\left(\frac{1}{\sqrt{2}} - \frac{1}{\sqrt{2}}i\right) = 2\sqrt{2} - 2\sqrt{2}\,i.$$

Exercise 11.5

Graph each complex number, its conjugate, its negative, and the negative of its conjugate. Draw line segments joining each pair of these four points.

1. $2 + 3i$ 2. $-3 + 4i$ 3. $4 - i$

4. $-2 - i$ 5. $4i$ 6. $-3i$

Write without absolute-value notation.

Examples   a. $|-3|$ b. $|2 + 5i|$ c. $|(-3, 2)|$

Solutions   By Definition 11.12,

a. $\sqrt{(-3)^2} = 3$ b. $\sqrt{2^2 + 5^2} = \sqrt{29}$

c. $\sqrt{(-3)^2 + (2)^2} = \sqrt{13}$

7. $|4|$ 8. $|-2|$ 9. $|3 + 2i|$

10. $|4 - i|$ 11. $|(2, 0)|$ 12. $|(3, -5)|$

13. $|(-2, -1)|$ 14. $|(-7, -1)|$

Write each complex number in the form r cis θ.

15. $3 + 3i$ 16. $2 - 2i$ 17. 5

18. $-7i$ 19. $2\sqrt{3} - 2i$ 20. $-3\sqrt{3} - 3i$

Write each complex number in the form a + bi.

21. 4 cis 240°   22. 3 cis 300°   23. 6 cis $(-30°)$

24. 5 cis 180°   25. 12 cis 420°   26. 10 cis $(-480°)$

For each given pair of complex numbers z_1 and z_2, find (a) $z_1 \cdot z_2$, and (b) z_1/z_2. Express each result in the form a + bi. Use Table VI as needed.

27. $z_1 = 3$ cis 90° and $z_2 = \sqrt{2}$ cis 45°   28. $z_1 = 4$ cis 30° and $z_2 = 2$ cis 60°

29. $z_1 = 6$ cis 150° and $z_2 = 18$ cis 570°   30. $z_1 = 14$ cis 210° and $z_2 = 2$ cis 120°

31. $z_1 = -3 + i$ and $z_2 = -2 - 4i$ 32. $z_1 = 2 + 3i$ and $z_2 = 2 + 3i$

33. Write $\left(-\dfrac{1}{2} - \dfrac{i\sqrt{3}}{2}\right)^3$ in the form $a + bi$.

34. Write $\left(-\dfrac{1}{2} + \dfrac{i\sqrt{3}}{2}\right)^3$ in the form $a + bi$.

35. Prove that the sum and product of two conjugate complex numbers are both real.

36. Prove that $|z_1 \cdot z_2| = |z_1| \cdot |z_2|$.

37. Show that the conjugate of the complex number r cis θ is r cis $(-\theta)$.

38. Show that if $a + bi = r$ cis θ, then $(a + bi)^2 = r^2$ cis 2θ.

39. Use the result of Problem 38 to show that if $a + bi = r$ cis θ, then

$$(a + bi)^3 = r^3 \text{ cis } 3\theta.$$

40. Use the trigonometric form to show that if z_1 and z_2 are complex numbers, then $z_1 \cdot z_2 = 0 + 0i$ if and only if either z_1 or z_2 or both are equal to $0 + 0i$.

11.6

De Moivre's theorem—powers and roots

Powers of complex numbers

Since $a + bi = r \text{ cis } \theta$, an application of Theorem 11.8-I to $(a + bi)^2$ results in

$$(a + bi)^2 = (r \text{ cis } \theta)(r \text{ cis } \theta) = r^2 \text{ cis } 2\theta. \tag{1}$$

Because

$$(a + bi)^3 = (a + bi)^2(a + bi),$$

from (1) we have

$$(a + bi)^3 = (r^2 \text{ cis } 2\theta)(r \text{ cis } \theta) = r^3 \text{ cis } 3\theta.$$

In a similar way, we can show that

$$(a + bi)^4 = (r^3 \text{ cis } 3\theta)(r \text{ cis } \theta) = r^4 \text{ cis } 4\theta,$$

and it seems plausible to make the following assertion, known as **De Moivre's theorem**. The proof is omitted.

Theorem 11.9 *If $z \in C$, $z = r \text{ cis } \theta$ and $n \in N$, then*

$$z^n = r^n \text{ cis } n\theta.$$

Example

Write $(\sqrt{3} + i)^7$ in the form $a + bi$.

Solution

For the modulus, we have $r = \sqrt{(\sqrt{3})^2 + 1^2} = 2$. For an argument, from $\tan \theta = 1/\sqrt{3}$, we obtain $\theta = 30°$. Thus, $(\sqrt{3} + i)^7 = (2 \text{ cis } 30°)^7$. Then, by De Moivre's theorem,

$$(2 \text{ cis } 30°)^7 = 2^7 \text{ cis } (7 \cdot 30)° = 128 \text{ cis } 210°.$$

Converting to the form $a + bi$, we find

$$128(\cos 210° + i \sin 210°) = 128\left(-\frac{\sqrt{3}}{2} - \frac{1}{2}i\right) = -64\sqrt{3} - 64i;$$

so

$$(\sqrt{3} + i)^7 = -64\sqrt{3} - 64i.$$

By appropriately defining z^0 and z^{-n}, we can extend De Moivre's theorem to include as exponents all $n \in J$.

Definition 11.14   *If $z \neq 0 + 0i$, then*

$$\text{I}\quad z^0 = 1 + 0i, \qquad \text{II}\quad z^{-n} = \frac{1}{z^n}, \quad \text{for } n \in J.$$

Theorem 11.10   *If $z \in C$, $z \neq 0 + 0i$, $z = r \text{ cis } \theta$ and $n \in J$, then*

$$z^n = r^n \text{ cis } n\theta.$$

Example   Write $(1 + i)^{-6}$ in the form $a + bi$.

Solution   Since $r = \sqrt{1^2 + 1^2} = \sqrt{2}$ and $\tan \theta = 1/1$, we have $\theta = 45°$. Hence,

$$(1 + i)^{-6} = (\sqrt{2} \text{ cis } 45°)^{-6}.$$

By Theorem 11.10,

$$(\sqrt{2} \text{ cis } 45°)^{-6} = (\sqrt{2})^{-6} \text{ cis } (-6 \cdot 45)°$$

$$= \frac{1}{8} \text{ cis } (-270°)$$

$$= \frac{1}{8} [\cos (-270°) + i \sin (-270°)].$$

Since $\cos (-270°) = 0$ and $\sin (-270°) = 1$, we obtain

$$(1 + i)^{-6} = \frac{1}{8} (0 + i) = \frac{1}{8} i.$$

Roots of   Yet another extension of De Moivre's theorem is possible if
complex numbers   we make the following definition.

Definition 11.15   *For $z \in C$, $n \in N$, w is an nth root of z provided*

$$w^n = z.$$

Theorem 11.11   *If $z \in C$, $z = r \text{ cis } \theta$ and $n \in N$, then*

$$w = r^{1/n} \text{ cis } \left(\frac{\theta}{n}\right)$$

is an nth root of z.

This theorem follows directly from De Moivre's theorem. The fact that

$$\text{cis } \theta = \text{cis } (\theta + k360°),$$

for $k \in J$, enables us to find n distinct complex nth roots for each $z \in C$, $z \neq 0 + 0i$, as illustrated in the following example.

Example   Write each of the four fourth roots of $z = 2 + 2\sqrt{3}\,i$ in the form r cis θ.

Solution   Since $r = \sqrt{2^2 + (2\sqrt{3})^2} = 4$ and $\theta = \text{Tan}^{-1} 2\sqrt{3}/2 = 60°$, we have

$$z = 2 + 2\sqrt{3}\,i = 4 \text{ cis } 60°.$$

From Theorem 11.11 and the periodic property of cosine and sine, each number

$$4^{1/4} \text{ cis } \left(\frac{60° + k360°}{4} \right),$$

for $k \in J$, is a fourth root of z. Taking $k = 0, 1, 2,$ and 3, in turn, gives the roots

$$w_0 = \sqrt{2} \text{ cis } 15°, \quad w_1 = \sqrt{2} \text{ cis } 105°, \quad w_2 = \sqrt{2} \text{ cis } 195°, \quad w_3 = \sqrt{2} \text{ cis } 285°.$$

The substitution of any other integer for k will produce one of these four complex numbers.

Exercise 11.6

Use Theorem 11.9 or 11.10 as appropriate to write each of the given expressions as a complex number of the form $a + bi$. Use Table VI as necessary.

1. $[2 \text{ cis } (-30°)]^7$

2. $(4 \text{ cis } 36°)^5$

3. $\left(-\frac{1}{2} + \frac{1}{2}\sqrt{3}\,i \right)^3$

4. $(1 + i)^{12}$

5. $(\sqrt{3} \text{ cis } 5°)^{12}$

6. $(\sqrt{2} \text{ cis } 30°)^{-7}$

7. $(\sqrt{3} - i)^{-5}$

8. $(1 - i)^{-6}$

9. $\dfrac{(1 + i)^3}{(1 + \sqrt{3}\,i)^5 (1 - i)^2}$

10. $\dfrac{4(\sqrt{3} + i)^3}{(1 - i)^3}$

11. $\dfrac{(1 - i)^5}{(1 + i)^6}$

12. $\dfrac{(1 + \sqrt{3}\,i)^{-4}}{(\sqrt{3} + i)^{-6}}$

Find the nth roots of z by applying Theorem 11.11. Leave the results in trigonometric form and list all n of the nth roots.

13. $z = 32 \text{ cis } 45°, \quad n = 5$

14. $z = 27, \quad n = 3$

15. $z = -16\sqrt{3} + 16i, \quad n = 5$

16. $z = 1 - i, \quad n = 4$

17. $z = -i, \quad n = 6$

18. $z = 2 + 2\sqrt{3}\,i, \quad n = 3$

Solve each of the following equations over C.

19. $x^5 = 16 - 16\sqrt{3}\,i$ 20. $x^3 + 4i = 4\sqrt{3}$

21. $x^7 + 1 = 0$ 22. $x^7 - 1 = 0$

23. Factor $x^4 + 16$ into linear factors.

24. Factor $x^5 - 1$ into linear factors.

25. Show that the sum of the four fourth roots of 1 is $0 + 0i$.

26. Explain why the sum of the nth roots of any complex number is zero for all $n > 1$.

27. Use De Moivre's theorem to prove that

$$\cos 3\theta = 4\cos^3 \theta - 3\cos \theta \quad \text{and} \quad \sin 3\theta = 3\sin \theta - 4\sin^3 \theta.$$

11.7
Two-dimensional vectors

Ordered pairs as vectors   In Chapter 10, we considered the set of $m \times n$ matrices, for m and n fixed, as a vector space over the field of real numbers. In particular, we referred to a $1 \times n$ matrix $[a_1\, a_2 \cdots a_n]$ as a row vector, and its transpose $[a_1\, a_2 \cdots a_n]^t$ as a column vector. The two operations involved in a vector space S, it will be recalled, are (i) the addition of vectors in S and (ii) the multiplication of vectors in S by elements of R. In order to develop a vector space using ordered pairs of real numbers as vectors, we begin with a definition.

Definition 11.16   *A two-dimensional vector* **v** *is an ordered pair of real numbers. That is, if $a, b \in R$, then*

$$\mathbf{v} = (a, b),$$

and the set of all such vectors is the set

$$V = \{(a, b)\,|\,a, b \in R\}.$$

Note that we use **boldface** type to denote vectors. In handwritten form, vectors are frequently identified by arrows drawn above symbols. Thus, $\vec{v}$ denotes a vector. Since we shall be dealing only with two-dimensional vectors, that is, vectors with only two components, we shall refer to these simply as *vectors*. Most of the concepts developed here, however, extend to sets of ordered triples, ordered quadruples, or, indeed, to ordered n-tuples of real numbers. These, in turn, can be considered as sets of vectors of three, four, or n dimensions.

Geometric
representation
Because every ordered pair (a, b) can be associated with a directed line segment, or **geometric vector**, originating (having **initial point**) at the origin and terminating (having **terminal point**) at the point in the plane corresponding to (a, b), there is a geometric interpretation that can be made of the vector-space algebra of ordered pairs. See Figure 11.5.

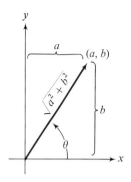

Figure 11.5

Definition 11.17   *For each vector* $\mathbf{v} = (a, b)$, *the norm, or magnitude, of* $\mathbf{v}$ *is the real number*

$$\|\mathbf{v}\| = \sqrt{a^2 + b^2}.$$

It is evident from the Pythagorean theorem that the norm of a vector is just the length of the associated geometric vector.

Definition 11.18   *For each vector* $\mathbf{v} = (a, b)$, *with* $\|\mathbf{v}\| \neq 0$, *the direction angle of* $\mathbf{v}$ *is the angle* θ *satisfying*

$$\cos \theta = \frac{a}{\|\mathbf{v}\|}, \qquad \sin \theta = \frac{b}{\|\mathbf{v}\|}, \qquad -180° < \theta \leq 180°.$$

Thus the direction angle of $\mathbf{v}$ is the angle θ, $-180° < \theta \leq 180°$, from the positive x-axis to the geometric vector associated with $\mathbf{v}$.

Example
Find the norm and direction angle, to the nearest 10', of $\mathbf{v} = (4, -5)$.

Solution
$\|\mathbf{v}\| = \sqrt{4^2 + (-5)^2} = \sqrt{41}$. Since $P(4, -5)$ is in the fourth quadrant,

$$\theta = \operatorname{Sin}^{-1} \frac{-5}{\sqrt{41}} \approx \operatorname{Sin}^{-1} \left(\frac{-5}{6.403}\right) \approx \operatorname{Sin}^{-1}(-0.7808) \approx -51° \, 20'.$$

Definition 11.19   *If* $\mathbf{v}_1 = (a_1, b_1)$ *and* $\mathbf{v}_2 = (a_2, b_2)$, *then*

$$\mathbf{v}_1 = \mathbf{v}_2 \quad \text{if and only if} \quad a_1 = a_2 \text{ and } b_1 = b_2.$$

Accordingly, vectors are equal if and only if they correspond to the same geometric vector.

Operations on   Now, let us define two operations, as follows.
vectors

Definition 11.20 *If* $v_1 = (a_1, b_1)$, $v_2 = (a_2, b_2)$, *and* $c \in R$, *then*

$$\text{I}\quad v_1 + v_2 = (a_1, b_1) + (a_2, b_2) = (a_1 + a_2, b_1 + b_2),$$

$$\text{II}\quad cv_1 = c(a_1, b_1) = (ca_1, cb_1).$$

From the vector viewpoint of ordered pairs, real numbers, such as c in the foregoing definition, are called **scalars**, and the operation defined in II is called the multiplication of a vector by a scalar. Graphically, operations I and II can be interpreted in terms of geometric vectors, as shown in Figures 11.6 and 11.7.

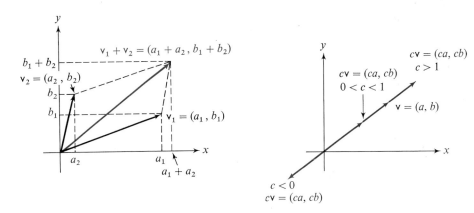

Figure 11.6   Figure 11.7

In Figure 11.6, the vectors v_1 and v_2 are said to be **noncollinear** because they correspond to noncollinear geometric vectors. It is evident that the sum of two such noncollinear vectors v_1 and v_2 corresponds to the diagonal of the parallelogram with adjacent sides corresponding to v_1 and v_2. This statement can be considered valid, in a limiting way, even if the geometric vectors corresponding to v_1 and v_2 are collinear. Accordingly, we say that vectors are added according to the **parallelogram law**.

In Figure 11.7, the vectors v_1 and cv_1 are said to be collinear, since they correspond to geometric vectors on the same line. For $c > 0$, the product cv corresponds to a geometric vector with the *same direction* as the one corresponding to v. If $0 < c < 1$, then $\|cv\| < \|v\|$. If $c < 0$, then cv corresponds to a geometric vector of *direction opposite* to the geometric vector corresponding to v.

V as a vector space   The set V of vectors constitutes a vector space. To see this, we need two definitions.

Definition 11.21 *The zero vector, $\mathbf{0}$, or identity vector for addition, is*

$$\mathbf{0} = (0, 0).$$

The zero vector has the property that $\mathbf{v} + \mathbf{0} = \mathbf{0} + \mathbf{v} = \mathbf{v}$ for each vector $\mathbf{v}$. The norm of $\mathbf{0}$ is, of course, 0. No particular direction is assigned to the geometric vector corresponding to $\mathbf{0}$; for this reason, *any and every direction might be assigned to it*, as suits our convenience.

Definition 11.22 *If $\mathbf{v} = (a, b)$ is a vector, then the negative of $\mathbf{v}$ is*

$$-\mathbf{v} = (-a, -b).$$

With Definitions 11.19–11.22 we now have exactly the same algebra, as regards addition and the multiplication by a real-number scalar, for the set of real-number pairs (a, b), whether they are regarded as 1×2 matrices, as 2×1 matrices, or as vectors. Since we have shown that

$$\{(a, b) \,|\, a, b \in R\}$$

is a vector space when viewed as the set of 1×2 matrices, it follows that the set is also a vector space when viewed as a set of vectors. We can state the result for vectors $\mathbf{v} = (a, b)$ as follows:

Theorem 11.12 *The set V of vectors $\mathbf{v} = (a, b)$, $a, b \in R$, with the set R of real numbers as scalar multipliers, is a vector space over the field R of real numbers. That is, if $\mathbf{v}_1$, $\mathbf{v}_2$, and $\mathbf{v}_3$ are vectors, and c and d are real scalars, then*

I $\mathbf{v}_1 + \mathbf{v}_2 \in V$,

II $(\mathbf{v}_1 + \mathbf{v}_2) + \mathbf{v}_3 = \mathbf{v}_1 + (\mathbf{v}_2 + \mathbf{v}_3)$,

III $\mathbf{v}_1 + \mathbf{0} = \mathbf{v}_1$ and $\mathbf{0} + \mathbf{v}_1 = \mathbf{v}_1$,

IV $\mathbf{v}_1 + (-\mathbf{v}_1) = \mathbf{0}$,

V $\mathbf{v}_1 + \mathbf{v}_2 = \mathbf{v}_2 + \mathbf{v}_1$,

VI $c\mathbf{v}_1 \in V$,

VII $c(d\mathbf{v}_1) = (cd)\mathbf{v}_1$,

VIII $(c + d)\mathbf{v}_1 = c\mathbf{v}_1 + d\mathbf{v}_1$,

IX $c(\mathbf{v}_1 + \mathbf{v}_2) = c\mathbf{v}_1 + c\mathbf{v}_2$,

X $1 \cdot \mathbf{v}_1 = \mathbf{v}_1$,

XI $(-1)\mathbf{v}_1 = -\mathbf{v}_1$,

XII $0 \cdot \mathbf{v}_1 = \mathbf{0}$,

XIII $c \cdot \mathbf{0} = \mathbf{0}$.

Some of these are again considered in the following exercise set, but this time from a vector viewpoint. The proofs of the various parts of this theorem were considered in Chapter 10.

In accord with Definition 11.22, we can also define the *difference* of two vectors.

Definition 11.23 *If* $\mathbf{v}_1$ *and* $\mathbf{v}_2$ *are vectors, then the difference* $\mathbf{v}_1 - \mathbf{v}_2$ *is*

$$\mathbf{v}_1 - \mathbf{v}_2 = \mathbf{v}_1 + (-\mathbf{v}_2),$$

and is viewed as the result of subtracting $\mathbf{v}_2$ *from* $\mathbf{v}_1$.

The geometric vector associated with $\mathbf{v}_1 - \mathbf{v}_2$ is shown in Figure 11.8.

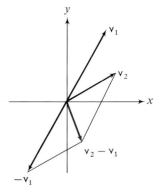

Figure 11.8

Exercise 11.7

Write each of the following in the form (a, b), *represent the vector operation(s) graphically, and find the norm and direction angle of* (a, b).

Example $(5, 3) - (2, -1)$

Solution $(5, 3) - (2, -1)$

$$= (5, 3) + (-2, 1)$$

$$= (3, 4);$$

$$\|(3, 4)\| = \sqrt{3^2 + 4^2} = 5,$$

$$\theta = \text{Cos}^{-1}\frac{3}{5} \approx 53° \, 10'.$$

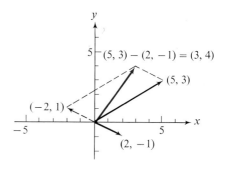

1. $(2, 4) + (6, 1)$

2. $(3, -1) + (4, 2)$

3. $(-3, 4) + (2, -1)$

4. $(-2, -4) + (7, -2)$

5. $(0, 3) + (5, 6)$

6. $(5, 2) + (3, 0)$

7. $(2, 0) + (3, 5) + (3, 1)$

8. $(-2, 4) + (3, 0) + (4, 3)$

9. $(6, -1) + (2, 0) + (0, 2)$

10. $(0, 4) + (2, -1) + (5, 4)$

11. $(5, 3) - (2, 2)$

12. $(7, 8) - (3, 4)$

13. $(-2, 4) - (2, -1)$

14. $(8, -3) - (2, -1)$

15. $(2, 4) + (0, 3) - (2, -1)$

16. $(-3, 0) + (2, 1) - (-1, 0)$

17. $(0, 2) - (3, 1) + (2, 0)$

18. $(4, 2) - (0, -2) + (4, 4)$

Example $2(3, 5) + 3(1, -2)$

Solution $2(3, 5) + 3(1, -2)$

$$= (6, 10) + (3, -6)$$

$$= (9, 4);$$

$$\|(9, 4)\| = \sqrt{9^2 + 4^2}$$

$$= \sqrt{97},$$

$$\theta = \text{Cos}^{-1} \frac{9}{\sqrt{97}} \approx \text{Cos}^{-1} 0.914 \approx 24°.$$

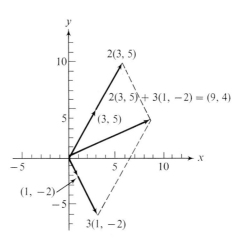

19. $3(1, 1) + 4(2, 2)$

20. $3(1, -4) + 2(3, 5)$

21. $5(0, 2) + 3(-2, -3)$

22. $2(3, 1) + 5(-6, 0)$

23. $(5, 6) - 2(1, -2)$

24. $(6, 2) - 3(-1, 2)$

25. $3(0, 4) - 4(2, 0)$

26. $7(5, 0) - 2(-3, 0)$

27. $2(1, 3) - 4(-2, 1) + 2(0, 1)$

28. $3(0, 2) + 5(1, -2) - 3(0, 1)$

29. $3(4, -1) - 2(0, 1) + (3, 0)$

30. $(4, 0) - 2(1, 3) + 3(0, -2)$

For Problems 31–40, let $\mathbf{v}_1 = (a_1, b_1)$, $\mathbf{v}_2 = (a_2, b_2)$, and $\mathbf{v}_3 = (a_3, b_3)$; $c, d \in R$. Show that each of the following statements is true, using appropriate definitions and properties of real numbers.

31. $\mathbf{v}_1 + \mathbf{v}_2 \in V$.

32. $\mathbf{v}_1 + \mathbf{v}_2 = \mathbf{v}_2 + \mathbf{v}_1$.

33. $(\mathbf{v}_1 + \mathbf{v}_2) + \mathbf{v}_3 = \mathbf{v}_1 + (\mathbf{v}_2 + \mathbf{v}_3)$

34. $\mathbf{v}_1 + \mathbf{0} = \mathbf{v}_1$

35. $\mathbf{v}_1 + (-\mathbf{v}_1) = \mathbf{0}$

36. $c\mathbf{v}_1 \in V$

37. $c(d\mathbf{v}_1) = (cd)\mathbf{v}_1$

38. $(c + d)\mathbf{v}_1 = c\mathbf{v}_1 + d\mathbf{v}_1$

39. $c(\mathbf{v}_1 + \mathbf{v}_2) = c\mathbf{v}_1 + c\mathbf{v}_2$

40. $1 \cdot \mathbf{v}_1 = \mathbf{v}_1$

11.8

Applications of vectors

"Free" geometric vectors In applications, it is convenient to consider all directed line segments—or geometric vectors—in the plane, not just those with initial point at the origin.

Definition 11.24 *Two geometric vectors are equivalent if they have the same length (magnitude) and the same direction.*

Thus the geometric vector from $(0, 0)$ to $(2, 3)$ and the geometric vector from $(4, 2)$ to $(6, 5)$ are equivalent. See Figure 11.9, where these and other geometric vectors equivalent to them are pictured.

Of course, every geometric vector in the plane is equivalent to infinitely many other such vectors, and, in particular, every geometric vector is equivalent to one in **standard position**, that is, with initial point at the origin.

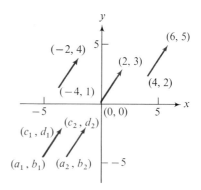

Figure 11.9

If you imagine a geometric vector as free to "slide" (parallel to its initial position) in the plane, then you can always "slide" any geometric vector in the plane onto any vector equivalent to it, as shown in Figure 11.10.

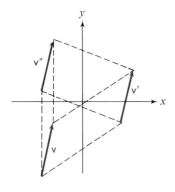

Figure 11.10

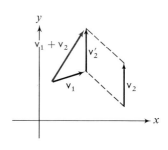

Figure 11.11

Using this notion, you can visualize the sum of two geometric vectors $\mathbf{v}_1$ and $\mathbf{v}_2$ as being determined by sliding one vector, say $\mathbf{v}_2$, onto an equivalent vector $\mathbf{v}_2'$, the initial point of which lies on the terminal point of $\mathbf{v}_1$. The sum $\mathbf{v}_1 + \mathbf{v}_2$ is then the vector with initial point the initial point of $\mathbf{v}_1$ and terminal point the terminal point of $\mathbf{v}_2'$, as shown in Figure 11.11.

Test for equivalence of geometric vectors To express analytically the conditions for the equivalence of two geometric vectors in the plane, we have the following result, proof of which is left as an exercise.

Theorem 11.13 *The geometric vector from (a_1, b_1) to (c_1, d_1) and the geometric vector from (a_2, b_2) to (c_2, d_2) are equivalent if and only if*

$$c_1 - a_1 = c_2 - a_2 \quad and \quad d_1 - b_1 = d_2 - b_2.$$

If they are equivalent, then both are geometric representations of the vector

$$(c_1 - a_1, d_1 - b_1).$$

Vectors representing physical quantities Many physical quantities, such as displacement, force, velocity and acceleration, are specified by a magnitude and a direction. Accordingly, vector methods are extensively used in treating these quantities.

Example An airplane flies 340 miles per hour in a direction 60° clockwise from north for two hours and then flies due north at the same speed for one hour. How far from the starting point is the airplane at the end of this time?

Solution The conditions can be represented with geometric vectors as illustrated in Figures a and b.

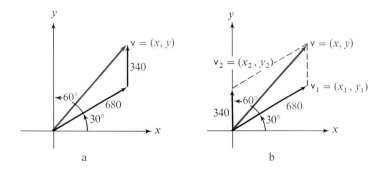

(Solution continued overleaf.)

Since $\sin 30° = \dfrac{1}{2} = \dfrac{y_1}{680}$, we have $y_1 = 340$;

since $\cos 30° = \dfrac{\sqrt{3}}{2} = \dfrac{x_1}{680}$,

we have $x_1 = 340\sqrt{3}$; hence $\mathbf{v}_1 = (340\sqrt{3}, 340)$.

Since $\mathbf{v}_2 = (x_2, y_2) = (0, 340)$, we have

$$\mathbf{v}_3 = \mathbf{v}_1 + \mathbf{v}_2 = (340\sqrt{3}, 340) + (0, 340)$$

$$= (340\sqrt{3}, 680),$$

$$\|\mathbf{v}_3\| = \sqrt{340^2 \cdot 3 + 680^2} = \sqrt{809{,}200}.$$

Hence, the airplane is $\sqrt{809{,}200}$, or about 900, miles from its starting point.

Consider next an application of a different sort, namely to a set of forces that are in balance.

Example

A force of magnitude 10 lbs is applied to an iron ring in a direction making an angle of 30° with the positive x-direction, and a second force, of magnitude 18 lbs, is applied to the ring in a direction making an angle of 45° with the positive x-direction. Determine the magnitude and direction of the force on the ring that will just balance these two given forces.

Solution

Since

$$\sin 30° = \frac{1}{2} = \frac{b}{10} \quad \text{and} \quad \cos 30° = \frac{\sqrt{3}}{2} = \frac{a}{10},$$

the first force can be represented by the vector $(a, b) = (5\sqrt{3}, 5)$. Again, since

$$\sin 45° = \cos 45° = \frac{1}{\sqrt{2}} = \frac{c}{18} = \frac{d}{18},$$

the second force can be represented by the vector $(c, d) = (9\sqrt{2}, 9\sqrt{2})$. If the force $F(x, y)$ is to balance these, then the sum of the three forces must vanish:

$$(5\sqrt{3}, 5) + (9\sqrt{2}, 9\sqrt{2}) + (x, y) = (0, 0).$$

Thus $F(x,y)$ is

$$(-5\sqrt{3} - 9\sqrt{2}, \ -5 - 9\sqrt{2}),$$

or approximately

$$(-21.4, \ -17.7).$$

The magnitude of the force F therefore is

$$\|F\| = \sqrt{x^2 + y^2}$$

$$\approx \sqrt{(-21.4)^2 + (-17.7)^2} \approx 27.8.$$

Since $x < 0$ and $y < 0$, F must be directed into the third quadrant; its angle θ from the positive x-direction is given by

$$\sin \theta = \frac{y}{\sqrt{x^2 + y^2}}$$

$$\approx \frac{-17.7}{27.8}$$

$$\approx -0.637, \quad -180° < \theta < 180°,$$

so that $\theta \approx -140° \ 30'$.

Exercise 11.8

Specify each vector as an equivalent vector with initial point at the origin.

Example $\mathbf{v} : (4, -1)$ to $(5, 3)$

Solution   By Theorem 11.13, this vector is equivalent to

$$[5 - 4, 3 - (-1)] \quad \text{or} \quad (1, 4).$$

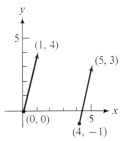

1. $\mathbf{v} : (2, 3)$ to $(6, 7)$ 2. $\mathbf{v} : (4, -3)$ to $(6, 2)$

3. $\mathbf{v} : (-2, 5)$ to $(7, -1)$ 4. $\mathbf{v} : (4, -2)$ to $(-3, 0)$

5. $\mathbf{v} : (-5, -3)$ to $(0, 0)$ 6. $\mathbf{v} : (-5, 0)$ to $(0, -3)$

The figure shows several geometric vectors. Express each of the following combinations of the associated vectors as an ordered pair (a, b).

Example $\mathbf{v}_2 + \mathbf{v}_7$

Solution $\mathbf{v}_2$ is equivalent to

$(5 - 8, 14 - 9)$

$= (-3, 5),$

$\mathbf{v}_7$ is equivalent to

$(11 - 3, -6 - (-6))$

$= (8, 0)$

Hence,

$\mathbf{v}_2 + \mathbf{v}_7$

$= (-3, 5) + (8, 0)$

$= (-3 + 8, 5 + 0)$

$= (5, 5).$

7. $\mathbf{v}_1 + \mathbf{v}_5$ 8. $\mathbf{v}_3 + \mathbf{v}_7$ 9. $\mathbf{v}_6 - \mathbf{v}_2$

10. $\mathbf{v}_3 - \mathbf{v}_4$ 11. $\mathbf{v}_2 + \mathbf{v}_3 + \mathbf{v}_4$ 12. $\mathbf{v}_6 + \mathbf{v}_1 + \mathbf{v}_5$

13. $\mathbf{v}_2 + \mathbf{v}_4 - \mathbf{v}_1$ 14. $\mathbf{v}_4 - \mathbf{v}_7 + \mathbf{v}_2$

15. An airplane flies 800 miles due south and then flies 600 miles 60° clockwise from north. How far from the starting point is the airplane at the end of this time?

16. An airplane is flying at 20,000 ft on a heading of 300° clockwise from north at 450 mph. The wind velocity at this level is 90 mph from 240° clockwise from north. Find the ground speed and the direction of the flight over the ground.

17. A weight of 160 lbs is placed on a smooth plane inclined at an angle of 30° with the horizontal (see figure). What is the minimum force that must be exerted against the weight and parallel to the plane to prevent the weight from slipping? *Hint*: Resolve the vertical force into components in the direction of the plane and perpendicular to it.

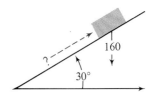

18. A force of 40 lbs is directed to the right. What vertical force must be applied so that the magnitude of the resultant force is 80 lbs? What angle does the resultant force make with the horizontal?

19. Forces having magnitudes of 12 and 18 lbs are applied to an iron ring in directions making clockwise angles measuring 30° and 120°, respectively, with the positive x-direction. Determine the magnitude and direction of the force on the ring that will just balance these two forces.

20. Determine the magnitude and direction of the force required for equilibrium in Problem 19 if the force making an angle of 30° has a magnitude of 18 lbs and the force making an angle of 120° has a magnitude of 10 lbs.

12

Sequences and series

12.1

Mathematical induction

The material in the present section depends on a special property of the set of natural numbers, or positive integers: $N = \{1, 2, 3, \ldots\}$.

The set N has the following properties:

$$a.\ 1 \in S. \qquad b.\ \text{If } k \in S, \text{ then } k + 1 \in S.$$

In addition, N *contains no elements not implied by properties a and b*. These properties underlie the following theorem, called the **principle of mathematical induction**, which we state without proof.

Theorem 12.1 *If a given sentence involving natural numbers n is true for $n = 1$, and if its truth for $n = k$ implies its truth for $n = k + 1$, then it is true for every natural number n.*

> Requirements of a proof by mathematical induction

We can exploit Theorem 12.1 to prove a number of assertions. Although the technique we shall use is called **proof by mathematical induction**, the argument we shall employ is deductive, as have been all of the other arguments in this book. Proofs by mathematical induction require two things:

a. A demonstration that the assertion to be proved is true for the natural number 1.
b. A demonstration that the truth of the assertion for a natural number k implies its truth for $k + 1$.

When these two demonstrations have been made, the principle of mathematical induction assures us that the assertion is true for every natural number.

Example   Prove that the sum of the first n natural numbers is $\dfrac{n(n+1)}{2}$.

Solution   In symbols, we wish to show that

$$1 + 2 + 3 + \cdots + n = \frac{n(n+1)}{2}.$$

We must do two things:

     1. We must first show that the assertion is true for $n = 1$, that is, that

$$1 = \frac{1(1+1)}{2},$$

which is true.

     2. We must next show that the truth of

$$1 + 2 + 3 + \cdots + k = \frac{k(k+1)}{2}$$

implies the truth of

$$1 + 2 + 3 + \cdots + k + (k+1) = \frac{(k+1)[(k+1)+1]}{2}.$$

That is, we must show that the truth of the assertion for $n = k$ implies its truth for $n = k + 1$. Now, assume the truth of

$$1 + 2 + 3 + \cdots + k = \frac{k(k+1)}{2}.$$

Then, by adding $k + 1$ to each member of this equation, we obtain

$$1 + 2 + 3 + \cdots + k + (k+1) = \frac{k(k+1)}{2} + (k+1)$$

$$= (k+1)\left(\frac{k}{2} + 1\right)$$

$$= (k+1)\left(\frac{k+2}{2}\right)$$

$$= \frac{(k+1)[(k+1)+1]}{2}.$$

Thus the second fact necessary for our proof is established. By the principle of mathematical induction, the assertion is true for every natural number n.

    This method of proof is often compared to lining up a row of dominoes, with the assumption that whenever one domino is toppled, the one following will topple. One then needs only to topple the first domino ($n = 1$) and the whole row behind it will topple, as indicated in Figure 12.1 on page 362.

Figure 12.1

Exercise 12.1

By mathematical induction, prove the validity of the formulas in Problems 1–10 for all positive integral values of n.

1. $\dfrac{1}{2} + \dfrac{2}{2} + \dfrac{3}{2} + \cdots + \dfrac{n}{2} = \dfrac{n(n+1)}{4}$

2. $1 + 3 + 5 + \cdots + (2n-1) = n^2$

3. $2 + 4 + 6 + \cdots + 2n = n(n+1)$

4. $2 + 6 + 10 + \cdots + (4n-2) = 2n^2$

5. $1^2 + 2^2 + 3^2 + \cdots + n^2 = \dfrac{n(n+1)(2n+1)}{6}$

6. $2 + 2^2 + 2^3 + \cdots + 2^n = 2^{n+1} - 2$

7. $1^3 + 3^3 + 5^3 + \cdots + (2n-1)^3 = n^2(2n^2 - 1)$

8. $\dfrac{1}{1 \cdot 2} + \dfrac{1}{2 \cdot 3} + \dfrac{1}{3 \cdot 4} + \cdots + \dfrac{1}{n(n+1)} = \dfrac{n}{n+1}$

9. $1 \cdot 2 + 2 \cdot 3 + 3 \cdot 4 + \cdots + n(n+1) = \dfrac{n(n+1)(n+2)}{3}$

10. $1 \cdot 4 + 2 \cdot 9 + 3 \cdot 16 + \cdots + n(n+1)^2 = \dfrac{1}{12} n(n+1)(n+2)(3n+5)$

11. Show that if $2 + 4 + 6 + \cdots + 2n = n(n+1) + 2$ is true for $n = k$, then it is true for $n = k+1$. Is it true for every $n \in N$?

12. Show that $n^3 + 11n = 6(n^2 + 1)$ is true for $n = 1, 2, \text{and } 3$. Is it true for every $n \in N$?

12.2

Sequences

Let us consider a class of functions in which each function has as its domain either the set N of positive integers or a subset of successive members of N.

Definition 12.1 *A sequence function is a function having as its domain the set N of positive integers 1, 2, 3, ... A finite-sequence function has as its domain the set of positive integers 1, 2, 3, . . . , n, for some fixed n.*

For example, the function defined by

$$s(n) = n + 3, \quad n \in \{1, 2, 3, \ldots\}, \tag{1}$$

is a sequence function. The elements in the range of such a function, considered in the order

$$s(1), s(2), s(3), s(4), \ldots,$$

are said to form a **sequence**. Similarly, the elements of a finite-sequence function, considered in order, constitute a **finite sequence**.

For example, the sequence associated with (1) is found by successively substituting the numbers 1, 2, 3, . . . , for n:

$$s(1) = 1 + 3 = 4,$$
$$s(2) = 2 + 3 = 5,$$
$$s(3) = 3 + 3 = 6,$$
$$s(4) = 4 + 3 = 7,$$

and so on. Thus the first four terms of (1) are 4, 5, 6, and 7. The nth term, or general term, is $n + 3$. As another example, the first five terms of the sequence defined by the equation

$$s(n) = \frac{3}{2n - 1}, \quad n \in \{1, 2, 3, \ldots\},$$

are 3/1, 3/3, 3/5, 3/7, and 3/9; and the twenty-fifth term is

$$s(25) = \frac{3}{2(25) - 1} = \frac{3}{49}.$$

Given several terms in a sequence, we are often able to construct an expression for a general term of a sequence to which they belong. Such a general term is not unique. Thus, if the first three terms in a sequence are 2, 4, 6, . . . , we may *surmise* that the general term is $s(n) = 2n$. Note, however, that the sequences for both

$$s(n) = 2n$$

and

$$s(n) = 2n + (n - 1)(n - 2)(n - 3)$$

start with 2, 4, 6, but that the two sequences differ for terms following the third.

> Sequence notation The notation ordinarily used for the terms in a sequence is not function notation as such. It is customary to denote a term in a sequence by means of a subscript. Thus, the sequence $s(1)$, $s(2)$, $s(3)$, $s(4)$, ... would appear as s_1, s_2, s_3, s_4, ...

Let us next consider two special kinds of sequences that have many applications. The first kind can be defined as follows:

Definition 12.2 *An arithmetic progression is a sequence defined by equations of the form*

$$s_1 = a, \qquad s_{n+1} = s_n + d,$$

where $a, d \in R$, and $n \in N$.

> General term of an arithmetic progression Since each term in such a sequence is obtained from the preceding term by adding d, d is called the **common difference**. Thus $3, 7, 11, 15, \ldots$ is an arithmetic progression, in which $s_1 = 3$ and $d = 4$.

For each arithmetic progression, the general term is established by the following.

Theorem 12.2 *The nth term in the sequence defined by*

$$s_1 = a, \qquad s_{n+1} = s_n + d,$$

where $a, d \in R$, and $n \in N$, is

$$s_n = a + (n - 1)d. \tag{2}$$

Proof We shall use mathematical induction. That (2) is true for the natural number 1 is evident by direct substitution of 1 in (2):

$$s_1 = a + (1 - 1)d = a.$$

If now we assume that (2) is true for the natural number k, then we have

$$s_k = a + (k - 1)d.$$

By the defining equation, we accordingly have

$$s_{k+1} = s_k + d.$$

Replacing s_k in this expression with $a + (k - 1)d$, we obtain

$$s_{k+1} = a + (k - 1)d + d$$
$$= a + kd$$
$$= a + [(k + 1) - 1]d,$$

and the principle of mathematical induction assures us that the relationship (2) is valid for all natural numbers.

The second kind of sequence we shall consider can be defined as follows:

Definition 12.3 *A geometric progression is a sequence defined by equations of the form*

$$s_1 = a, \qquad s_{n+1} = rs_n,$$

where $a, r \in R$, $a \neq 0$, $r \neq 0$, and $n \in N$.

Thus, $3, 9, 27, 81, \ldots$ is a geometric progression in which each term except the first is obtained by multiplying the preceding term by 3. Since the effect of multiplying the terms in this way is to produce a fixed ratio between any two successive terms, the multiplier, r, is called the **common ratio**.

General term of a geometric progression   The general term for a geometric progression is that established by the following theorem. The proof by induction is left as an exercise.

Theorem 12.3 *The nth term in the sequence defined by*

$$s_1 = a, \qquad s_{n+1} = rs_n,$$

where $a, r \in R$, $a \neq 0$, $r \neq 0$, and $n \in N$, is

$$s_n = ar^{n-1}.$$

Exercise 12.2

Find the first four terms in the sequence with the general term as given.

Examples   a. $s_n = \dfrac{n(n + 1)}{2}$ b. $s_n = (-1)^n 2^n$

(Solution overleaf.)

Solutions   a. $s_1 = \dfrac{1(1+1)}{2} = 1$ b. $s_1 = (-1)^1 2^1 = -2$

$$s_2 = \dfrac{2(2+1)}{2} = 3 \qquad\qquad s_2 = (-1)^2 2^2 = 4$$

$$s_3 = \dfrac{3(3+1)}{2} = 6 \qquad\qquad s_3 = (-1)^3 2^3 = -8$$

$$s_4 = \dfrac{4(4+1)}{2} = 10; \qquad\qquad s_4 = (-1)^4 2^4 = 16;$$

$$1, 3, 6, 10 \qquad\qquad\qquad\qquad -2, 4, -8, 16$$

1. $s_n = n - 5$ 2. $s_n = 2n - 3$ 3. $s_n = \dfrac{n^2 - 2}{2}$

4. $s_n = \dfrac{3}{n^2 + 1}$ 5. $s_n = 1 + \dfrac{1}{n}$ 6. $s_n = \dfrac{n}{2n - 1}$

7. $s_n = \dfrac{n(n-1)}{2}$ 8. $s_n = \dfrac{5}{n(n+1)}$ 9. $s_n = (-1)^n$

10. $s_n = (-1)^{n+1}$ 11. $s_n = \dfrac{(-1)^n(n-2)}{n}$ 12. $s_n = (-1)^{n-1} 3^{n+1}$

Write the next three terms in each of the following arithmetic progressions.

Examples   a. $5, 9, \ldots$ b. $x, x - a, \ldots$

Solutions   Find the common difference and then continue the sequence.

    a. $d = 9 - 5 = 4;$ b. $d = (x - a) - x = -a;$

       $13, 17, 21$ $x - 2a, x - 3a, x - 4a$

13. $3, 7, \ldots$ 14. $-6, -1, \ldots$ 15. $x, x + 1, \ldots$

16. $a, a + 5, \ldots$ 17. $2x + 1, 2x + 4, \ldots$ 18. $3a, 5a, \ldots$

Write the next four terms in each of the following geometric progressions.

Examples   a. $3, 6, \ldots$ b. $x, 2, \ldots$

Solutions   Find the common ratio, and then continue the sequence.

a. $r = \dfrac{6}{3} = 2$;

12, 24, 48, 96

b. $r = \dfrac{2}{x}$ $(x \neq 0)$;

$\dfrac{4}{x}, \dfrac{8}{x^2}, \dfrac{16}{x^3}, \dfrac{32}{x^4}$

19. $2, 8, \ldots$

20. $4, 8, \ldots$

21. $\dfrac{2}{3}, \dfrac{4}{3}, \ldots$

22. $\dfrac{1}{2}, -\dfrac{3}{2}, \ldots$

23. $\dfrac{a}{x}, -1, \ldots$

24. $\dfrac{a}{b}, \dfrac{a}{bc}, \ldots$

Example   Find the general term and the fourteenth term of the arithmetic progression $-6, -1, \ldots$.

Solution   Find the common difference.

$$d = -1 - (-6) = 5$$

Use $s_n = a + (n-1)d$.

$$s_n = -6 + (n-1)5 = 5n - 11$$
$$s_{14} = 5(14) - 11 = 59$$

25. Find the general term and the seventh term in the arithmetic progression $7, 11, \ldots$.

26. Find the general term and the twelfth term in the arithmetic progression $2, \dfrac{5}{2}, \ldots$.

27. Find the general term and the twentieth term in the arithmetic progression $3, -2, \ldots$.

28. Find the general term and the ninth term in the arithmetic progression $\dfrac{3}{4}, 2, \ldots$.

Example   Find the general term and also the ninth term of the geometric progression $-24, 12, \ldots$.

Solution   Find the common ratio.

$$r = \frac{12}{-24} = -\frac{1}{2}$$

Use $s_n = ar^{n-1}$.

$$s_n = -24\left(-\frac{1}{2}\right)^{n-1}$$

$$s_9 = -24\left(-\frac{1}{2}\right)^8 = -\frac{3}{32}$$

29. Find the general term and the sixth term in the geometric progression 48, 96, ...

30. Find the general term and the eighth term in the geometric progression $-3, \dfrac{3}{2}, \ldots$.

31. Find the general term and the seventh term in the geometric progression $-\dfrac{1}{3}, 1, \ldots$.

32. Find the general term and the ninth term in the geometric progression $-81, -27, \ldots$.

33. If the third term in an arithmetic progression is 7 and the eighth term is 17, find the common difference. What are the first and the twentieth terms?

34. If the fifth term of an arithmetic progression is -16 and the twentieth term is -46, what is the twelfth term?

35. Which term in the arithmetic progression 4, 1, ... is -77?

36. What is the twelfth term in an arithmetic progression in which the second term is x and the third term is y?

37. Find the first term of a geometric progression with fifth term 48 and ratio 2.

38. Find two different values for x so that $-\dfrac{3}{2}, x, -\dfrac{8}{27}$ will be in geometric progression.

39. By mathematical induction, prove Theorem 12.3.

12.3

Series

Associated with any sequence is a *series*.

Definition 12.4   *A series is the indicated sum of the terms in a sequence.*

For example, with the finite sequence

$$4, 7, 10, \ldots, 3n + 1,$$

for a given counting number n, there is associated the finite series

$$S_n = 4 + 7 + 10 + \cdots + (3n + 1);$$

similarly, with the finite sequence

$$x, x^2, x^3, x^4, \ldots, x^n,$$

there is associated the finite series

$$S_n = x + x^2 + x^3 + x^4 + \cdots + x^n.$$

Since the terms in the series are the same as those in the sequence, we can refer to the first term or the second term or the general term of a series in the same manner as we do for a sequence.

Sum of the first n terms of an arithmetic progression   Consider the series S_n of the first n terms of the general arithmetic progression,

$$S_n = a + (a + d) + (a + 2d) + \cdots + [a + (n-1)d], \quad (1)$$

and then consider the same series written as

$$S_n = s_n + (s_n - d) + (s_n - 2d) + \cdots + [s_n - (n - 1)d], \quad (2)$$

where the terms are displayed in reverse order. Adding (1) and (2) term by term, we have

$$S_n + S_n = (a + s_n) + (a + s_n) + (a + s_n) + \cdots + (a + s_n),$$

where the term $(a + s_n)$ occurs n times. Then

$$2S_n = n(a + s_n),$$

$$S_n = \frac{n}{2}(a + s_n). \quad (3)$$

If (3) is rewritten as

$$S_n = n\left(\frac{a + s_n}{2}\right),$$

we observe that the sum is given by the product of the number of terms in the series and the average of the first and last terms. The validity of (3) can be established by mathematical induction and is left as an exercise.

An alternate form for (3) is obtained by substituting $a + (n - 1)d$ for s_n in (3) to obtain

$$S_n = \frac{n}{2}(a + [a + (n - 1)d]),$$

$$S_n = \frac{n}{2}[2a + (n - 1)d], \quad (3a)$$

where the sum is now expressed in terms of a, n, and d.

Sum of the first n terms of a geometric progression   To find an explicit representation for the sum of a given number of terms in a geometric progression in terms of a, r, and n, we employ a device somewhat similar to the one used in finding the sum in an arithmetic progression. Consider the geometric series (4) containing n terms, and the series (5) obtained by multiplying both members of (4) by r:

$$S_n = a + ar + ar^2 + ar^3 + \cdots + ar^{n-2} + ar^{n-1}, \tag{4}$$

$$rS_n = ar + ar^2 + ar^3 + ar^4 + \cdots + ar^{n-1} + ar^n. \tag{5}$$

When we subtract (5) from (4), all terms in the right-hand members except the first term in (4) and the last term in (5) vanish, yielding

$$S_n - rS_n = a - ar^n.$$

Factoring S_n from the left-hand member gives

$$(1 - r)S_n = a - ar^n,$$

$$S_n = \frac{a - ar^n}{1 - r}, \tag{6}$$

if $r \neq 1$, and we have a formula for the sum of the first n terms of a geometric progression. Establishing the validity of Equation (6), which can be accomplished by mathematical induction, is left as an exercise.

An alternative expression for (6) can be obtained by first writing

$$S_n = \frac{a - r(ar^{n-1})}{1 - r},$$

and then, since $s_n = ar^{n-1}$, expressing this as

$$S_n = \frac{a - rs_n}{1 - r}, \tag{7}$$

where the sum is now given in terms of a, s_n, and r.

Sigma notation   A series for which the general term is known can be represented in a very convenient, compact way by means of what is called **sigma**, or **summation**, **notation**. The Greek letter $\sum$ (sigma) is used to denote a sum. For example,

$$S_n = 4 + 7 + 10 + \cdots + (3n + 1)$$

can be written

$$S_n = \sum_{j=1}^{n} (3j + 1),$$

where we understand that S_n is the series having terms obtained by replacing j in the expression $3j + 1$ with the numbers $1, 2, 3, \ldots, n$, successively. Similarly,

$$S = \sum_{j=3}^{6} j^2$$

appears in expanded form as

$$S = 3^2 + 4^2 + 5^2 + 6^2,$$

where the first value for j is 3 and the last is 6.

The variable used in conjunction with summation notation is called the **index of summation**, and the set of integers over which we sum (in this case, $\{3, 4, 5, 6\}$) is called the **range of summation**.

Notation for an infinite sum   To indicate that a series has an infinite number of terms, we cannot use the notation S_n for the sum, because there is no value to substitute for n. We therefore adopt notation such as

$$S_\infty = \sum_{j=1}^{\infty} \frac{1}{2^j} \tag{8}$$

to indicate that there is no last term in a series. In expanded form, the infinite series (8) is given by

$$S_\infty = \frac{1}{2} + \frac{1}{4} + \frac{1}{8} + \cdots.$$

The meaning, if any, of such an infinite sum will be discussed in Section 12.4.

Exercise 12.3

Write each series in expanded form.

Examples   a. $\displaystyle\sum_{j=2}^{5} (j^2 + 1)$ b. $\displaystyle\sum_{k=1}^{\infty} (-1)^k 2^{k+1}$

Solutions   a. $j = 2$, $2^2 + 1 = 5$;   b. $k = 1$, $(-1)^1 2^{1+1} = (-1)(4) = -4$;
 $j = 3$, $3^2 + 1 = 10$; $k = 2$, $(-1)^2 2^{2+1} = (1)(8) = 8$;
 $j = 4$, $4^2 + 1 = 17$; $k = 3$, $(-1)^3 2^{3+1} = (-1)(16) = -16$.
 $j = 5$, $5^2 + 1 = 26$. $\displaystyle\sum_{k=1}^{\infty} (-1)^k 2^{k+1} = -4 + 8 - 16 + \cdots$

$\displaystyle\sum_{j=2}^{5} (j^2 + 1) = 5 + 10 + 17 + 26$

1. $\displaystyle\sum_{j=1}^{4} j^2$ 2. $\displaystyle\sum_{j=1}^{4} (3j - 2)$ 3. $\displaystyle\sum_{j=1}^{3} \frac{(-1)^j}{2^j}$

4. $\displaystyle\sum_{i=3}^{5} \frac{(-1)^{i+1}}{i-2}$ 5. $\displaystyle\sum_{k=0}^{\infty} \frac{1}{2^k}$ 6. $\displaystyle\sum_{k=0}^{\infty} \frac{k}{1+k}$

Write each series in sigma notation.

Examples   a. $5 + 8 + 11 + 14$ b. $x^2 + x^4 + x^6$ c. $\dfrac{3}{5} + \dfrac{5}{7} + \dfrac{7}{9} + \cdots$

Solutions   Find an expression for the general term and write in sigma notation.

a. $3j + 2$ b. x^{2j} c. $\dfrac{2j+1}{2j+3}$

$\displaystyle\sum_{j=1}^{4} (3j + 2)$ $\displaystyle\sum_{j=1}^{3} x^{2j}$ $\displaystyle\sum_{j=1}^{\infty} \dfrac{2j+1}{2j+3}$

7. $x + x^3 + x^5 + x^7$ 8. $x^3 + x^5 + x^7 + x^9 + x^{11}$

9. $1 + 4 + 9 + 16 + 25$ 10. $\dfrac{1}{3} + \dfrac{1}{9} + \dfrac{1}{27} + \dfrac{1}{81}$

11. $1 \cdot 2 + 2 \cdot 3 + 3 \cdot 4 + 4 \cdot 5 + \cdots$ 12. $\dfrac{1}{2} + \dfrac{2}{3} + \dfrac{3}{4} + \dfrac{4}{5} + \cdots$

13. $\dfrac{2}{1} + \dfrac{3}{2} + \dfrac{4}{3} + \dfrac{5}{4} + \cdots$ 14. $\dfrac{1}{1} + \dfrac{2}{3} + \dfrac{3}{5} + \dfrac{4}{7} + \cdots$

Find each of the following sums.

Example $\displaystyle\sum_{j=1}^{12} (4j + 1)$

Solution   Write the first two or three terms in expanded form:

$$5 + 9 + 13 + \cdots.$$

This is an arithmetic series. The first term is 5 and the common difference is 4. Therefore we can use

$$S_n = \frac{n}{2} [2a + (n - 1)d]$$

to obtain

$$S_{12} = \frac{12}{2} [2(5) + (12 - 1)4] = 324.$$

15. $\displaystyle\sum_{j=1}^{7} (2j + 1)$ 16. $\displaystyle\sum_{j=1}^{21} (3j - 2)$ 17. $\displaystyle\sum_{j=3}^{15} (7j - 1)$

18. $\displaystyle\sum_{j=10}^{20} (2j - 3)$ 19. $\displaystyle\sum_{k=1}^{8} \left(\dfrac{1}{2}k - 3\right)$ 20. $\displaystyle\sum_{k=1}^{100} k$

Example $\displaystyle\sum_{j=2}^{5}\left(\frac{1}{3}\right)^{j}$

Solution   Write the first two terms in expanded form:

$$\left(\frac{1}{3}\right)^{2}+\left(\frac{1}{3}\right)^{3}+\cdots.$$

This is a geometric series in which the first term is $\frac{1}{9}$, the ratio is $\frac{1}{3}$, and $n=4$. Therefore we can use $S_{n}=\dfrac{a-ar^{n}}{1-r}$ to obtain

$$S_{4}=\frac{\dfrac{1}{9}-\dfrac{1}{9}\left(\dfrac{1}{3}\right)^{4}}{1-\dfrac{1}{3}}=\frac{40}{243}.$$

21. $\displaystyle\sum_{j=1}^{6} 3^{j}$ 22. $\displaystyle\sum_{j=1}^{4}(-2)^{j}$ 23. $\displaystyle\sum_{k=3}^{7}\left(\frac{1}{2}\right)^{k-2}$

24. $\displaystyle\sum_{j=3}^{12} 2^{j-5}$ 25. $\displaystyle\sum_{j=1}^{6}\left(\frac{1}{3}\right)^{j}$ 26. $\displaystyle\sum_{j=1}^{4}(3+2^{j})$

27. Find the sum of all even integers n, for $13<n<29$.

28. Find the sum of all integral multiples of 7 between 8 and 110.

29. How many bricks will there be in a pile one brick in thickness if there are 27 bricks in the bottom row, 25 in the second row, and so forth, to the top row, which has one brick?

30. If there are a total of 256 bricks in a pile arranged in the manner of those in Problem 29, how many bricks are there in the third row from the bottom of the pile?

31. Find $\displaystyle\sum_{j=1}^{n}\left(\frac{1}{2}\right)^{j}$ for $n=2, 3, 4,$ and 5. What value do you think $\displaystyle\sum_{j=1}^{n}\left(\frac{1}{2}\right)^{j}$ approximates as n becomes larger and larger?

32. Find p if $\displaystyle\sum_{j=1}^{5} pj=14$.

33. Find p and q if $\displaystyle\sum_{j=1}^{4}(pj+q)=28$ and $\displaystyle\sum_{j=2}^{5}(pj+q)=44$.

34. Consider $S_{n}=\displaystyle\sum_{i=1}^{n} f(i)$. Explain why this equation defines a sequence function. What is the variable denoting an element in the domain? The range?

35. By mathematical induction, prove that for an arithmetic progression, $S_n = \dfrac{n}{2}(a + s_n)$ for all positive integral values of n.

36. Show that the sequence formed by adding the corresponding terms in two arithmetic progressions is an arithmetic progression.

37. Show that the sum of the terms in two series with terms in arithmetic progression can be written as a series with terms in arithmetic progression.

38. By mathematical induction, prove that, for a geometric progression, $S_n = \dfrac{a - ar^n}{1 - r}$ for all positive integral values of n, provided $r \neq 1$.

12.4

Limits of sequences and series

A sequence that is strictly increasing but bounded

Consider the sequence function defined by

$$S_n = \frac{n}{n + 1}, \quad n \in N. \tag{1}$$

If we write the range of (1) in the form

$$\frac{1}{2}, \frac{2}{3}, \frac{3}{4}, \frac{4}{5}, \cdots, \frac{n}{n + 1}, \cdots,$$

then it is clear that each of the terms is greater than the preceding term; indeed, the difference of consecutive terms is

$$\frac{n + 1}{n + 2} - \frac{n}{n + 1} = \frac{(n^2 + 2n + 1) - (n^2 + 2n)}{(n + 1)(n + 2)} = \frac{1}{(n + 1)(n + 2)} > 0.$$

Such a sequence is said to be **strictly increasing**. On the other hand, it is also clear that, no matter how large a value is assigned to n, we have

$$\frac{n}{n + 1} < 1,$$

because the denominator is one larger than the numerator, in fact, we have

$$1 - \frac{n}{n + 1} = \frac{(n + 1) - n}{n + 1} = \frac{1}{n + 1} > 0.$$

Thus we have a sequence in which each term is greater than the preceding term and yet no term is equal to or greater than 1.

Limit of a sequence　We note, however—and this is a very basic consideration—that the value of $n/(n + 1)$ is as close to 1 as we please if n is large enough. For example, the difference satisfies

$$1 - \frac{n}{n + 1} = \frac{1}{n + 1}, \quad \text{and we have} \quad \frac{1}{n + 1} < \frac{1}{1000}$$

provided $n + 1 > 1000$—that is, $n > 999$. If it is true that the nth term in a sequence differs from the number L by as little as we please for all sufficiently large n, we say that **the sequence approaches the number L as a limit**. The symbolism

$$\lim_{n \to \infty} s_n = L$$

(read "the limit, as n increases without bound, of s_n is L") is used to denote this situation. A thorough discussion of the notion of a limit is included in courses in calculus, and will not be attempted here. A few elementary ideas, however, are in order.

A sequence in which the nth term approaches a number L as $n \to \infty$ is said to be **a convergent sequence**, and the sequence is said to **converge** to L. It is not necessary for convergence that a sequence be strictly increasing. For example,

$$1, \frac{1}{2}, \frac{1}{3}, \frac{1}{4}, \ldots, \frac{1}{n}, \ldots$$

converges to 0, but each term in the sequence is less than, instead of greater than, the term that precedes it. Again, the sequence

$$-1, \frac{1}{2}, -\frac{1}{3}, \frac{1}{4}, \ldots, \frac{(-1)^n}{n}, \ldots$$

converges to 0 but is neither increasing nor decreasing. We can rephrase the definition of convergence of a sequence as follows:

Definition 12.5　*A sequence $s_1, s_2, \ldots, s_n, \ldots$ converges to the number L,*

$$\lim_{n \to \infty} s_n = L,$$

if and only if the absolute value of the difference between the nth term in the sequence and the number L is as small as we please for all sufficiently large n. Thus the sequence converges to the number L if and only if

$$\lim_{n \to \infty} |L - s_n| = 0.$$

For example, the **alternating** (because the signs alternate) **sequence**

$$\frac{-2}{3}, \frac{4}{9}, \frac{-8}{27}, \ldots, \left(\frac{-2}{3}\right)^n, \ldots$$

converges to 0 since the absolute value of the difference between $(-2/3)^n$ and 0, that is, $|0 - (-2/3)^n|$, is as small as we please for n large enough. We express this by writing

$$\lim_{n \to \infty} \left(\frac{-2}{3}\right)^n = 0.$$

On the other hand, the alternating sequence

$$\frac{1}{2}, \; -\frac{2}{3}, \frac{3}{4}, \; -\frac{4}{5}, \; \ldots, (-1)^{n+1} \frac{n}{n+1}, \; \ldots$$

does not converge. As n increases, the nth term oscillates back and forth from the neighborhood of $+1$ to the neighborhood of -1, and we cannot find a number L such that $\lim\limits_{n \to \infty} |L - s_n| = 0$. Such a sequence is said to **diverge**. A sequence such as

$$1, 2, 3, \ldots, n, \ldots$$

also is said to diverge. An answer to the logical question of what we mean by "enough" when we say "n large enough" requires a more precise definition of limit than we have given here. As remarked earlier, a course in the calculus will treat this in detail.

Sequence of partial   For an infinite series,
sums of a series

$$S_\infty = \sum_{j=1}^{\infty} s_j,$$

we can consider the infinite sequence of **partial sums**:

$$S_1 = s_1,$$
$$S_2 = s_1 + s_2,$$
$$\cdot \quad \cdot \quad \cdot \quad \cdot$$
$$S_n = s_1 + s_2 + \cdots + s_n,$$
$$\cdot \quad \cdot \quad \cdot \quad \cdot \quad \cdot \quad \cdot$$

Definition 12.6   *An infinite series*

$$S_\infty = \sum_{j=1}^{\infty} s_j$$

converges if and only if $S_1, S_2, \ldots, S_n, \ldots$, *the corresponding sequence of partial sums, converges.*

If the sequence of partial sums converges to the number L,

$$\lim_{n \to \infty} S_n = L,$$

then L is said to be the **sum** of the infinite series, and we write

$$S_\infty = \sum_{j=1}^{\infty} s_j = L.$$

If the sequence of partial sums diverges, then the series is said to **diverge**.

Sum of an infinite geometric progression We recall from Section 12.3 that the sum of n terms (the nth partial sum) of a geometric progression is given, for $r \neq 1$, by

$$S_n = \frac{a - ar^n}{1 - r}. \tag{2}$$

If $|r| < 1$, that is, if $-1 < r < 1$, then $|r|^n$ becomes smaller and smaller for increasingly large n. For example, if $r = 1/2$, then

$$r^2 = \frac{1}{4}, \quad r^3 = \frac{1}{8}, \quad r^4 = \frac{1}{16},$$

and so forth; and $(1/2)^n$ is as small as we please if n is sufficiently large. Writing (2) as

$$S_n = \frac{a}{1 - r}(1 - r^n) \tag{3}$$

we see that the value of the factor $(1 - r^n)$ is as close as we please to 1 provided $|r| < 1$ and n is taken large enough. Since this argument shows that the sequence of partial sums (3) converges to

$$\frac{a}{1 - r},$$

we have the following result.

Theorem 12.4 *The sum of an infinite geometric progression, $a + ar + ar^2 + \cdots + ar^n + \cdots$, with $|r| < 1$, is*

$$S_\infty = \lim_{n \to \infty} S_n = \frac{a}{1 - r}.$$

An interesting application of this sum arises in connection with repeating decimals—that is, decimal numerals that, after a finite number of decimal places, have endlessly repeating groups of digits. For example,

$$0.21\overline{21},$$

$$0.138\overline{512}512$$

are repeating decimals. The bar denotes that the numerals appearing under it are repeated endlessly. Consider the problem of expressing such a decimal fraction as an arithmetic fraction. We illustrate the process involved with the first example above. The decimal $0.21\overline{21}$ can be written as

$$0.21 + 0.0021 + 0.000021 + \cdots, \tag{4}$$

which is a geometric progression with ratio $r = 0.01$. Since the ratio is less than 1 in absolute value, we can use Theorem 12.4 to find the sum of the infinite series (4). Thus

$$S_\infty = \frac{a}{1-r} = \frac{0.21}{1-0.01} = \frac{21}{99} = \frac{7}{33},$$

and the given decimal fraction is equivalent to 7/33.

Power series for circular functions The function values in the ranges of some important nonalgebraic functions are not easily evaluated for all x in the domain. Fortunately, infinite **power series** in x, or series of the form

$$a_0 + a_1 x + \cdots + a_n x^n + \cdots,$$

for these functions can be determined by means of calculus. For example, by means of calculus, it can be shown that, for all $x \in R$,

$$\cos x = 1 - \frac{x^2}{1 \cdot 2} + \frac{x^4}{1 \cdot 2 \cdot 3 \cdot 4} - \cdots, \tag{5}$$

$$\sin x = x - \frac{x^3}{1 \cdot 2 \cdot 3} + \frac{x^5}{1 \cdot 2 \cdot 3 \cdot 4 \cdot 5} - \cdots. \tag{6}$$

Example Find an approximation for $\cos 0.1$, using the first two terms in the series (5), above.

Solution $\cos (0.1) \approx 1 - \dfrac{(0.1)^2}{2 \cdot 1} = 1.00000 - 0.00500$

$$= 0.99500.$$

Power series for e^x We shall consider one more series, namely the power series for e^x. It can be shown that, for all $x \in R$,

$$e^x = 1 + x + \frac{x^2}{1 \cdot 2} + \frac{x^3}{1 \cdot 2 \cdot 3} + \cdots, \tag{7}$$

and for all $z \in C$,

$$e^z = 1 + z + \frac{z^2}{1 \cdot 2} + \frac{z^3}{1 \cdot 2 \cdot 3} + \cdots.$$

In particular, for $z = ix$, we have

$$e^{ix} = 1 + ix - \frac{x^2}{1 \cdot 2} - i \frac{x^3}{1 \cdot 2 \cdot 3} + \cdots .$$

Now the terms in the series for e^{ix} are alternately the terms in the series for $\cos x$ and i times the terms in the series for $\sin x$. Accordingly, we have the celebrated identity

$$e^{ix} = \cos x + i \sin x,$$

which is known as **Euler's formula** in honor of its discoverer, the prolific Swiss mathematician Leonhard Euler (1701–1783).

Exercise 12.4

Discuss the limiting behavior of each expression as $n \to \infty$.

Example $\dfrac{n^2 + 3}{n^2}$

Solution   By writing $\dfrac{n^2 + 3}{n^2}$ as $\dfrac{n^2}{n^2} + \dfrac{3}{n^2}$ and then as $1 + \dfrac{3}{n^2}$, we observe that

$$\lim_{n \to \infty} \frac{n^2 + 3}{n^2} = \lim_{n \to \infty} \left(1 + \frac{3}{n^2} \right) = 1 + 0 = 1.$$

1. $\dfrac{1}{n}$ 2. $1 + \dfrac{1}{n^2}$ 3. $\dfrac{n+1}{n}$ 4. $\dfrac{n+3}{n^2}$

5. $2n$ 6. $(-1)^n$ 7. $\dfrac{1}{2^n}$ 8. $(-1)^n \dfrac{1}{n}$

State which of the following sequences are convergent.

Example $1, \dfrac{3}{2}, \dfrac{7}{4}, \dfrac{15}{8}, \ldots, \dfrac{2^n - 1}{2^{n-1}}, \ldots$

Solution   Writing the general term as $\dfrac{2^n}{2^{n-1}} - \dfrac{1}{2^{n-1}}$, or $2 - \dfrac{1}{2^{n-1}}$, we observe that

$$\lim_{n \to \infty} \frac{2^n - 1}{2^{n-1}} = \lim_{n \to \infty} \left(2 - \frac{1}{2^{n-1}} \right) = 2 - 0 = 2.$$

The sequence is convergent.

9. $\dfrac{1}{2}, \dfrac{1}{4}, \dfrac{1}{8}, \dfrac{1}{16}, \ldots, \dfrac{1}{2^n}, \ldots$

10. $2, \dfrac{3}{2}, \dfrac{4}{3}, \dfrac{5}{4}, \ldots, \dfrac{n+1}{n}, \ldots$

11. $1, 2, 3, 4, 5, \ldots, n, \ldots$

12. $2, 4, 6, 8, \ldots, 2n, \ldots$

13. $1, -\dfrac{1}{2}, \dfrac{1}{4}, -\dfrac{1}{8}, \ldots, (-1)^{n-1}\dfrac{1}{2^{n-1}}, \ldots$

14. $1, -1, 1, -1, \ldots, (-1)^{n+1}, \ldots$

Find the sum of each of the following infinite geometric series. If the series has no sum, so state.

Examples

a. $3 + 2 + \cdots$

b. $\dfrac{1}{81} - \dfrac{1}{54} + \cdots$

Solutions

a. $r = \dfrac{2}{3}$; series has a sum since $|r| < 1$.

$$S_\infty = \frac{a}{1-r} = \frac{3}{1 - \dfrac{2}{3}} = 9$$

b. $r = -\dfrac{1}{54} \div \dfrac{1}{81} = -\dfrac{3}{2}$; series does not have a sum since $|r| > 1$.

15. $12 + 6 + \cdots$

16. $2 + 1 + \cdots$

17. $\dfrac{1}{36} + \dfrac{1}{30} + \cdots$

18. $\dfrac{1}{16} - \dfrac{1}{8} + \cdots$

19. $\displaystyle\sum_{j=1}^{\infty} \left(\dfrac{2}{3}\right)^j$

20. $\displaystyle\sum_{j=1}^{\infty} \left(-\dfrac{1}{4}\right)^j$

Find an arithmetic fraction equal to each of the given decimal numerals.

Example

$0.81\overline{81}$

Solution

Rewrite as a series: $0.81 + 0.0081 + 0.000081 + \cdots$.

Find the common ratio: $r = 0.01$. Use $S_\infty = \dfrac{a}{1-r}$.

$$S_\infty = \frac{0.81}{1 - 0.01} = \frac{81}{99} = \frac{9}{11}.$$

21. $0.31\overline{31}$

22. $0.45\overline{45}$

23. $2.41\overline{0410}$

24. $3.027\overline{027}$

25. $0.128\overline{88}$

26. $0.8\overline{3333}$

27. A force is applied to a particle moving in a straight line in such a fashion that each second it moves only one half of the distance it moved the preceding second. If the particle moves ten centimeters the first second, approximately how far will it move before coming to rest?

28. The arc length through which the bob on a pendulum moves is nine-tenths of its preceding arc length. Approximately how far will the bob move before coming to rest if the first arc length is 12 inches?

In Problems 29–36, use the first two terms of Equation (5), (6), or (7), in this section to find an approximation for the given expression.

29. $\cos 0.2$ 30. $\cos 0.3$ 31. $\cos 0.4$ 32. $\sin 0.1$

33. $\sin 0.3$ 34. $\sin 0.4$ 35. e^0 36. $e^{0.1}$

37. Use Euler's formula to find a value for $e^{i\pi}$.

38. Use Euler's formula to show that $e^{-ix} = \cos x - i \sin x$.

12.5

The binomial theorem

There are situations, as in Equations (5), (6), and (7) on page 378 and in the binomial expansion below, in which it is necessary to write the product of several consecutive positive integers. To facilitate writing products of this type, we use a special symbol $n!$ (read "n factorial" or "factorial n"), which is defined by

$$n! = n(n-1)(n-2) \cdots (3)(2)(1).$$

Thus

$$5! = 5 \cdot 4 \cdot 3 \cdot 2 \cdot 1 \quad (\text{read "five factorial"}),$$

and

$$8! = 8 \cdot 7 \cdot 6 \cdot 5 \cdot 4 \cdot 3 \cdot 2 \cdot 1 \quad (\text{read "eight factorial"}).$$

Factorial notation can also be used to represent products of consecutive positive integers, beginning with integers different from 1. For example,

$$8 \cdot 7 \cdot 6 \cdot 5 = \frac{8!}{4!},$$

because

$$\frac{8!}{4!} = \frac{8 \cdot 7 \cdot 6 \cdot 5 \cdot 4 \cdot 3 \cdot 2 \cdot 1}{4 \cdot 3 \cdot 2 \cdot 1} = 8 \cdot 7 \cdot 6 \cdot 5.$$

Since

$$n! = n(n - 1)(n - 2)(n - 3) \cdots 5 \cdot 4 \cdot 3 \cdot 2 \cdot 1$$

and

$$(n - 1)! = (n - 1)(n - 2)(n - 3) \cdots 5 \cdot 4 \cdot 3 \cdot 2 \cdot 1,$$

for $n > 1$ we can write the recursive relationship

$$\boldsymbol{n! = n(n - 1)!.} \tag{1}$$

For example,

$$7! = 7 \cdot 6!,$$
$$27! = 27 \cdot 26!,$$
$$(n + 2)! = (n + 2)(n + 1)!.$$

If (1) is to hold also for $n = 1$, then we must have

$$1! = 1 \cdot (1 - 1)!$$

or

$$1! = 1 \cdot 0!.$$

Therefore, for consistency, we define $0!$ by

$$\boldsymbol{0! = 1.}$$

A special case of the use of factorial notation occurs in the form

$$\binom{n}{r} = \frac{n!}{r!\,(n - r)!}.$$

The symbol $\binom{n}{r}$ is read "the number of combinations of n things taken r at a time"; it denotes the number of r-element subsets of a set containing n members. Combinations will be discussed in Section 13.2. As examples, we have

$$\binom{5}{3} = \frac{5!}{3!(5 - 3)!} = \frac{5!}{3!\,2!} = \frac{5 \cdot 4 \cdot 3!}{3!(2 \cdot 1)} = 10,$$

$$\binom{5}{1} = \frac{5!}{1!(5 - 1)!} = \frac{5!}{4!} = \frac{5 \cdot 4!}{4!} = 5,$$

and

$$\binom{5}{0} = \frac{5!}{0!(5 - 0)!} = \frac{5!}{5!} = 1.$$

The series obtained by expanding a binomial of the form

$$(a + b)^n$$

is particularly useful in certain branches of mathematics. Starting with familiar examples, where n takes the values 1, 2, 3, 4, and 5 in turn, we can show by direct multiplication that

$$(a + b)^1 = a + b$$
$$(a + b)^2 = a^2 + 2ab + b^2$$
$$(a + b)^3 = a^3 + 3a^2b + 3ab^2 + b^3$$
$$(a + b)^4 = a^4 + 4a^3b + 6a^2b^2 + 4ab^3 + b^4$$
$$(a + b)^5 = a^5 + 5a^4b + 10a^3b^2 + 10a^2b^3 + 5ab^4 + b^5.$$

We observe that in each case:

1. The first term is a^n.

2. The variable factors of the second term are $a^{n-1}b^1$, and the coefficient is n, which can be written in the form

$$\frac{n}{1!}.$$

3. The variable factors of the third term are $a^{n-2}b^2$, and the coefficient can be written in the form

$$\frac{n(n-1)}{2!}.$$

4. The variable factors of the fourth term are $a^{n-3}b^3$, and the coefficient can be written in the form

$$\frac{n(n-1)(n-2)}{3!}.$$

The foregoing expansions suggest the following result, known as the **binomial theorem**. Since its proof, which requires the use of mathematical induction, is quite lengthy, is omitted.

Theorem 12.5 *For each natural number n,*

$$(a + b)^n = a^n + \frac{n}{1!} a^{n-1}b + \frac{n(n-1)}{2!} a^{n-2}b^2 + \frac{n(n-1)(n-2)}{3!} a^{n-3}b^3$$

$$+ \cdots + \frac{n(n-1)(n-2)\cdots(n-r+2)}{(r-1)!} a^{n-r+1}b^{r-1} + \cdots + b^n, \quad (2)$$

where r is the number of the term.

For example,

$$(x - 2)^4 = x^4 + \frac{4}{1!} x^3(-2)^1 + \frac{4 \cdot 3}{2!} x^2(-2)^2 + \frac{4 \cdot 3 \cdot 2}{3!} x(-2)^3 + \frac{4 \cdot 3 \cdot 2 \cdot 1}{4!} (-2)^4$$

$$= x^4 - 8x^3 + 24x^2 - 32x + 16.$$

In this case, $a = x$ and $b = -2$ in the binomial expansion.

Observe that the coefficients of the terms in the binomial expansion (2) can be represented as follows.

1st term: $\quad \dbinom{n}{0} = \dfrac{n!}{0!n!} = 1,$

2nd term: $\quad \dbinom{n}{1} = \dfrac{n!}{1!(n-1)!} = \dfrac{n \cdot (n-1)!}{1 \cdot (n-1)!} = n,$

3rd term: $\quad \dbinom{n}{2} = \dfrac{n!}{2!(n-2)!} = \dfrac{n(n-1)(n-2)!}{2!(n-2)!} = \dfrac{n(n-1)}{2!},$

rth term: $\quad \dbinom{n}{r-1} = \dfrac{n!}{(r-1)!(n-r+1)!}$

$$= \dfrac{n(n-1)(n-2) \cdots (n-r+2)(n-r+1)!}{(r-1)!(n-r+1)!}$$

$$= \dfrac{n(n-1)(n-2) \cdots (n-r+2)}{(r-1)!}.$$

Hence, the binomial expansion (2) can be written as

$$(a + b)^n = \dbinom{n}{0} a^n + \dbinom{n}{1} a^{n-1}b + \dbinom{n}{2} a^{n-2}b^2 + \dbinom{n}{3} a^{n-3}b^3 + \cdots$$

$$+ \dbinom{n}{r-1} a^{n-r+1}b^{r-1} + \cdots + \dbinom{n}{n} b^n. \quad (3)$$

For example,

$$(x - 2)^4 = \dbinom{4}{0} x^4 + \dbinom{4}{1} x^3(-2)^1 + \dbinom{4}{2} x^2(-2)^2 + \dbinom{4}{3} x(-2)^3 + \dbinom{4}{4} (-2)^4.$$

Simplifying the coefficients, we obtain the same result as above,

$$(x - 2)^4 = x^4 - 8x^3 + 24x^2 - 32x + 16.$$

Note that the rth term in a binomial expansion is given by

$$\dbinom{n}{r-1} a^{n-r+1}b^{r-1} = \dfrac{n(n-1)(n-2)(n-r+2)}{(r-1)!} a^{n-r+1}b^{r-1}. \quad (4)$$

For example, by the left-hand member of (4), the seventh term of $(x - 2)^{10}$ is

$$\binom{10}{6} x^4 (-2)^6 = \frac{10!}{6!\,4!} x^4 (-2)^6 = \frac{10 \cdot 9 \cdot 8 \cdot 7 \cdot 6!}{6! \cdot 4 \cdot 3 \cdot 2 \cdot 1} x^4 (64)$$

$$= 13440\, x^4,$$

while, by the right-hand member of (4), we have

$$\frac{10 \cdot 9 \cdot 8 \cdot 7}{4 \cdot 3 \cdot 2 \cdot 1} x^4 (64) = 13440\, x^4.$$

Exercise 12.5

1. Write $(2n)!$ in expanded form for $n = 4$.

2. Write $2n!$ in expanded form for $n = 4$.

3. Write $n(n - 1)!$ in expanded form for $n = 6$.

4. Write $2n(2n - 1)!$ in expanded form for $n = 2$.

Write in expanded form and simplify.

Examples	a. $\dfrac{7!}{4!}$	b. $\dfrac{4!\,6!}{8!}$
Solutions	a. $\dfrac{7 \cdot 6 \cdot 5 \cdot 4!}{4!}$	b. $\dfrac{4 \cdot 3 \cdot 2 \cdot 1 \cdot 6!}{8 \cdot 7 \cdot 6!}$
	210	$\dfrac{3}{7}$

5. $5!$ 6. $7!$ 7. $\dfrac{9!}{7!}$ 8. $\dfrac{12!}{11!}$

9. $\dfrac{5!\,7!}{8!}$ 10. $\dfrac{12!\,8!}{16!}$ 11. $\dfrac{8!}{2!\,(8 - 2)!}$ 12. $\dfrac{10!}{4!\,(10 - 4)!}$

Write each product in factorial notation.

Examples	a. $1 \cdot 2 \cdot 3 \cdot 4 \cdot 5 \cdot 6$	b. $11 \cdot 12 \cdot 13 \cdot 14$	c. 150
Solutions	a. $6!$	b. $\dfrac{14!}{10!}$	c. $\dfrac{150!}{149!}$

13. $1 \cdot 2 \cdot 3$ 14. $1 \cdot 2 \cdot 3 \cdot 4 \cdot 5$ 15. $3 \cdot 4 \cdot 5 \cdot 6$

16. 7 17. $8 \cdot 7 \cdot 6$ 18. $28 \cdot 27 \cdot 26 \cdot 25 \cdot 24$

Write each expression in factorial notation and simplify.

Examples   a. $\dbinom{6}{2}$ b. $\dbinom{4}{4}$

Solutions   a. $\dbinom{6}{2} = \dfrac{6!}{2!(6-2)!} = \dfrac{6!}{2!4!}$ b. $\dbinom{4}{4} = \dfrac{4!}{4!(4-4)!}$

$$= \dfrac{6 \cdot 5 \cdot 4!}{2 \cdot 1 \cdot 4!} = 15 \qquad\qquad = \dfrac{4!}{4!0!} = 1$$

19. $\dbinom{6}{5}$ 20. $\dbinom{4}{2}$ 21. $\dbinom{3}{3}$ 22. $\dbinom{5}{5}$

23. $\dbinom{7}{0}$ 24. $\dbinom{2}{0}$ 25. $\dbinom{5}{2}$ 26. $\dbinom{5}{3}$

Write each expression in factored form and show the first three factors and the last three factors.

Example $(2n + 1)!$

Solution $(2n + 1)(2n)(2n - 1) \cdot \cdots \cdot 3 \cdot 2 \cdot 1$

27. $n!$ 28. $(n + 4)!$ 29. $(3n)!$

30. $3n!$ 31. $(n - 2)!$ 32. $(3n - 2)!$

Simplify each expression.

Examples   a. $\dfrac{(n-1)!}{(n-3)!}$ b. $\dfrac{(n-1)!(2n)!}{2n!(2n-2)!}$

Solutions   a. $\dfrac{(n-1)(n-2)(n-3)!}{(n-3)!}$ b. $\dfrac{(n-1)!(2n)(2n-1)(2n-2)!}{2(n)(n-1)!(2n-2)!}$

              $(n-1)(n-2)$ $2n - 1$

33. $\dfrac{(n+2)!}{n!}$ 34. $\dfrac{(n+2)!}{(n-1)!}$ 35. $\dfrac{(n+1)(n+2)!}{(n+3)!}$

36. $\dfrac{(2n+4)!}{(2n+2)!}$ 37. $\dfrac{(2n)!(n-2)!}{4(2n-2)!(n)!}$ 38. $\dfrac{(2n+1)!(2n-1)!}{[(2n)!]^2}$

Expand.

Example $(a-3b)^4$

Solution   From the binomial expansion (2),

$$(a-3b)^4 = a^4 + \frac{4}{1!}a^3(-3b) + \frac{4\cdot 3}{2!}a^2(-3b)^2 + \frac{4\cdot 3\cdot 2}{3!}a(-3b)^3$$

$$+ \frac{4\cdot 3\cdot 2\cdot 1}{4!}(-3b)^4$$

$$= a^4 - 12a^3b + 54a^2b^2 - 108ab^3 + 81b^4.$$

Alternatively, from the binomial expansion (3),

$$(a-3b)^4 = \binom{4}{0}a^4 + \binom{4}{1}a^3(-3b) + \binom{4}{2}a^2(-3b)^2 + \binom{4}{3}a(-3b)^3$$

$$+ \binom{4}{4}(-3b)^4,$$

which also simplifies to the expression obtained above.

39. $(x+3)^5$ 40. $(2x+y)^4$ 41. $(x-3)^4$ 42. $(2x-1)^5$

43. $\left(2x-\dfrac{y}{2}\right)^3$ 44. $\left(\dfrac{x}{3}+3\right)^5$ 45. $\left(\dfrac{x}{2}+2\right)^6$ 46. $\left(\dfrac{2}{3}-a^2\right)^4$

Write the first four terms in each expansion. Do not simplify the terms.

Example $(x+2y)^{15}$

Solution   You may write either

$$(x+2y)^{15} = x^{15} + \frac{15}{1!}x^{14}(2y) + \frac{15\cdot 14}{2!}x^{13}(2y)^2 + \frac{15\cdot 14\cdot 13}{3!}x^{12}(2y)^3$$

or

$$(x+2y)^{15} = \binom{15}{0}x^{15} + \binom{15}{1}x^{14}(2y) + \binom{15}{2}x^{13}(2y)^2 + \binom{15}{3}x^{12}(2y)^3.$$

47. $(x + y)^{20}$ 48. $(x - y)^{15}$ 49. $(a - 2b)^{12}$

50. $(2a - b)^{12}$ 51. $(x - \sqrt{2})^{10}$ 52. $\left(\dfrac{x}{2} + 2\right)^{8}$

Find each power to the nearest hundredth.

Example $(0.97)^{7}$

Solution   Either form of the binomial expansion (2) or (3) on pages 383 and 384 can be used. We shall use (2).

$$(0.97)^7 = (1 - 0.03)^7$$

$$= 1^7 + \frac{7}{1!}(1)^6(-0.03)^1 + \frac{7 \cdot 6}{2!}(1)^5(-0.03)^2$$

$$+ \frac{7 \cdot 6 \cdot 5}{3!}(1)^4(-0.03)^3 + \cdots$$

$$= 1 - 0.21 + 0.0189 - 0.000945 + \cdots$$

$$= 0.807955^{+}$$

Hence, to the nearest hundredth, $(0.97)^7 = 0.81$.

53. $(1.02)^{10}$ *Hint:* $1.02 = (1 + 0.02)$

54. $(1.01)^{15}$

55. If an amount of money (P) is invested at 4% compounded annually, the amount (A) present at the end of (n) years is given by $A = P(1 + 0.04)^n$. Find the amount A (to the nearest dollar) if \$1000 was invested for 10 years.

56. In Problem 55, find the amount present at the end of 20 years.

Find each specified term.

Example $(x - 2y)^{12}$, the seventh term.

Solution   In Formula (4) page 384, use $n = 12$ and $r = 6$.

$$\frac{12!}{6!(12 - 6)!} x^6(-2y)^6 = 59{,}136x^6y^6$$

57. $(a - b)^{15}$, the sixth term.   58. $(x + 2)^{12}$, the fifth term.

59. $(x - 2y)^{10}$, the fifth term.   60. $(a^3 - b)^9$, the seventh term.

61. Given that the binomial formula holds for $(1 + x)^n$ where n is a negative integer:

    a. Write the first four terms of $(1 + x)^{-1}$.

    b. Find the first four terms of the quotient $1/(1 + x)$ by dividing $(1 + x)$ into 1. Compare the results of (a) and (b).

62. Given that the binomial formula holds as an infinite "sum" for $(1 + x)^n$, where n is a rational number and $|x| < 1$, find to two decimal places.

    a. $\sqrt{1.02}$;   b. $\sqrt{0.99}$.

13

Probability

13.1

Basic counting principles; permutations

Associated with each finite set A is a nonnegative integer n, namely the number of elements in A. Hence, we have a function from the set of all finite sets to the set of nonnegative integers. The symbolism $n(A)$ is used to denote elements in the range of this set function n. For example, if

$$A = \{5, 7, 9\}, \quad B = \{1/2, 0, 3, -5, 7\}, \quad C = \emptyset,$$

then

$$n(A) = 3, \quad n(B) = 5, \quad \text{and} \quad n(C) = 0.$$

Counting properties All the sets with which we shall hereafter be concerned are assumed to be finite sets. We then have the following properties, called **counting properties**, for the function n.

I $n(A \cup B) = n(A) + n(B), \quad \text{if } A \cap B = \emptyset.$

Thus, if A and B are disjoint sets then the number of elements in their union is the sum of the number of elements in A and the number of elements in B.

For example, suppose there are five roads from town R to town S, and two railroads from town R to town S. If A is the set of roads and B the set of railroads from R to S, then $n(A) = 5$, $n(B) = 2$, and $n(A \cup B) = 5 + 2 = 7$; thus there are seven ways one can go from town R to town S by driving or riding on a train.

II $n(A \cup B) = n(A) + n(B) - n(A \cap B),$

$$\text{if } A \cap B \neq \emptyset.$$

That is, if A and B overlap, then to count the number of elements in $A \cup B$, we might add the number of elements in A to the number of elements in B. But since any elements in the intersection of A and B are counted twice in this process (once in A and once in B), we must subtract the number of

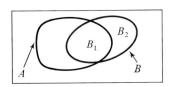

Figure 13.1

such elements from the sum $n(A) + n(B)$ to obtain the number of elements in $A \cup B$, as suggested in Figure 13.1.

For example, suppose there are fifteen unrelated girls and seventeen unrelated boys in a mathematics class, and suppose that there are precisely two brother-sister pairs in the class. If A denotes the set of different families represented by the girls and B denotes the set of different families represented by the boys, then the number of different families represented by all of the members of the class is $n(A) + n(B) - n(A \cap B) = 15 + 17 - 2 = 30$. Actually, Property II is a consequence of Property I.

III $n(A \times B) = n(A) \cdot n(B)$.

This asserts that the number of elements in the Cartesian product of sets A and B is the product of the number of elements in A and the number of elements in B.

For example, suppose again that there are five roads from town R to town S (set A), and further suppose that there are three roads from town S to town T (set B). Then for each element of A there are three elements of B, and the total possible ways one can drive from R to T via S is

$$n(A \times B) = n(A) \cdot n(B) = 5 \cdot 3 = 15.$$

Given the set of digits $A = \{1, 2, 3\}$, how many different three-digit numerals can be constructed from the members of A if no member is used more than once? The answer to this question can be obtained by simply listing the different three-digit numerals, 1 2 3, 1 3 2, 2 1 3, 2 3 1, 3 1 2, 3 2 1, and counting them. Such a procedure would be quite impracticable, however, if the number of members of the given set of digits were very large. Another way to arrive at the same conclusion is by applying the third counting property. If we let A denote the set of possible first digits in the foregoing numerals, then $n(A) = 3$. Since no numeral may be used more than once, and since one numeral has already been used for a first digit, there remain but two possibilities for the second digit. If B denotes the set of possible second digits after the first digit has been chosen, then $n(B) = 2$. By similar reasoning, if C is the set of possible third digits after the first two have been chosen, then $n(C) = 1$. By the third counting property (applied twice), we find that

$$n(A \times B \times C) = [n(A) \cdot n(B)] \cdot n(C) = 3 \cdot 2 \cdot 1 = 6.$$

Permutations   Each of the three-digit numerals discussed above is called a **permutation** of the elements of the set of numerals $\{1, 2, 3\}$.

Definition 13.1   *A permutation of a set A is an ordering (first, second, etc.) of the members of A.*

With Definition 13.1, we can state the following result.

Theorem 13.1 *Let $P_{n,n}$ denote the number of distinct permutations of the members of a set A containing n members. Then*

$$P_{n,n} = n!. \tag{1}$$

The symbol $P_{n,n}$ (or sometimes $_nP_n$, or P_n^n) is read "the number of permutations of n things taken n at a time."

Proof Let A_1 denote the set of possible selections for the first member. Then $A_1 = A$, and $n(A_1) = n(A) = n$. Having made a first selection, let A_2 denote the set of possible second selections. Then $A_2 \subset A$, and $n(A_2) = n(A_1) - 1 = n - 1$. A continuation of this procedure, together with successive application of the third counting principle, leads to

$$P_{n,n} = n(A_1) \cdot n(A_2) \cdot \cdots \cdot n(A_n) = n \cdot (n-1)(n-2) \cdots 1 = n!,$$

as was to be proved.

Example   In how many ways can nine men be assigned positions to form distinct baseball teams?

Solution   Let A denote the set of men, so that $n(A) = 9$. The total number of ways in which 9 men can be assigned 9 positions on a team, or, in other words, the number of possible permutations of the members of a 9-element set is, by Equation (1),

$$P_{9,9} = 9! = 9 \cdot 8 \cdot 7 \cdots 1 = 362,880.$$

Theorem 13.2 *Let $P_{n,r}$ denote the number of permutations of the members, taken r at a time, of a set A containing n members; that is, let $P_{n,r}$ be the number of distinct orderings of r elements when there is a set A of n elements from which to choose. Then*

$$P_{n,r} = n(n-1)(n-2) \cdots [n - (r-1)]$$
$$= n(n-1)(n-2) \cdots (n-r+1). \tag{2}$$

The proof follows the proof of Theorem 13.1, except that the last subset considered is such that $n(A_r) = n - r + 1$.

Example   In how many ways can a basketball team be formed by choosing players for the five positions from a set of ten players?

Solution   Let A denote the set of players, so that $n(A) = 10$. Then from (2) and the fact that a basketball team consists of 5 players, we have

$$P_{10,5} = 10 \cdot 9 \cdot 8 \cdots (10 - 5 + 1) = 10 \cdot 9 \cdot 8 \cdot 7 \cdot 6 = 30,240.$$

An alternative expression for $P_{n,r}$ can be obtained by observing that

$$P_{n,r} = n(n-1)(n-2) \cdots (n-r+1)$$

$$= \frac{n(n-1)(n-2) \cdots (n-r+1)(n-r)!}{(n-r)!},$$

so that

$$P_{n,r} = \frac{n!}{(n-r)!}. \tag{3}$$

Distinguishable permutations   The problem of finding the number of distinguishable permutations of n objects taken n at a time, if some of the objects are identical, requires a little more careful analysis. As an example, consider the number of permutations of the letters of the word *DIVISIBLE*. We can make a distinction between the three *I*'s by assigning subscripts to each so that we have nine distinct letters,

$$D, I_1, V, I_2, S, I_3, B, L, E.$$

The number of permutations of these nine letters is of course 9!. If the letters other than I_1, I_2, and I_3 are retained in the position they occupy in a permutation of the above nine letters, I_1, I_2, and I_3 can be permuted among themselves 3! ways. Thus, if P is the number of *distinguishable* permutations of the letters

$$D, I, V, I, S, I, B, L, E,$$

then, since for each of these there are 3! ways in which the *I*'s can be permuted without otherwise changing the order of the other letters, it follows that

$$3! \cdot P = 9!,$$

from which

$$P = \frac{9!}{3!}.$$

As another example, consider the letters of the word *MISSISSIPPI*. There would exist 11! distinguishable permutations of the letters in this word if each letter were distinct. Note, however, that the letters S and I each appear four times and the letter P appears twice. Reasoning as we did in the previous example, we see that the number P of distinguishable permutations of the letters in *MISSISSIPPI* is given by

$$4!4!2!P = 11!,$$

from which

$$P = \frac{11!}{4!4!2!}.$$

Exercise 13.1

Given the following sets, find n(A ∩ B), n(A ∪ B), and n(A × B).

Example $A = \{a, b, c\}, \quad B = \{c, d\}$

Solution $A \cap B = \{c\}.$ Therefore $n(A \cap B) = 1.$

$A \cup B = \{a, b, c, d\}.$ Therefore $n(A \cup B) = 4.$

$A \times B = \{(a, c), (a, d), (b, c), (b, d), (c, c), (c, d)\}.$ Therefore $n(A \times B) = 6.$

1. $A = \{d, e\}, \quad B = \{e, f, g, h\}$ 2. $A = \{e\}, \quad B = \{a, b, c, d\}$

3. $A = \{1, 2, 3, 4\}, \quad B = \{3, 4, 5, 6\}$ 4. $A = \{1, 2\}, \quad B = \{3, 4, 5\}$

5. $A = \{1, 2\}, \quad B = \{1, 2\}$ 6. $A = \emptyset, \quad B = \{2, 3, 4\}$

Example   In how many ways can three members of a class be assigned a grade of A, B, C, or D?

Solution   Sometimes a simple diagram, such as ____, ____, ____, designating a sequence, is a helpful preliminary device. Since each of the students may receive any one of four different grades, the sequence would appear as

$$4, \, 4, \, 4.$$

From counting property III, there are $4 \cdot 4 \cdot 4$, or 64, possible ways the grades may be assigned.

Example   In how many different ways can three members of a class be assigned a grade of A, B, C, or D so that no two members receive the same grade?

Solution   Since the first student may receive any one of four different grades, the second student may then receive any one of three different grades, and the third student may then receive any one of two different grades, the sequence would appear as

$$4, \, 3, \, 2.$$

From counting property III, there are $4 \cdot 3 \cdot 2$, or 24, possible ways the grades may be assigned. In this case, we could have obtained the same result directly from Theorem 13.2, since $P_{4, \, 3} = 4 \cdot 3 \cdot 2 = 24.$

In each problem, a digit or letter may be used more than once unless stated otherwise.

7. How many different two-digit numerals can be formed from the digits 5 and 6?

8. How many different two-digit numerals can be formed from the digits 7, 8, and 9?

9. In how many different ways can four students be seated in a row?

10. In how many different ways can five students be seated in a row?

11. In how many different ways can four questions on a true-false test be answered?

12. In how many different ways can five questions on a true-false test be answered?

13. In how many ways can you write different three-digit numerals from $\{2, 3, 4, 5\}$?

14. How many different seven-digit telephone numbers can be formed from the set of digits $\{1, 2, 3, 4, 5, 6, 7, 8, 9, 0\}$?

15. In how many ways can you write different three-digit numerals, using $\{2, 3, 4, 5\}$, if no digit is to be used more than once in each numeral?

16. How many different seven-digit telephone numbers can be formed from the set of digits $\{1, 2, 3, 4, 5, 6, 7, 8, 9, 0\}$ if no digit is to be used more than once in any number?

17. How many three-letter arrangements can be formed from $\{A, N, S, W, E, R\}$?

18. How many different three-letter arrangements can be formed from $\{A, N, S, W, E, R\}$ if no letter is to be used more than once in any arrangement?

19. How many four-digit numerals for positive odd integers can be formed from $\{1, 2, 3, 4, 5\}$?

20. How many four-digit numerals for positive even integers can be formed from the set $\{1, 2, 3, 4, 5\}$?

21. How many numerals for positive integers less than 500 can be formed from $\{3, 4, 5\}$?

22. How many numerals for positive odd integers less than 500 can be formed from $\{3, 4, 5\}$?

23. How many numerals for positive even integers less than 500 can be formed from $\{3, 4, 5\}$?

24. How many numerals for positive even integers between 400 and 500, inclusive, can be formed from $\{3, 4, 5\}$?

25. How many permutations of the elements of $\{P, R, I, M, E\}$ end in a vowel?

26. How many permutations of the elements of $\{P, R, O, D, U, C, T\}$ end in a vowel?

27. Find the number of distinguishable permutations of the letters in the word *LIMIT*.

28. Find the number of distinguishable permutations of the letters in the word

COMBINATION.

29. Find the number of distinguishable permutations of the letters in the word

$$COLORADO.$$

30. Find the number of distinguishable permutations of the letters in the word

$$TALLAHASSEE.$$

31. Show that $P_{5,3} = 5(P_{4,2})$.   32. Show that $P_{5,r} = 5(P_{4,r-1})$.

33. Show that $P_{n,3} = n(P_{n-1,2})$.   34. Show that $P_{n,3} - P_{n,2} = (n-3)(P_{n,2})$.

35. Solve for n: $P_{n,5} = 5(P_{n,4})$.   36. Solve for n: $P_{n,5} = 9(P_{n-1,4})$.

Example   In how many ways can four students be seated around a circular table?

Solution   In any such arrangement (which is called a **circular permutation**), there is no first position. Each person can take four different initial positions without affecting the arrangement. Thus, there are

$$\frac{4!}{4} = 6 \quad \text{arrangements.}$$

(In general, there are $n!/n$, or $(n-1)!$, circular permutations of n things taken n at a time).

37. In how many ways can five students be seated around a circular table?

38. In how many ways can six students be seated around a circular table?

39. In how many ways can six students be seated around a circular table if a certain two must be seated together?

40. In how many ways can three different keys be arranged on a key ring? *Hint*: Arrangements should be considered identical if one can be obtained from the other by turning the ring over. In general, there are only $(1/2)(n-1)!$ distinct arrangements of n keys on a ring $(n \geq 3)$.

13.2

Combinations

An additional counting concept is needed before we turn our attention to probability—namely, finding the number of distinct r-element subsets of an n-element set with no reference to relative order of the elements in the subset. For example, five different playing cards can be arranged in 5! permutations, but to a poker player they represent the same hand. The set of five cards (with no reference to the arrangement of the cards) is called a **combination**.

Definition 13.2 *A subset of an n-element set A is called a combination.*

The counting of combinations is related to the counting of permutations. From Theorem 13.2, we know that the number of distinct permutations of n elements of a set A taken r at a time is given by

$$P_{n,r} = \frac{n!}{(n-r)!}.$$

With this in mind, consider the following result concerning the number $\binom{n}{r}$ that was introduced in Section 12.5.

Theorem 13.3 *Let* $\binom{n}{r}$ *denote the number of distinct combinations of the members, taken r at a time, of a set A containing n members. Then*

$$\binom{n}{r} = \frac{P_{n,r}}{r!}. \tag{1}$$

Proof There are, by definition, $\binom{n}{r}$ r-element subsets of the set A, where $n(A) = n$. Also, from Theorem 13.1, each of these subsets has $r!$ permutations of its members. There are therefore $\binom{n}{r}r!$ permutations of n elements of A taken r at a time. Thus

$$P_{n,r} = \binom{n}{r}r!,$$

from which we obtain

$$\binom{n}{r} = \frac{P_{n,r}}{r!},$$

as was to be shown.

Thus, to find the number of r-element subsets of an n-element set A, we count the number of permutations of the elements of A taken r at a time, and then divide by the number of possible permutations of an r-element set. This seems very much like counting a set of people by counting the number of arms and legs and dividing the result by 4, but this approach gives us a very useful expression for the number we seek, $\binom{n}{r}$. Since

$$P_{n,r} = n(n-1)(n-2)\cdots(n-r+1),$$

it follows that

$$\binom{n}{r} = \frac{P_{n,r}}{r!} = \frac{n(n-1)(n-2)\cdots(n-r+1)}{r!}. \tag{2}$$

Example
In how many ways can a committee of five be selected from a set of twelve persons?

Solution
What we wish here is the number of 5-element subsets of a 12-element set. From (2), we have

$$\binom{12}{5} = \frac{12 \cdot 11 \cdot 10 \cdot 9 \cdot 8}{5 \cdot 4 \cdot 3 \cdot 2 \cdot 1} = 792.$$

By Equation (3) on page 393, we have the alternative expression

$$\binom{n}{r} = \frac{P_{n,r}}{r!} = \frac{n!}{r!(n-r)!}. \tag{3}$$

Since the numbers $\binom{n}{r}$ are the coefficients in the binomial expansion, and since these coefficients are symmetric, we have the following plausible assertion.

Theorem 13.4 $\binom{n}{r} = \binom{n}{n-r}.$

Proof From (3), we have

$$\binom{n}{r} = \frac{n!}{r!(n-r)!}$$

and

$$\binom{n}{n-r} = \frac{n!}{(n-r)![n-(n-r)]!} = \frac{n!}{(n-r)!r!},$$

and the theorem is proved.

Theorem 13.4 is plausible also since each time a distinct set of r objects is chosen, a distinct set of $n - r$ objects remains unchosen.

Exercise 13.2

Example
How many different amounts of money can be formed from a penny, a nickel, a dime, and a quarter?

Solution   We want to find the total number of combinations that can be formed by taking the coins 1, 2, 3, and 4 at a time. By (3) we have,

$$\binom{4}{1} = \frac{4!}{1!\,3!} = 4, \quad \binom{4}{2} = \frac{4!}{2!\,2!} = 6, \quad \binom{4}{3} = \frac{4!}{3!\,1!} = 4, \quad \binom{4}{4} = \frac{4!}{4!\,0!} = 1,$$

and the total number of combinations is 15. Clearly each combination gives a diferent amount.

1. How many different amounts of money can be formed from a penny, a nickel, and a dime?

2. How many different amounts of money can be formed from a penny, a nickel, a dime, a quarter, and a half-dollar?

3. How many different committees of four persons each can be chosen from a group of six persons?

4. How many different committees of four persons each can be chosen from a group of ten persons?

5. In how many different ways can a set of five cards be selected from a standard bridge deck containing 52 cards?

6. In how many different ways can a set of 13 cards be selected from a standard bridge deck of 52 cards?

7. In how many different ways can a hand consisting of five spades, five hearts, and three diamonds be selected from a standard bridge deck of 52 cards?

8. In how many different ways can a hand consisting of ten spades, one heart, one diamond, and one club be selected from a standard bridge deck of 52 cards?

9. In how many different ways can a hand consisting of either five spades, five hearts, five diamonds, or five clubs be selected from a standard bridge deck?

10. In how many different ways can a hand consisting of three aces and two cards that are not aces be selected from a standard bridge deck?

11. A combination of three balls is picked at random from a box containing five red, four white, and three blue balls. In how many ways can the set chosen contain at least one white ball?

12. In Problem 11, in how many ways can the set chosen contain at least one white and one blue ball?

13. A set of five distinct points lies on a circle. How many inscribed triangles can be drawn having all their vertices in this set?

14. A set of ten distinct points lies on a circle. How many inscribed quadrilaterals can be drawn having all their vertices in this set?

15. A set of ten distinct points lies on a circle. How many inscribed hexagons can be drawn having all their vertices in this set?

16. Given $\binom{n}{3} = \binom{50}{47}$, find n.

17. Given $\binom{n}{7} = \binom{n}{5}$, find n.

18. Expand $(a + b)^5$. Write the coefficient of each term in the form $\binom{n}{r}$.

19. Write the first four terms of the expansion of $(a + b)^{10}$. Write the coefficient of each term in the form $\binom{n}{r}$.

20. Write the first eight terms of the expansion of $(a + b)^{12}$. Write the coefficient of each term in the form $\binom{n}{r}$.

13.3

Probability functions

Sample spaces, outcomes, and events

When an experiment of some kind is undertaken, associated with the experiment is a set of possible results. For example, when a die is rolled, it will come to a stop with the number of spots on its upper face corresponding to one of the numerals 1, 2, 3, 4, 5, 6. This exhausts all possibilities. Of course, in the absence of chicanery, exactly which one of these random results will occur cannot be precisely specified in advance of the experiment. A question of great practical importance regarding the result of casting a die, and the result of any comparable experiment in which the outcome is uncertain, is "Can we assign some kind of measure to the degree of uncertainty involved?" The answer is "Yes," but before we assign a measure, let us define some necessary terms.

Definition 13.3 *The set of all possible results of an experiment is called a sample space for the experiment.*

Definition 13.4 *Each element of a sample space is called an outcome or sample point.*

Definition 13.5 *Any subset of a sample space is called an event, and is commonly denoted by the letter E.*

The reason for the terminology in Definition 13.5 is that, in conducting an experiment, one may be interested in sets of outcomes rather than in individual outcomes. In the tossing of a die, for example, if the sample space is taken as $\{1, 2, 3, 4, 5, 6\}$, then the event that an outcome (a numeral) denotes an even integer is the set $\{2, 4, 6\}$, which is a subset of the sample space. The event that an outcome denotes an odd integer is the set $\{1, 3, 5\}$. These two events are complements of each other, and are examples of **complimentary events**, that is, two events whose intersection is $\emptyset$ and whose union is the entire sample space.

The number of possible events in an n-element sample space is the number of possible subsets of an n-element set, namely 2^n, where both the null set (impossible event) and the entire sample space (certain event) are included. That is,

$$\binom{n}{0} + \binom{n}{1} + \binom{n}{2} + \cdots + \binom{n}{n} = (1 + 1)^n = 2^n,$$

where $\binom{n}{0}$ represents the single event that is contributed by the null set.

We can define a *probability function P* on a sample space S by means of either a priori or a posteriori considerations, as follows.

A priori considerations involve physical, geometrical, and other inherent properties of the experiment in question. They involve no sampling of outcomes. One way to assign a probability to an event is simply to use the ratio of the number of outcomes in the event to the number of possible outcomes. Thus, when a die is cast and we admit as outcomes the die's stopping with any of its six different faces uppermost, then without making any trial throws we would assign the value $1/6$ as the probability of each of the six possible outcomes.

More generally, we have the following definition.

Definition 13.6   *If E is any subset containing n(E) members (outcomes) of a sample space containing n(S) equally likely outcomes, then the probability of the occurrence of E, P(E), is given by*

$$P(E) = \frac{n(E)}{n(S)}. \tag{1}$$

Example   If two dice are cast, what is the (a priori) probability that the sum of the number of dots showing on the top faces of the dice is less than 6?

Solution   For our sample spaces, let us consider A the set of possible outcomes for one die and B the set for the other. Then

$$n(A) = 6 \quad \text{and} \quad n(B) = 6.$$

The possible outcomes for both would be $S = A \times B$, the Cartesian product of A and B, and $n(A \times B) = n(S) = 36$. Each outcome here is

(*Solution continued overleaf.*)

an ordered pair (a, b), where a is the numeral on the upper face of the first die, and b is that of the second die. The event we seek is the event $\{(a, b) \mid a + b < 6\}$. The lattice in the figure shows the situation schematically. Since

$$E = \{(1, 1), (1, 2), (1, 3), (1, 4), (2, 1),$$
$$(2, 2), (2, 3), (3, 1), (3, 2), (4, 1)\},$$

we have $n(E) = 10$, and

$$P(E) = \frac{n(E)}{n(S)} = \frac{10}{36} = \frac{5}{18}.$$

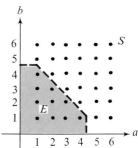

Care must be taken in interpreting the meaning of "the probability of an event." In the foregoing example, 5/18 does not assure us, for instance, that E will occur 5 times out of 18 casts, or, indeed, that even one cast out of 18 will produce the event described. What it does imply, however, is that if you cast the dice a very great number of times, then you can *expect* the sum on the exposed faces to be less than 6 about 5/18 of the time.

A posteriori considerations involve testing the experiment a certain number of times. Mortality tables give probability functions of this sort. Actually, (1) can still be used in defining such probability functions, provided we interpret the function $n(E)$ as being the number of times the event E occurred in the test and $n(S)$ as being the total number of times the experiment was performed in the test.

Although diagrams are often useful in illustrating an event and a sample space, the number of outcomes $N(E)$ in the event and the number $N(S)$ of possible outcomes in the sample space ordinarily are computed directly.

Example

If three marbles are drawn at random from an urn containing six white marbles and four blue marbles, what is the probability that the three marbles are all white?

Solution

Since the number of ways of drawing three white marbles from six white marbles is $\binom{6}{3}$ and the number of ways of drawing three marbles from the total number of marbles is $\binom{10}{3}$, we have

$$P(E) = \frac{N(E)}{N(S)} = \frac{\binom{6}{3}}{\binom{10}{3}} = \frac{\dfrac{6!}{3!\,3!}}{\dfrac{10!}{3!\,7!}} = \frac{1}{6}.$$

Exercise 13.3

1. A die is cast. List the outcomes in the sample space. List the outcomes in the event E that the number on the upper face of the die is greater than 2. Determine $P(E)$.

2. A coin is tossed. List the outcomes in the sample space. List the outcomes in the event E that a head appears. Determine $P(E)$.

3. Two coins are tossed. List the outcomes in the sample space. List the outcomes in the event E that both coins show the same face. Determine $P(E)$.

4. A die is cast and a coin is tossed. List the outcomes in the sample space. List the outcomes in the event E that the coin shows a head and the die shows a numeral greater than 4. Determine $P(E)$.

In Exercises 5–10, consider an experiment in which two dice are cast. Determine the probability that the specified event occurs.

5. The sum of the numbers of dots shown is 7.

6. The sum of the numbers of dots shown is 8.

7. The sum of the numbers of dots shown is 3.

8. The sum of the numbers of dots shown is 12.

9. At least one of the numbers of dots shown is less than 3.

10. Both dice show the same number of dots.

In Exercises 11–14, consider an experiment in which a single card is drawn at random from a standard pack of 52 cards. Determine the probability that the specified event occurs.

11. The card is the King of hearts.   12. The card is an ace.

13. The card is a black face card.   14. The card is a 5, 6, 7, or 8.

In Exercises 15–18, consider an experiment in which two cards are drawn at random from a standard pack of 52 cards. Determine the probability that the specified event occurs.

15. Both cards are spades.

16. Both cards are red.

17. Both cards are face cards (Jack, Queen, or King).

18. Both cards are aces.

In Exercises 19–22, consider an experiment in which 2 marbles are drawn at random from an urn containing 8 red, 6 blue, 4 green, and 2 white marbles. Determine the probability that the specified event occurs.

19. Both marbles are white.   20. Both marbles are green.

21. Both marbles are blue.   22. Both marbles are red.

13.4
Probability of the union of events

Probability of the complement of an event   If we denote the complement of an event E in a sample space S by E', then $P(E')$ denotes the probability of the occurrence of E'. Since an outcome in S must lie in either E or E', but not both, and since $P(S) = 1$, it follows that

$$P(E) + P(E') = 1,$$

or

$$P(E') = 1 - P(E).$$

We can often use this fact to simplify the computing of a probability.

Example   If two marbles are drawn at random from an urn containing 6 white, 2 red, and 5 green marbles, what is the probability that not both are white, that is, that at least one is not white?

Solution   Rather than consider all of the possible pairs of marbles in which not both are white, we simply compute the probability of the event E that they *are* both white and subtract the result from 1. We have

$$P(E) = \frac{\binom{6}{2}}{\binom{13}{2}} = \frac{15}{78} = \frac{5}{26}.$$

Then

$$P(E') = 1 - \frac{5}{26} = \frac{21}{26},$$

and this is just the probability that not both marbles are white.

   The ratio of the probability of an event to the probability of its complement is given a special name.

Definition 13.7   *The odds that an experiment with sample space S will result in an event E are given by*

$$\frac{P(E)}{P(E')} = \frac{P(E)}{1 - P(E)}, \quad P(E') \neq 0.$$

Example   Find the odds that the sum of the numbers determined by a single cast
                 of two dice will be less than 6.

Solution   From the example on pages 401 and 402, the probability of the event
                 is 5/18. Hence

$$P(E') = 1 - P(E) = 1 - \frac{5}{18} = \frac{13}{18}.$$

Then, by Definition 13.7,

$$\frac{P(E)}{P(E')} = \frac{5/18}{13/18} = \frac{5}{13},$$

and the odds that the sum of the numbers will be less than 6 are 5/13,
or 5 to 13.

Probability of the   By the second counting principle on page 390, if E_1 and E_2 are
union of events   events in a sample space S, then

$$n(E_1 \cup E_2) = n(E_1) + n(E_2) - n(E_1 \cap E_2).$$

This leads directly to the following theorem.

Theorem 13.5   *If S is a sample space, and E_1 and E_2 are any events in S, then*

$$P(E_1 \text{ or } E_2) = P(E_1 \cup E_2) = P(E_1) + P(E_2) - P(E_1 \cap E_2).$$

Example   If two cards are drawn from a standard deck of playing cards, what is
                 the probability that either both are red or both are Jacks?

Solution   A deck of cards contains 52 cards, and an outcome here consists of two
                 cards. Hence, the number of elements of the sample space is the number
                 of ways (combinations) one can draw two cards from 52, which is
                 $\binom{52}{2}$. Let E_1 be the event that both are red. Since there are 26 red
                 cards in a deck, the number of outcomes (combinations) in the event
                 E_1 is

$$n(E_1) = \binom{26}{2}.$$

Let E_2 be the event that both cards are Jacks. Then, because there are
four Jacks in a deck,

$$n(E_2) = \binom{4}{2}.$$

(Solution continued overleaf.)

Since there is only one pair of red Jacks, $n(E_1 \cap E_2) = 1$. We then have

$$P(E_1 \cup E_2) = P(E_1) + P(E_2) - P(E_1 \cup E_2)$$

$$= \frac{\binom{26}{2}}{\binom{52}{2}} + \frac{\binom{4}{2}}{\binom{52}{2}} - \frac{1}{\binom{52}{2}} = \frac{\binom{26}{2} + \binom{4}{2} - 1}{\binom{52}{2}}$$

$$= \frac{\frac{26 \cdot 25}{1 \cdot 2} + \frac{4 \cdot 3}{1 \cdot 2} - 1}{\frac{52 \cdot 51}{1 \cdot 2}} = \frac{325 + 6 - 1}{1326} = \frac{330}{1326} = \frac{165}{663} = \frac{55}{221}.$$

Probability of the union of disjoint events

Of course, if E_1 and E_2 are *disjoint*, then $E_1 \cap E_2 = \emptyset$, so that $P(E_1 \cap E_2) = 0$, and the equation in Theorem 13.5 reduces to the equation in Definition 13.6, namely,

$$P(E_1 \cup E_2) = P(E_1) + P(E_2).$$

Disjoint events are said to be **mutually exclusive**.

Example

A card is drawn at random from a standard deck of 52 cards. What is the probability that the card is either a face card (Jack, Queen, or King) or a four?

Solution

Let E_1 be the event that the card is a four. There are four such cards in a deck, so that $n(E_1) = 4$. Let E_2 be the event that the card is a Jack, Queen, or King. There are twelve such cards in a deck. Hence $n(E_2) = 12$. Since the sample space is just the entire deck, $n(S) = 52$, and since E_1 and E_2 are mutually exclusive,

$$P(E_1 \cup E_2) = P(E_1) + P(E_2)$$

$$= \frac{4}{52} + \frac{12}{52} = \frac{16}{52} = \frac{4}{13}.$$

Exercise 13.4

Two dice are cast. Let E be the event that both dice show the same numeral. Let F be the event that the sum of the numbers thrown is greater than eight. Find the probabilities of the following events.

1. $P(E)$　　　　　　2. $P(F)$　　　　　　3. $P(E \cup F)$

4. $P(E')$　　　　　　5. $P(F')$　　　　　　6. $P(E' \cup F')$

*A box contains five red, four white, and three blue marbles. Two marbles are drawn from the box. Let RR be the event that both marbles are red, WW that both marbles are white, **BB** that both marbles are blue, and RW, RB, BW that a red and a white, a red and a blue, and a blue and a white are drawn, respectively. Find the probability of each event.*

7. $P(RR)$ 8. $P(BB)$ 9. $P(WW)$

10. $P(RW)$ 11. $P(RB)$ 12. $P(BW)$

13. What is the probability that neither is white?

14. What is the probability that neither is blue?

15. What is the probability that at least one is red?

16. What is the probability that either one is red or else both are white?

17. What is the probability that by drawing a single card from a standard deck of 52 cards, one will get a 2, 3, or 4?

18. What is the probability that if two cards are drawn from a standard deck of 52 cards they will be of the same suit? Different suits?

19. One box contains three red and eight white marbles, and a second box contains five red and two white marbles. If one marble is drawn from each box, what is the probability of drawing:
    a. Two red marbles?
    b. Two white marbles?
    c. One red and one white marble?

20. In Problem 19, what are the odds of drawing:
    a. Two red marbles?
    b. Two white marbles?
    c. One red and one white marble?

*If the probability of the event E that a person will receive k dollars is P(E), then the person's **mathematical expectation** relative to this event is kP(E).*

21. A lottery offers a prize of $50, and 70 tickets are sold. What is the mathematical expectation of a person who buys three tickets? If each ticket costs $1, is the person's expectation greater or less than his outlay?

22. The odds that a certain horse will win the Irish Sweepstakes are 2 to 7. If you hold a ticket on this horse to pay $100,000 if he wins, what is your mathematical expectation?

If E_1, E_2, E_3, etc., are mutually exclusive events, and the return to you is k_1 if E_1 occurs, k_2 if E_2 occurs, etc., then your mathematical expection is $\sum_{i=1}^{n} k_i P(E_i)$.

23. One coin is selected at random from a collection containing a penny, a nickel, and a dime. What is the expectation?

24. One coin is selected at random from a collection containing a dime, a quarter, and half-dollar. What is the expectation?

25. Three $1 bills and four $5 bills are hidden from view. What is the expectation on a single selection?

26. Three $1 bills, four $5 bills, and one $10 bill are hidden from view. What is the expectation on a single draw?

13.5

Probability of the intersection of events

In some experiments, we may be interested in events that are not dependent on each other, in the sense that the occurrence of one may have no effect on the probability of the occurrence or nonoccurrence of the other.

Independent events   Consider an experiment in which two cards are drawn at random, one after the other, from a deck of ten cards, six of which are red and four blue. We can inquire into the probability that the first card drawn is red and the second blue. The simplest such situation would be one in which the first card is drawn, observed, and returned to the deck, which is then shuffled thoroughly before the second card is drawn. In this case, the sample space would consist of a set of ordered pairs (x, y), where x is the result of the first draw and y the result of the second draw. Since there are ten possibilities in each case, the sample space would consist of $10 \times 10 = 100$ ordered pairs. In the Cartesian graph of Figure 13.2, r_i and b_i are used to designate the drawing of red and blue cards, respectively.

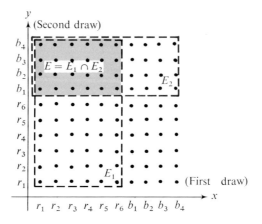

Figure 13.2

The events E_1 that the first card drawn is a red card and E_2 that the second card drawn is a blue card are outlined in the figure. The event E that both E_1 and E_2 occur is the intersection of E_1 and E_2; that is, $E = E_1 \cap E_2$. By inspection,

$$P(E_1) = \frac{60}{100} = \frac{3}{5}, \qquad P(E_2) = \frac{40}{100} = \frac{2}{5},$$

and

$$P(E) = P(E_1 \cap E_2) = \frac{24}{100} = \frac{6}{25}.$$

Moreover, in this example, it is evident that

$$P(E) = P(E_1 \cap E_2) = P(E_1) \cdot P(E_2).$$

Definition 13.8   *If E_1 and E_2 are events in a sample space, and if*

$$P(E_1 \cap E_2) = P(E_1) \cdot P(E_2),$$

then E_1 and E_2 are independent events. If two events are not independent, then they are said to be dependent.

Dependent events   Now consider the same experiment, except that this time the first card is not returned to the deck before the second is taken.
Then there will be ten possible first draws, but only nine possible second draws. The sample space will therefore contain 10×9 ordered pairs, such that no ordered pair with first and second components the same remains in the set.

Figure 13.3 shows a graph of the sample space, which is the same as that in the preceding figure except that one diagonal is missing. Sets with graphs that have a

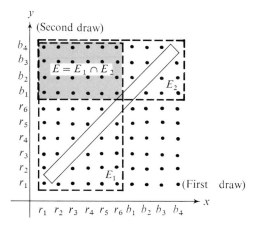

Figure 13.3

missing diagonal are called **deleted Cartesian sets**. Again, the figure shows E_1 and E_2, the events that a red and a blue are obtained on the first and second draw, respectively. The event that both occur is $E = E_1 \cap E_2$, which is also shown in the figure. By inspection,

$$P(E_1) = \frac{54}{90} = \frac{3}{5}, \quad P(E_2) = \frac{36}{90} = \frac{2}{5}, \quad \text{and} \quad P(E) = P(E_1 \cap E_2) = \frac{24}{90} = \frac{4}{15}.$$

This time,

$$P(E_1 \cap E_2) \neq P(E_1) \cdot P(E_2),$$

so that the events are dependent.

Conditional probability   If we write 4/15 as

$$\frac{4}{15} = \frac{3}{5} \cdot \frac{4}{9},$$

then we have

$$P(E_1 \cap E_2) = P(E_1) \cdot \frac{4}{9},$$

where 4/9 can be interpreted as the probability of the occurrence of E_2, *given the occurrence of* E_1. This probability is denoted by the symbol $P(E_2 | E_1)$.

Definition 13.9 *If E_1 and E_2 are events in a sample space, and $P(E_1) \neq 0$, then the conditional probability $P(E_2 | E_1)$ of E_2 given E_1 is*

$$P(E_2 | E_1) = \frac{P(E_1 \cap E_2)}{P(E_1)}.$$

If E_1 and E_2 are independent and $P(E_1) \neq 0$, then by Definitions 13.8 and 13.9,

$$P(E_2 | E_1) = \frac{P(E_1 \cap E_2)}{P(E_1)} = \frac{P(E_1) \cdot P(E_2)}{P(E_1)} = P(E_2).$$

Exercise 13.5

1. A bag contains four red marbles and ten blue marbles. If two marbles are drawn in succession, and if the first is not replaced, what is the probability that the first is red and the second is blue? Are the two draws independent events?

2. A red die and a green die are cast. What is the probability of obtaining a sum greater than 9, given that the green die shows 4?

3. A bag contains four white and six red marbles. Two marbles are drawn from the bag and replaced, and two more marbles are then drawn from the bag. What is the probability of drawing:
   a. Two red marbles on the first draw and two white ones on the second draw?
   b. A total of two white marbles?
   c. Four white marbles?
   d. Four red marbles?

4. In Problem 3, what is the probability of drawing:
   a. Exactly three white marbles in the two draws?
   b. At least three white marbles in the two draws?
   c. Exactly two red marbles in the two draws?
   d. At least two red marbles in the two draws?

5. A red and a green die are cast. Let E_1 be the event that at least one die shows 3 and E_2 be the event that the sum of the two numbers thrown is 8.
   a. Find $P(E_1)$.   b. Find $P(E_2)$.
   c. Find $P(E_2|E_1)$.   d. Are E_1 and E_2 independent?

6. In Problem 5, let E_1 be the event that neither die shows a result larger than 4 and let E_2 be the event that the dice do not show the same number.
   a. Find $P(E_1)$.   b. Find $P(E_2)$.
   c. Find $P(E_2|E_1)$.   d. Are E_1 and E_2 independent?

7. A coin is tossed three consecutive times. What is the probability that:
   a. The second toss is a head?
   b. The third toss is a head?
   c. Both the second and third tosses are heads?
   d. The first and third tosses are heads?
   e. The first and third tosses are heads but the second is not?

8. In Problem 7, state whether each of the following pairs of events is independent.
   a. a and b b. a and c c. a and d
   d. a and e e. d and e

9. The probability that A will pass a course in college algebra is 5/6, that B will pass, 3/4, and that C will pass, 2/3. What is the probability of each of the following events?
   a. At least one of the three will pass.
   b. At least A and C will pass.
   c. A and C will pass but B will not.
   d. At least two of the three will pass.

10. In Problem 9, what is the probability of c, given the occurrence of a? Are the events a and c independent?

11. A day is selected at random in some fashion such that any day of the week is an equally likely choice. Let the probability be 1/30 that a day selected at random will be a rainy day.
    a. What is the probability that a rainy Wednesday will be selected?
    b. What is the probability that a dry Thursday will be selected?

c. What is the probability that either Monday, Tuesday, or Wednesday will be selected, and that it will not rain that day?

d. Let E_1 be the selection of Sunday, and let E_2 be the event that it does not rain on the day selected. What is the conditional probability of E_2, given the occurrence of E_1? Are E_1 and E_2 independent?

12. Two identical urns contain marbles. One urn contains all red marbles, while half of the contents of the other urn are white and half are red. If a marble is drawn at random from one of the urns and found to be red, what is the probability that it was drawn from the urn containing only red marbles?

13. Of two dice, one is normal, but the other has three faces showing 4 spots and four faces showing 3 spots. If one of the dice is chosen at random and a 4 is thrown with it, what is the probability that the normal die was chosen?

14. Argue that if E_1 and E_2 are mutually exclusive events with nonzero probabilities, then E_1 and E_2 are not independent.

Appendix

Table 1   Squares, square roots, and prime factors

No.	Sq.	Sq. Root	Prime Factors	No.	Sq.	Sq. Root	Prime Factors
1	1	1.000		51	2,601	7.141	$3 \cdot 17$
2	4	1.414	2	52	2,704	7.211	$2^2 \cdot 13$
3	9	1.732	3	53	2,809	7.280	53
4	16	2.000	2^2	54	2,916	7.348	$2 \cdot 3^3$
5	25	2.236	5	55	3,025	7.416	$5 \cdot 11$
6	36	2.449	$2 \cdot 3$	56	3,136	7.483	$2^3 \cdot 7$
7	49	2.646	7	57	3,249	7.550	$3 \cdot 19$
8	64	2.828	2^3	58	3,364	7.616	$2 \cdot 29$
9	81	3.000	3^2	59	3,481	7.681	59
10	100	3.162	$2 \cdot 5$	60	3,600	7.746	$2^2 \cdot 3 \cdot 5$
11	121	3.317	11	61	3,721	7.810	61
12	144	3.464	$2^2 \cdot 3$	62	3,844	7.874	$2 \cdot 31$
13	169	3.606	13	63	3,969	7.937	$3^2 \cdot 7$
14	196	3.742	$2 \cdot 7$	64	4,096	8.000	2^6
15	225	3.873	$3 \cdot 5$	65	4,225	8.062	$5 \cdot 13$
16	256	4.000	2^4	66	4,356	8.124	$2 \cdot 3 \cdot 11$
17	289	4.123	17	67	4,489	8.185	67
18	324	4.243	$2 \cdot 3^2$	68	4,624	8.246	$2^2 \cdot 17$
19	361	4.359	19	69	4,761	8.307	$3 \cdot 23$
20	400	4.472	$2^2 \cdot 5$	70	4,900	8.367	$2 \cdot 5 \cdot 7$
21	441	4.583	$3 \cdot 7$	71	5,041	8.426	71
22	484	4.690	$2 \cdot 11$	72	5,184	8.485	$2^3 \cdot 3^2$
23	529.	4.796	23	73	5,329	8.544	73
24	576	4.899	$2^3 \cdot 3$	74	5,476	8.602	$2 \cdot 37$
25	625	5.000	5^2	75	5,625	8.660	$3 \cdot 5^2$
26	676	5.099	$2 \cdot 13$	76	5,776	8.718	$2^2 \cdot 19$
27	729	5.196	3^3	77	5,929	8.775	$7 \cdot 11$
28	784	5.292	$2^2 \cdot 7$	78	6,084	8.832	$2 \cdot 3 \cdot 13$
29	841	5.385	29	79	6,241	8.888	79
30	900	5.477	$2 \cdot 3 \cdot 5$	80	6,400	8.944	$2^4 \cdot 5$
31	961	5.568	31	81	6,561	9.000	3^4
32	1,024	5.657	2^5	82	6,724	9.055	$2 \cdot 41$
33	1,089	5.745	$3 \cdot 11$	83	6,889	9.110	83
34	1,156	5.831	$2 \cdot 17$	84	7,056	9.165	$2^2 \cdot 3 \cdot 7$
35	1,225	5.916	$5 \cdot 7$	85	7,225	9.220	$5 \cdot 17$
36	1,296	6.000	$2^2 \cdot 3^2$	86	7,396	9.274	$2 \cdot 43$
37	1,369	6.083	37	87	7,569	9.327	$3 \cdot 29$
38	1,444	6.164	$2 \cdot 19$	88	7,744	9.381	$2^3 \cdot 11$
39	1,521	6.245	$3 \cdot 13$	89	7,921	9.434	89
40	1,600	6.325	$2^3 \cdot 5$	90	8,100	9.487	$2 \cdot 3^2 \cdot 5$
41	1,681	6.403	41	91	8,281	9.539	$7 \cdot 13$
42	1,764	6.481	$2 \cdot 3 \cdot 7$	92	8,464	9.592	$2^2 \cdot 23$
43	1,849	6.557	43	93	8,649	9.644	$3 \cdot 31$
44	1,936	6.633	$2^2 \cdot 11$	94	8,836	9.695	$2 \cdot 47$
45	2,025	6.708	$3^2 \cdot 5$	95	9,025	9.747	$5 \cdot 19$
46	2,116	6.782	$2 \cdot 23$	96	9,216	9.798	$2^5 \cdot 3$
47	2,209	6.856	47	97	9,409	9.849	97
48	2,304	6.928	$2^4 \cdot 3$	98	9,604	9.899	$2 \cdot 7^2$
49	2,401	7.000	7^2	99	9,801	9.950	$3^2 \cdot 11$
50	2,500	7.071	$2 \cdot 5^2$	100	10,000	10.000	$2^2 \cdot 5^2$

Table II   Common logarithms

x	0	1	2	3	4	5	6	7	8	9
1.0	.0000	.0043	.0086	.0128	.0170	.0212	.0253	.0294	.0334	.0374
1.1	.0414	.0453	.0492	.0531	.0569	.0607	.0645	.0682	.0719	.0755
1.2	.0792	.0828	.0864	.0899	.0934	.0969	.1004	.1038	.1072	.1106
1.3	.1139	.1173	.1206	.1239	.1271	.1303	.1335	.1367	.1399	.1430
1.4	.1461	.1492	.1523	.1553	.1584	.1614	.1644	.1673	.1703	.1732
1.5	.1761	.1790	.1818	.1847	.1875	.1903	.1931	.1959	.1987	.2014
1.6	.2041	.2068	.2095	.2122	.2148	.2175	.2201	.2227	.2253	.2279
1.7	.2304	.2330	.2355	.2380	.2405	.2430	.2455	.2480	.2504	.2529
1.8	.2553	.2577	.2601	.2625	.2648	.2672	.2695	.2718	.2742	.2765
1.9	.2788	.2810	.2833	.2856	.2878	.2900	.2923	.2945	.2967	.2989
2.0	.3010	.3032	.3054	.3075	.3096	.3118	.3139	.3160	.3181	.3201
2.1	.3222	.3243	.3263	.3284	.3304	.3324	.3345	.3365	.3385	.3404
2.2	.3424	.3444	.3464	.3483	.3502	.3522	.3541	.3560	.3579	.3598
2.3	.3617	.3636	.3655	.3674	.3692	.3711	.3729	.3747	.3766	.3784
2.4	.3802	.3820	.3838	.3856	.3874	.3892	.3909	.3927	.3945	.3962
2.5	.3979	.3997	.4014	.4031	.4048	.4065	.4082	.4099	.4116	.4133
2.6	.4150	.4166	.4183	.4200	.4216	.4232	.4249	.4265	.4281	.4298
2.7	.4314	.4330	.4346	.4362	.4378	.4393	.4409	.4425	.4440	.4456
2.8	.4472	.4487	.4502	.4518	.4533	.4548	.4564	.4579	.4594	.4609
2.9	.4624	.4639	.4654	.4669	.4683	.4698	.4713	.4728	.4742	.4757
3.0	.4771	.4786	.4800	.4814	.4829	.4843	.4857	.4871	.4886	.4900
3.1	.4914	.4928	.4942	.4955	.4969	.4983	.4997	.5011	.5024	.5038
3.2	.5051	.5065	.5079	.5092	.5105	.5119	.5132	.5145	.5159	.5172
3.3	.5185	.5198	.5211	.5224	.5237	.5250	.5263	.5276	.5289	.5302
3.4	.5315	.5328	.5340	.5353	.5366	.5378	.5391	.5403	.5416	.5428
3.5	.5441	.5453	.5465	.5478	.5490	.5502	.5514	.5527	.5539	.5551
3.6	.5563	.5575	.5587	.5599	.5611	.5623	.5635	.5647	.5658	.5670
3.7	.5682	.5694	.5705	.5717	.5729	.5740	.5752	.5763	.5775	.5786
3.8	.5798	.5809	.5821	.5832	.5843	.5855	.5866	.5877	.5888	.5899
3.9	.5911	.5922	.5933	.5944	.5955	.5966	.5977	.5988	.5999	.6010
4.0	.6021	.6031	.6042	.6053	.6064	.6075	.6085	.6096	.6107	.6117
4.1	.6128	.6138	.6149	.6160	.6170	.6180	.6191	.6201	.6212	.6222
4.2	.6232	.6243	.6253	.6263	.6274	.6284	.6294	.6304	.6314	.6325
4.3	.6335	.6345	.6355	.6365	.6375	.6385	.6395	.6405	.6415	.6425
4.4	.6435	.6444	.6454	.6464	.6474	.6484	.6493	.6503	.6513	.6522
4.5	.6532	.6542	.6551	.6561	.6571	.6580	.6590	.6599	.6609	.6618
4.6	.6628	.6637	.6646	.6656	.6665	.6675	.6684	.6693	.6702	.6712
4.7	.6721	.6730	.6739	.6749	.6758	.6767	.6776	.6785	.6794	.6803
4.8	.6812	.6821	.6830	.6839	.6848	.6857	.6866	.6875	.6884	.6893
4.9	.6902	.6911	.6920	.6928	.6937	.6946	.6955	.6964	.6972	.6981
5.0	.6990	.6998	.7007	.7016	.7024	.7033	.7042	.7050	.7059	.7067
5.1	.7076	.7084	.7093	.7101	.7110	.7118	.7126	.7135	.7143	.7152
5.2	.7160	.7168	.7177	.7185	.7193	.7202	.7210	.7218	.7226	.7235
5.3	.7243	.7251	.7259	.7267	.7275	.7284	.7292	.7300	.7308	.7316
5.4	.7324	.7332	.7340	.7348	.7356	.7364	.7372	.7380	.7388	.7396
x	0	1	2	3	4	5	6	7	8	9

Table II (*continued*)

x	0	1	2	3	4	5	6	7	8	9
5.5	.7404	.7412	.7419	.7427	.7435	.7443	.7451	.7459	.7466	.7474
5.6	.7482	.7490	.7497	.7505	.7513	.7520	.7528	.7536	.7543	.7551
5.7	.7559	.7566	.7574	.7582	.7589	.7597	.7604	.7612	.7619	.7627
5.8	.7634	.7642	.7649	.7657	.7664	.7672	.7679	.7686	.7694	.7701
5.9	.7709	.7716	.7723	.7731	.7738	.7745	.7752	.7760	.7767	.7774
6.0	.7782	.7789	.7796	.7803	.7810	.7818	.7825	.7832	.7839	.7846
6.1	.7853	.7860	.7868	.7875	.7882	.7889	.7896	.7903	.7910	.7917
6.2	.7924	.7931	.7938	.7945	.7952	.7959	.7966	.7973	.7980	.7987
6.3	.7993	.8000	.8007	.8014	.8021	.8028	.8035	.8041	.8048	.8055
6.4	.8062	.8069	.8075	.8082	.8089	.8096	.8102	.8109	.8116	.8122
6.5	.8129	.8136	.8142	.8149	.8156	.8162	.8169	.8176	.8182	.8189
6.6	.8195	.8202	.8209	.8215	.8222	.8228	.8235	.8241	.8248	.8254
6.7	.8261	.8267	.8274	.8280	.8287	.8293	.8299	.8306	.8312	.8319
6.8	.8325	.8331	.8338	.8344	.8351	.8357	.8363	.8370	.8376	.8382
6.9	.8388	.8395	.8401	.8407	.8414	.8420	.8426	.8432	.8439	.8445
7.0	.8451	.8457	.8463	.8470	.8476	.8482	.8488	.8494	.8500	.8506
7.1	.8513	.8519	.8525	.8531	.8537	.8543	.8549	.8555	.8561	.8567
7.2	.8573	.8579	.8585	.8591	.8597	.8603	.8609	.8615	.8621	.8627
7.3	.8633	.8639	.8645	.8651	.8657	.8663	.8669	.8675	.8681	.8686
7.4	.8692	.8698	.8704	.8710	.8716	.8722	.8727	.8733	.8739	.8745
7.5	.8751	.8756	.8762	.8768	.8774	.8779	.8785	.8791	.8797	.8802
7.6	.8808	.8814	.8820	.8825	.8831	.8837	.8842	.8848	.8854	.8859
7.7	.8865	.8871	.8876	.8882	.8887	.8893	.8899	.8904	.8910	.8915
7.8	.8921	.8927	.8932	.8938	.8943	.8949	.8954	.8960	.8965	.8971
7.9	.8976	.8982	.8987	.8993	.8998	.9004	.9009	.9015	.9020	.9025
8.0	.9031	.9036	.9042	.9047	.9053	.9058	.9063	.9069	.9074	.9079
8.1	.9085	.9090	.9096	.9101	.9106	.9112	.9117	.9122	.9128	.9133
8.2	.9138	.9143	.9149	.9154	.9159	.9165	.9170	.9175	.9180	.9186
8.3	.9191	.9196	.9201	.9206	.9212	.9217	.9222	.9227	.9232	.9238
8.4	.9243	.9248	.9253	.9258	.9263	.9269	.9274	.9279	.9284	.9289
8.5	.9294	.9299	.9304	.9309	.9315	.9320	.9325	.9330	.9335	.9340
8.6	.9345	.9350	.9355	.9360	.9365	.9370	.9375	.9380	.9385	.9390
8.7	.9395	.9400	.9405	.9410	.9415	.9420	.9425	.9430	.9435	.9440
8.8	.9445	.9450	.9455	.9460	.9465	.9469	.9474	.9479	.9484	.9489
8.9	.9494	.9499	.9504	.9509	.9513	.9518	.9523	.9528	.9533	.9538
9.0	.9542	.9547	.9552	.9557	.9562	.9566	.9571	.9576	.9581	.9586
9.1	.9590	.9595	.9600	.9605	.9609	.9614	.9619	.9624	.9628	.9633
9.2	.9638	.9643	.9647	.9652	.9657	.9661	.9666	.9671	.9675	.9680
9.3	.9685	.9689	.9694	.9699	.9703	.9708	.9713	.9717	.9722	.9727
9.4	.9731	.9736	.9741	.9745	.9750	.9754	.9759	.9763	.9768	.9773
9.5	.9777	.9782	.9786	.9791	.9795	.9800	.9805	.9809	.9814	.9818
9.6	.9823	.9827	.9832	.9836	.9841	.9845	.9850	.9854	.9859	.9863
9.7	.9868	.9872	.9877	.9881	.9886	.9890	.9894	.9899	.9903	.9908
9.8	.9912	.9917	.9921	.9926	.9930	.9934	.9939	.9943	.9948	.9952
9.9	.9956	.9961	.9965	.9969	.9974	.9978	.9983	.9987	.9991	.9996
x	0	1	2	3	4	5	6	7	8	9

Table III Exponential functions

x	e^z	e^{-z}	x	e^z	e^{-z}
0.00	1.0000	1.0000	1.5	4.4817	0.2231
0.01	1.0101	0.9901	1.6	4.9530	0.2019
0.02	1.0202	0.9802	1.7	5.4739	0.1827
0.03	1.0305	0.9705	1.8	6.0496	0.1653
0.04	1.0408	0.9608	1.9	6.6859	0.1496
0.05	1.0513	0.9512	2.0	7.3891	0.1353
0.06	1.0618	0.9418	2.1	8.1662	0.1225
0.07	1.0725	0.9324	2.2	9.0250	0.1108
0.08	1.0833	0.9331	2.3	9.9742	0.1003
0.09	1.0942	0.9139	2.4	11.023	0.0907
0.10	1.1052	0.9048	2.5	12.182	0.0821
0.11	1.1163	0.8958	2.6	13.464	0.0743
0.12	1.1275	0.8869	2.7	14.880	0.0672
0.13	1.1388	0.8781	2.8	16.445	0.0608
0.14	1.1503	0.8694	2.9	18.174	0.0550
0.15	1.1618	0.8607	3.0	20.086	0.0498
0.16	1.1735	0.8521	3.1	22.198	0.0450
0.17	1.1853	0.8437	3.2	24.533	0.0408
0.18	1.1972	0.8353	3.3	27.113	0.0369
0.19	1.2092	0.8270	3.4	29.964	0.0334
0.20	1.2214	0.8187	3.5	33.115	0.0302
0.21	1.2337	0.8106	3.6	36.598	0.0273
0.22	1.2461	0.8025	3.7	40.447	0.0247
0.23	1.2586	0.7945	3.8	44.701	0.0224
0.24	1.2712	0.7866	3.9	49.402	0.0202
0.25	1.2840	0.7788	4.0	54.598	0.0183
0.30	1.3499	0.7408	4.1	60.340	0.0166
0.35	1.4191	0.7047	4.2	66.686	0.0150
0.40	1.4918	0.6703	4.3	73.700	0.0136
0.45	1.5683	0.6376	4.4	81.451	0.0123
0.50	1.6487	0.6065	4.5	90.017	0.0111
0.55	1.7333	0.5769	4.6	99.484	0.0101
0.60	1.8221	0.5488	4.7	109.95	0.0091
0.65	1.9155	0.5220	4.8	121.51	0.0082
0.70	2.0138	0.4966	4.9	134.29	0.0074
0.75	2.1170	0.4724	5.0	148.41	0.0067
0.80	2.2255	0.4493	5.5	244.69	0.0041
0.85	2.3396	0.4274	6.0	403.43	0.0025
0.90	2.4596	0.4066	6.5	665.14	0.0015
0.95	2.5857	0.3867	7.0	1096.6	0.0009
1.0	2.7183	0.3679	7.5	1808.0	0.0006
1.1	3.0042	0.3329	8.0	2981.0	0.0003
1.2	3.3201	0.3012	8.5	4914.8	0.0002
1.3	3.6693	0.2725	9.0	8103.1	0.0001
1.4	4.0552	0.2466	10.0	22026	0.00005

Table IV Natural logarithms of numbers

n	$\log_e n$	n	$\log_e n$	n	$\log_e n$
	*	4.5	1.5041	9.0	2.1972
0.1	7.6974	4.6	1.5261	9.1	2.2083
0.2	8.3906	4.7	1.5476	9.2	2.2192
0.3	8.7960	4.8	1.5686	9.3	2.2300
0.4	9.0837	4.9	1.5892	9.4	2.2407
0.5	9.3069	5.0	1.6094	9.5	2.2513
0.6	9.4892	5.1	1.6292	9.6	2.2618
0.7	9.6433	5.2	1.6487	9.7	2.2721
0.8	9.7769	5.3	1.6677	9.8	2.2824
0.9	9.8946	5.4	1.6864	9.9	2.2925
1.0	0.0000	5.5	1.7047	10	2.3026
1.1	0.0953	5.6	1.7228	11	2.3979
1.2	0.1823	5.7	1.7405	12	2.4849
1.3	0.2624	5.8	1.7579	13	2.5649
1.4	0.3365	5.9	1.7750	14	2.6391
1.5	0.4055	6.0	1.7918	15	2.7081
1.6	0.4700	6.1	1.8083	16	2.7726
1.7	0.5306	6.2	1.8245	17	2.8332
1.8	0.5878	6.3	1.8405	18	2.8904
1.9	0.6419	6.4	1.8563	19	2.9444
2.0	0.6931	6.5	1.8718	20	2.9957
2.1	0.7419	6.6	1.8871	25	3.2189
2.2	0.7885	6.7	1.9021	30	3.4012
2.3	0.8329	6.8	1.9169	35	3.5553
2.4	0.8755	6.9	1.9315	40	3.6889
2.5	0.9163	7.0	1.9459	45	3.8067
2.6	0.9555	7.1	1.9601	50	3.9120
2.7	0.9933	7.2	1.9741	55	4.0073
2.8	1.0296	7.3	1.9879	60	4.0943
2.9	1.0647	7.4	2.0015	65	4.1744
3.0	1.0986	7.5	2.0149	70	4.2485
3.1	1.1314	7.6	2.0281	75	4.3175
3.2	1.1632	7.7	2.0412	80	4.3820
3.3	1.1939	7.8	2.0541	85	4.4427
3.4	1.2238	7.9	2.0669	90	4.4998
3.5	1.2528	8.0	2.0794	100	4.6052
3.6	1.2809	8.1	2.0919	110	4.7005
3.7	1.3083	8.2	2.1041	120	4.7875
3.8	1.3350	8.3	2.1163	130	4.8676
3.9	1.3610	8.4	2.1282	140	4.9416
4.0	1.3863	8.5	2.1401	150	5.0106
4.1	1.4110	8.6	2.1518	160	5.0752
4.2	1.4351	8.7	2.1633	170	5.1358
4.3	1.4586	8.8	2.1748	180	5.1930
4.4	1.4816	8.9	2.1861	190	5.2470

* Subtract 10 for $n < 1$. Thus $\log_e 0.1 = 7.6974 - 10 = -2.3026$.

Table V Values of circular functions

Real Number x or θ radians	θ degrees	sin x or sin θ	csc x or csc θ	tan x or tan θ	cot x or cot θ	sec x or sec θ	cos x or cos θ
0.00	0° 00′	0.0000	No value	0.0000	No value	1.000	1.000
.01	0° 34′	.0100	100.0	.0100	100.0	1.000	1.000
.02	1° 09′	.0200	50.00	.0200	49.99	1.000	0.9998
.03	1° 43′	.0300	33.34	.0300	33.32	1.000	0.9996
.04	2° 18′	.0400	25.01	.0400	24.99	1.001	0.9992
0.05	2° 52′	0.0500	20.01	0.0500	19.98	1.001	0.9988
.06	3° 26′	.0600	16.68	.0601	16.65	1.002	.9982
.07	4° 01′	.0699	14.30	.0701	14.26	1.002	.9976
.08	4° 35′	.0799	12.51	.0802	12.47	1.003	.9968
.09	5° 09′	.0899	11.13	.0902	11.08	1.004	.9960
0.10	5° 44′	0.0998	10.02	0.1003	9.967	1.005	0.9950
.11	6° 18′	.1098	9.109	.1104	9.054	1.006	.9940
.12	6° 53′	.1197	8.353	.1206	8.293	1.007	.9928
.13	7° 27′	.1296	7.714	.1307	7.649	1.009	.9916
.14	8° 01′	.1395	7.166	.1409	7.096	1.010	.9902
0.15	8° 36′	0.1494	6.692	0.1511	6.617	1.011	0.9888
.16	9° 10′	.1593	6.277	.1614	6.197	1.013	.9872
.17	9° 44′	.1692	5.911	.1717	5.826	1.015	.9856
.18	10° 19′	.1790	5.586	.1820	5.495	1.016	.9838
.19	10° 53′	.1889	5.295	.1923	5.200	1.018	.9820
0.20	11° 28′	0.1987	5.033	0.2027	4.933	1.020	0.9801
.21	12° 02′	.2085	4.797	.2131	4.692	1.022	.9780
.22	12° 36′	.2182	4.582	.2236	4.472	1.025	.9759
.23	13° 11′	.2280	4.386	.2341	4.271	1.027	.9737
.24	13° 45′	.2377	4.207	.2447	4.086	1.030	.9713
0.25	14° 19′	0.2474	4.042	0.2553	3.916	1.032	0.9689
.26	14° 54′	.2571	3.890	.2660	3.759	1.035	.9664
.27	15° 28′	.2667	3.749	.2768	3.613	1.038	.9638
.28	16° 03′	.2764	3.619	.2876	3.478	1.041	.9611
.29	16° 37′	.2860	3.497	.2984	3.351	1.044	.9582
0.30	17° 11′	0.2955	3.384	0.3093	3.233	1.047	0.9553
.31	17° 46′	.3051	3.278	.3203	3.122	1.050	.9523
.32	18° 20′	.3146	3.179	.3314	3.018	1.053	.9492
.33	18° 54′	.3240	3.086	.3425	2.920	1.057	.9460
.34	19° 29′	.3335	2.999	.3537	2.827	1.061	.9428
0.35	20° 03′	0.3429	2.916	0.3650	2.740	1.065	0.9394
.36	20° 38′	.3523	2.839	.3764	2.657	1.068	.9359
.37	21° 12′	.3616	2.765	.3879	2.578	1.073	.9323
.38	21° 46′	.3709	2.696	.3994	2.504	1.077	.9287
.39	22° 21′	.3802	2.630	.4111	2.433	1.081	.9249
0.40	22° 55′	0.3894	2.568	0.4228	2.365	1.086	0.9211
.41	23° 29′	.3986	2.509	.4346	2.301	1.090	.9171
.42	24° 04′	.4078	2.452	.4466	2.239	1.095	.9131
.43	24° 38′	.4169	2.399	.4586	2.180	1.100	.9090
.44	25° 13′	.4259	2.348	.4708	2.124	1.105	.9048
0.45	25° 47′	0.4350	2.299	0.4831	2.070	1.111	0.9004

Table V (*continued*)

Real Number x or θ radians	θ degrees	sin x or sin θ	csc x or csc θ	tan x or tan θ	cot x or cot θ	sec x or sec θ	cos x or cos θ
0.45	25° 47′	0.4350	2.299	0.4831	2.070	1.111	0.9004
.46	26° 21′	.4439	2.253	.4954	2.018	1.116	.8961
.47	26° 56′	.4529	2.208	.5080	1.969	1.122	.8916
.48	27° 30′	.4618	2.166	.5206	1.921	1.127	.8870
.49	28° 04′	.4706	2.125	.5334	1.875	1.133	.8823
0.50	28° 39′	0.4794	2.086	0.5463	1.830	1.139	0.8776
.51	29° 13′	.4882	2.048	.5594	1.788	1.146	.8727
.52	29° 48′	.4969	2.013	.5726	1.747	1.152	.8678
.53	30° 22′	.5055	1.978	.5859	1.707	1.159	.8628
.54	30° 56′	.5141	1.945	.5994	1.668	1.166	.8577
0.55	31° 31′	0.5227	1.913	0.6131	1.631	1.173	0.8525
.56	32° 05′	.5312	1.883	.6269	1.595	1.180	.8473
.57	32° 40′	.5396	1.853	.6410	1.560	1.188	.8419
.58	33° 14′	.5480	1.825	.6552	1.526	1.196	.8365
.59	33° 48′	.5564	1.797	.6696	1.494	1.203	.8309
0.60	34° 23′	0.5646	1.771	0.6841	1.462	1.212	0.8253
.61	34° 57′	.5729	1.746	.6989	1.431	1.220	.8196
.62	35° 31′	.5810	1.721	.7139	1.401	1.229	.8139
.63	36° 06′	.5891	1.697	.7291	1.372	1.238	.8080
.64	36° 40′	.5972	1.674	.7445	1.343	1.247	.8021
0.65	37° 15′	0.6052	1.652	0.7602	1.315	1.256	0.7961
.66	37° 49′	.6131	1.631	.7761	1.288	1.266	.7900
.67	38° 23′	.6210	1.610	.7923	1.262	1.276	.7838
.68	38° 58′	.6288	1.590	.8087	1.237	1.286	.7776
.69	39° 32′	.6365	1.571	.8253	1.212	1.297	.7712
0.70	40° 06′	0.6442	1.552	0.8423	1.187	1.307	0.7648
.71	40° 41′	.6518	1.534	.8595	1.163	1.319	.7584
.72	41° 15′	.6594	1.517	.8771	1.140	1.330	.7518
.73	41° 50′	.6669	1.500	.8949	1.117	1.342	.7452
.74	42° 24′	.6743	1.483	.9131	1.095	1.354	.7385
0.75	42° 58′	0.6816	1.467	0.9316	1.073	1.367	0.7317
.76	43° 33′	.6889	1.452	.9505	1.052	1.380	.7248
.77	44° 07′	.6961	1.436	.9697	1.031	1.393	.7179
.78	44° 41′	.7033	1.422	.9893	1.011	1.407	.7109
.79	45° 16′	.7104	1.408	1.009	.9908	1.421	.7038
0.80	45° 50′	0.7174	1.394	1.030	0.9712	1.435	0.6967
.81	46° 25′	.7243	1.381	1.050	.9520	1.450	.6895
.82	46° 59′	.7311	1.368	1.072	.9331	1.466	.6822
.83	47° 33′	.7379	1.355	1.093	.9146	1.482	.6749
.84	48° 08′	.7446	1.343	1.116	.8964	1.498	.6675
0.85	48° 42′	0.7513	1.331	1.138	0.8785	1.515	0.6600
.86	49° 16′	.7578	1.320	1.162	.8609	1.533	.6524
.87	49° 51′	.7643	1.308	1.185	.8437	1.551	.6448
.88	50° 25′	.7707	1.297	1.210	.8267	1.569	.6372
.89	51° 00′	.7771	1.287	1.235	.8100	1.589	.6294
0.90	51° 34′	0.7833	1.277	1.260	0.7936	1.609	0.6216
.91	52° 08′	.7895	1.267	1.286	.7774	1.629	.6137
.92	52° 43′	.7956	1.257	1.313	.7615	1.651	.6058
.93	53° 17′	.8016	1.247	1.341	.7458	1.673	.5978
.94	53° 51′	.8076	1.238	1.369	.7303	1.696	.5898
0.95	54° 26′	0.8134	1.229	1.398	0.7151	1.719	0.5817

Table V (*continued*)

Real Number x or θ radians	θ degrees	sin x or sin θ	csc x or csc θ	tan x or tan θ	cot x or cot θ	sec x or sec θ	cos x or cos θ
0.95	54° 26′	0.8134	1.229	1.398	0.7151	1.719	0.5817
.96	55° 00′	.8192	1.221	1.428	.7001	1.744	.5735
.97	55° 35′	.8249	1.212	1.459	.6853	1.769	.5653
.98	56° 09′	.8305	1.204	1.491	.6707	1.795	.5570
.99	56° 43′	.8360	1.196	1.524	.6563	1.823	.5487
1.00	57° 18′	0.8415	1.188	1.557	0.6421	1.851	0.5403
1.01	57° 52′	.8468	1.181	1.592	.6281	1.880	.5319
1.02	58° 27′	.8521	1.174	1.628	.6142	1.911	.5234
1.03	59° 01′	.8573	1.166	1.665	.6005	1.942	.5148
1.04	59° 35′	.8624	1.160	1.704	.5870	1.975	.5062
1.05	60° 10′	0.8674	1.153	1.743	0.5736	2.010	0.4976
1.06	60° 44′	.8724	1.146	1.784	.5604	2.046	.4889
1.07	61° 18′	.8772	1.140	1.827	.5473	2.083	.4801
1.08	61° 53′	.8820	1.134	1.871	.5344	2.122	.4713
1.09	62° 27′	.8866	1.128	1.917	.5216	2.162	.4625
1.10	63° 02′	0.8912	1.122	1.965	0.5090	2.205	0.4536
1.11	63° 36′	.8957	1.116	2.014	.4964	2.249	.4447
1.12	64° 10′	.9001	1.111	2.066	.4840	2.295	.4357
1.13	64° 45′	.9044	1.106	2.120	.4718	2.344	.4267
1.14	65° 19′	.9086	1.101	2.176	.4596	2.395	.4176
1.15	65° 53′	0.9128	1.096	2.234	0.4475	2.448	0.4085
1.16	66° 28′	.9168	1.091	2.296	.4356	2.504	.3993
1.17	67° 02′	.9208	1.086	2.360	.4237	2.563	.3902
1.18	67° 37′	.9246	1.082	2.427	.4120	2.625	.3809
1.19	68° 11′	.9284	1.077	2.498	.4003	2.691	.3717
1.20	68° 45′	0.9320	1.073	2.572	0.3888	2.760	0.3624
1.21	69° 20′	.9356	1.069	2.650	.3773	2.833	.3530
1.22	69° 54′	.9391	1.065	2.733	.3659	2.910	.3436
1.23	70° 28′	.9425	1.061	2.820	.3546	2.992	.3342
1.24	71° 03′	.9458	1.057	2.912	.3434	3.079	.3248
1.25	71° 37′	0.9490	1.054	3.010	0.3323	3.171	0.3153
1.26	72° 12′	.9521	1.050	3.113	.3212	3.270	.3058
1.27	72° 46′	.9551	1.047	3.224	.3102	3.375	.2963
1.28	72° 20′	.9580	1.044	3.341	.2993	3.488	.2867
1.29	73° 55′	.9608	1.041	3.467	.2884	3.609	.2771
1.30	74° 29′	0.9636	1.038	3.602	0.2776	3.738	0.2675
1.31	75° 03′	.9662	1.035	3.747	.2669	3.878	.2579
1.32	75° 38′	.9687	1.032	3.903	.2562	4.029	.2482
1.33	76° 12′	.9711	1.030	4.072	.2456	4.193	.2385
1.34	76° 47′	.9735	1.027	4.256	.2350	4.372	.2288
1.35	77° 21′	0.9757	1.025	4.455	0.2245	4.566	0.2190
1.36	77° 55′	.9779	1.023	4.673	.2140	4.779	.2092
1.37	78° 30′	.9799	1.021	4.913	.2035	5.014	.1994
1.38	79° 04′	.9819	1.018	5.177	.1931	5.273	.1896
1.39	79° 38′	.9837	1.017	5.471	.1828	5.561	.1798
1.40	80° 13′	0.9854	1.015	5.798	0.1725	5.883	0.1700
1.41	80° 47′	.9871	1.013	6.165	.1622	6.246	.1601
1.42	81° 22′	.9887	1.011	6.581	.1519	6.657	.1502
1.43	81° 56′	.9901	1.010	7.055	.1417	7.126	.1403
1.44	82° 30′	.9915	1.009	7.602	.1315	7.667	.1304
1.45	83° 05′	0.9927	1.007	8.238	0.1214	8.299	0.1205

Table V (*continued*)

Real Number x or θ radians	θ degrees	$\sin x$ or $\sin \theta$	$\csc x$ or $\csc \theta$	$\tan x$ or $\tan \theta$	$\cot x$ or $\cot \theta$	$\sec x$ or $\sec \theta$	$\cos x$ or $\cos \theta$
1.45	83° 05′	0.9927	1.007	8.238	0.1214	8.299	0.1205
1.46	83° 39′	.9939	1.006	8.989	.1113	9.044	.1106
1.47	84° 13′	.9949	1.005	9.887	.1011	9.938	.1006
1.48	84° 48′	.9959	1.004	10.98	.0910	11.03	.0907
1.49	85° 22′	.9967	1.003	12.35	.0810	12.39	.0807
1.50	85° 57′	0.9975	1.003	14.10	0.0709	14.14	0.0707
1.51	86° 31′	.9982	1.002	16.43	.0609	16.46	.0608
1.52	87° 05′	.9987	1.001	19.67	.0508	19.69	.0508
1.53	87° 40′	.9992	1.001	24.50	.0408	24.52	.0408
1.54	88° 14′	.9995	1.000	32.46	.0308	32.48	.0308
1.55	88° 49′	0.9998	1.000	48.08	0.0208	48.09	0.0208
1.56	89° 23′	.9999	1.000	92.62	.0108	92.63	.0108
1.57	89° 57′	1.000	1.000	1256	.0008	1256	.0008

Table VI Values of trigonometric functions

Angle θ									
Degrees	Radians	sin θ	csc θ	tan θ	cot θ	sec θ	cos θ		
0° 00′	.0000	.0000	No value	.0000	No value	1.000	1.0000	1.5708	90° 00′
10	029	029	343.8	029	343.8	000	000	679	50
20	058	058	171.9	058	171.9	000	000	650	40
30	087	087	114.6	087	114.6	000	1.0000	621	30
40	116	116	85.95	116	85.94	000	.9999	592	20
50	145	145	68.76	145	68.75	000	999	563	10
1° 00′	.0175	.0175	57.30	.0175	57.29	1.000	.9998	1.5533	89° 00′
10	204	204	49.11	204	49.10	000	998	504	50
20	233	233	42.98	233	42.96	000	997	475	40
30	262	262	38.20	262	38.19	000	997	446	30
40	291	291	34.38	291	34.37	000	996	417	20
50	320	320	31.26	320	31.24	001	995	388	10
2° 00′	.0349	.0349	28.65	.0349	28.64	1.001	.9994	1.5359	88° 00′
10	378	378	26.45	378	26.43	001	993	330	50
20	407	407	24.56	407	24.54	001	992	301	40
30	436	436	22.93	437	22.90	001	990	272	30
40	465	465	21.49	466	21.47	001	989	243	20
50	495	494	20.23	495	20.21	001	988	213	10
3° 00′	.0524	.0523	19.11	.0524	19.08	1.001	.9986	1.5184	87° 00′
10	553	552	18.10	553	18.07	002	985	155	50
20	582	581	17.20	582	17.17	002	983	126	40
30	611	610	16.38	612	16.35	002	981	097	30
40	640	640	15.64	641	15.60	002	980	068	20
50	669	669	14.96	670	14.92	002	978	039	10
4° 00′	.0698	.0698	14.34	.0699	14.30	1.002	.9976	1.5010	86° 00′
10	727	727	13.76	729	13.73	003	974	981	50
20	756	756	13.23	758	13.20	003	971	952	40
30	785	785	12.75	787	12.71	003	969	923	30
40	814	814	12.29	816	12.25	003	967	893	20
50	844	843	11.87	846	11.83	004	964	864	10
5° 00′	.0873	.0872	11.47	.0875	11.43	1.004	.9962	1.4835	85° 00′
10	902	901	11.10	904	11.06	004	959	806	50
20	931	929	10.76	934	10.71	004	957	777	40
30	960	958	10.43	963	10.39	005	954	748	30
40	.0989	.0987	10.13	.0992	10.08	005	951	719	20
50	.1018	.1016	9.839	.1022	9.788	005	948	690	10
6° 00′	.1047	.1045	9.567	.1051	9.514	1.006	.9945	1.4661	84° 00′
10	076	074	9.309	080	9.255	006	942	632	50
20	105	103	9.065	110	9.010	006	939	603	40
30	134	132	8.834	139	8.777	006	936	573	30
40	164	161	8.614	169	8.556	007	932	544	20
50	193	190	8.405	198	8.345	007	929	515	10
7° 00′	.1222	.1219	8.206	.1228	8.144	1.008	.9925	1.4486	83° 00′
10	251	248	8.016	257	7.953	008	922	457	50
20	280	276	7.834	287	7.770	008	918	428	40
30	309	305	7.661	317	7.596	009	914	399	30
40	338	334	7.496	346	7.429	009	911	370	20
50	367	363	7.337	376	7.269	009	907	341	10
8° 00′	.1396	.1392	7.185	.1405	7.115	1.010	.9903	1.4312	82° 00′
		cos θ	sec θ	cot θ	tan θ	csc θ	sin θ	Radians	Degrees
								Angle θ	

Table VI (*continued*)

Degrees	Radians	sin θ	csc θ	tan θ	cot θ	sec θ	cos θ		
8° 00′	.1396	.1392	7.185	.1405	7.115	1.010	.9903	1.4312	82° 00′
10	425	421	7.040	435	6.968	010	899	283	50
20	454	449	6.900	465	827	011	894	254	40
30	484	478	765	495	691	011	890	224	30
40	513	507	636	524	561	012	886	195	20
50	542	536	512	554	435	012	881	166	10
9° 00′	.1571	.1564	6.392	.1584	6.314	1.012	.9877	1.4137	81° 00′
10	600	593	277	614	197	013	872	108	50
20	629	622	166	644	6.084	013	868	079	40
30	658	650	6.059	673	5.976	014	863	050	30
40	687	679	5.955	703	871	014	858	1.4021	20
50	716	708	855	733	769	015	853	1.3992	10
10° 00′	.1745	.1736	5.759	.1763	5.671	1.015	.9848	1.3963	80° 00′
10	774	765	665	793	576	016	843	934	50
20	804	794	575	823	485	016	838	904	40
30	833	822	487	853	396	017	833	875	30
40	862	851	403	883	309	018	827	846	20
50	891	880	320	914	226	018	822	817	10
11° 00′	.1920	.1908	5.241	.1944	5.145	1.019	.9816	1.3788	79° 00′
10	949	937	164	.1974	5.066	019	811	759	50
20	.1978	965	089	.2004	4.989	020	805	730	40
30	.2007	.1994	5.016	035	915	020	799	701	30
40	036	.2022	4.945	065	843	021	793	672	20
50	065	051	876	095	773	022	787	643	10
12° 00′	.2094	.2079	4.810	.2126	4.705	1.022	.9781	1.3614	78° 00′
10	123	108	745	156	638	023	775	584	50
20	153	136	682	186	574	024	769	555	40
30	182	164	620	217	511	024	763	526	30
40	211	193	560	247	449	025	757	497	20
50	240	221	502	278	390	026	750	468	10
13° 00′	.2269	.2250	4.445	.2309	4.331	1.026	.9744	1.3439	77° 00′
10	298	278	390	339	275	027	737	410	50
20	327	306	336	370	219	028	730	381	40
30	356	334	284	401	165	028	724	352	30
40	385	363	232	432	113	029	717	323	20
50	414	391	182	462	061	030	710	294	10
14° 00′	.2443	.2419	4.134	.2493	4.011	1.031	.9703	1.3265	76° 00′
10	473	447	086	524	3.962	031	696	235	50
20	502	476	4.039	555	914	032	689	206	40
30	531	504	3.994	586	867	033	681	177	30
40	560	532	950	617	821	034	674	148	20
50	589	560	906	648	776	034	667	119	10
15° 00′	.2618	.2588	3.864	.2679	3.732	1.035	.9659	1.3090	75° 00′
10	647	616	822	711	689	036	652	061	50
20	676	644	782	742	647	037	644	032	40
30	705	672	742	773	606	038	636	1.3003	30
40	734	700	703	805	566	039	628	1.2974	20
50	763	728	665	836	526	039	621	945	10
16° 00′	.2793	.2756	3.628	.2867	3.487	1.040	.9613	1.2915	74° 00′
		cos θ	sec θ	cot θ	tan θ	csc θ	sin θ	Radians	Degrees

Angle θ

Table VI (*continued*)

Degrees	Radians	sin θ	csc θ	tan θ	cot θ	sec θ	cos θ		
16° 00′	.2793	.2756	3.628	.2867	3.487	1.040	.9613	1.2915	74° 00′
10	822	784	592	899	450	041	605	886	50
20	851	812	556	931	412	042	596	857	40
30	880	840	521	962	376	043	588	828	30
40	909	868	487	.2944	340	044	580	799	20
50	938	896	453	.3026	305	045	572	770	10
17° 00′	.2967	.2924	3.420	.3057	3.271	1.046	.9563	1.2741	73° 00′
10	.2996	952	388	089	237	047	555	712	50
20	.3025	.2979	357	121	204	048	546	683	40
30	054	.3007	326	153	172	048	537	654	30
40	083	035	295	185	140	049	528	625	20
50	113	062	265	217	108	050	520	595	10
18° 00′	.3142	.3090	3.236	.3249	3.078	1.051	.9511	1.2566	72° 00′
10	171	118	207	281	047	052	502	537	50
20	200	145	179	314	3.018	053	492	508	40
30	229	173	152	346	2.989	054	483	479	30
40	258	201	124	378	960	056	474	450	20
50	287	228	098	411	932	057	465	421	10
19° 00′	.3316	.3256	3.072	.3443	2.904	1.058	.9455	1.2392	71° 00′
10	345	283	046	476	877	059	446	363	50
20	374	311	3.021	508	850	060	436	334	40
30	403	338	2.996	541	824	061	426	305	30
40	432	365	971	574	798	062	417	275	20
50	462	393	947	607	773	063	407	246	10
20° 00′	.3491	.3420	2.924	.3640	2.747	1.064	.9397	1.2217	70° 00′
10	520	448	901	673	723	065	387	188	50
20	549	475	878	706	699	066	377	159	40
30	578	502	855	739	675	068	367	130	30
40	607	529	833	772	651	069	356	101	20
50	636	557	812	805	628	070	346	072	10
21° 00′	.3665	.3584	2.790	.3839	2.605	1.071	.9336	1.2043	69° 00′
10	694	611	769	872	583	072	325	1.2014	50
20	723	638	749	906	560	074	315	1.1985	40
30	752	665	729	939	539	075	304	956	30
40	782	692	709	.3973	517	076	293	926	20
50	811	719	689	.4006	496	077	283	897	10
22° 00′	.3840	.3746	2.669	.4040	2.475	1.079	.9272	1.1868	68° 00′
10	869	773	650	074	455	080	261	839	50
20	898	800	632	108	434	081	250	810	40
30	927	827	613	142	414	082	239	781	30
40	956	854	595	176	394	084	228	752	20
50	985	881	577	210	375	085	216	723	10
23° 00′	.4014	.3907	2.559	.4245	2.356	1.086	.9205	1.1694	67° 00′
10	043	934	542	279	337	088	194	665	50
20	072	961	525	314	318	089	182	636	40
30	102	.3987	508	348	300	090	171	606	30
40	131	.4014	491	383	282	092	159	577	20
50	160	041	475	417	264	093	147	548	10
24° 00′	.4189	.4067	2.459	.4452	2.246	1.095	.9135	1.1519	66° 00′
		cos θ	sec θ	cot θ	tan θ	csc θ	sin θ	Radians	Degrees

Angle θ

Table VI (*continued*)

Angle θ									Angle θ	
Degrees	Radians	sin θ	csc θ	tan θ	cot θ	sec θ	cos θ			
24° 00′	.4189	.4067	2.459	.4452	2.246	1.095	.9135	1.1519	66° 00′	
10	218	094	443	487	229	096	124	490	50	
20	247	120	427	522	211	097	112	461	40	
30	276	147	411	557	194	099	100	432	30	
40	305	173	396	592	177	100	088	403	20	
50	334	200	381	628	161	102	075	374	10	
25° 00′	.4363	.4226	2.366	.4663	2.145	1.103	.9063	1.1345	65° 00′	
10	392	253	352	699	128	105	051	316	50	
20	422	279	337	734	112	106	038	286	40	
30	451	305	323	770	097	108	026	257	30	
40	480	331	309	806	081	109	013	228	20	
50	509	358	295	841	066	111	.9001	199	10	
26° 00′	.4538	.4384	2.281	.4877	2.050	1.113	.8988	1.1170	64° 00′	
10	567	410	268	913	035	114	975	141	50	
20	596	436	254	950	020	116	962	112	40	
30	625	462	241	.4986	2.006	117	949	083	30	
40	654	488	228	.5022	1.991	119	936	054	20	
50	683	514	215	059	977	121	923	1.1025	10	
27° 00′	.4712	.4540	2.203	.5095	1.963	1.122	.8910	1.0996	63° 00′	
10	741	566	190	132	949	124	897	966	50	
20	771	592	178	169	935	126	884	937	40	
30	800	617	166	206	921	127	870	908	30	
40	829	643	154	243	907	129	857	879	20	
50	858	669	142	280	894	131	843	850	10	
28° 00′	.4887	.4695	2.130	.5317	1.881	1.133	.8829	1.0821	62° 00′	
10	916	720	118	354	868	134	816	792	50	
20	945	746	107	392	855	136	802	763	40	
30	.4974	772	096	430	842	138	788	734	30	
40	.5003	797	085	467	829	140	774	705	20	
50	032	823	074	505	816	142	760	676	10	
29° 00′	.5061	.4848	2.063	.5543	1.804	1.143	.8746	1.0647	61° 00′	
10	091	874	052	581	792	145	732	617	50	
20	120	899	041	619	780	147	718	588	40	
30	149	924	031	658	767	149	704	559	30	
40	178	950	020	696	756	151	689	530	20	
50	207	.4975	010	735	744	153	675	501	10	
30° 00′	.5236	.5000	2.000	.5774	1.732	1.155	.8660	1.0472	60° 00′	
10	265	025	1.990	812	720	157	646	443	50	
20	294	050	980	851	709	159	631	414	40	
30	323	075	970	890	698	161	616	385	30	
40	352	100	961	930	686	163	60!	356	20	
50	381	125	951	.5969	675	165	587	327	10	
31° 00′	.5411	.5150	1.942	.6009	1.664	1.167	.8572	1.0297	59° 00′	
10	440	175	932	048	653	169	557	268	50	
20	469	200	923	088	643	171	542	239	40	
30	498	225	914	128	632	173	526	210	30	
40	527	250	905	168	621	175	511	181	20	
50	556	275	896	208	611	177	496	152	10	
32° 00′	.5585	.5299	1.887	.6249	1.600	1.179	.8480	1.0123	58° 00′	
		cos θ	sec θ	cot θ	tan θ	csc θ	sin θ	Radians	Degrees	
									Angle θ	

427

Table VI (*continued*)

Angle θ									
Degrees	Radians	sin θ	csc θ	tan θ	cot θ	sec θ	cos θ		
32° 00′	.5585	.5299	1.887	.6249	1.600	1.179	.8480	1.0123	58° 00′
10	614	324	878	289	590	181	465	094	50
20	643	348	870	330	580	184	450	065	40
30	672	373	861	371	570	186	434	036	30
40	701	398	853	412	560	188	418	1.0007	20
50	730	422	844	453	550	190	403	.9977	10
33° 00′	.5760	.5446	1.836	.6494	1.540	1.192	.8387	.9948	57° 00′
10	789	471	828	536	530	195	371	919	50
20	818	495	820	577	520	197	355	890	40
30	847	519	812	619	511	199	339	861	30
40	876	544	804	661	501	202	323	832	20
50	905	568	796	703	492	204	307	803	10
34° 00′	.5934	.5592	1.788	.6745	1.483	1.206	.8290	.9774	56° 00′
10	963	616	781	787	473	209	274	745	50
20	.5992	640	773	830	464	211	258	716	40
30	.6021	664	766	873	455	213	241	687	30
40	050	688	758	916	446	216	225	657	20
50	080	712	751	.6959	437	218	208	628	10
35° 00′	.6109	.5736	1.743	.7002	1.428	1.221	.8192	.9599	55° 00′
10	138	760	736	046	419	223	175	570	50
20	167	783	729	089	411	226	158	541	40
30	196	807	722	133	402	228	141	512	30
40	225	831	715	177	393	231	124	483	20
50	254	854	708	221	385	233	107	454	10
36° 00′	.6283	.5878	1.701	.7265	1.376	1.236	.8090	.9425	54° 00′
10	312	901	695	310	368	239	073	396	50
20	341	925	688	355	360	241	056	367	40
30	370	948	681	400	351	244	039	338	30
40	400	972	675	445	343	247	021	308	20
50	429	.5995	668	490	335	249	.8004	279	10
37° 00′	.6458	.6018	1.662	.7536	1.327	1.252	.7986	.9250	53° 00′
10	487	041	655	581	319	255	969	221	50
20	516	065	649	627	311	258	951	192	40
30	545	088	643	673	303	260	934	163	30
40	574	111	636	720	295	263	916	134	20
50	603	134	630	766	288	266	898	105	10
38° 00′	.6632	.6157	1.624	.7813	1.280	1.269	.7880	.9076	52° 00′
10	661	180	618	860	272	272	862	047	50
20	690	202	612	907	265	275	844	.9018	40
30	720	225	606	.7954	257	278	826	.8988	30
40	749	248	601	.8002	250	281	808	959	20
50	778	271	595	050	242	284	790	930	10
39° 00′	.6807	.6293	1.589	.8098	1.235	1.287	.7771	.8901	51° 00′
10	836	316	583	146	228	290	753	872	50
20	865	338	578	195	220	293	735	843	40
30	894	361	572	243	213	296	716	814	30
40	923	383	567	292	206	299	698	785	20
50	952	406	561	342	199	302	679	756	10
40° 00′	.6981	.6428	1.556	.8391	1.192	1.305	.7660	.8727	50° 00′
		cos θ	sec θ	cot θ	tan θ	csc θ	sin θ	Radians	Degrees
									Angle θ

Table VI (*continued*)

Angle θ									
Degrees	Radians	sin θ	csc θ	tan θ	cot θ	sec θ	cos θ		
40° 00′	.6981	.6428	1.556	.8391	1.192	1.305	.7660	.8727	50° 00′
10	.7010	450	550	441	185	309	642	698	50
20	039	472	545	491	178	312	623	668	40
30	069	494	540	541	171	315	604	639	30
40	098	517	535	591	164	318	585	610	20
50	127	539	529	642	157	322	566	581	10
41° 00′	.7156	.6561	1.524	.8693	1.150	1.325	.7547	.8552	49° 00′
10	185	583	519	744	144	328	528	523	50
20	214	604	514	796	137	332	509	494	40
30	243	626	509	847	130	335	490	465	30
40	272	648	504	899	124	339	470	436	20
50	301	670	499	.8952	117	342	451	407	10
42° 00′	.7330	.6691	1.494	.9004	1.111	1.346	.7431	.8378	48° 00′
10	359	713	490	057	104	349	412	348	50
20	389	734	485	110	098	353	392	319	40
30	418	756	480	163	091	356	373	290	30
40	447	777	476	217	085	360	353	261	20
50	476	799	471	271	079	364	333	232	10
43° 00′	.7505	.6820	1.466	.9325	1.072	1.367	.7314	.8203	47° 00′
10	534	841	462	380	066	371	294	174	50
20	563	862	457	435	060	375	274	145	40
30	592	884	453	490	054	379	254	116	30
40	621	905	448	545	048	382	234	087	20
50	650	926	444	601	042	386	214	058	10
44° 00′	.7679	.6947	1.440	.9657	1.036	1.390	.7193	.8029	46° 00′
10	709	967	435	713	030	394	173	.7999	50
20	738	.6988	431	770	024	398	153	970	40
30	767	.7009	427	827	018	402	133	941	30
40	796	030	423	884	012	406	112	912	20
50	825	050	418	.9942	006	410	092	883	10
45° 00′	.7854	.7071	1.414	1.000	1.000	1.414	.7071	.7854	45° 00′
		cos θ	sec θ	cot θ	tan θ	csc θ	sin θ	Radians	Degrees
								Angle θ	

ANSWERS

Exercise 1.1 (page 4)

1. {3, 4, 5, 6}   **3.** { }

5. {Sunday, Monday, Tuesday, Wednesday, Thursday, Friday, Saturday}

7. {natural numbers less than 3} = {1, 2}   **9.** {2} ≠ {−2}   **11.** $\emptyset$ ≠ {0}

13. 3 ∈ {2, 3, 4}   **15.** {2} ∉ {2, 3, 4}   **17.** 5 ∉ {4, 5, 6}   **19.** $\emptyset$ ⊂ {4, 5, 6}

21. a. {5, 6, 7} **b.** {5, 6}, {5, 7}, {6, 7} **c.** {5}, {6}, {7} **d.** $\emptyset$

23. a. $A \subset U$ **b.** $C \not\subset A$ **c.** $A \not\subset B$ **d.** $C \subset B$

25. $\{x \mid x$ is an odd natural number$\}$ **27.** $\{x \mid 2^x = 5\}$ **29.** $\{x \mid x \notin A\}$

Exercise 1.2 (page 7)

1. $A' = \{1, 3, 5, 7, 9\} = C$ **3.** $C' = \{2, 4, 6, 8, 10\} = A$

5. $A \cup B = \{1, 2, 3, 4, 5, 6, 8, 10\}$ **7.** $A \cap C = \emptyset$

9. $A' \cup C' = U$ or $\{1, 2, 3, 4, 5, 6, 7, 8, 9, 10\}$ **11.** $A' \cup C = C$ or $\{1, 3, 5, 7, 9\}$

13. 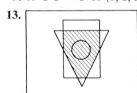   **15.** 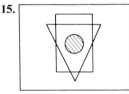   **17.**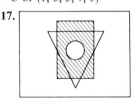

19. 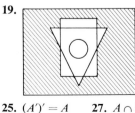   **21.** 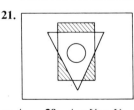   **23.**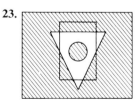

25. $(A')' = A$ **27.** $A \cap U = A$ **29.** $A \cup U = U$

31. $A \cup \emptyset = A$ **33.** $\emptyset' \cap \emptyset = \emptyset$ **35.** $\emptyset \cup \emptyset = \emptyset$

37. a. $A \cup B = \emptyset$ if $A = B = \emptyset$ **b.** $A \cup \emptyset = \emptyset$ if $A = \emptyset$ **c.** $A \cap U = U$ if $A = U$
 d. $A \cup B = A$ if $B \subset A$ **e.** $A \cup \emptyset = U$ if $A = U$ **f.** $A' \cap U = U$ if $A = \emptyset$
 g. $A \cap B = A$ if $A \subset B$ **h.** $A' \cup \emptyset = \emptyset$ if $A = U$ **i.** $A \cup B = A \cap B$ if $A = B$

Exercise 1.3 (page 12)

1. False   **3.** True   **5.** True

7. False   **9.** True   **11.** True

13. {1, 2, 3, 4, 5}   **15.** {1, 2, 3, 4, 5, 6}   **17.** {−9, −8, −7, −6}

19. $\{x \mid x \in N\}$ **21.** $\{x \mid x \in R\}$ **23.** $\{x \mid x \in R, x$ between -4 and $3\}$

25. a. {4}   **b.** {4, −2, 0}   **c.** {4, −2, $\frac{2}{5}$, 0, −$\frac{3}{4}$}   **d.** {4, −2, $\frac{2}{5}$, 0, −$\frac{3}{4}$, $\sqrt{2}$, $\sqrt{7}$}

27. Substitution law for equality, E-4   **29.** Symmetric law for equality, E-2

31. Transitive law for equality, E-3   **33.** Closure law for multiplication, F-6

35. Associative law for multiplication, F-7   **37.** Additive-inverse law, F-4

39. Commutative law for multiplication, F-10   **41.** Commutative law for addition, F-5

43. Distributive law, F-8   **45.** Commutative law for addition, F-5

47. Commutative law for multiplication, F-10   **49.** Distributive law, F-8

51. Closed   **53.** Closed

55. Closed   **57.** Not closed

Exercise 1.4 (page 18)

1. 1.1   **3.** 1.8-V   **5.** 1.11-III   **7.** 1.7   **9.** 1.9

11. 1.11-VII   **13.** 1.8-III   **15.** 1.11-IV   **17.** 1.2   **19.** 1.5

For Problems 21–43, the proofs given are not unique. They are here presented in a precise statement-reason format.

21. 1. $a, b, c \in R$ and $a = b$ 1. Hyp.
2. $ac, bc \in R$ 2. F-6
3. $ac = ac$ and $ca = ca$ 3. E-1
4. $ac = bc$ and $ca = cb$ 4. E-4

23. 1. Let $a + b = a$ and
$a + b' = a$ 1. Hyp.
2. $a + b = a + b'$ 2. E-3
3. $b = b'$ 3. Th. 1.4
4. $a + 0 = a$ 4. F-3
5. $a + 0 = a + b$ 5. E-4
6. $0 = b$ 6. Th. 1.4
7. $b' = 0$ 7. E-4

25. 1. $a + c = b + c$ 1. Hyp.
2. $(a + c) + (^-c) = (b + c) + (^-c)$ 2. Th. 1.1
3. $a + (c + (^-c)) = b + (c + (^-c))$ 3. F-2
4. $a + 0 = b + 0$ 4. F-4
5. $a = b$ 5. F-3

27. 1. $0 + 0 = 0$ 1. F-3
2. $a(0 + 0) = a \cdot 0$ 2. Th. 1.2
3. $a \cdot 0 + a \cdot 0 = a \cdot 0$ 3. F-8
4. $-(a \cdot 0) + [a \cdot 0 + a \cdot 0] = -(a \cdot 0) + a \cdot 0$ 4. Th. 1.1
5. $[-(a \cdot 0) + a \cdot 0] + a \cdot 0 = -(a \cdot 0) + a \cdot 0$ 5. F-2
6. $0 + a \cdot 0 = 0$ 6. F-4
7. $a \cdot 0 = 0$ 7. F-3

29.
1. $(a + b) + [-(a + b)] = 0$ 1. F-4
2. $-a + (a + b) + [-(a + b)] = -a + 0$ 2. Th. 1.1
3. $(-a + a) + b + [-(a + b)] = -a + 0$ 3. F-2
4. $0 + b + [-(a + b)] = -a + 0$ 4. F-4
5. $b + [-(a + b)] = -a$ 5. F-3
6. $-b + (b + [-(a + b)]) = -b + (-a)$ 6. Th. 1.1
7. $(-b + b) + [-(a + b)] = -b + (-a)$ 7. F-2
8. $0 + [-(a + b)] = -b + (-a)$ 8. F-4
9. $-(a + b) = -b + (-a)$ 9. F-3
10. $-(a + b) = (-a) + (-b)$ 10. F-5

31.
1. $b + (-b) = 0$ 1. F-4
2. $(-a)[b + (-b)] = (-a) \cdot 0$ 2. Th. 1.2
3. $(-a) \cdot b + (-a)(-b) = (-a) \cdot 0$ 3. F-8
4. $(-a) \cdot b + (-a)(-b) = 0$ 4. Th. 1.6
5. $-(ab) + ab = 0$ 5. F-4, F-5
6. $-(ab) + (-a)(-b) = 0$ 6. E-4 (Prob. 30)
7. $-(ab) + (-a)(-b) = -(ab) + ab$ 7. E-3
8. $(-a)(-b) = ab$ 8. Th. 1.4

33.
1. $\dfrac{-a}{-b} = -a \cdot \dfrac{1}{-b}$ 1. Def. 1.13

2. $= -\left(a \cdot \dfrac{1}{-b}\right)$ 2. Th. 1.8-III

3. $= -\left(\dfrac{a}{-b}\right)$ 3. Def. 1.13

4. $= -\left(-\dfrac{a}{b}\right)$ 4. Th. 1.8-V

5. $\dfrac{-a}{-b} = \dfrac{a}{b}$ 5. Th. 1.8-I; E-3

35.
1. $abc = abc$ 1. E-1
2. $(ac)b = a(bc)$ 2. Def. 1.11, F-10, F-7

3. $\dfrac{ac}{bc} = \dfrac{a}{b}$ 3. Th. 1.9

37.
1. $\dfrac{a}{b} \cdot \dfrac{c}{d} = \left(a \cdot \dfrac{1}{b}\right)\left(c \cdot \dfrac{1}{d}\right)$ 1. Def. 1.13

2. $= ac \cdot \dfrac{1}{b} \cdot \dfrac{1}{d}$ 2. F-10

3. $= ac \cdot \dfrac{1}{bd}$ 3. Th. 1.11-I

4. $\dfrac{a}{b} \cdot \dfrac{c}{d} = \dfrac{ac}{bd}$ 4. Def. 1.13; E-3

39. 1. $\dfrac{a}{b}+\dfrac{c}{d}=\dfrac{ad}{bd}+\dfrac{bc}{bd}$ 1. Th. 1.10

     2. $\dfrac{a}{b}+\dfrac{c}{d}=\dfrac{ad+bc}{bd}$ 2. Th. 1.11-III

41. 1. $\dfrac{\dfrac{1}{a}}{\dfrac{b}{b}}=\dfrac{1\cdot b}{\dfrac{a}{b}\cdot b}$ 1. Th. 1.10

     2. $=\dfrac{b}{a\cdot\dfrac{1}{b}\cdot b}$ 2. Def. 1.13

     3. $=\dfrac{b}{a}$ 3. F-11

43. (A) 1. $\dfrac{a}{b}=q$ 1. Hyp.

     2. $b\cdot\dfrac{a}{b}=bq$ 2. Th. 1.2

     3. $\left(b\cdot\dfrac{1}{b}\right)a=bq$ 3. Def. 1.13

     4. $a=bq$ 4. F-11

     (B) 1. $a=bq$ 1. Hyp.

     2. $\dfrac{1}{b}\cdot a=\dfrac{1}{b}\cdot bq$ 2. Th. 1.2

     3. $\dfrac{a}{b}=\left(\dfrac{1}{b}\cdot b\right)q$ 3. Def. 1.13

     4. $\dfrac{a}{b}=q$ 4. F-11

Exercise 1.5 (page 22)

1. I   **3.** III   **5.** II   **7.** IV   **9.** $7>3$ **11.** $-4<-3$ **13.** $-1\le x\le1$

15. $x>0$ **17.** $x\ge0$ **19.** 3   **21.** 7   **23.** -2

25. $3x$ if $x\ge0,$ $-3x$ if $x<0$ **27.** $x+1$ if $x\ge-1,$ $-(x+1)$ if $x<-1$

29. $y-3$ if $y\ge3,$ $-(y-3)$ if $y<3$ **31.** $|-2|<|-5|$ **33.** $-7<|-1|$

35. $|-3|>0$ **37.** $2<5$ **39.** $7<8$ **41.** $|x|<3$

Exercise 2.1 (page 27)

1. 3 in y **3.** 4 in x,y; 3 in x; 2 in y **5.** Not a polynomial   **7.** $-1, 1, -2, 20$

9. $1, 1, 0, 1, 0$ **11.** $1, 1$ **13.** $39, 6$ **15.** $2x+1, 4x-5$

17. $3x^2-4x+2, -x^2+4$ **19.** $3x^2-5x+5$ **21.** $3x^2-4x+4$

23. x^2-2x+6 **25.** n, n

Exercise 2.2 (page 30)

1. $-6x^3y^4$ **3.** $-8x^3+4x^2-4x$ **5.** $a^2bc-ab^2c+2abc^2$ **7.** $x^2+7x+10$

9. $x^2-4xy+4y^2$ **11.** $10x^2+17x+3$ **13.** $9a^2-4b^2$ **15.** x^3+6x^2+7x-4

17. $2x^2+8x+6$ **19.** $-2ac+6ad+bc-3bd$ **21.** a^4-ab^3 **23.** $4a+4$

25. $20x^2+4x$ **27.** $x^2-5x+11, x^2-x+5$

29. $x^4+6x^3+9x^2, a^8-6a^6+8a^4+3a^2$ **31.** $2ab; 0>2ab; 0<2ab$

Exercise 2.3 (page 33)

1. $xy(y + x)$ **3.** $3x^3y(3x^2 - x + 2)$ **5.** $(x - 6)(x - 2)$ **7.** $(x - 5)(x + 5)$

9. $(4x - 3)(3x + 1)$ **11.** $3(x + 2)(x + 2)$ **13.** $(y^2 + 2)(y^2 + 1)$

15. $(2a^2 + 1)(a + 1)(a - 1)$ **17.** $(3x - y)(y - x)[x^2 + (y - 2x)^2]$

19. $(x + y)(x + a)$ **21.** $(a - 2b)(a^2 + 2b^2)$ **23.** $(y - 3x)(y^2 + 3xy + 9x^2)$

25. $(2x - y)(x^2 - xy + y^2)$ **27.** $(a^n - 2)(a^n + 2)$ **29.** $(x^n - y^n)(x^n + y^n)(x^{2n} + y^{2n})$

31. $(3x^{2n} - 1)(x^{2n} - 3)$ **33.** $2(y^n - 30)(y^n + 24)$

35. $ac - ad + bd - bc = a(c - d) - b(c - d) = (a - b)(c - d)$;
$ac - ad + bd - bc = bd - ad - bc + ac = d(b - a) - c(b - a) = (b - a)(d - c)$

37. $(x^2 + xy + y^2)(x^2 - xy + y^2)$

Exercise 2.4 (page 41)

1. $4ay^2$ $(a, y \neq 0)$ **3.** xyz $(x, y \neq 0)$ **5.** $2x + 1$ $(x \neq 0)$

7. $4a^2 + 2a + 2$ $(a \neq 0)$ **9.** $x^2 - 4x - 3$ $(x \neq 0)$ **11.** $3x^2 - 2x + \dfrac{3}{5x}$ $(x, y \neq 0)$

13. $2y^2 - y + \dfrac{13}{2} + \dfrac{-3/2}{2y + 1}$ $\left(y \neq -\dfrac{1}{2}\right)$ **15.** $2y^3 - y^2 - \dfrac{1}{2}y - \dfrac{5}{4} + \dfrac{-13y/4 + 9/4}{2y^2 + y + 1}$

17. $x^3 - x^2 - \dfrac{1}{x - 2}$ $(x \neq 2)$ **19.** $2x^2 - 2x + 3 + \dfrac{-8}{x + 1}$ $(x \neq -1)$

21. $2x^3 + 10x^2 + 50x + 249 + \dfrac{1251}{x - 5}$ $(x \neq 5)$ **23.** $x^2 + 2x - 3 + \dfrac{4}{x + 2}$ $(x \neq -2)$

25. $x^5 + x^4 + 2x^3 + 2x^2 + 2x + 1 + \dfrac{1}{x - 1}$ $(x \neq 1)$ **27.** $x^4 + x^3 + x^2 + x + 1$ $(x \neq 1)$

29. 3, 17, 47   **31.** 56, 12, 326   **33.** $-685, 95, 719$

Exercise 2.5 (page 45)

1. $\dfrac{a}{b}$ $(a, b, c \neq 0)$ **3.** 2 $(x + y \neq 0)$ **5.** -1 $(b - a \neq 0)$ **7.** $-(x + 1)$ $(x \neq 1)$

9. $x^2 - 2x - \dfrac{3}{2}$ $(x \neq 0)$ **11.** $y + 7$ $(y \neq 2)$ **13.** $-(2y + 5)$ $\left(y \neq \dfrac{1}{2}\right)$

15. $\dfrac{x^2 + xy + y^2}{x + y}$ $(x^2 - y^2 \neq 0)$ **17.** $\dfrac{y^2 + 4}{y^2 + 3}$ $(y^2 - 4 \neq 0)$

19. $\dfrac{x - y + 1}{-(x + y)}$ $(x \neq y, x \neq -y)$ **21.** $\dfrac{9}{12}$ **23.** $\dfrac{ab^2}{a^2b}$ $(a, b \neq 0)$

25. $\dfrac{3(y - 3)}{y^2 - y - 6}$ $(y \neq -2, 3)$ **27.** $\dfrac{3(a^2 - 3a + 9)}{a^3 + 27}$ $(a \neq -3)$

29. Yes; $x = 2$ **31.** For $N > 0$: $a - b < 0$ or $a < b$;   for $N < 0$: $a - b > 0$ or $a > b$

Exercise 2.6 (page 48)

1. $\dfrac{2x-1}{y}$ **3.** 1 **5.** $\dfrac{2a+9}{9}$ **7.** $\dfrac{x+7}{30}$ **9.** $\dfrac{5}{2a+2b}$ **11.** $-\dfrac{1}{2(y-3)}$

13. $\dfrac{3}{3-x}$ **15.** $\dfrac{-1}{(a+2)(a+3)}$ **17.** $\dfrac{-3(x^2+y^2)}{(2x-y)(x-2y)}$ **19.** $\dfrac{y(y-11)}{(y-4)(y+4)(y-1)}$

21. $\dfrac{x^3-2x^2+2x-2}{(x-1)^2}$ **23.** $\dfrac{4x^3-4x^2+4x+4}{(2x+1)(2x-1)}$ **25.** $\dfrac{x(5x-11)}{(x-4)(x-1)^2}$

27. $\dfrac{4}{(y+1)(y+3)}$ **29.** $\dfrac{y^2+xz}{(y-x)(y-z)}$

Exercise 2.7 (page 51)

1. $\dfrac{-b^2}{a}$ **3.** $\dfrac{1}{ax^2y}$ **5.** 5 **7.** $\dfrac{a^2}{2b}$ **9.** $\dfrac{(x+1)(x-3)}{2(x-1)}$ **11.** $\dfrac{2x}{x+5}$

13. $(5ab-4)(4ab+3)$ **15.** $x-y$ **17.** $\dfrac{x^2-1}{x^2}$ **19.** $\dfrac{x+5}{x(x+1)}$

21. $\dfrac{2y^2-6y+3}{2}$ **23.** $\dfrac{7}{10a+2}$ **25.** $\dfrac{4a^2-3a}{4a+1}$ **27.** $\dfrac{-1}{y-3}$ **29.** $\dfrac{a-2b}{a+2b}$

Exercise 2.8 (page 57)

1. 2 **3.** $\dfrac{1}{27}$ **5.** 32 **7.** $\dfrac{8}{27}$ **9.** x^2 **11.** $a^{17/12}$ **13.** $x^{1/3}$ **15.** $\dfrac{1}{x^{16/15}}$

17. $\dfrac{y^2}{x}$ **19.** $\left(\dfrac{x}{y}\right)^{5/4}$ **21.** $x^n \cdot y^4$ **23.** $x^{3n/2}$ **25.** $x^{5n/2} \cdot y^{3m/2-1}$ **27.** $x - x^{2/3}$

29. $x - 2x^{1/2}y^{-1/2} + y^{-1}$ **31.** $x + y - (x+y)^{3/2}$ **33.** $x + y$ **35.** $x(x^{1/2}+1)$

37. $x^{-1/2}(x^{-1}+1)$ **39.** $y^{-3/2}(1-y^{5/2})$ **41.** 5 **43.** $2|x|$ **45.** $\dfrac{2}{|x|(x+1)^{1/2}}$

Exercise 2.9 (page 61)

1. $\sqrt[3]{a^2}$ **3.** $3\sqrt[3]{x}$ **5.** $-6\sqrt[3]{(xy)^2}$ **7.** $\sqrt[7]{(x-y)^4}$ **9.** $x^{2/3}$ **11.** $(2xy^2)^{1/5}$

13. $-3(a^3b)^{1/4}$ **15.** $3(x^2-y)^{1/3}$ **17.** 12 **19.** -3 **21.** x^2y **23.** $\dfrac{2}{3}x^3y^5$

25. $2x^2\sqrt{x}$ **27.** $xy\sqrt[4]{3xy}$ **29.** $3\sqrt[4]{3}$ **31.** $\sqrt{3}$ **33.** $\sqrt[3]{2b}$ **35.** xy **37.** b

39. $\dfrac{\sqrt{3}}{3}$ **41.** $\dfrac{\sqrt{2x}}{2}$ **43.** $\dfrac{\sqrt[3]{3x}}{x}$ **45.** $\dfrac{1}{\sqrt{7}}$ **47.** $\dfrac{x}{\sqrt{xy}}$ **49.** $3\sqrt{3}$ **51.** $-4\sqrt{2}$

53. $5\sqrt[3]{2}$ **55.** $3\sqrt{2}+\sqrt{6}$ **57.** $1-\sqrt{5}$ **59.** $x - \sqrt{3x} - 6$ **61.** $2(1-\sqrt{3})$

63. $\dfrac{x(\sqrt{x}+3)}{x-9}$ **65.** $\dfrac{-1}{2(1+\sqrt{2})}$ **67.** $2|x|$ **69.** $3|x|\sqrt{x-1}$

Exercise 3.1 (page 67)

1. $\{-2\}$ **3.** $\{1\}$ **5.** $\{-7\}$ **7.** $\{15\}$ **9.** $\emptyset$ **11.** $\left\{-\dfrac{14}{5}\right\}$ **13.** $k = v - gt$

15. $c = \dfrac{2A - bh}{h}$ **17.** $n = \dfrac{l - a + d}{d}$ **19.** $y' = \dfrac{3x + 1}{x^2 - 2y^3}$ **21.** $x_1 = \dfrac{x_4}{x_2 - 2x_3}$

23. $y = 6(x - x_1) + y_1$ **25.** $-\dfrac{3}{5}$ **27.** 6 in. **29.** \$600 at 7%, \$1400 at 5%

31. 32 and 64 mph

Exercise 3.2 (page 72)

1. $\{0, -2\}$ **3.** $\{2, -7\}$ **5.** $\left\{\dfrac{1}{2}, 1\right\}$ **7.** $\left\{1, -\dfrac{10}{3}\right\}$ **9.** $\{2, -2\}$

11. $\{\sqrt{5}, -\sqrt{5}\}$ **13.** $\{6 + \sqrt{5}, 6 - \sqrt{5}\}$ **15.** $\{2, -6\}$ **17.** $\{-4, -5\}$

19. $\left\{\dfrac{1}{2}, -2\right\}$ **21.** $(x - 2)^2 + (y - 2)^2 = 5^2$ **23.** $[x - (-3)]^2 + (y - 1)^2 = 2^2$

25. $\left(x - \dfrac{1}{2}\right)^2 + [y - (-1)]^2 = 2^2$ **27.** $\{1, 2\}$ **29.** $\left\{2, \dfrac{3}{2}\right\}$ **31.** $\left\{3, -\dfrac{3}{2}\right\}$

33. $\{\sqrt{5}\}$ **35.** $\left\{\sqrt{3}, -\dfrac{\sqrt{3}}{2}\right\}$ **37.** $\{2k, -k\}$ **39.** $\left\{\dfrac{1 + \sqrt{1 - 4ac}}{2a}, \dfrac{1 - \sqrt{1 - 4ac}}{2a}\right\}$

41. 4 **43.** $k < -2$ **45.** 6, 7 **47.** 12 in. **49.** $2\frac{1}{2}$ sec, $\dfrac{5}{4}\sqrt{6}$ sec

Exercise 3.3 (page 77)

1. $\{64\}$ **3.** $\{-25\}$ **5.** $\{4\}$ **7.** $\{5\}$ **9.** $\{16\}$ **11.** $\{4\}$ **13.** $\{1, 3\}$

15. $A = \pi r^2$ **17.** $y = \dfrac{1}{x^3}$ **19.** $y = \sqrt{a^2 - x^2}$ **21.** $\{25\}$ **23.** $\left\{\dfrac{\sqrt{2}}{2}, -\dfrac{\sqrt{2}}{2}\right\}$

25. $\{2, -7, -3, -2\}$ **27.** $\{64, -8\}$ **29.** $\left\{\dfrac{1}{4}, -\dfrac{1}{3}\right\}$

Exercise 3.4 (page 82)

1. $\{x \mid x > 1\}$ **3.** $\{x \mid x > 3\}$ **5.** $\left\{x \mid x \le \dfrac{13}{2}\right\}$

7. $\{x \mid -2 < x < 5\}$

9. $\{x \mid x \geq 13\}$

11. $\{x \mid x < -1 \text{ or } x > 2\}$

13. $\{x \mid x < -1 \text{ or } x > 4\}$

15. $\{x \mid -\sqrt{5} < x < \sqrt{5}\}$

17. $\left\{x \mid x < 0 \text{ or } x \geq \dfrac{1}{2}\right\}$

19. $\left\{x \mid -\dfrac{8}{3} < x < -2\right\}$

21. $\{x \mid x < 0 \text{ or } 2 < x \leq 4\}$

23. $\{x \mid -3 < x < 0 \text{ or } x > 2\}$

25. At least 48% but less than 98%.

Exercise 3.5 (page 86)

1. $\{6, -6\}$ **3.** $\{-3, 5\}$ **5.** $\left\{1, \dfrac{1}{3}\right\}$

7. $\{x \mid -2 < x < 2\}$ **9.** $\{x \mid x \leq -7 \text{ or } x \geq 1\}$ **11.** $\{x \mid x \leq 1 \text{ or } x \geq 4\}$

13. $|x - 2| < 1$ **15.** $|x + 8| \leq 1$ **17.** $|4x - 5| \leq 19$

19. (I) If $a \geq 0$, then $-a \leq 0$ and, by Def. 1.15, $|-a| = -(-a) = a$; also $|a| = a$;
hence $|-a| = |a|$.
(II) If $a < 0$, then $-a > 0$ and, by Def. 1.15, $|-a| = -a$; also $|a| = -a$;
hence $|-a| = |a|$.

21. Since $a^2 \geq 0$ for all $a \in R$, $|a^2| = a^2$(Def. 1.15). If $a \geq 0$, then $|a| = a$, so $|a|^2 = a^2$;
while if $a < 0$, then $|a| = -a$, so $|a|^2 = (-a)(-a) = a^2$. Hence $|a^2| = |a|^2 = a^2$.

Exercise 4.1 (page 92)

1. $\left\{(0, 6), (1, 4), (2, 2), (-3, 12), \left(\dfrac{2}{3}, \dfrac{14}{3}\right)\right\}$

3. $\left\{(0, 0), (1, -3), (2, 3), \left(-3, -\dfrac{9}{7}\right), \left(\dfrac{2}{3}, -\dfrac{9}{7}\right)\right\}$

5. $\left\{(0, \sqrt{11}), (1, \sqrt{14}), (2, \sqrt{17}), (-3, \sqrt{2}), \left(\dfrac{2}{3}, \sqrt{13}\right)\right\}$

7. (a) Domain: $\{2, 5, 7\}$, Range: $\{3, 7, 8\}$ (b) A function

9. (a) Domain: $\{2, 3\}$, Range: $\{-1, 4, 6\}$ (b) Not a function

11. (a) Domain: $\{5, 6, 7\}$, Range: $\{5, 6, 7\}$ (b) A function

13. Domain: $\{x \mid x \in R\}$ **15.** Domain: $\{x \mid x \in R\}$ **17.** Domain: $\{x \mid x \in R, x \neq 2\}$

19. Domain: $\{x \mid x \in R, x \geq 0\}$ **21.** Domain: $\{x \mid x \in R, -2 \leq x \leq 2\}$

23. Domain: $\{x \mid x \in R, x \neq 0, 1\}$ **25.** Yes   **27.** Yes   **29.** No   **31.** No
33. 2   **35.** -1 **37.** 9   **39.** a^2 **41.** 1   **43.** $a - 2$ **45.** ± 1 **47.** ± 3
49. a. 2   **b.** 0   **c.** 2   **d.** x **51. a.** even   **b.** odd   **53.** $x = y + 2$ or $y = x - 2$; yes

55. $2y - 3x = 6$ or $y = \dfrac{3}{2}x + 3$; yes   **57.** $x = y^2 + 3$ or $y = \pm\sqrt{x + 3}$; no

Exercise 4.2 (page 101)

1.

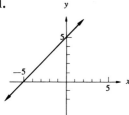

3.

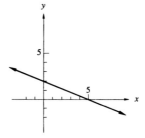

5.

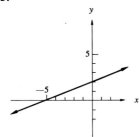

7.

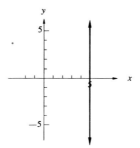

9.

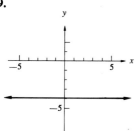

11. Distance, 5; slope, $\dfrac{4}{3}$ **13.** Distance, 13; slope $\dfrac{12}{5}$ **15.** Distance, $\sqrt{2}$; slope, 1

17. $7, 2\sqrt{17}, \sqrt{89}$ **19.** 10, 21, 17   **21.** $x - 2y - 1 = 0$ **23.** $x - y + 1 = 0$

25. $y - 2 = 0$ **27.** $y = -\dfrac{3}{2}x + \dfrac{1}{2}$; slope, $-\dfrac{3}{2}$; intercept, $\dfrac{1}{2}$

29. $y = \dfrac{1}{3}x - \dfrac{2}{3}$; slope, $\dfrac{1}{3}$; intercept, $-\dfrac{2}{3}$ **31.** $y = \dfrac{8}{3}x$; slope, $\dfrac{8}{3}$; intercept, 0

33. $x - 2y = 0$

35. Since $m = \dfrac{y_2 - y_1}{x_2 - x_1}$ and $y - y_1 = m(x - x_1)$, we have by substitution

$$y - y_1 = \left(\dfrac{y_2 - y_1}{x_2 - x_1}\right)(x - x_1).$$

37. $F(x) = \dfrac{-x + 11}{3}$ **39. a.** $x - 3y - 3 = 0$ **b.** $x - y + 2 = 0$

41. Let $M(x, y)$ be midpoint of P_1P_2; then from triangles P_1AM and P_1BP_2,

$$\dfrac{x - x_1}{x_2 - x_1} = \dfrac{P_1M}{P_1P_2} = \dfrac{1}{2} \quad \text{and} \quad x = \dfrac{x_1 + x_2}{2};$$

similarly, $\dfrac{y - y_1}{y_2 - y_1} = \dfrac{1}{2}$ and $y = \dfrac{y_1 + y_2}{2}$.

43.

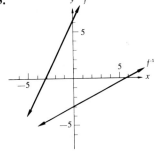

45.

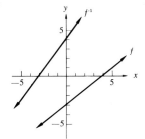

Exercise 4.3 (page 105)

1. a. $3x + y = 16$ **b.** $x - 3y = 2$ **3. a.** $3x - 2y = 0$ **b.** $2x + 3y = 0$

5. a. $2x + 3y = -7$ **b.** $3x - 2y = -4$ **7. a.** $x + 2y = 5$ **b.** $2x - y = -10$

9. $x + y = 12$ **11.** $7x - 3y = -8$ **13.** $ax - by = \dfrac{1}{2}(a^2 - b^2)$

15. Slope of $\overline{AB}$: $m_1 = -1$; slope of $\overline{AC}$: $m_2 = 1$; $m_1 m_2 = -1$; thus, by Th. 1.2,

$\overline{AB} \perp \overline{AC}$. Hence, $\triangle ABC$ is a right triangle.

17. Slope of $\overline{PQ}$: $m_4 = \dfrac{1}{3}$; slope of $\overline{QR}$: $m_2 = \dfrac{-2}{3}$; slope of $\overline{RS}$: $m_3 = \dfrac{1}{3}$;

slope of $\overline{SP}$: $m_4 = \dfrac{-2}{3}$; $m_1 = m_3$ and $m_2 = m_4$, thus, by Th. 1.1, $\overline{PQ}$ is parallel to

$\overline{RS}$, and $\overline{QR}$ is parallel to $\overline{SP}$. Hence, $PQRS$ is a parallelogram.

19. From geometry, if L_1 is parallel to L_2, then $\angle P_3 P_1 Q_1$ is congruent to $\angle P_4 P_2 Q_2$.
Also, $\angle P_1 Q_1 P_3$ is congruent to $\angle P_2 Q_2 P_4$, since each is a right angle. Hence, $\triangle P_1 Q_1 P_3$
is similar to $\triangle P_2 Q_2 P_4$. Because the lengths of corresponding sides of similar triangles

are proportional, it follows that $\dfrac{Q_1 P_3}{P_1 Q_1} = \dfrac{Q_2 P_4}{P_2 Q_2}$.

But $\dfrac{Q_1 P_3}{P_1 Q_1} = \dfrac{y_3}{x_3 - x_1} = m_1$ and $\dfrac{Q_2 P_4}{P_2 Q_2} = \dfrac{y_4}{x_4 - x_2} = m_2$. Hence, $m_1 = m_2$.

21. $\dfrac{y_1 - y_2}{x_2 - x_1} = -\dfrac{y_2 - y_1}{x_2 - x_1} = -m_2$; $\dfrac{x_2 - x_1}{y_3 - y_1} = \dfrac{1}{m_1}$.

Hence, $-m_2 = \dfrac{1}{m_1}$, or $m_1 m_2 = -1$.

Exercise 4.4 (page 111)

1. $(4, 0)$; $(1, 0)$; $\left(\dfrac{5}{2}, -\dfrac{9}{4}\right)$, minimum   **3.** $(7, 0)$; $(-1, 0)$; $(3, -16)$, minimum

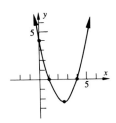

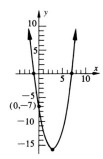

5. $(4, 0)$; $(1, 0)$; $\left(\dfrac{5}{2}, \dfrac{9}{4}\right)$, maximum   **7.** No intercepts; $(0, 2)$, minimum

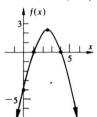   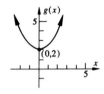

9.   **11.** 4; 4

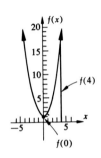

13.   **15.**

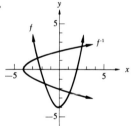   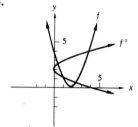

17. Varying k has the effect of translating the graph along the y-axis.

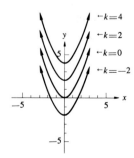

Exercise 4.5 (page 116)

1. Circle

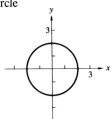

3. Ellipse

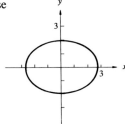

5. Hyperbola   **7.** Pair of straight lines through origin

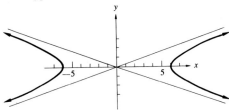

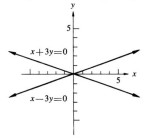

9. A point (the origin)   **11.** The null set, Ø

13. Two parallel lines   **15.** The points of intersection are exactly
                                                                those points common to both graphs.

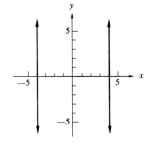

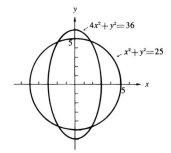

Exercise 4.6 (page 119)

1. $x^2 + y^2 - 2x = 0$ **3.** $x^2 + y^2 + 6x - 2y + 6 = 0$

5. $x^2 + y_1^2 - 2hx - 2ky + h^2 + k^2 - r^2 = 0$ **7.** $x^2 + y^2 - 6x - 10y + 9 = 0$

9. $x - y = 3$ **11.** $3x^2 + 3y^2 - 46x + 131 = 0$ **13.** $y^2 - 8x = 0$

15. $x^2 + 6x + 4y + 13 = 0$ **17.** $3x^2 + 4y^2 = 48$

19. $25x^2 + 16y^2 - 200x - 160y + 400 = 0$ **21.** $3x^2 - y^2 = 3$

23. $16y^2 - 9x^2 + 36x - 180 = 0$ **25.** $x^2 - 2py + p^2 = 0$

Exercise 4.7 (page 122)

1.
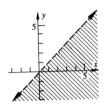

3.
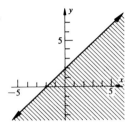

5.
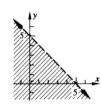

7.
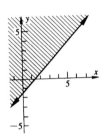

9.
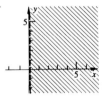

11.
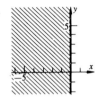

13.
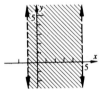

15.
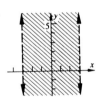

17.

19.
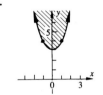

21.
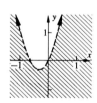

23.
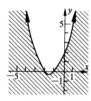

25.
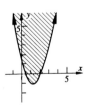

27.
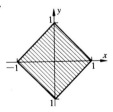

Exercises 4.8 (page 127)

1. a. $(2, 1, 1)$ **b.** $(-1, 0, 3)$ **c.** $(6, 2, -2)$

3. a. $(4, 0, -16)$ **b.** $(0, \sqrt{2}, -4), (0, -\sqrt{2}, -4)$ **c.** No point exists.

5. a. -1 **b.** -4 **c.** -2 **7. a.** $(0, 0, 2)$ **b.** $(0, 3, 0)$ **c.** $(6, 0, 0)$

9. a. $(0, 0, 2)$, $(0, 0, -2)$ **b.** $(0, 2, 0)$, $(0, -2, 0)$ **c.** $(\sqrt{2}, 0, 0)$, $(-\sqrt{2}, 0, 0)$

11. a. No point exists.   **b.** $(0, -9, 0)$ **c.** $(3, 0, 0)$, $(-3, 0, 0)$

13. a. $(0, 0, 0)$ **b.** $(0, 0, 0)$ **c.** $(0, 0, 0)$ **15.** $(2, -2, z)$

17. $(\sqrt{13}, y, 2)$, $(-\sqrt{13}, y, 2)$ **19.** $(x, 3, z)$ **21.** $(-2, y, z)$

23. 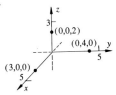   **25.**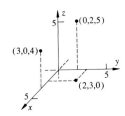

27. 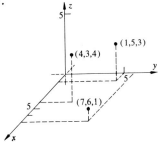   **29.**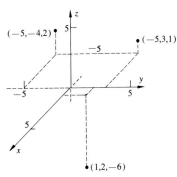

31. $\sqrt{19}$ **33.** $\sqrt{14}$ **35.** $\sqrt{22}$

Exercise 4.9 (page 134)

1. 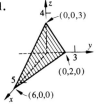   **3.** 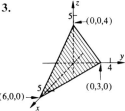   **5.**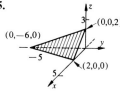

7. 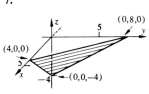   **9.** 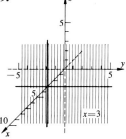   **11.**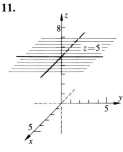

13.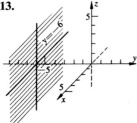

15.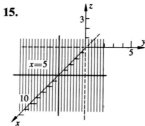

17.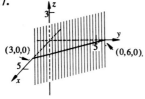

19.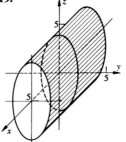

21.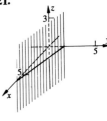

23.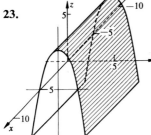

25.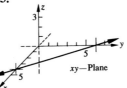

27.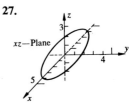

29.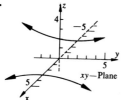

31. No trace exists.

33.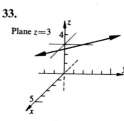

35.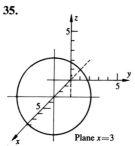

37.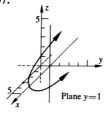

39.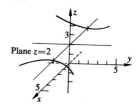

41.

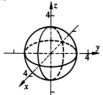

43.

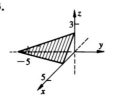

45.

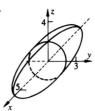

47.

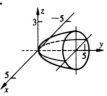

Exercise 5.1 (page 139)

1.

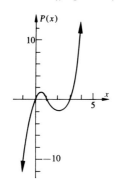

3.

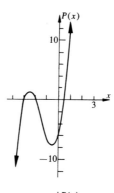

5.

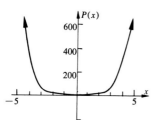

7.

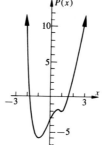

Exercise 5.2 (page 143)

1. 2 positive; 2 negative   **3.** No positive; 1 negative   **5.** No positive; no negative

7. Upper bound 3; lower bound -4 **9.** Upper bound 4; lower bound -3

11. Upper bound 2; lower bound -3 **13.** Upper bound 1; lower bound -2

21.

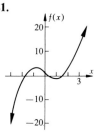

23.

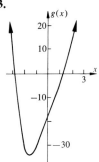

25.

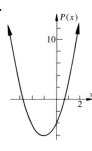

Exercise 5.3 (page 146)

1. $-2, 1, 2$ **3.** $-1, 1$ **5.** $-4, -1, 1, 2$ **7.** $-\dfrac{1}{2}, 2, \dfrac{5}{2}$ **9.** $-\dfrac{2}{3}, \dfrac{1}{2}, \dfrac{5}{2}$

11. $-2, -\dfrac{1}{2}, \dfrac{1}{2}, 3$

13.

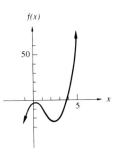

15.

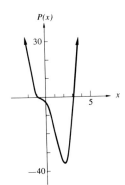

17.

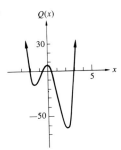

Exercise 5.4 (page 149)

1. 1.5   **3.** 2.1   **5.** 1.7   **7.** 0.7   **9.** 1.71

Exercise 5.5 (page 154)

1. $x = 3$ **3.** $x = 3; x = -2$ **5.** $x = -1; x = -4$ **7.** $x = -1$

9.

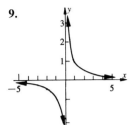

11.

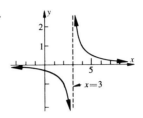

13.

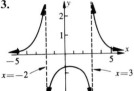

15.

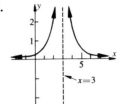

17. $x = 2; x = -2; y = 0$ **19.** $x = 4; y = x + 4$ **21.** $x = 4; x = -1; y = 1$

23.

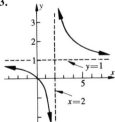

25.

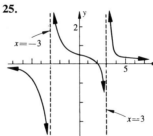

27.

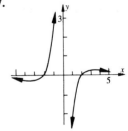

29.

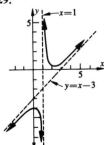

31.

Exercise 6.1 (page 159)

1. $(0, 1), (1, 3), (2, 9)$ **3.** $(0, -1), (1, -5), (2, -25)$ **5.** $(-3, 8), (0, 1), \left(3, \dfrac{1}{8}\right)$

7. $\left(-2, \dfrac{1}{100}\right)$, $(1, 10)$, $(0, 1)$

9.

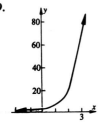

11.

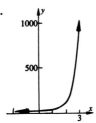

13.

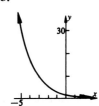

15.

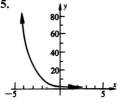

17.

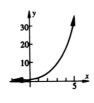

19.

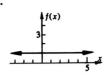

No; a constant function

21. a. $\{-2\}$ **b.** $\{-4\}$ **c.** $\left\{\dfrac{3}{4}\right\}$

Exercise 6.2 (page 162)

1. $\log_4 16 = 2$ **3.** $\log_3 27 = 3$ **5.** $\log_{1/2} \dfrac{1}{4} = 2$ **7.** $\log_8 \dfrac{1}{2} = -\dfrac{1}{3}$

9. $\log_{10} 100 = 2$ **11.** $\log_{10}(0.1) = -1$ **13.** $2^6 = 64$ **15.** $3^2 = 9$

17. $\left(\dfrac{1}{3}\right)^{-2} = 9$ **19.** $10^3 = 1000$ **21.** $10^{-2} = 0.01$ **23.** 2 **25.** 3 **27.** $\dfrac{1}{2}$

29. -1 **31.** 1 **33.** 2 **35.** -1 **37.** $\{2\}$ **39.** $\{2\}$ **41.** $\{64\}$ **43.** $\{-3\}$

45. $\{100\}$ **47.** $\{4\}$

49. By definition, $\log_b 1$ is a number such that $b^{\log_b 1} = 1$; therefore $\log_b 1 = 0$.

51. $\log_b b^x = x$ implies $b^x = b^x$, which is true for all $x \in R$. **53.** $\log_b x + \log_b y$

55. $\log_b x - \log_b y$ **57.** $5 \log_b x$ **59.** $\dfrac{1}{3} \log_b x$ **61.** $\dfrac{1}{2}(\log_b x - \log_b z)$

63. $\dfrac{1}{3}(\log_{10} x + 2 \log_{10} y - \log_{10} z)$ **65.** $\log_{10} 2 + \log_{10} \pi + \dfrac{1}{2} \log_{10} l - \dfrac{1}{2} \log_{10} g$

67. $\log_b xy$ **69.** $\log_b x^2 y^3$ **71.** $\log_b \dfrac{x^3 y}{z^2}$ **73.** $\log_{10} \dfrac{x(x-2)}{z^2}$

75. By Th. 6.3-III, $\dfrac{1}{4} \log_{10} 8 = \dfrac{1}{4} \log_{10} 2^3 = \dfrac{3}{4} \log_{10} 2$;

hence, $\dfrac{1}{4} \log_{10} 8 + \dfrac{1}{4} \log_{10} 2 = \dfrac{3}{4} \log_{10} 2 + \dfrac{1}{4} \log_{10} 2 = \log_{10} 2$.

Exercise 6.3 (page 169)

1. 2 **3.** -3 or $7-10$ **5.** -4 or $6-10$ **7.** 4 **9.** 0.8280
11. $9.9101-10$ **13.** $8.9031-10$ **15.** 2.3945 **17.** 4.10 **19.** 3.67
21. 0.0642 **23.** 5480 **25.** 0.000718 **27.** 0.6246 **29.** 3.1824 **31.** 4.5695
33. $9.7095-10$ **35.** 3.225 **37.** 10.52 **39.** 0.05075 **41.** 0.7495
43. $\log_{10} 3.751$; 0.751 is closer to a tabulated value **45.** 9.1 **47.** 5000 **49.** 113

Exercise 6.4 (page 175)

1. 4.014 **3.** 2.299 **5.** 0.000461 **7.** 64.34 **9.** 2.010 **11.** 3.435×10^{-10}
13. 0.04582 **15.** 0.2776 **17.** 9.872 **19.** 4.746 **21.** 1.394 **23.** 3.484
25. $\left\{\dfrac{\log_{10} 7}{\log_{10} 2}\right\}$ **27.** $\left\{\dfrac{\log_{10} 8}{\log_{10} 3} - 1\right\}$ **29.** $\left\{\dfrac{1}{2}\left(\dfrac{\log_{10} 3}{\log_{10} 7} + 1\right)\right\}$
31. $\left\{\sqrt{\dfrac{\log_{10} 15}{\log_{10} 4}}, -\sqrt{\dfrac{\log_{10} 15}{\log_{10} 4}}\right\}$ **33.** $\left\{\dfrac{-1}{\log_{10} 3}\right\}$ **35.** $\left\{1 - \dfrac{\log_{10} 15}{\log_{10} 3}\right\}$
37. $n = \dfrac{\log_{10} y}{\log_{10} \dot{x}}$ **39.** $t = \dfrac{\log_{10} y}{k \log_{10} e}$ **41.** 1.343 **43.** 5% **45.** 2.5%
47. 20 yrs **49.** 12 yrs **51.** $7400, $7430 **53.** 7 **55.** 7.7 **57.** 6.2
59. 1.0×10^{-3} **61.** 2.5×10^{-6} **63.** 6.3×10^{-8} **65.** 1.11 sec **67.** 12.05 gr
69. 30.0 in., 16.1 in.

Exercise 6.5 (page 181)

1. 3.32 **3.** 3.41 **5.** 1.08 **7.** 0.79 **9.** 1.0986 **11.** 2.8332 **13.** 5.7900
15. 6.1093 **17.** 1.6487 **19.** 29.964 **21.** 1.260 **23.** 0.08607 **25.** 0.0821
27. 0.7600 **29.** $\dfrac{1}{3}$ **31.** 2.10 **33.** 2.86

35. Since $\log_{10} 4 = \log_{10} 2^2 = 2 \log_{10} 2$ and $\log_2 10 = \dfrac{\log_{10} 10}{\log_{10} 2}$ and $\log_{10} 10 = 1$,

$(\log_{10} 4 - \log_{10} 2)(\log_2 10) = (2 \log_{10} 2 - \log_{10} 2)\left(\dfrac{1}{\log_{10} 2}\right) = 1.$

Exercise 7.1 (page 186)

1. $\dfrac{\pi}{3}$, pos. **3.** $\dfrac{3\pi}{7}$, pos. **5.** $\dfrac{\pi}{5}$, pos. **7.** $\dfrac{2\pi}{3}$, pos. **9.** $\dfrac{\pi}{4}$, pos. **11.** $\dfrac{4}{5}$, I
13. $-\dfrac{12}{13}$, IV **15.** $\dfrac{\sqrt{5}}{3}$, IV **17.** $-\dfrac{1}{2}$, III **19.** $\dfrac{4}{5}$, IV **21.** $\cos(\pi - s) = -\cos s$

23. $\sin\left(\dfrac{\pi}{2}+s\right)=\cos s$

25. Because the circle is symmetric to the horizontal axis, it follows that if $\dfrac{\pi}{2}<s<\pi$, so that $(\cos s,\sin s)=(-x,y)$. Then $(\cos(-s),\sin(-s))=(-x,-y)$, from which the theorem follows.

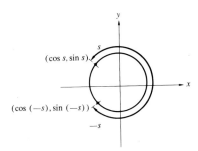

Exercise 7.2 (page 193)

1. $\dfrac{1}{\sqrt{2}}$ **3.** $-\dfrac{1}{2}$ **5.** 1 **7.** -1 **9.** 0 **11.** 0 **13.** $\dfrac{1}{\sqrt{2}}$ **15.** $-\dfrac{1}{2}$

17. $\dfrac{\pi}{4}$ **19.** $\dfrac{5\pi}{6}$ **21.** $\dfrac{2\pi}{3}$ **23.** $\dfrac{4\pi}{3}$ **25.** $\dfrac{\pi}{4}<s\le\pi$ **27.**

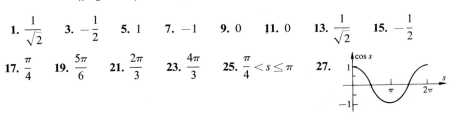

29. Suppose that $0<a<2\pi$. Then if a is a period of the cosine function,
$$\cos(s+a)=\cos s,\text{ for all }s\in R.$$
Let $s=0$; then
$$\cos(s+a)=\cos(0+a)=\cos a=\cos 0=1.$$
But $\cos a$ must correspond to the first coordinate of a point on the unit circle, and the only point with first coordinate 1 is $(1,0)$. This implies $a=2k\pi$, contrary to hypothesis.

31. Let $f(s)=\cos s+\sin s$. Now we know that $\cos s=\cos(s+2\pi)$ and $\sin s=\sin(s+2\pi)$, so that
$$f(s)=\cos s+\sin s=\cos(s+2\pi)+\sin(s+2\pi)=f(s+2\pi).$$
Thus $f(s)=f(s+2\pi)$ and 2π is indeed a period of $f(s)$.

Exercise 7.3 (page 197)

1. $-\cos\dfrac{2\pi}{5}$ **3.** $-\cos\dfrac{\pi}{8}$ **5.** $\cos\dfrac{\pi}{7}$ **7.** $-\cos\dfrac{\pi}{6}$ **9.** $-\cos\dfrac{3\pi}{7}$ **11.** $-\cos\dfrac{2\pi}{5}$

13. $-\cos\dfrac{2\pi}{11}$ **15.** $\cos\dfrac{2\pi}{9}$ **17.** 0.8309 **19.** 0.2482 **21.** 0.1403 **23.** 0.3474

25. -0.4267 **27.** 0.4801 **29.** -0.5312 **31.** -0.9950 **33.** 0.26 **35.** 0.79

37. 0.535

39. $\cos(x_1 - x_2) = \cos(x_1 + (-x_2))$
$$= \cos x_1 \cos(-x_2) - \sin x_1 \sin(-x_2)$$
$$= \cos x_1 \cos x_2 + \sin x_1 \sin x_2$$

41. $\cos(2\pi - x) = \cos 2\pi \cos x + \sin 2\pi \sin x$
$$= 1 \cdot \cos x + 0 \cdot \sin x = \cos x$$

43. $\cos\left(\dfrac{\pi}{2} + x\right) = \cos\dfrac{\pi}{2}\cos x - \sin\dfrac{\pi}{2}\sin x$
$$= 0 \cdot \cos x - 1 \cdot \sin x = -\sin x$$

45. $\cos\left(\dfrac{3\pi}{2} + x\right) = \cos\dfrac{3\pi}{2}\cos x - \sin\dfrac{3\pi}{2}\sin x$
$$= 0 \cdot \cos x - (-1) \cdot \sin x = \sin x$$

Exercise 7.4 (page 201)

1. $-\sin\dfrac{2\pi}{5}$ **3.** $-\sin\dfrac{\pi}{8}$ **5.** $-\sin\dfrac{4\pi}{11}$ **7.** $\sin\dfrac{\pi}{7}$ **9.** $\sin\dfrac{\pi}{6}$ **11.** $-\sin\dfrac{5\pi}{11}$

13. $\sin\dfrac{2\pi}{5}$ **15.** $-\sin\dfrac{\pi}{5}$ **17.** 0.3051 **19.** 0.9356 **21.** 0.9975 **23.** 0.8634

25. 0.8016 **27.** -0.7243 **29.** -0.9391 **31.** 0.1889 **33.** 1.50 **35.** 1.03

37. 0.412

39. $\sin(x_1 - x_2) = \sin(x_1 + (-x_2))$
$$= \sin x_1 \cos(-x_2) + \cos x_1 \sin(-x_2)$$
$$= \sin x_1 \cos x_2 - \cos x_1 \sin x_2$$

41. $\sin(\pi + x) = \sin \pi \cos x + \cos \pi \sin x$
$$= 0 \cdot \cos x + (-1) \cdot \sin x = -\sin x$$

43. $\sin\left(\dfrac{\pi}{2} + x\right) = \sin\dfrac{\pi}{2}\cos x + \cos\dfrac{\pi}{2}\sin x$
$$= 1 \cdot \cos x + 0 \cdot \sin x = \cos x$$

45. $\sin\left(\dfrac{3\pi}{2} + x\right) = \sin\dfrac{3\pi}{2}\cos x + \cos\dfrac{3\pi}{2}\sin x$
$$= (-1) \cdot \cos x + 0 \cdot \sin x = -\cos x$$

Exercise 7.5 (page 210)

1.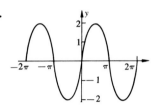

3.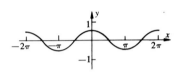

5.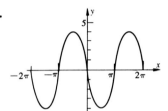

7.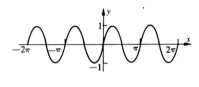

9.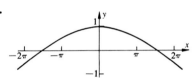

11.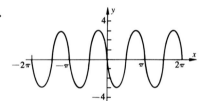

13.

15.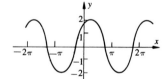

17.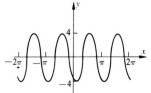

19.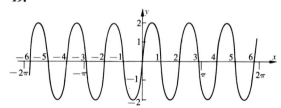

NOTE: LONG MARKS π UNITS; SHORT MARKS INTEGERS

21.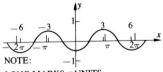

NOTE:
LONG MARKS π UNITS;
SHORT MARKS INTEGERS

23. $\left\{ -2\pi, -\dfrac{3\pi}{2}, -\pi, -\dfrac{\pi}{2}, 0, \dfrac{\pi}{2}, \pi, \dfrac{3\pi}{2}, 2\pi \right\}$

25. $\left\{ -\dfrac{3\pi}{2}, \dfrac{3\pi}{2} \right\}$

27. $\{-6, -5, -4, -3, -2, -1, 0, 1, 2, 3, 4, 5, 6\}$

29.

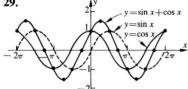

31.

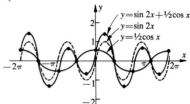

33.

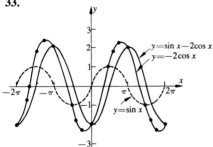

35.

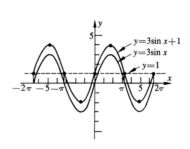

37.

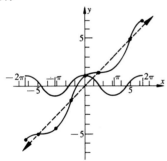

Exercise 7.6 (page 213)

1. $\dfrac{1}{2}\sqrt{2-\sqrt{2}}$ **3.** $\dfrac{1}{2}\sqrt{2+\sqrt{2}}$ **5.** $\dfrac{1}{2}\sqrt{2-\sqrt{2}}$ **7.** $\dfrac{1}{2}\sqrt{2-\sqrt{2-\sqrt{3}}}$

9. a. $-\dfrac{7}{25}$ **b.** $\dfrac{24}{25}$ **c.** $-\dfrac{2\sqrt{5}}{5}$ **d.** $\dfrac{\sqrt{5}}{5}$

11. a. 0.96 **b.** -0.28 **c.** 0.45 **d.** -0.89 **13.** $\dfrac{1}{2}$ **15.** $-\dfrac{\sqrt{3}}{2}$

17. By Formula (3), $\cos 2x = 1 - 2\sin^2 x$, hence $\sin^2 x = \tfrac{1}{2}(1 - \cos 2x)$.

19. Use $\cos^3 x = \cos(2x + x)$, Th. 7.3, Th. 7.9; then $\cos^3 x = \cos(2x + x) = \cos^2 x \cos x - \sin^2 x \sin x = (\cos^2 x - \sin^2 x) \cos x - (2 \sin x \cos x) \sin x = \cos^3 x - \sin^2 x \cos x - 2 \sin^2 x \cos x = \cos^3 x - 3 \sin^2 x \cos x$. From Th. 7.1, with

$s = x$, $\cos^2 x + \sin^2 x = 1$ and $\sin^2 x = 1 - \cos^2 x$; hence $\cos^3 x - 3 \sin^2 x \cos x = \cos^3 x - 3(1 - \cos^2 x) \cos x = \cos^3 x - 3 \cos x + 3 \cos^3 x = 4 \cos^3 x - 3 \cos x.$

21. Let $x = x + \pi$; then use Th. 7.6 and substitution $\cos 2\pi = 1$, $\sin 2\pi = 0$; then $\sin 2x = \sin [2(x + \pi)] = \sin (2x + 2\pi) = \sin 2x$; hence, π is a period of $\sin 2x$. To show π is fundamental period: suppose there exists a, $0 < a < \pi$, such that $\sin 2(x + a) = \sin 2x$. Then $\sin 2(x + a) = \sin (2x + 2a) = \sin 2x$ implies $a = 0$, but assumption was $0 < a < \pi$, so no a exists; hence π is the fundamental period of $\sin 2x$.

23. Let $x = x + 4\pi$; then use substitution $\cos 2\pi = 1$ and $\sin 2\pi = 0$; then $\sin \dfrac{x}{2} = \sin \left[\dfrac{1}{2}(x + 4\pi) \right] = \sin \left[\dfrac{x}{2} + 2\pi \right] = \sin \dfrac{x}{2}$; hence 4π is a period for $\sin \dfrac{x}{2}$. To show π is the fundamental period, use procedure of Problem 21 above assuming $0 < a < 4\pi$

25. By Formula (4) with $x = \dfrac{x}{2}$, $\cos 2 \left(\dfrac{x}{2} \right) = 2 \cos^2 \left(\dfrac{x}{2} \right) - 1$. Then $\cos x = 2 \cos^2 \left(\dfrac{x}{2} \right) - 1$ and $\cos^2 \left(\dfrac{x}{2} \right) = \dfrac{1 + \cos x}{2}$; hence $\cos \dfrac{x}{2} = \pm \sqrt{\dfrac{1 + \cos x}{2}}$.

Exercise 7.7 (page 221)

1. $\cos x = -\dfrac{15}{17}$, $\sec x = -\dfrac{17}{15}$, $\csc x = -\dfrac{17}{8}$, $\tan x = \dfrac{8}{15}$, $\cot x = \dfrac{15}{8}$

3. $\sin x = \dfrac{5}{13}$, $\cos x = \dfrac{12}{13}$, $\sec x = \dfrac{13}{12}$, $\csc x = \dfrac{13}{5}$, $\cot x = \dfrac{12}{5}$

5. $\sin x = -\dfrac{1}{\sqrt{10}}$, $\cos x = \dfrac{3}{\sqrt{10}}$, $\sec x = \dfrac{\sqrt{10}}{3}$, $\csc x = -\sqrt{10}$, $\tan x = -\dfrac{1}{3}$

7. $\tan (\pi - x) = \dfrac{\tan \pi - \tan x}{1 + \tan \pi \tan x} = \dfrac{0 - \tan x}{1 + 0 \cdot \tan x}$

$\qquad = -\tan x \quad \left(x \neq \dfrac{(2n+1)\pi}{2}, n \in J \right)$

9. $\tan \left(\dfrac{\pi}{2} - x \right) = \dfrac{\sin \left(\dfrac{\pi}{2} - x \right)}{\cos \left(\dfrac{\pi}{2} - x \right)} = \dfrac{\sin \dfrac{\pi}{2} \cos x - \cos \dfrac{\pi}{2} \sin x}{\cos \dfrac{\pi}{2} \cos x + \sin \dfrac{\pi}{2} \sin x}$

$\qquad = \dfrac{1 \cdot \cos x - 0 \cdot \sin x}{0 \cdot \cos x + 1 \cdot \sin x} = \dfrac{\cos x}{\sin x}$

$\qquad = \cot x \quad (x \neq n\pi, n \in J)$

11. $\tan \dfrac{\pi}{7}$ **13.** $-\tan \dfrac{\pi}{11}$ **15.** $-\tan \dfrac{5\pi}{13}$ **17.** $\tan \dfrac{3\pi}{7}$ **19.** 5.177   **21.** 2.360

23. -0.0601 **25.** 0.1003   **27.** 1.247   **29.** -1.381 **31.** 0.97   **33.** 1.15   **35.** 1.55

Exercise 7.8 (page 226)

1.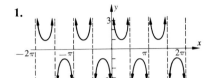

3.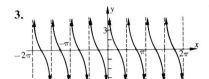

5.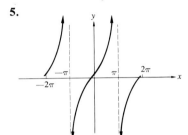

7.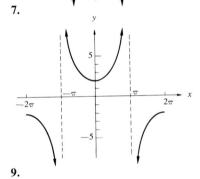

9.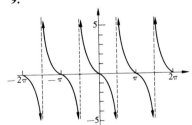

11.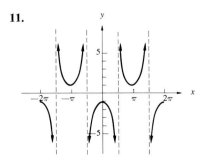

13.

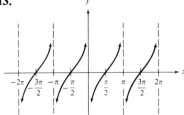

15.

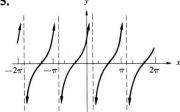

17.

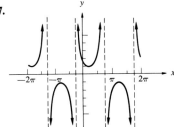

Exercise 7.9 (page 230)

1. $-\tan x$ **3.** $\tan x$ **5.** $\tan (x_1 + x_2)$ **7.** $\cos 2x$ **9.** $\sin (x_1 - x_2)$

11. $\tan 6x$ **13.** $\sin^2 x$ **15.** $\cos 2x_1$

17. By Identity 12, $\tan x = \dfrac{\sin x}{\cos x}$; hence $\cos x \tan x = \cos x \left(\dfrac{\sin x}{\cos x}\right) = \sin x.$

19. By Identity 18, $\cot x = \dfrac{\cos x}{\sin x}$; then $\cot^2 x = \dfrac{\cos^2 x}{\sin^2 x}$; hence $\sin^2 x \cot^2 x =$

$\sin^2 x \left(\dfrac{\cos^2 x}{\sin^2 x}\right) = \cos^2 x$

21. By Identity 14, $1 + \tan^2 x = \sec^2 x$; and by Identity 16, $\sec x = \dfrac{1}{\cos x}$. Then

$\sec^2 x = \dfrac{1}{\cos^2 x}$; hence $\cos^2 x(1 + \tan^2 x) = \cos^2 x(\sec^2 x) = \cos^2 x \left(\dfrac{1}{\cos^2 x}\right) = 1.$

23. By Identity 1, $\sin^2 x + \cos^2 x = 1$; hence $1 - \cos^2 x = \sin^2 x$. Using Identities 16, 17, and 18 with this yields $\sec x \csc x - \cot x = \left(\dfrac{1}{\cos x}\right)\left(\dfrac{1}{\sin x}\right) - \dfrac{\cos x}{\sin x} = \dfrac{1 - \cos^2 x}{\cos x \sin x}$

$= \dfrac{\sin^2 x}{\cos x \sin x} = \dfrac{\sin x}{\cos x} = \tan x.$

25. By Identities 16 and 12, $\dfrac{\sin x \sec x}{\tan x} = \dfrac{(\sin x)\left(\dfrac{1}{\cos x}\right)}{\left(\dfrac{\sin x}{\cos x}\right)} = \dfrac{\sin x}{\sin x} = 1.$

27. By Identity 14, $\tan^2 x + 1 = \sec^2 x$; hence $\sec^2 x - 1 = \tan^2 x$; also, by Identity 1, $\sin^2 x + \cos^2 x = 1$, and if $\sin^2 x \neq 0$. Then $\dfrac{\sin^2 x}{\sin^2 x} + \dfrac{\cos^2 x}{\sin^2 x} = \dfrac{1}{\sin^2 x}$, or $1 + \cot^2 x$ $= \csc^2 x$; hence $\csc^2 x - 1 = \cot^2 x$. Using these results with Identity 19 yields $(\sec^2 x - 1)(\csc^2 x - 1) = (\tan^2 x)(\cot^2 x) = (\tan^2 x)\left(\dfrac{1}{\tan^2 x}\right) = 1.$

29. From Identity 1, since $\sin^2 x + \cos^2 x = 1$, $1 - \sin^2 x = \cos^2 x$. From this result and Identity 16, $\dfrac{1}{1 + \sin x} + \dfrac{1}{1 - \sin x} = \dfrac{1 - \sin x + 1 + \sin x}{(1 + \sin x)(1 - \sin x)} = \dfrac{2}{1 - \sin^2 x} = \dfrac{2}{\cos^2 x}$

$= 2\left(\dfrac{1}{\cos x}\right)^2 = 2 \sec^2 x.$

31. From Identity 18, $\sin x \cot x = \sin x \left(\dfrac{\cos x}{\sin x}\right) = \cos x.$

33. From Identities 16 and 12, $\sec x - \cos x = \dfrac{1}{\cos x} - \cos x = \dfrac{1 - \cos^2 x}{\cos x} = \dfrac{\sin^2 x}{\cos x}$

$= \sin x \left(\dfrac{\sin x}{\cos x}\right) = \sin x \tan x.$

35. From Problem 27 above, $1 + \cot^2 x = \csc^2 x$; also, from Identity 19, $\cot x = \dfrac{1}{\tan x}$;

hence $\tan x \cot x = 1$ and $\tan x = \dfrac{1}{\cot x}$, $x \neq 0$. Then $\dfrac{1 + \tan^2 x}{\tan^2 x} = \dfrac{1}{\tan^2 x} + 1 =$ $\cot^2 x + 1 = 1 + \cot^2 x = \csc^2 x.$

37. From Identity 12 and the fact that $1 - \cos^2 x = \sin^2 x$ (from Problem 23 above) $\tan^2 x - \sin^2 x = \dfrac{\sin^2 x}{\cos^2 x} - \sin^2 x = \dfrac{\sin^2 x - \sin^2 x \cos^2 x}{\cos^2 x} = \dfrac{\sin^2 x(1 - \cos^2 x)}{\cos^2 x} =$

$\left(\dfrac{\sin x}{\cos x}\right)^2 (1 - \cos^2 x) = \tan^2 x \sin^2 x = \sin^2 x \tan^2 x.$

39. From Identity 14, $\dfrac{1}{\sec x - \tan x} \cdot \dfrac{\sec x + \tan x}{\sec x + \tan x} = \dfrac{\sec x + \tan x}{\sec^2 x - \tan^2 x} =$

$\dfrac{\sec x + \tan x}{(\tan^2 x + 1) - \tan^2 x} = \sec x + \tan x.$

41. From Identities 9, 12, 14, and 16, $\dfrac{2 \tan x}{1 + \tan^2 x} = \dfrac{2\left(\dfrac{\sin x}{\cos x}\right)}{\sec^2 x} = \dfrac{2\left(\dfrac{\sin x}{\cos x}\right)}{\left(\dfrac{1}{\cos x}\right)^2} = 2 \sin x \cos x$

$= \sin 2x$.

43. From Problem 27 above, $1 + \cot^2 x = \csc^2 x$, and from Identities 19 and 22, $\cot 2x = \dfrac{\cot^2 x - 1}{2 \cot x}$. From these results and Identities 17 and 18, $\cot x - \cot 2x =$

$\cot x - \dfrac{\cot x^2 - 1}{2 \cot x} = \dfrac{2 \cot^2 x - \cot^2 x + 1}{2 \cot x} = \dfrac{\cot^2 x + 1}{2 \cot x} = \dfrac{\csc^2 x}{2 \cot x} = \dfrac{\csc x}{2}\left(\dfrac{\csc x}{\cot x}\right) =$

$\dfrac{\csc x}{2}\left(\dfrac{\dfrac{1}{\sin x}}{\dfrac{\cos x}{\sin x}}\right) = \dfrac{\csc x}{2 \cos x} = \dfrac{1}{2 \cos x \sin x} = \dfrac{1}{\sin 2x} = \csc 2x$.

45. From Identities 8c, 9, and the fact that $1 - \sin^2 x = \cos^2 x$ derived from Identity 1, followed by Identity 18, $\dfrac{1 + \cos 2x}{\sin 2x} = \dfrac{1 + (1 - 2 \sin^2 x)}{2 \sin x \cos x} = \dfrac{2(1 - \sin^2 x)}{2 \sin x \cos x} =$

$\dfrac{2 \cos^2 x}{2 \sin x \cos x} = \dfrac{\cos x}{\sin x} = \cot x$.

47. $\sin^2 x = \dfrac{1 - \cos 2x}{2}$

Exercise 8.1 (page 236)

1. a. $0°$ **b.** $90°$ **c.** $180°$ **d.** $270°$ **e.** $360°$ **3.** $40°$ **5.** $252°$ **7.** $17.2°$

9. $207.5°$ **11.** 0.35 **13.** 2.27 **15.** 7.33 **17.** 13.09 **19.** $57.3°$

21. $390°, -330°$; $30° + 360°k$, $k \in J$ **23.** $120°, 480°$; $-240° + 360°k$, $k \in J$

25. $60°, -300°$; $420° + 360°k$, $k \in J$ **27.** $30°, 390°$; $-330° + 360°k$, $k \in J$

29. $s \approx 2.09''$ **31.** $s \approx 0.72'$ **33.** $s \approx 4.71''$

Exercise 8.2 (page 243)

1. $\sin \alpha = \dfrac{4}{5}$, $\cos \alpha = \dfrac{3}{5}$, $\tan \alpha = \dfrac{4}{3}$, $\cot \alpha = \dfrac{3}{4}$, $\sec \alpha = \dfrac{5}{3}$, $\csc \alpha = \dfrac{5}{4}$

3. $\sin \alpha = -\dfrac{\sqrt{2}}{2}$, $\cos \alpha = \dfrac{\sqrt{2}}{2}$, $\tan \alpha = -1$, $\cot \alpha = -1$, $\sec \alpha = \sqrt{2}$,

$\csc \alpha = -\sqrt{2}$

5. $\sin \alpha = -\dfrac{4}{5}, \quad \cos \alpha = -\dfrac{3}{5}, \quad \tan \alpha = \dfrac{4}{3}, \quad \cot \alpha = \dfrac{3}{4}, \quad \sec \alpha = -\dfrac{5}{3}, \quad \csc \alpha = -\dfrac{5}{4}$

7. $\sin \alpha = 1, \quad \cos \alpha = 0, \quad \tan \alpha$ not defined, $\quad \cot \alpha = 0, \quad \sec \alpha$ not defined, $\quad \csc \alpha = 1$

9. IV   **11.** I   **13.** IV   **15.** $-\dfrac{1}{\sqrt{3}}$ **17.** $-\dfrac{1}{\sqrt{2}}$ **19.** $-\dfrac{1}{2}$ **21.** $-\dfrac{1}{2}$

23. $\dfrac{\sqrt{3}}{2}$ **25.** $\sqrt{3}$ **27.** 0.5299   **29.** 0.6494   **31.** 1.046   **33.** 0.808

35. 0.152   **37.** 1.015   **39.** 0.7431   **41.** -0.8391 **43.** -0.6428 **45.** 0.8772

47. 1.827   **49.** -0.0610

51. If the coordinates are associated with lengths of line segments as indicated in the figure, then by the Pythagorean theorem, $OP_1 = \sqrt{x_1{}^2 + y_1{}^2}$ and $OP_2 = \sqrt{x_2{}^2 + y_2{}^2}$. Since $\triangle OAP$ and $\triangle OBP$ are similar, their corresponding sides are proportional and the desired results follow by substituting in the following ratios:

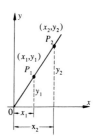

$$\frac{OA}{AP_1} = \frac{OB}{BP_2}; \quad \frac{OA}{OP_1} = \frac{OB}{OP_2}; \quad \text{and} \quad \frac{AP_1}{OP_1} = \frac{BP_2}{OP_2}.$$

Exercise 8.3 (page 247)

1. $A = 36° \, 50', B = 53° \, 10', c = 10, \mathscr{A} = 24.0$ **3.** $A = 36°, a = 87.2, c = 148.3,$
$\mathscr{A} = 5231$

5. $B = 68°, a = 6.0, b = 14.8, \mathscr{A} = 44.4$ **7.** 16

9. $\cos \theta = \dfrac{-\sqrt{3}}{2}, \tan \theta = \dfrac{1}{\sqrt{3}}, \cot \theta = \sqrt{3}, \csc \theta = -2, \sec \theta = \dfrac{-2}{\sqrt{3}}$

11. $\sin \theta = \dfrac{2\sqrt{10}}{7}, \tan \theta = -\dfrac{2\sqrt{10}}{3}, \cot \theta = -\dfrac{3}{2\sqrt{10}}, \sec \theta = -\dfrac{7}{3}, \csc \theta = \dfrac{7}{2\sqrt{10}}$

13. $\cos \theta = \dfrac{-4}{5}, \tan \theta = \dfrac{-3}{4}, \cot \theta = \dfrac{-4}{3}, \csc \theta = \dfrac{5}{3}, \sec \theta = \dfrac{-5}{4}$

15. $\sin \theta = \dfrac{-5}{13}, \cos \theta = \dfrac{-12}{13}, \cot \theta = \dfrac{12}{5}, \csc \theta = \dfrac{-13}{5}, \sec \theta = \dfrac{-13}{12}$

17. $\sin \theta = \dfrac{-\sqrt{3}}{2}, \cos \theta = \dfrac{-1}{2}, \tan \theta = \sqrt{3}, \cot \theta = \dfrac{1}{\sqrt{3}}, \csc \theta = \dfrac{-2}{\sqrt{3}}$

19. $\dfrac{3+\sqrt{2}}{2}$ **21.** 20.6   **23.** Altitude, 14.1; $\mathscr{A}=394.8$ **25.** 18.4 inches

27. 38° 40′, 51° 20′   **29.** 82.6 feet

Exercise 8.4 (page 253)

1. $C=70°, a=19.7, c=18.8$ **3.** $B=58°, a=74.9, b=74.2$

5. $B=47° 40′, b=66.2, c=34.0$ **7.** One triangle possible

9. One triangle possible   **11.** Two triangles possible   **13.** A right triangle

15. $B=18° 40′, C=48° 20′, b=1.7$ **17.** $A=108° 40′, C=28° 40′, a=8.7$

19. $A=107° 20′, B=40° 10′, a=2.7$; or $A=7° 40′, B=139° 50′, a=0.4$

21. No possible solution   **23.** $B=25° 40′, C=94° 20′, c=15.9$

25. $A=93° 40′, C=53° 00′, a=763$; or $A=19° 40′, C=127° 00′, a=257$

27. $\mathscr{A}=21$ **29.** $\mathscr{A}=3.8$ **31.** $\mathscr{A}=2.6$ **33.** 197.6 feet

35. From the law of sines and the properties of proportion, $\dfrac{a}{\sin\alpha}=\dfrac{b}{\sin\beta}; \dfrac{a}{b}=\dfrac{\sin\alpha}{\sin\beta};$

$\dfrac{a}{b}+1=\dfrac{\sin\alpha}{\sin\beta}+1$; then $\dfrac{a+b}{b}=\dfrac{\sin\alpha+\sin\beta}{\sin\beta}$.

37. Dividing result from Problem 36 by the result of Problem 35 gives

$\dfrac{a-b}{a+b}=\dfrac{\sin\alpha-\sin\beta}{\sin\alpha+\sin\beta}$; from Theorems 7.6 and 7.7 we see that $\sin x_1 \cos x_2=$

$\dfrac{1}{2}[\sin(x_1+x_2)+\sin(x_1-x_2)]$; hence $\sin(x_1+x_2)+\sin(x_1-x_2)=2\sin x_1 \cos x_2$.

Let $\alpha=x_1+x_2$ and $\beta=x_1-x_2$; then $2x_1=\alpha+\beta$, $x_1=\dfrac{1}{2}(\alpha+\beta)$; $2x_2=\alpha-\beta$,

$x_2=\dfrac{1}{2}(\alpha-\beta)$; so $\sin\alpha+\sin\beta=2\sin\dfrac{1}{2}(\alpha+\beta)\cos\dfrac{1}{2}(\alpha-\beta)$. Using this result and

replacing β by $-\beta$ yields $\sin\alpha-\sin\beta=2\sin\dfrac{1}{2}(\alpha-\beta)\cos\dfrac{1}{2}(\alpha+\beta)$. Then $\dfrac{a-b}{a+b}=$

$\dfrac{\sin\alpha-\sin\beta}{\sin\alpha+\sin\beta}=\dfrac{2\sin\frac{1}{2}(\alpha-\beta)\cos\frac{1}{2}(\alpha+\beta)}{2\sin\frac{1}{2}(\alpha+\beta)\cos\frac{1}{2}(\alpha-\beta)}=\left[\dfrac{\sin\frac{1}{2}(\alpha-\beta)}{\cos\frac{1}{2}(\alpha-\beta)}\right]\cdot\left[\dfrac{\cos\frac{1}{2}(\alpha+\beta)}{\sin\frac{1}{2}(\alpha+\beta)}\right]$

$=\tan\dfrac{1}{2}(\alpha-\beta)\cdot\cot\dfrac{1}{2}(\alpha+\beta)=\dfrac{\tan\frac{1}{2}(\alpha-\beta)}{\tan\frac{1}{2}(\alpha+\beta)}$.

Exercise 8.5 (page 257)

1. $c=6.8, \alpha=132° 50′, \beta=17° 10′$ **3.** $b=10.3, \alpha=23° 40′, \gamma=34°$

5. $a=14.9, \beta=75° 10′, \gamma=24° 10′$ **7.** $\alpha=38° 10′, \beta=81° 50′, \gamma=60°$

9. 71° 40′   **11.** $\mathscr{A}=9.8$

13. From the law of cosines, $a^2 = b^2 + c^2 - 2bc \cos \alpha$; $\cos \alpha = \dfrac{b^2 + c^2 - a^2}{2bc}$; then

$$1 + \cos \alpha = 1 + \frac{b^2 + c^2 - a^2}{2bc} = \frac{(b^2 + 2bc + c^2) - a^2}{2bc} = \frac{(b + c)^2 - a^2}{2bc} =$$

$$\frac{(b + c + a)(b + c - a)}{2bc}.$$

15. Since $s = \dfrac{a + b + c}{2}$, $s - a = \dfrac{a + b + c}{2} - a = \dfrac{b + c - a}{2}$. From Problem 13 above,

$$1 + \cos \alpha = \frac{(b + c + a)(b + c - a)}{2bc}; \quad \frac{1 + \cos \alpha}{2} = \left(\frac{b + c + a}{2}\right)\left(\frac{b + c - a}{2}\right) \cdot \left(\frac{1}{bc}\right) =$$

$$\frac{s(s - a)}{bc}; \text{ hence } \cos \frac{1}{2} \alpha = \sqrt{\frac{1 + \cos \alpha}{2}} = \sqrt{\frac{s(s - a)}{bc}}.$$

17. $87° \ 30'$

19. Multiplying the results of Problems 13 and 14 gives $1 - \cos^2 \alpha$

$$= \frac{(b + c + a)(b + c - a)(a - b + c)(a + b - c)}{4b^2c^2}; \text{ since } s = \frac{a + b + c}{2}, s - a =$$

$$\frac{b + c - a}{2}, s - b = \frac{a - b + c}{2}, \text{ and } s - c = \frac{a + b - c}{2}. \text{ Using the above results with}$$

the fact that $\mathscr{A} = \dfrac{1}{2} bc \sin \alpha$ and $\sin \alpha = \sqrt{1 - \cos^2 \alpha}$ yields

$$\mathscr{A} = \frac{1}{2} bc \sqrt{\frac{(b + c + a)(b + c - a)(a - b + c)(a + b - c)}{4b^2c^2}} =$$

$$\sqrt{\left(\frac{b + c + a}{2}\right)\left(\frac{b + c - a}{2}\right)\left(\frac{a - b + c}{2}\right)\left(\frac{a + b - c}{2}\right)} = \sqrt{s(s - a)(s - b)(s - c)}.$$

21. Let α be the right angle. Then a is the length of the hypotenuse in the right triangle; from the law of cosines and the fact that $\cos 90° = 0$, $a^2 = b^2 + c^2 - 2bc \cos 90° = b^2 + c^2$. **23.** $\mathscr{A} \approx 16$

Exercise 8.6 (page 262)

1. $(6, 125°), (6, -235°), (-6, 305°), (-6, -55°)$
3. $(-2, -90°), (-2, 270°), (2, 90°), (2, -270°)$
5. $(6, -240°), (6, 120°), (-6, -60°), (-6, 300°)$ 7. $\left(\dfrac{5}{\sqrt{2}}, \dfrac{5}{\sqrt{2}}\right)$ 9. $\left(\dfrac{\sqrt{3}}{4}, -\dfrac{1}{4}\right)$
11. $\left(\dfrac{-10}{\sqrt{2}}, \dfrac{-10}{\sqrt{2}}\right)$ 13. $(6, 45°), (6, -315°)$ 15. $(2, 240°), (2, -120°)$
17. $(0, \theta°), (0, -\theta°)$, for any $\theta > 0$ 19. $r = 5$ 21. $r \sin \theta = -4$
23. $r^2(\cos^2 \theta + 9 \sin^2 \theta) = 9$ 25. $x^2 + y^2 = 25$ 27. $x^2 + y^2 - 9x = 0$
29. $y^2 = 4x + 4$

31. $\sec \theta/2 = \dfrac{1}{\cos (\theta/2)}$ and $\cos^2 (\theta/2) = \dfrac{1+\cos\theta}{2}$; hence $r = \sec^2 (\theta/2) = \dfrac{1}{\cos^2 (\theta/2)} =$

$\dfrac{2}{1+\cos\theta}$. Now $\cos\theta = \dfrac{x}{r}$ and $r^2 = x^2+y^2$; so $r = \dfrac{2}{1+\dfrac{x}{r}}$ and $r\left(1+\dfrac{x}{r}\right) = 2$;

hence $r+x=2$, or $r=2-x$. Squaring both sides $r^2 = 4-4x+x^2$ and substitu-

ting for r^2 gives $x = 1 - \dfrac{1}{4}\,y^2$, whose graph is a parabola.

33.

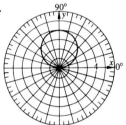

Exercise 9.1 (page 268)

1. $\dfrac{\pi}{6}$ **3.** $\dfrac{\pi}{4}$ **5.** Does not exist.   **7.** $\dfrac{-\pi}{6}$ **9.** 0.10   **11.** 0.39   **13.** -1.04

15. 1.29   **17.** $\dfrac{\pi}{4}$ **19.** $\dfrac{\pi}{3}$ **21.** $\dfrac{\sqrt{3}}{2}$ **23.** $\dfrac{-\sqrt{3}}{2}$ **25.** $\dfrac{\sqrt{3}}{2}$ **27.** $\dfrac{2}{3}$

29. $\dfrac{3+4\sqrt{3}}{10}$ **31.** $\dfrac{3\pi}{4}$ **33.** $\sqrt{1-x^2}$ **35.** $\dfrac{y}{\sqrt{1-y^2}}$ **37.** $\sqrt{\dfrac{1+x}{2}}$

39. $x = \dfrac{1}{2}\cos\dfrac{y}{3}$ **41.** $x = \tan 2y - \pi$ **43.** Since $-\dfrac{\pi}{2} \le \text{Arc}\sin\dfrac{2}{5} \le \dfrac{\pi}{2}$,

$\tan\left(\text{Arc}\sin\dfrac{2}{5}\right) = \dfrac{2}{\sqrt{21}}$; hence $\text{Arc}\sin\dfrac{2}{5} = \text{Arc}\tan\dfrac{2}{\sqrt{21}}$.

45. Not always. Let Arc cos $(\cos x) = \alpha$. Then $0 \le \alpha \le \pi$ and $\cos\alpha = \cos x$; so $\alpha = x$ if and only if $0 \le x \le \pi$.

Exercise 9.2 (page 274)

1. $\left\{p\,|\,p = \dfrac{\pi}{6} + 2k\pi, k \in J\right\} \cup \left\{p\,|\,p = \dfrac{5\pi}{6} + 2k\pi, k \in J\right\}$ **3.** $\emptyset$

5. $\left\{r\,|\,r = \dfrac{\pi}{4} + 2k\pi, k \in J\right\} \cup \left\{r\,|\,r = \dfrac{7\pi}{4} + 2k\pi, k \in J\right\}$

7. $\left\{x\,|\,x = \dfrac{4\pi}{3} + 2k\pi, k \in J\right\} \cup \left\{x\,|\,x = \dfrac{5\pi}{3} + 2k\pi, k \in J\right\}$

9. a. $\left\{\dfrac{\pi}{3}+2k\pi\right\}\cup\left\{\dfrac{5\pi}{3}+2k\pi\right\},\ k\in J$ **b.** $\left\{\left(\dfrac{\pi}{3}+2k\pi\right)^{R}\right\}\cup\left\{\left(\dfrac{5\pi}{3}+2k\pi\right)^{R}\right\},\ k\in J;$

$\{(60+360k)^{\circ}\}\cup\{(300+360k)^{\circ}\},\ k\in J$

11. a. $\left\{\dfrac{\pi}{3}+k\pi\right\},\ k\in J$ **b.** $\left\{\left(\dfrac{\pi}{3}+k\pi\right)^{R}\right\},\ k\in J;\ \{(60+180k)^{\circ}\},\ k\in J$

13. a. $\left\{\dfrac{\pi}{4}+2k\pi\right\}\cup\left\{\dfrac{7\pi}{4}+2k\pi\right\},\ k\in J$ **b.** $\left\{\left(\dfrac{\pi}{4}+2k\pi\right)^{R}\right\}\cup\left\{\left(\dfrac{7\pi}{4}+2k\pi\right)^{R}\right\},\ k\in J;$

$\{(45+360k)^{\circ}\}\cup\{(315+360k)^{\circ}\},\ k\in J$

15. a. $\{0.25+2k\pi\}\cup\{2.89+2k\pi\},\ k\in J$ **b.** $\{(0.25+2k\pi)^{R}\}\cup\{2.89+2k\pi)^{R}\},\ k\in J;$

$\{14°\ 30'+360°k\}\cup\{165°\ 30'+360°k\},\ k\in J$

17. a. $\{1.25+k\pi\},\ k\in J$ **b.** $\{(1.25+k\pi)^{R}\},\ k\in J;\ \{71°\ 30'+180°k\},\ k\in J$

19. a. $\{1.16+2k\pi\}\cup\{5.12+2k\pi\},\ k\in J$ **b.** $\{(1.16+2k\pi)^{R}\}\cup\{5.12+2k\pi)^{R}\},\ k\in J;$

$\{66°\ 30'+360°k\}\cup\{293°\ 30'+360°k\},\ k\in J$

21. $\left\{\dfrac{\pi}{3}+k\pi\right\},\ k\in J$ **23.** $\left\{\dfrac{\pi}{4}+\dfrac{k\pi}{2}\right\},\ k\in J$ **25.** $\left\{\dfrac{\pi}{2}+2k\pi\right\},\ k\in J$

27. $\{30°,\ 45°,\ 135°,\ 150°,\ 225°,\ 315°\}$ **29.** $\{0°,\ 120°,\ 180°,\ 240°,\ 360°\}$

31. $\{30°,\ 150°,\ 180°\}$ **33.** $\left\{\dfrac{\pi^{R}}{4},\ \dfrac{5\pi^{R}}{4}\right\}$ **35.** $\{0^{R},\ 2\pi^{R}\}$

37. $\{1.11^{R},\ 1.77^{R},\ 4.25^{R},\ 4.91^{R}\}$ **39.** $\{0.67^{R},\ 2.48^{R}\}$

41. $\{1.9,\ -1.9\}$

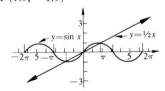

Exercise 9.3 (page 278)

1. $\{22°\ 30',\ 157°\ 30',\ 202°\ 30',\ 337°\ 30'\}$ **3.** $\{60°,\ 300°\}$

5. $\{0°,\ 60°,\ 120°,\ 180°,\ 240°,\ 300°\}$ **7.** $\{45°,\ 225°\}$

9. $\{67°\ 30',\ 157°\ 30',\ 247°\ 30',\ 337°\ 30'\}$ **11.** $\{30°,\ 90°,\ 150°,\ 210°,\ 270°,\ 330°\}$

13. $\left\{\dfrac{\pi}{2}+k\pi,\ \dfrac{\pi}{6}+2k\pi,\ \dfrac{5\pi}{6}+2k\pi\right\},\ k\in J$ **15.** $\left\{\dfrac{\pi}{2}+k\pi,\ \dfrac{\pi}{3}+2k\pi,\ \dfrac{5\pi}{3}+2k\pi\right\},\ k\in J$

17. $\left\{k\pi,\ \dfrac{\pi}{2}+k\pi\right\},\ k\in J$ **19.** $\left\{\dfrac{k\pi}{2}\right\},\ k\in J$

21. $\left\{\dfrac{\pi}{6}+2k\pi,\ \dfrac{5\pi}{6}+2k\pi,\ \pi+2k\pi\right\},\ k\in J$

Exercise 10.1 (page 283)

1. 2×2, $\begin{bmatrix} 6 & 2 \\ -1 & 3 \end{bmatrix}$ **3.** 2×3, $\begin{bmatrix} 2 & 1 \\ -7 & 4 \\ 3 & 0 \end{bmatrix}$ **5.** 3×3, $\begin{bmatrix} 2 & 4 & -2 \\ 3 & 0 & 3 \\ -1 & 1 & 1 \end{bmatrix}$

7. 2×4, $\begin{bmatrix} 4 & 2 \\ -3 & 1 \\ -1 & 1 \\ 0 & 6 \end{bmatrix}$ **9.** 1×5, $\begin{bmatrix} 0 \\ 0 \\ 0 \\ 0 \\ 0 \end{bmatrix}$ **11.** $\begin{bmatrix} 7 & -2 & 3 \\ 6 & 1 & -2 \end{bmatrix}$

13. $\begin{bmatrix} 1 & 1 & 6 & 10 \end{bmatrix}$ **15.** $\begin{bmatrix} 0 & 3 \\ -2 & 0 \\ 3 & 1 \end{bmatrix}$ **17.** $0_{3 \times 2}$

19. From the results of Problem 18 and Def. 10.3,

$$X + A = \begin{bmatrix} b_{11} - a_{11} & b_{12} - a_{12} \\ b_{21} - a_{21} & b_{22} - a_{22} \end{bmatrix} + \begin{bmatrix} a_{11} & a_{12} \\ a_{21} & a_{22} \end{bmatrix} = \begin{bmatrix} b_{11} & b_{12} \\ b_{21} & b_{22} \end{bmatrix} = B;$$

hence X is the solution for $X + A = B$. Since $B - A = B + (-A)$, then by Def. 10.3, Def. 10.5, and the results of Problem 18,

$$B - A = B + (-A) = \begin{bmatrix} b_{11} & b_{12} \\ b_{21} & b_{22} \end{bmatrix} + \begin{bmatrix} -a_{11} & -a_{12} \\ -a_{21} & -a_{22} \end{bmatrix} = \begin{bmatrix} b_{11} - a_{11} & b_{12} - a_{12} \\ b_{21} - a_{21} & b_{22} - a_{22} \end{bmatrix} = X;$$

hence X is also the solution for $X = B - A$. Therefore $X + A = B$ and $X = B - A$ are equivalent matrix equations.   **21.** $\begin{bmatrix} 2 & 2 \\ -1 & 1 \end{bmatrix}$ **23.** $\begin{bmatrix} 1 & 1 \\ 0 & 3 \end{bmatrix}$

25. Let $A_{m \times n}$, $B_{m \times n}$, and $C_{m \times n}$ have entries a_{ij}, b_{ij}, and c_{ij} respectively; then, we have by Def. 10.3, each entry of $(A + B) + C$ is $a_{ij} + b_{ij} + c_{ij}$ and each entry of $A + (B + C)$ is $a_{ij} + b_{ij} + c_{ij}$; hence $(A + B) + C = A + (B + C)$.

Exercise 10.2 (page 290)

1. $\begin{bmatrix} 0 & -5 & 5 \\ -15 & 5 & -10 \end{bmatrix}$ **3.** $\begin{bmatrix} -1 \end{bmatrix}$ **5.** $\begin{bmatrix} 1 & -13 \\ 4 & -7 \end{bmatrix}$

7. $\begin{bmatrix} 30 & -39 \\ 29 & 14 \end{bmatrix}$ **9.** $\begin{bmatrix} -5 & -1 \\ 8 & -1 \end{bmatrix}$ **11.** $\begin{bmatrix} -1 & 0 & -2 \\ 1 & 2 & 8 \\ 0 & 1 & 3 \end{bmatrix}$

13. $\begin{bmatrix} 1 & 0 & 0 \\ 0 & 1 & 0 \\ 0 & 0 & 1 \end{bmatrix}$ **15.** $\begin{bmatrix} 1 & 0 \\ -1 & 2 \end{bmatrix}$ **17.** $\begin{bmatrix} 1 & -2 \\ 1 & 2 \end{bmatrix}$

19. $\begin{bmatrix} -2 & 3 \\ 2 & -4 \end{bmatrix}$ **21.** $\begin{bmatrix} -1 & 1 \\ -1 & -1 \end{bmatrix}$ **23.** $\begin{bmatrix} 1 & -1 \\ 1 & 0 \end{bmatrix}$

25. We have $A + B = \begin{bmatrix} 0 & 2 \\ -1 & 3 \end{bmatrix}$, $A - B = \begin{bmatrix} -2 & 2 \\ 1 & -1 \end{bmatrix}$, $A^2 = \begin{bmatrix} 1 & 0 \\ 0 & 1 \end{bmatrix}$, $B^2 = \begin{bmatrix} 1 & 0 \\ -3 & 4 \end{bmatrix}$, and $AB = \begin{bmatrix} -3 & 4 \\ -1 & 2 \end{bmatrix}$. Then **(a)** $(A+B)(A+B) = \begin{bmatrix} 0 & 2 \\ -1 & 3 \end{bmatrix} \cdot \begin{bmatrix} 0 & 2 \\ -1 & 3 \end{bmatrix} = \begin{bmatrix} -2 & 6 \\ -3 & 7 \end{bmatrix}$

and $A^2 + 2AB + B^2 = \begin{bmatrix} 1 & 0 \\ 0 & 1 \end{bmatrix} + 2\begin{bmatrix} -3 & 4 \\ -1 & 2 \end{bmatrix} + \begin{bmatrix} 1 & 0 \\ -3 & 4 \end{bmatrix} = \begin{bmatrix} -4 & 8 \\ -5 & 9 \end{bmatrix}$;

hence $(A+B)(A+B) \neq A^2 + 2AB + B^2$. **(b)** $(A+B)(A-B) = \begin{bmatrix} 0 & 2 \\ -1 & 3 \end{bmatrix} \cdot$

$\begin{bmatrix} -2 & 2 \\ 1 & -1 \end{bmatrix} = \begin{bmatrix} 2 & -2 \\ 5 & -5 \end{bmatrix}$ and $A^2 - B^2 = \begin{bmatrix} 1 & 0 \\ 0 & 1 \end{bmatrix} - \begin{bmatrix} 1 & 0 \\ -3 & 4 \end{bmatrix} = \begin{bmatrix} 0 & 0 \\ 3 & -3 \end{bmatrix}$;

hence $(A+B)(A-B) \neq A^2 - B^2$.

For Problems 27 and 29, let $A = \begin{bmatrix} a_{11} & a_{12} \\ a_{21} & a_{22} \end{bmatrix}$, $B = \begin{bmatrix} b_{11} & b_{12} \\ b_{21} & b_{22} \end{bmatrix}$, and $C = \begin{bmatrix} c_{11} & c_{12} \\ c_{21} & c_{22} \end{bmatrix}$

27. Since $AB = \begin{bmatrix} a_{11}b_{11} + a_{12}b_{21} & a_{11}b_{12} + a_{12}b_{22} \\ a_{21}b_{11} + a_{22}b_{21} & a_{21}b_{12} + a_{22}b_{22} \end{bmatrix}$, $(AB)C =$

$\begin{bmatrix} (a_{11}b_{11}+a_{12}b_{21})c_{11}+(a_{11}b_{12}+a_{12}b_{22})c_{21} & (a_{11}b_{11}+a_{12}b_{21})c_{12}+(a_{11}b_{12}+a_{12}b_{22})c_{22} \\ (a_{21}b_{11}+a_{22}b_{21})c_{11}+(a_{21}b_{12}+a_{22}b_{22})c_{21} & (a_{21}b_{11}+a_{22}b_{21})c_{12}+(a_{21}b_{12}+a_{22}b_{22})c_{22} \end{bmatrix}$.

The element in the first row, first column is given by
$a_{11}b_{11}c_{11} + a_{12}b_{21}c_{11} + a_{11}b_{12}c_{21} + a_{12}b_{22}c_{21} =$
$(a_{11}b_{11}c_{11} + a_{11}b_{12}c_{21}) + (a_{12}b_{21}c_{11} + a_{12}b_{22}c_{21}) =$
$a_{11}(b_{11}c_{11} + b_{12}c_{21}) + a_{12}(b_{21}c_{11} + b_{22}c_{21})$.

When the remaining elements are treated in a similar manner, $(AB)C =$

$\begin{bmatrix} a_{11}(b_{11}c_{11}+b_{12}c_{21})+a_{12}(b_{21}c_{11}+b_{22}c_{21}) & a_{11}(b_{11}c_{12}+b_{12}c_{22})+a_{12}(b_{21}c_{12}+b_{22}c_{22}) \\ a_{21}(b_{11}c_{11}+b_{12}c_{21})+a_{22}(b_{21}c_{11}+b_{22}c_{21}) & a_{21}(b_{11}c_{12}+b_{12}c_{22})+a_{22}(b_{21}c_{12}+b_{22}c_{22}) \end{bmatrix} =$

$\begin{bmatrix} a_{11} & a_{12} \\ a_{21} & a_{22} \end{bmatrix} \cdot \begin{bmatrix} b_{11}c_{11} + b_{12}c_{21} & b_{11}c_{12} + b_{12}c_{22} \\ b_{21}c_{11} + b_{22}c_{21} & b_{21}c_{12} + b_{22}c_{22} \end{bmatrix} =$

$\begin{bmatrix} a_{11} & a_{12} \\ a_{21} & a_{22} \end{bmatrix} \cdot \left(\begin{bmatrix} b_{11} & b_{12} \\ b_{21} & b_{22} \end{bmatrix} \cdot \begin{bmatrix} c_{11} & c_{12} \\ c_{21} & c_{22} \end{bmatrix} \right) = A(BC)$.

29. $B + C = \begin{bmatrix} b_{11} + c_{11} & b_{12} + c_{12} \\ b_{21} + c_{21} & b_{21} + c_{22} \end{bmatrix}$;

hence $(B+C)A = \begin{bmatrix} (b_{11}+c_{11})a_{11}+(b_{12}+c_{12})a_{21} & (b_{11}+c_{11})a_{12}+(b_{12}+c_{12})a_{22} \\ (b_{21}+c_{21})a_{11}+(b_{22}+c_{22})a_{21} & (b_{21}+c_{21})a_{12}+(b_{22}+c_{22})a_{22} \end{bmatrix}$.

The element in the first row, first column is given by
$b_{11}a_{11} + c_{11}a_{11} + b_{12}a_{21} + c_{12}a_{21} = (b_{11}a_{11} + b_{12}a_{21}) + (c_{11}a_{11} + c_{22}a_{21})$.

When the remaining elements are treated in a similar manner, $(B+C)A$

$= \begin{bmatrix} (b_{11}a_{11}+b_{12}a_{21})+(c_{11}a_{11}+c_{12}a_{21}) & (b_{11}a_{12}+b_{12}a_{22})+(c_{11}a_{12}+c_{12}a_{22}) \\ (b_{21}a_{11}+b_{22}a_{21})+(c_{21}a_{11}+c_{22}a_{21}) & (b_{21}a_{12}+b_{22}a_{22})+(c_{21}a_{12}+c_{22}a_{22}) \end{bmatrix}$

$= \begin{bmatrix} b_{11}a_{11}+b_{12}a_{21} & b_{11}a_{12}+b_{12}a_{22} \\ b_{21}a_{11}+b_{22}a_{21} & b_{21}a_{12}+b_{22}a_{22} \end{bmatrix} + \begin{bmatrix} c_{11}a_{11}+c_{12}a_{21} & c_{11}a_{12}+c_{12}a_{22} \\ c_{21}a_{11}+c_{22}a_{21} & c_{21}a_{12}+c_{22}a_{22} \end{bmatrix}$

$= \begin{bmatrix} b_{11} & b_{12} \\ b_{21} & b_{22} \end{bmatrix} \cdot \begin{bmatrix} a_{11} & a_{12} \\ a_{21} & a_{22} \end{bmatrix} + \begin{bmatrix} c_{11} & c_{12} \\ c_{21} & c_{22} \end{bmatrix} \cdot \begin{bmatrix} a_{11} & a_{12} \\ a_{21} & a_{22} \end{bmatrix} = BC + CA$.

31. $(A_{2\times2} \cdot B_{2\times2}) = \begin{bmatrix} a_{11}b_{11} + a_{12}b_{21} & a_{11}b_{12} + a_{12}b_{22} \\ a_{21}b_{11} + a_{22}b_{21} & a_{21}b_{12} + a_{22}b_{22} \end{bmatrix}.$

Hence, $(A_{2\times2} \cdot B_{2\times2})^t = \begin{bmatrix} a_{11}b_{11} + a_{12}b_{21} & a_{21}b_{11} + a_{22}b_{21} \\ a_{11}b_{12} + a_{12}b_{22} & a_{21}b_{12} + a_{22}b_{22} \end{bmatrix}.$

By the commutative property of multiplication for real numbers, $(A_{2\times2} \cdot B_{2\times2})^t$

$= \begin{bmatrix} b_{11}a_{11} + b_{21}a_{12} & b_{11}a_{21} + b_{21}a_{22} \\ b_{12}a_{11} + b_{22}a_{12} & b_{12}a_{21} + b_{22}a_{22} \end{bmatrix} = \begin{bmatrix} b_{11} & b_{21} \\ b_{12} & b_{22} \end{bmatrix} \cdot \begin{bmatrix} a_{11} & a_{21} \\ a_{12} & a_{22} \end{bmatrix} = B_2^t{}_{\times2} \cdot A_2^t{}_{\times2}.$

33. Let $A = \begin{bmatrix} a_{11} & a_{12} \\ a_{21} & a_{22} \end{bmatrix}$ and c be a scalar; then by Def. 10.6 and the closure property of

multiplication of real numbers, $c\begin{bmatrix} a_{11} & a_{12} \\ a_{21} & a_{22} \end{bmatrix} = \begin{bmatrix} ca_{11} & ca_{12} \\ ca_{21} & ca_{22} \end{bmatrix}$, which is a 2×2 matrix.

35. By Def. 10.6 and the distributive property for real numbers, and by Def. 10.3,

$(c+d)\begin{bmatrix} a_{11} & a_{12} \\ a_{21} & a_{22} \end{bmatrix} = \begin{bmatrix} (c+d)a_{11} & (c+d)a_{12} \\ (c+d)a_{21} & (c+d)a_{22} \end{bmatrix} = \begin{bmatrix} ca_{11} + da_{11} & ca_{12} + da_{12} \\ ca_{21} + da_{21} & ca_{22} + da_{22} \end{bmatrix}$

$= \begin{bmatrix} ca_{11} & ca_{12} \\ ca_{21} & ca_{22} \end{bmatrix} + \begin{bmatrix} da_{11} & da_{12} \\ da_{21} & da_{22} \end{bmatrix} = c\begin{bmatrix} a_{11} & a_{12} \\ a_{21} & a_{22} \end{bmatrix} + d\begin{bmatrix} a_{11} & a_{12} \\ a_{21} & a_{22} \end{bmatrix} = cA + dA.$

37. By Defs. 10.6 and 10.5, $(-1)A = -1\begin{bmatrix} a_{11} & a_{12} \\ a_{21} & a_{22} \end{bmatrix} = \begin{bmatrix} -a_{11} & -a_{12} \\ -a_{21} & -a_{22} \end{bmatrix} = -A.$

39. $0_{2\times2} = \begin{bmatrix} 0 & 0 \\ 0 & 0 \end{bmatrix}$; hence, by Def. 10.6,

$c \cdot 0_{2\times2} = c\begin{bmatrix} 0 & 0 \\ 0 & 0 \end{bmatrix} = \begin{bmatrix} c \cdot 0 & c \cdot 0 \\ c \cdot 0 & c \cdot 0 \end{bmatrix} = \begin{bmatrix} 0 & 0 \\ 0 & 0 \end{bmatrix} = 0_{2\times2}.$

Exercise 10.3 (page 296)

1. 0   **3.** -6 **5.** -2

7. $M_{11} = \begin{vmatrix} 0 & 3 & -1 \\ 1 & 2 & 2 \\ -1 & 3 & 1 \end{vmatrix}$, $A_{11} = \begin{vmatrix} 0 & 3 & -1 \\ 1 & 2 & 2 \\ -1 & 3 & 1 \end{vmatrix}$

9. $M_{23} = \begin{vmatrix} 2 & 1 & 0 \\ -2 & 1 & 2 \\ 1 & -1 & 1 \end{vmatrix}$, $A_{23} = -\begin{vmatrix} 2 & 1 & 0 \\ -2 & 1 & 2 \\ 1 & -1 & 1 \end{vmatrix}$

11. $M_{31} = \begin{vmatrix} 1 & -2 & 0 \\ 0 & 3 & -1 \\ -1 & 3 & 1 \end{vmatrix}$, $A_{31} = \begin{vmatrix} 1 & -2 & 0 \\ 0 & 3 & -1 \\ -1 & 3 & 1 \end{vmatrix}$

13. $M_{44} = \begin{vmatrix} 2 & 1 & -2 \\ 1 & 0 & 3 \\ -2 & 1 & 2 \end{vmatrix}$, $A_{44} = \begin{vmatrix} 2 & 1 & -2 \\ 1 & 0 & 3 \\ -2 & 1 & 2 \end{vmatrix}$

15. 1   **17.** 0   **19.** -30 **21.** $3x + 1$ **23.** 3   **25.** 0   **27.** -1 **29.** 0

31. x^3 **33.** $x = 5$ **35.** $x = 3$

37. Expanding by elements in the first row yields $\begin{vmatrix} 0 & 1 & 0 & 0 \\ 1 & 0 & 3 & 2 \\ 5 & -1 & 2 & 1 \\ 1 & 0 & 1 & 1 \end{vmatrix} = -\begin{vmatrix} 1 & 3 & 2 \\ 5 & 2 & 1 \\ 1 & 1 & 1 \end{vmatrix};$

expanding by elements in the third row yields $-\begin{vmatrix} 3 & 2 \\ 2 & 1 \end{vmatrix} + \begin{vmatrix} 1 & 2 \\ 2 & 1 \end{vmatrix} - \begin{vmatrix} 1 & 3 \\ 5 & 2 \end{vmatrix}.$

$= 1 - 9 + 13 = 5$ **39.** 2, 6, 24

41. Let $A = \begin{bmatrix} a_{11} & a_{12} \\ a_{21} & a_{22} \end{bmatrix}$; by Def. 10.6 and the fact that $\delta(A) = a_{11}a_{22} - a_{12}a_{21}$, it follows that

$aA = \begin{bmatrix} aa_{11} & aa_{12} \\ aa_{21} & aa_{22} \end{bmatrix}$ and $\delta(aA) = a^2 a_{11}a_{22} - a^2 a_{12}a_{21} = a^2(a_{11}a_{22} - a_{12}a_{21}) = a^2\delta(A).$

43. Let $A = \begin{bmatrix} a_{11} & a_{12} \\ a_{21} & a_{22} \end{bmatrix}$ and $B = \begin{bmatrix} b_{11} & b_{12} \\ b_{21} & b_{22} \end{bmatrix}$; then $\delta(A = a_{11}a_{22} - a_{12}a_{21}$ and

$\delta(B) = b_{11}b_{22} - b_{12}b_{21}$, $AB = \begin{bmatrix} a_{11}b_{11} + a_{12}b_{21} & a_{11}b_{12} + a_{12}b_{22} \\ a_{21}b_{11} + a_{22}b_{21} & a_{21}b_{12} + a_{22}b_{22} \end{bmatrix}$

and $\delta(AB) = (a_{11}b_{11} + a_{12}b_{21})(a_{21}b_{12} + a_{22}b_{22}) - (a_{11}b_{12} + a_{12}b_{22})(a_{21}b_{11} + a_{22}b_{21})$,
which simplifies to $a_{11}b_{11}a_{22}b_{22} - a_{11}b_{12}a_{22}b_{21} - a_{12}b_{22}a_{21}b_{11} + a_{12}b_{21}a_{21}b_{12}$
$= a_{11}a_{22}(b_{11}b_{22} - b_{12}b_{21}) - a_{12}a_{21}(b_{11}b_{22} - b_{12}b_{21})$
$= (a_{11}a_{22} - a_{12}a_{21})(b_{11}b_{22} - b_{12}b_{21}) = \delta(A) \cdot \delta(B).$

Exercise 10.4 (page 302)

1. Theorem 10.8   **3.** Theorem 10.10   **5.** Theorem 10.11

7. Theorem 10.11   **9.** Theorem 10.13   **11.** Theorem 10.13

13. $\begin{vmatrix} 1 & 3 \\ 0 & -4 \end{vmatrix}$ **15.** $\begin{vmatrix} 1 & -2 & 1 \\ 0 & 7 & 1 \\ 0 & 2 & 1 \end{vmatrix}$ **17.** $\begin{vmatrix} 0 & 1 & -3 & -2 \\ 0 & 2 & 1 & 2 \\ 1 & 1 & 2 & 3 \\ 0 & 1 & 1 & 1 \end{vmatrix}$

19. $-1\begin{vmatrix} 2 & 1 \\ -1 & 2 \end{vmatrix} = -5$ **21.** $\begin{vmatrix} -1 & -5 \\ 2 & -2 \end{vmatrix} = 12$ **23.** $\begin{vmatrix} 4 & 4 \\ 3 & 7 \end{vmatrix} = 16$

25. $\begin{vmatrix} -1 & 1 \\ 4 & -1 \end{vmatrix} = -3$ **27.** $\begin{vmatrix} 3 & 1 \\ 5 & 4 \end{vmatrix} = 7$ **29.** $\begin{vmatrix} 3 & 6 \\ 0 & 0 \end{vmatrix} = 0$

31. $\begin{vmatrix} 6 & 1 \\ 0 & 3 \end{vmatrix} = 18$ **33.** $-16\begin{vmatrix} 1 & 2 \\ 2 & 3 \end{vmatrix} = 16$ **35.** $\begin{vmatrix} 4 & -4 \\ 3 & -9 \end{vmatrix} = -24$

37. Let $A = \begin{vmatrix} x & y & 1 \\ x_1 & y_1 & 1 \\ x_2 & y_2 & 1 \end{vmatrix} = 0$. Expanding about the first row gives

$\delta(A) = x\begin{vmatrix} y_1 & 1 \\ y_2 & 1 \end{vmatrix} - y\begin{vmatrix} x_1 & 1 \\ x_2 & 1 \end{vmatrix} + 1\begin{vmatrix} x_1 & y_1 \\ x_2 & y_2 \end{vmatrix} = 0;$

hence $(y_1 - y_2)x + (x_2 - x_1)y + (x_1y_2 - y_1x_2) = 0$. Further, y_1, y_2, x_1, x_2 are real numbers; hence there exist real numbers a, b, c such that $(y_1 - y_2) = a$, $(x_2 - x_1) = b$, and $(x_1y_2 - y_1x_2) = c$. Substituting yields $ax + by + c = 0$, which is the equation of a straight line.

39. Multiply column 1 by $(-a)$ and add result to column 2; also, multiply column 1 by $(-a^2)$ and add result to column 3, to obtain,

$$\begin{vmatrix} 1 & a & a^2 \\ 1 & b & b^2 \\ 1 & c & c^2 \end{vmatrix} = \begin{vmatrix} 1 & 0 & 0 \\ 1 & b-a & b^2-a^2 \\ 1 & c-a & c^2-a^2 \end{vmatrix}.$$ Expand about the first row to obtain

$$1 \begin{vmatrix} b-a & b^2-a^2 \\ c-a & c^2-a^2 \end{vmatrix}$$

$$= (b-a)[c^2-a^2] - (c-a)[b^2-a^2]$$
$$= (b-a)[(c-a)(c+a)] - (c-a)\cdot[(b-a)(b+a)]$$
$$= -(a-b)[(c-a)(c+a)] + (c-a)[(a-b)(a+b)]$$
$$= (a-b)(c-a)[-(c+a)+(a+b)]$$
$$= (a-b)(c-a)(b-c) = (b-c)(c-a)(a-b).$$

41. Let $A = \begin{bmatrix} a_{11} & a_{12} \\ a_{21} & a_{22} \end{bmatrix}$ and $B = \begin{bmatrix} b_{11} & b_{12} \\ b_{21} & b_{22} \end{bmatrix}$.

Then, by Def. 10.7, $AB = \begin{bmatrix} a_{11}b_{11} + a_{12}b_{21} & a_{11}b_{12} + a_{12}b_{22} \\ a_{21}b_{11} + a_{22}b_{21} & a_{21}b_{12} + a_{22}b_{22} \end{bmatrix}$,

and, by Def. 10.6, $a(AB) = \begin{bmatrix} a(a_{11}b_{11} + a_{12}b_{21}) & a(a_{11}b_{12} + a_{12}b_{22}) \\ a(a_{21}b_{11} + a_{22}b_{21}) & a(a_{21}b_{12} + a_{22}b_{22}) \end{bmatrix}$;

also $aA = \begin{bmatrix} aa_{11} & aa_{12} \\ aa_{21} & aa_{22} \end{bmatrix}$ and $(aA)B = \begin{bmatrix} aa_{11}b_{11} + aa_{12}b_{21} & aa_{11}b_{12} + aa_{12}b_{22} \\ aa_{21}b_{11} + aa_{22}b_{21} & aa_{21}b_{12} + aa_{22}b_{22} \end{bmatrix}$

$$= \begin{bmatrix} a(a_{11}b_{11} + a_{12}b_{21}) & a(a_{11}b_{12} + a_{12}b_{22}) \\ a(a_{21}b_{11} + a_{22}b_{21}) & a(a_{21}b_{12} + a_{22}b_{22}) \end{bmatrix} = a(AB);$$

further, $aB = \begin{bmatrix} ab_{11} & ab_{12} \\ ab_{21} & ab_{22} \end{bmatrix}$ and $A(aB) = \begin{bmatrix} a_{11}ab_{11} + a_{12}ab_{21} & a_{11}ab_{12} + a_{12}ab_{22} \\ a_{21}ab_{11} + a_{22}ab_{21} & a_{21}ab_{12} + a_{22}ab_{22} \end{bmatrix}$

$$= \begin{bmatrix} a(a_{11}b_{11} + a_{12}b_{21}) & a(a_{11}b_{12} + a_{12}b_{22}) \\ a(a_{21}b_{11} + a_{22}b_{21}) & a(a_{21}b_{12} + a_{22}b_{22}) \end{bmatrix} = a(AB).$$

43. Let $A = \begin{vmatrix} a_{11} & a_{12} \\ a_{21} & a_{22} \end{vmatrix}$; $\delta(A) = a_{11}a_{22} - a_{12}a_{21}$.

Interchange rows 1 and 2 and let $B = \begin{vmatrix} a_{21} & a_{22} \\ a_{11} & a_{12} \end{vmatrix}$;

then $\delta(B) = a_{21}a_{12} - a_{22}a_{11} = -(a_{11}a_{22} - a_{12}a_{21}) = -\delta(A)$.

Interchange columns 1 and 2 and let $C = \begin{vmatrix} a_{12} & a_{11} \\ a_{22} & a_{21} \end{vmatrix}$;

then $\delta(C) = a_{12}a_{21} - a_{11}a_{22} = -(a_{11}a_{22} - a_{12}a_{21}) = -\delta(A)$.

45. Let $A = \begin{vmatrix} a_{11} & a_{12} \\ a_{21} & a_{22} \end{vmatrix}$; then $\delta(A) = a_{11}a_{22} - a_{12}a_{21}$.

Let $B = \begin{vmatrix} ka_{11} & ka_{12} \\ a_{21} & a_{22} \end{vmatrix}$; then $\delta(B) = ka_{11}a_{22} - ka_{12}a_{21} = $

$k(a_{11}a_{22} - a_{12}a_{21}) = k\delta(A)$; similarly if $C = \begin{vmatrix} a_{11} & a_{12} \\ ka_{21} & ka_{22} \end{vmatrix}$.

Let $D = \begin{vmatrix} ka_{11} & a_{12} \\ ka_{21} & a_{22} \end{vmatrix}$; then $\delta(D) = ka_{11}a_{22} - a_{12}ka_{21} = $

$k(a_{11}a_{22} - a_{12}a_{21}) = k \, \delta(A)$; similarly if $E = \begin{vmatrix} a_{11} & ka_{12} \\ a_{21} & ka_{22} \end{vmatrix}$.

Exercise 10.5 (page 308)

1. $\begin{bmatrix} 3 & -2 \\ -1 & 1 \end{bmatrix}$ **3.** $\dfrac{1}{5}\begin{bmatrix} 1 & 3 \\ -1 & 2 \end{bmatrix}$ **5.** $|A| = 0$; no inverse

7. $-1\begin{bmatrix} 4 & -7 \\ -3 & 5 \end{bmatrix}$ **9.** $\dfrac{1}{2}\begin{bmatrix} -2 & -4 \\ 4 & 7 \end{bmatrix}$ **11.** $|A| = 0$; no inverse

13. $\dfrac{1}{6}\begin{bmatrix} 2 & 2 & -5 \\ -4 & 2 & 1 \\ 0 & 0 & 3 \end{bmatrix}$ **15.** $\dfrac{1}{3}\begin{bmatrix} -2 & 3 & -1 \\ -1 & 0 & 1 \\ 6 & -6 & 3 \end{bmatrix}$ **17.** $|A| = 0$; no inverse

19. $|A| = 0$; no inverse   **21.** $\begin{bmatrix} 2 & -3 & 11 \\ -2 & 4 & -13 \\ 1 & -2 & 7 \end{bmatrix}$ **23.** $-1\begin{bmatrix} 0 & 0 & -1 \\ 0 & -1 & 0 \\ -1 & 0 & 0 \end{bmatrix}$

25. Let $A \cdot B = \begin{bmatrix} 2 & 3 \\ 1 & -1 \end{bmatrix} \cdot \begin{bmatrix} 0 & 1 \\ 3 & 1 \end{bmatrix}$; $A \cdot B = \begin{bmatrix} 9 & 5 \\ -3 & 0 \end{bmatrix}$ and $\delta(AB) = 15$,

so $(A \cdot B)^{-1} = \dfrac{1}{15}\begin{bmatrix} 0 & -5 \\ 3 & 9 \end{bmatrix}$; also, since $\delta(A) = -5$ and $\delta(B) = -3$,

$B^{-1} = -\dfrac{1}{3}\begin{bmatrix} 1 & -1 \\ -3 & 0 \end{bmatrix}$ and $A^{-1} = \dfrac{1}{5}\begin{bmatrix} -1 & -3 \\ -1 & 2 \end{bmatrix}$; $B^{-1} \cdot A^{-1} = \dfrac{1}{15}\begin{bmatrix} 0 & -5 \\ 3 & 9 \end{bmatrix}$;

hence $(A \cdot B)^{-1} = B^{-1} \cdot A^{-1}$.

27. Let $A = \begin{bmatrix} 3 & 0 & 1 \\ 2 & 1 & 0 \\ 0 & 1 & 2 \end{bmatrix}$; then $\delta(A) = 8$ and $A^{-1} = \dfrac{1}{8}\begin{bmatrix} 2 & 1 & -1 \\ -4 & 6 & 2 \\ 2 & -3 & 3 \end{bmatrix}$.

Let $B = \begin{bmatrix} 2 & 1 & 0 \\ 1 & 1 & 2 \\ 0 & 1 & 0 \end{bmatrix}$; then $\delta(B) = -\dfrac{1}{4}$ and $B^{-1} = -\dfrac{1}{4}\begin{bmatrix} -2 & 0 & 2 \\ 0 & 0 & -4 \\ 1 & -2 & 1 \end{bmatrix}$;

$A \cdot B = \begin{bmatrix} 6 & 4 & 0 \\ 5 & 3 & 2 \\ 1 & 3 & 2 \end{bmatrix}$, $\delta(A \cdot B) = -32$, and $(A \cdot B)^{-1} = -\dfrac{1}{32}\begin{bmatrix} 0 & -8 & 8 \\ -8 & 12 & -12 \\ 12 & -14 & -2 \end{bmatrix}$;

$B^{-1} \cdot A^{-1} = -\dfrac{1}{4}\begin{bmatrix} -2 & 0 & 2 \\ 0 & 0 & -4 \\ 1 & -2 & 1 \end{bmatrix} \cdot \dfrac{1}{8}\begin{bmatrix} 2 & 1 & -1 \\ -4 & 6 & 2 \\ 2 & -3 & 3 \end{bmatrix} = -\dfrac{1}{32}\begin{bmatrix} 0 & -8 & 8 \\ -8 & 12 & -12 \\ 12 & -14 & -2 \end{bmatrix}$;

hence $(A \cdot B)^{-1} = B^{-1} \cdot A^{-1}$.

29. Let $A = \begin{bmatrix} a_{11} & a_{12} \\ a_{21} & a_{22} \end{bmatrix}$; then $\delta(A) = (a_{11}a_{22} - a_{12}a_{21}) \neq 0$ since A is nonsingular,

and hence $\dfrac{1}{\delta(A)}$ is defined and A^{-1} exists.

$$A^{-1} = \frac{1}{\delta(A)} \begin{bmatrix} a_{22} & -a_{12} \\ -a_{21} & a_{11} \end{bmatrix} = \begin{bmatrix} \dfrac{a_{22}}{\delta(A)} & \dfrac{-a_{12}}{\delta(A)} \\ \dfrac{-a_{21}}{\delta(A)} & \dfrac{a_{11}}{\delta(A)} \end{bmatrix}$$

and $\delta(A^{-1}) = \dfrac{a_{22}a_{11}}{[\delta(A)]^2} - \dfrac{a_{12}a_{21}}{[\delta(A)]^2} = \dfrac{a_{22}a_{11} - a_{12}a_{21}}{[\delta(A)]^2} = \dfrac{\delta(A)}{[\delta(A)]^2} = \dfrac{1}{\delta(A)}$.

31. From the results of Problem 43, Ex. 10.3, and Problem 29 above,

$$\delta[B^{-1}AB] = \delta[B^{-1}(AB)] = \delta(B^{-1}) \cdot \delta(AB) = \delta(B^{-1}) \cdot \delta(A) \cdot \delta(B)$$

$$= \frac{1}{\delta(B)} \cdot \delta(A) \cdot \delta(B) = \delta(A).$$

Exercise 10.6 (page 313)

1. $\{(1, 1)\}$ **3.** $\{(2, 2)\}$ **5.** $\{(6, 4)\}$ **7.** $\{(1, 1, 1)\}$ **9.** $\{(1, 1, 0)\}$

11. $\{(1, -2, 3)\}$ **13.** $\{(3, -1, -2)\}$

Exercise 10.7 (page 317)

1. $\left\{\left(\dfrac{13}{5}, \dfrac{3}{5}\right)\right\}$ **3.** $\left\{\left(\dfrac{22}{7}, \dfrac{20}{7}\right)\right\}$ **5.** Inconsistent

7. $\left\{\left(\dfrac{1}{a+b}, \dfrac{1}{a+b}\right)\right\} (a \neq -b)$ **9.** $\{(1, 1, 0)\}$ **11.** $\{(1, -2, 3)\}$

13. $\{(3, -1, -2)\}$ **15.** $\left\{\left(-\dfrac{1}{3}, -\dfrac{25}{24}, -\dfrac{5}{8}\right)\right\}$ **17.** $\{(-1, 1, 0, 2)\}$

19. $A = \begin{vmatrix} a_1 & b_1 \\ a_2 & b_2 \end{vmatrix}$ and $\delta(A) = a_1b_2 - a_2b_1$; $Ay = \begin{vmatrix} a_1 & c_1 \\ a_2 & c_2 \end{vmatrix}$ and $\delta(Ay) = a_1c_2 - a_2c_1 = 0$,

so $a_1c_2 = a_2c_1$; $Ax = \begin{vmatrix} c_1 & b_1 \\ c_2 & b_2 \end{vmatrix}$ and $\delta(Ax) = b_2c_1 - b_1c_2 = 0$, so $b_1c_2 = b_2c_1$;

hence $\dfrac{a_1c_2}{b_1c_2} = \dfrac{a_2c_1}{b_2c_1}; \dfrac{a_1}{b_1} = \dfrac{a_2}{b_2}$ and $a_1b_2 = a_2b_1$, so $a_1b_2 - a_2b_1 = 0$; therefore $\delta(A) = 0$.

Exercise 10.8 (page 322)

1. $\{(2, -1)\}$ **3.** $\{(5, 1)\}$ **5.** $\{(8, 1)\}$ **7.** Since $y = z$ and $x = 2$, there are an infinite number of answers of the form $\{(2, a, a)\}$, where $a \in R$.

9. $\left\{ \left(\frac{5}{4}, \frac{5}{2}, -\frac{1}{2} \right) \right\}$ **11.** $\left\{ \left(-\frac{77}{27}, -\frac{8}{27}, \frac{29}{27} \right) \right\}$

13. $\begin{bmatrix} k & 0 \\ 0 & 1 \end{bmatrix} \begin{bmatrix} a & b \\ c & d \end{bmatrix} = \begin{bmatrix} k \cdot a + 0 \cdot c & k \cdot b + 0 \cdot d \\ 0 \cdot a + 1 \cdot c & 0 \cdot b + 1 \cdot d \end{bmatrix}$

$$= \begin{bmatrix} ka & kb \\ c & d \end{bmatrix}$$

15. $\begin{bmatrix} 0 & 1 \\ 1 & 0 \end{bmatrix} \begin{bmatrix} a & b \\ c & d \end{bmatrix} = \begin{bmatrix} 0 \cdot a + 1 \cdot c & 0 \cdot b + 1 \cdot d \\ 1 \cdot a + 0 \cdot c & 1 \cdot b + 0 \cdot d \end{bmatrix}$

$$= \begin{bmatrix} c & d \\ a & b \end{bmatrix}$$

17. $\begin{bmatrix} 1 & 0 \\ k & 1 \end{bmatrix} \begin{bmatrix} a & b \\ c & d \end{bmatrix} = \begin{bmatrix} 1 \cdot a + 0 \cdot c & 1 \cdot b + 0 \cdot d \\ k \cdot a + 1 \cdot c & k \cdot b + 1 \cdot d \end{bmatrix}$

$$= \begin{bmatrix} a & b \\ ka + c & kb + d \end{bmatrix}$$

Exercise 11.1 (page 327)

1. $(5, 7)$ **3.** $(-6, -1)$ **5.** $(3, 7)$ **7.** $(3, 4)$ **9.** $(0, 2)$ **11.** $(5, 14)$

13. $(-2, 14)$ **15.** $(1, -5)$ **17.** $(-1, 7)$ **19.** $(a_1 + a_2, 0); (a_1 a_2, 0)$

21.

	If $a, b, c \in R$	If $(a, b), (c, d), (e, f) \in C$	
F-1	$a + b \in R$	$(a, b) + (c, d) \in C$	Closure law for addition
F-2	$(a + b) + c = a + (b + c)$	$[(a, b) + (c, d)] + (e, f) = (a, b) + [(c, d) + (e, f)]$	Associative law for addition
F-3	There exists $0 \in R$ such that $a + 0 = 0$ and $0 + a = a$	There exists $(0, 0) \in C$ such that $(a, b) + (0, 0) = (a, b)$ and $(0, 0) + (a, b) = (a, b)$	Additive-identity law
F-4	For each $a \in R$, there exists $-a \in R$ such that $a + (-a) = 0$ and $(-a) + a = 0$	For each $(a, b) \in C$, there exists $-(a, b) \in C$ such that $(a, b) + [-(a, b)] = (0, 0)$ and $[-(a, b)] + (a, b) = (0, 0)$	Additive-inverse law
F-5	$a + b = b + a$	$(a, b) + (c, d) = (c, d) + (a, b)$	Commutative law for addition

(Table continued overleaf.)

	If $a, b, c \in R$	If $(a, b), (c, d), (e, f) \in C$	
F-6	$ab \in R$	$(a, b) \cdot (c, d) \in C$	Closure law for multiplication
F-7	$(ab)c = a(bc)$	$[(a, b) \cdot (c, d)] \cdot (e, f)$ $= (a, b) \cdot [(c, d) \cdot (e, f)]$	Associative law for multiplication
F-8	There exists $1 \in R$ such that $a \cdot 1 = a$ and $1 \cdot a = a$	There exists $(1, 0) \in C$ such that $(a, b) \cdot (1, 0) = (a, b)$ and $(1, 0) \cdot (a, b) = (a, b)$	Multiplicative-identity law
F-9	For each $a \in R$, $a \neq 0$, there exists $a^{-1} \in R$, such that $aa^{-1} = 1$ and $a^{-1}a = 1$	For each $(a, b) \in C$, (a, b) $\neq (0, 0)$, there exists $(a, b)^{-1} \in C$, such that $(a, b) \cdot (a, b)^{-1} = (1, 0)$ and $(a, b)^{-1} \cdot (a, b) = (1, 0)$	Multiplicative-inverse law
F-10	$ab = ba$	$(a, b) \cdot (c, d) = (c, d) \cdot (a, b)$	Commutative law for multiplication
F-11	$a(b + c) = ab + ac$ and $(b + c)a = ba + ca$	$(a, b) \cdot [(c, d) + (e, f)]$ $= (a, b) \cdot (c, d) + (a, b) \cdot (e, f)$ and $[(c, d) + (e, f)] \cdot (a, b)$ $= (c, d) \cdot (a, b) + (e, f) \cdot (a, b)$	Distributive law

23. Let $z_1 = (a_1, b_1)$ and $z_2 = (a_2, b_2)$; then by Def. 11.3-I and F-5 for real numbers, $z_1 + z_2 = (a_1, b_1) + (a_2, b_2) = (a_1 + a_2, b_1 + b_2) = (a_2 + a_1, b_2 + b_1) = z_2 + z_1$.

25. Let $z_1 = (a_1, b_1)$, $z_2 = (a_2, b_2)$, and $z_3 = (a_3, b_3)$; then by Def. 11.3-II and F-11, F-2, F-5 for real numbers, $(z_1 \cdot z_2)z_3 = [(a_1, b_1) \cdot (a_2, b_2)] \cdot (a_3, b_3) = [(a_1a_2 - b_1b_2, a_1b_2 + a_2b_1)] \cdot (a_3, b_3) = [(a_1a_2 - b_1b_2)a_3 - (a_1b_2 + a_2b_1)b_3, (a_1a_2 - b_1b_2)b_3 + (a_1b_2 + a_2b_1)a_3] = [a_1(a_2a_3 - b_2b_3) - b_1(a_2b_3 + b_2a_3), a_1(a_2b_3 + b_2a_3) + b_1(a_2a_3 - b_2b_3)] = (a_1, b_1) \cdot [a_2a_3 - b_2b_3, a_2b_3 + b_2a_3] = (a_1, b_1) \cdot [(a_2, b_2) \cdot (a_3, b_3)]$.

27. $\left(\dfrac{a}{a^2 + b^2}, \dfrac{-b}{a^2 + b^2} \right)$, $(a^2 + b^2 \neq 0)$

Exercise 11.2 (page 331)

1. $(3, 1)$ 3. $(-9, 1)$ 5. $(8, -8)$ 7. $\left(\dfrac{7}{4}, -\dfrac{1}{4} \right)$ 9. $\left(\dfrac{1}{10}, -\dfrac{7}{10} \right)$

11. $(-2, 0)$ 13. $\left(\dfrac{23}{37}, \dfrac{27}{37} \right)$ 15. $\left(-\dfrac{2}{5}, -\dfrac{9}{5} \right)$ 17. $\left(-\dfrac{5}{2}, -\dfrac{1}{2} \right)$ 19. $(2, 0)$

21. Let $z = (a, b)$; then $\bar{z} = (a, -b)$. From this result and Ths. 11.3 and 11.4,
$$\frac{(a, b)}{(a, b)} = \frac{(a, b) \cdot (a, -b)}{(a, b) \cdot (a, -b)} = \frac{(a^2 + b^2, 0)}{(a^2 + b^2, 0)} = \left(\frac{a^2 + b^2}{a^2 + b^2}, \frac{0}{a^2 + b^2}\right) = (1, 0).$$

23. Let $z_1 = (a_1, b_1)$, $z_2 = (a_2, b_2)$; then $\bar{z}_1 = (a_1, -b_1)$, $\bar{z}_2 = (a_2, -b_2)$,
$\bar{z}_1 + \bar{z}_2 = (a_1, -b_1) + (a_2, -b_2) = [(a_1 + a_2), -(b_1 + b_2)] = \overline{z_1 + z_2}$.

25. For F-1: $(a, 0) + (b, 0) = (a + b, 0) \in C$.
For F-2: $[(a, 0) + (b, 0)] + (c, 0) = (a + b, 0) + (c, 0) = [(a + b) + c, 0]$
$= [a + (b + c), 0] = (a, 0) + [b + c, 0] = (a, 0) + [(b, 0) + (c, 0)]$.
For F-3: $(0, 0)$ is the additive identity, since $(a, 0) + (0, 0) = (a + 0, 0) = (a, 0)$
and $(0, 0) + (a, 0) = (0 + a, 0) = (a, 0)$.
For F-4: $(-a, 0)$ is the additive inverse, since $(a, 0) + (-a, 0)$
$= (a - a, 0) = (0, 0)$ and $(-a, 0) + (a, 0) = (-a + a, 0) = (0, 0)$.
For F-5: $(a, 0) + (b, 0) = (a + b, 0) = (b + a, 0) = (b, 0) + (a, 0)$.
For F-6: $(a, 0) \cdot (b, 0) = (ab - 0, 0 + 0) = (ab, 0) \in C$.
For F-7: $[(a, 0) \cdot (b, 0)] \cdot (c, 0) = (ab, 0) \cdot (c, 0) = [(ab)c, 0] = [a(bc), 0] = (a, 0) \cdot (bc, 0)$
$= (a, 0) \cdot [(b, 0) \cdot (c, 0)]$.
For F-8: $(1, 0)$ is the multiplicative identity since
$(a, 0) \cdot (1, 0) = (a, 0)$ and $(1, 0) \cdot (a, 0) = (a, 0)$.
For F-9: $\left(\dfrac{1}{a}, 0\right)$ is the multiplicative inverse, since if
$a \neq 0$, then $(a, 0) \cdot \left(\dfrac{1}{a}, 0\right) = (1, 0)$ and $\left(\dfrac{1}{a}, 0\right) \cdot (a, 0) = (1, 0)$.
For F-10: $(a, 0) \cdot (b, 0) = (ab, 0) = (ba, 0) = (b, 0) \cdot (a, 0)$.
For F-11: $(a, 0) \cdot [(b, 0) + (c, 0)] = (a, 0) \cdot (b + c, 0) = (ab + ac, 0) = (ab, 0) + (ac, 0)$
$= (a, 0) \cdot (b, 0) + (a, 0) \cdot (c, 0)$; $[(b, 0) + (c, 0)] \cdot (a, 0) = (b + c, 0) \cdot (a, 0)$.
$= (ba + ca, 0) = (ba, 0) + (ca, 0) = (b, 0) \cdot (a, 0) + (c, 0) \cdot (a, 0)$.

Exercise 11.3 (page 335)

1. $2 + 6i$ **3.** $5 - 2i$ **5.** $-7 - 3i$ **7.** $4 + 0i$ **9.** $(2, 3)$ **11.** $(-3, 1)$

13. $(0, 4)$ **15.** $(7, 0)$ **17.** $2 - 3i$ **19.** $-3 - i$ **21.** $0 - 4i$ **23.** $7 - 0i$

25. $x = \dfrac{3}{2}, y = -2$ **27.** $x = 2, y = -2; x = -2, y = 2$

29. $x = 3, y = 9; x = -3, y = -9$ **31.** $5 + 5i$ **33.** $5 - 3i$ **35.** $-1 - 2i$

37. $1 + i$ **39.** $-\dfrac{1}{10} + \dfrac{7}{10}i$ **41.** $-8 - 6i$ **43.** $2 - 2i$ **45.** $4 - i\sqrt{7}$

47. $5 - 2i$ **49.** $-\sqrt{7}$ **51.** $1 - i$ **53.** $\dfrac{4}{13} + \dfrac{7}{13}i$

55. $c(a + bi) = (c + 0i)(a + bi)$
$\qquad = (ca - 0b) + (cb + 0a)i$
$\qquad = ca + cbi$

Exercise 11.4 (page 338)

1. $\{i, -i\}$ **3.** $\left\{1, \dfrac{1}{2}\right\}$ **5.** $\left\{-\dfrac{1}{3}, 1 + \sqrt{5}, 1 - \sqrt{5}\right\}$

7. $\left\{-1, \dfrac{1}{2}, -\dfrac{1}{2} + i\dfrac{\sqrt{3}}{2}, \right.$ $\left. -\dfrac{1}{2} - i\dfrac{\sqrt{3}}{2}\right\}$ **9.** $-2i$ **11.** $-i$ **13.** $i, -2i$

15. $3i, \dfrac{-3 + \sqrt{29}}{2}, \dfrac{-3 - \sqrt{29}}{2}$

17. $1 - i; \ x^3 - 2x + 4 = 0$ **19.** $P(x) = (2x - 3)(x - 2 + 2i)(x - 2 - 2i)$

Exercise 11.5 (page 343)

1. 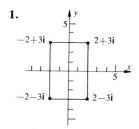   **3.** 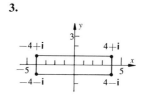   **5.**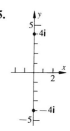

7. 4 **9.** $\sqrt{13}$ **11.** 2 **13.** $\sqrt{5}$ **15.** $3\sqrt{2}$ cis $45°$ **17.** 5 cis $0°$

19. 4 cis $330°$ **21.** $-2 - 2\sqrt{3}\,i$ **23.** $3\sqrt{3} - 3i$ **25.** $6 + 6\sqrt{3}\,i$

27. a. $-3 + 3i$ **b.** $\dfrac{3}{2} + \dfrac{3}{2}i$ **29. a.** 108 **b.** $\dfrac{1}{6} - \dfrac{\sqrt{3}}{6}i$

31. a. $10 + 10i$ **b.** $\dfrac{1}{10} - \dfrac{7}{10}i$ **33.** $1 + 0i$

35. Let $z = a + bi$; then $\bar{z} = a - bi$. $(a + bi) + (a - bi) = (a + a) + (b - b)i = 2a + 0i$
$= 2a \in R; (a + bi) \cdot (a - bi) = (a^2 + b^2) + (ab - ab)i = (a^2 + b^2) + 0i = (a^2 + b^2) \in R.$

37. r cis $\theta = r(\cos\theta + i\sin\theta) = r\cos\theta + (r\sin\theta)i = a + bi$, where $a = r\cos\theta$ and
$b = r\sin\theta$. For $(a + bi)$, $\bar{z} = a - bi = r\cos\theta - r\sin\theta i =$
$r\cos(-\theta) + [r\sin(-\theta)]i = r$ cis$(-\theta)$.

39. From the result of Problem 38, and from Th. 11.8-I,
$(a + bi)^3 = (a + bi)^2(a + bi) = (r^2$ cis $2\theta)(r$ cis $\theta) = r^3$ cis $(2\theta + \theta) = r^3$ cis 3θ.

Exercise 11.6 (page 347)

1. $-64\sqrt{3} + 64i$ **3.** $1 + 0i$ **5.** $729\left(\dfrac{1}{2} + \dfrac{\sqrt{3}}{2}i\right)$ **7.** $\dfrac{1}{64}\left(-\sqrt{3} + i\right)$

9. $\dfrac{1}{64}[(\sqrt{3}-1)-(\sqrt{3}+1)i]$ **11.** $-\dfrac{1}{2}-\dfrac{1}{2}i$

13. 2 cis 9°; 2 cis 81°; 2 cis 153°; 2 cis 225°; 2 cis 297°

15. 2 cis 30°; 2 cis 102°; 2 cis 174°; 2 cis 246°; 2 cis 318°

17. cis 45°; cis 105°; cis 165°; cis 225°; cis 285°; cis 345°

19. 2 cis 60°; 2 cis 132°; 2 cis 204°; 2 cis 276°; 2 cis 348°

21. cis $25\frac{5}{7}°$; cis $77\frac{1}{7}°$; cis $128\frac{4}{7}°$; cis 180°; cis $231\frac{3}{7}°$; cis $282\frac{6}{7}°$; cis $334\frac{2}{7}°$

23. $(x - 2 \text{ cis } 45°)(x - 2 \text{ cis } 135°)(x - 2 \text{ cis } 225°)(x - 2 \text{ cis } 315°)$

25. The four roots are $-1, 1, i, -i$; hence their sum is $0 + 0i$.

27. By De Moivre's theorem, $(\cos\theta + i\sin\theta)^3 = \cos 3\theta + i\sin 3\theta$. Since, $(\cos\theta + i\sin\theta)^3$
$= \cos^3\theta + 3i\cos^2\theta\sin\theta + 3i^2\cos\theta\sin^2\theta + i^3\sin^3\theta$, $i^2 = -1$, $i^3 = -i$, $\cos^2\theta$
$= (1 - \sin^2\theta)$, and $\sin^2\theta = (1 - \cos^2\theta)$, the left-hand member reduces to
$(4\cos^3\theta - 3\cos\theta) + (3\sin\theta - 4\sin^3\theta)i$.
Now, $(4\cos^3\theta - 3\cos\theta) + (3\sin\theta - 4\sin^3\theta)i = \cos 3\theta + (\sin 3\theta)i$.
Hence, by definition of equality of two complex numbers,
$\cos 3\theta = 4\cos^3\theta - 3\cos\theta$ and $\sin 3\theta = 3\sin\theta - 4\sin^3\theta$.

Exercise 11.7 (page 352)

1. $(8, 5), \sqrt{89}, \theta \approx 32°$ **3.** $(-1, 3), \sqrt{10}, \theta \approx 108° \, 30'$ **5.** $(5, 9), \sqrt{106}, \theta \approx 61°$

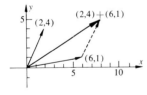   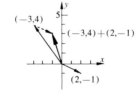   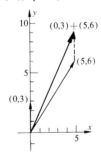

7. $(8, 6), 10, \theta \approx 36° \, 50'$ **9.** $(8, 1), \sqrt{65}, \theta \approx 7° \, 10'$ **11.** $(3, 1), \sqrt{10}, \theta \approx 18° \, 20'$

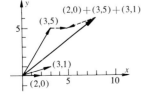   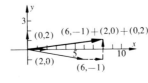   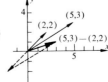

13. $(-4, 5)$, $\sqrt{41}$, $\theta \approx 128° \, 40'$ **15.** $(0, 8)$, 8, $\theta = 90°$ **17.** $(-1, 1)$, $\sqrt{2}$, $\theta = 135°$

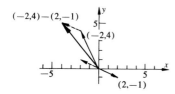

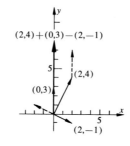

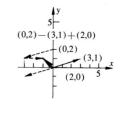

19. $(11, 11)$, $11\sqrt{2}$, $\theta = 45°$ **21.** $(-6, 1)$, $\sqrt{37}$, $\theta \approx 170° \, 30'$ **23.** $(3, 10)$, $\sqrt{109}$, $\theta \approx 73° \, 20'$

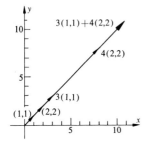

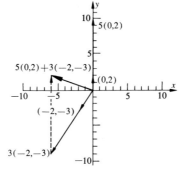

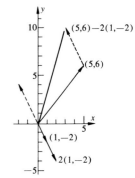

25. $(-8, 12)$, $4\sqrt{13}$, $\theta \approx 123° \, 40'$ **27.** $(10, 4)$, $2\sqrt{29}$, $\theta \approx 21° \, 50'$

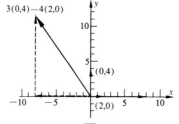

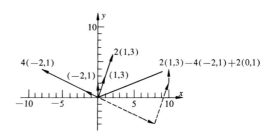

29. $(15, -5)$, $5\sqrt{10}$, $\theta \approx -18°30'$

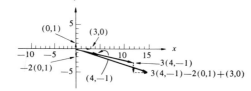

31. $\mathbf{v}_1 + \mathbf{v}_2 = (a_1, b_1) + (a_2, b_2)$ Hyp.
  $= (a_1 + a_2, b_1 + b_2)$ Def. 11.20-I

By the closure law for R, $a_1 + a_2 \in R$ and $b_1 + b_2 \in R$, so that by Def. 11.16 $(a_1 + a_2, b_1 + b_2) \in V$.

33. $(\mathbf{v}_1 + \mathbf{v}_2) + \mathbf{v}_3$

$\quad = [(a_1, b_1) + (a_2, b_2)] + (a_3, b_3)$ Hyp.

$\quad = (a_1 + a_2, b_1 + b_2) + (a_3, b_3)$ Def. 11.20-I

$\quad = [(a_1 + a_2) + a_3, (b_1 + b_2) + b_3]$ Def. 11.20-I

$\quad = [a_1 + (a_2 + a_3), b_1 + (b_2 + b_3)]$ Assoc. law of add. for R

$\quad = (a_1, b_1) + (a_2 + a_3, b_2 + b_3)$ Def. 11.20-I

$\quad = (a_1, b_1) + [(a_2, b_2) + (a_3, b_3)]$ Def. 11.20-I

$\quad = \mathbf{v}_1 + (\mathbf{v}_2 + \mathbf{v}_3)$ Hyp.

35. $\mathbf{v}_1 + (-\mathbf{v}_1)$

$\quad = (a_1, b_1) + (-a_1, -b_1)$ Hyp., Def. 11.22

$\quad = (a_1 + (-a_1), b_1 + (-b_1))$ Def. 11.20-I

$\quad = (a_1 - a_1, b_1 - b_1)$ Def. 1.12

$\quad = (0, 0)$ Add. inverse law for R

$\quad = \mathbf{0}$ Def. 11.21

37. $c(d\mathbf{v}_1)$

$\quad = c(d(a_1, b_1))$ Hyp.

$\quad = c(da_1, db_1)$ Def. 11.20-II

$\quad = (cda_1, cdb_1)$ Def. 11.20-II

$\quad = cd(a_1, b_1)$ Closure mult. law for R, Def. 11.20-II

$\quad = (cd)\mathbf{v}_1$ Def. 11.20-II

39. $c(\mathbf{v}_1 + \mathbf{v}_2)$

$\quad = c[(a_1, b_1) + (a_2, b_2)]$ Hyp.

$\quad = c(a_1 + a_2, b_1 + b_2)$ Def. 11.20-I

$\quad = [c(a_1 + a_2), c(b_1 + b_2)]$ Def. 11.20-II

$\quad = (ca_1 + ca_2, cb_1 + cb_2)$ Distrib. law for R

$\quad = (ca_1, cb_1) + (ca_2, cb_2)$ Def. 11.20-I

$\quad = c\mathbf{v}_1 + c\mathbf{v}_2$ Def. 11.20-II

Exercise 11.8 (page 357)

1. $(4, 4)$ **3.** $(9, -6)$ **5.** $(5, 3)$ **7.** $(2, -10)$ **9.** $(5, -12)$ **11.** $(10, 15)$

13. $(-8, 14)$ **15.** $\sqrt{520{,}000}$ miles   **17.** 80 lbs

19. Magnitude, 21.6 lbs; $\theta = -93° \, 40'$

Exercise 12.1 (page 362)

1. a. For $n = 1 : \dfrac{n}{2} = \dfrac{1}{2}$; $\dfrac{n(n+1)}{4} = \dfrac{1(1+1)}{4} = \dfrac{1}{2}$ **b.** For $n = k : \dfrac{1}{2} + \dfrac{2}{2} + \dfrac{3}{2} + \cdots + \dfrac{k}{2} =$

$\dfrac{k(k+1)}{4}$ and $(k+1)$th term $= \dfrac{k+1}{2}$; hence $\dfrac{1}{2} + \dfrac{2}{2} + \dfrac{3}{2} + \cdots + \dfrac{k}{2} + \dfrac{k+1}{2}$

$\quad = \dfrac{k(k+1)}{4} + \dfrac{k+1}{2} = \dfrac{k^2 + k + 2k + 2}{4} = \dfrac{k^2 + 3k + 2}{4} = \dfrac{(k+1)(k+2)}{4}$.

3. a. For $n = 1$: $2n = 2(1) = 2$; $n(n + 1) = 1(1 + 1) = 2$.

b. For $n = k$: $2 + 4 + 6 + \cdots + 2k = k(k + 1)$ and $(k + 1)$th term is $2(k + 1)$; hence $2 + 4 + 6 + \cdots + 2k + 2(k + 1) = k(k + 1) + 2(k + 1) = (k + 1)(k + 2)$.

5. a. For $n = 1$: $n^2 = 1^2 = 1$; $\dfrac{n(n + 1)(2n + 1)}{6} = \dfrac{1(2)(3)}{6} = 1$.

b. For $n = k$: $1^2 + 2^2 + 3^2 + \cdots + k^2 = \dfrac{k(k + 1)(2k + 1)}{6}$ and $(k + 1)$th term is

$(k + 1)^2$; hence $1^2 + 2^2 + 3^2 + \cdots + k^2 + (k + 1)^2 = \dfrac{k(k + 1)(2k + 1)}{6} + (k + 1)^2$

$= \dfrac{k(k + 1)(2k + 1) + 6(k + 1)^2}{6} = \dfrac{(k + 1)[k(2k + 1) + 6(k + 1)]}{6}$

$= \dfrac{(k + 1)(2k^2 + 7k + 6)}{6} = \dfrac{(k + 1)(k + 2)(2k + 3)}{6} = \dfrac{(k + 1)[(k + 1) + 1][2(k + 1) + 1]}{6}$.

7. a. For $n = 1$: $(2n - 1)^3 = (2 - 1)^3 = 1^3 = 1$; $n^2(2n^2 - 1) = 1(2 - 1) = 1(1) = 1$.

b. For $n = k$: $1^3 + 3^3 + 5^3 + \cdots + (2k - 1)^3 = k^2(2k^2 - 1)$ and the $(k + 1)$th term is $[2(k + 1) - 1]^3 = (2k + 1)^3$; hence $1^3 + 3^3 + 5^3 + \cdots + (2k - 1)^3 + (2k + 1)^3$ $= k^2(2k^2 - 1) + (2k + 1)^3 = 2k^4 + 8k^3 + 11k^2 + 6k + 1$; by use of the factor theorem and synthetic division, $2k^4 + 8k^3 + 11k^2 + 6k + 1 = (k + 1)(k + 1)$ $(2k^2 + 4k + 1)$; also, $2k^2 + 4k + 1 = 2(k^2 + 2k + 1) - 2 + 1 = 2(k + 1)^2 - 1$; hence $2k^4 + 8k^3 + 11k^2 + 6k + 1 = (k + 1)^2[2(k + 1)^2 - 1]$.

9. a. For $n = 1$: $n(n + 1) = 1(2) = 2$; $\dfrac{n(n + 1)(n + 2)}{3} = \dfrac{1(2)(3)}{3} = 2$.

b. For $n = k$: $1 \cdot 2 + 2 \cdot 3 + 3 \cdot 4 + \cdots + k(k + 1) = \dfrac{k(k + 1)(k + 2)}{3}$

and the $(k + 1)$th term is $(k + 1)[(k + 1) + 1] = (k + 1)(k + 2)$; hence

$1 \cdot 2 + 2 \cdot 3 + 3 \cdot 4 + \cdots + k(k + 1) + [(k + 1)(k + 2)] =$

$\dfrac{k(k + 1)(k + 2)}{3} + (k + 1)(k + 2) = \dfrac{[k(k + 1)(k + 2)] + [3(k + 1)(k + 2)]}{3} =$

$\dfrac{(k + 1)(k + 2)(k + 3)}{3} = \dfrac{(k + 1)[(k + 1) + 1][(k + 1) + 2]}{3}$.

11. For $n = k$: $2 + 4 + 6 + \cdots + 2k = k(k + 1) + 2$ and $(k + 1)$th term is $2(k + 1)$; hence $2 + 4 + 6 + \cdots + 2k + 2(k + 1) = k(k + 1) + 2 + 2(k + 1) = (k^2 + 3k + 2) + 2$ $= (k + 1)(k + 2) + 2 = (k + 1)[(k + 1) + 1] + 2$. However, for $n = 1$: $2n = 2(1) = 2$; $n(n + 1) + 2 = 1(2) + 2 = 4$. Hence not true for every $n \in N$.

Exercise 12.2 (page 365)

1. $-4, -3, -2, -1$ **3.** $-\dfrac{1}{2}, 1, \dfrac{7}{2}, 7$ **5.** $2, \dfrac{3}{2}, \dfrac{4}{3}, \dfrac{5}{4}$ **7.** $0, 1, 3, 6$

9. $-1, 1, -1, 1$ **11.** $1, 0, -\dfrac{1}{3}, \dfrac{1}{2}$ **13.** $11, 15, 19$ **15.** $x+2, x+3, x+4$

17. $2x+7, 2x+10, 2x+13$ **19.** $32, 128, 512, 2048$ **21.** $\dfrac{8}{3}, \dfrac{16}{3}, \dfrac{32}{3}, \dfrac{64}{3}$

23. $\dfrac{x}{a}, -\dfrac{x^2}{a^2}, \dfrac{x^3}{a^3}, -\dfrac{x^4}{a^4}$ **25.** $4n+3, 31$ **27.** $-5n+8, -92$ **29.** $48(2)^{n-1}, 1536$

31. $-\dfrac{1}{3}(-3)^{n-1}, -243$ **33.** $2; 3; 41$ **35.** 28th   **37.** 3

39. a. For $n=1: s_n = s_1 = a; ar^{n-1} = a(r)^0 = a(1) = a.$
b. For $n=k: s_k = ar^{k-1}$; to obtain s_{k+1}, multiply s_k by r; hence, from multiplying both sides by r, $s_{k(r)} = s_{k+1} = (ar^{k-1})(r) = ar^{k-1+1} = ar^{(k+1)-1}.$

Exercise 12.3 (page 371)

1. $1 + 4 + 9 + 16$ **3.** $-\dfrac{1}{2} + \dfrac{1}{4} - \dfrac{1}{8} + \dfrac{1}{16}$ **5.** $1 + \dfrac{1}{2} + \dfrac{1}{4} + \cdots$ **7.** $\displaystyle\sum_{j=1}^{4} x^{2j-1}$

9. $\displaystyle\sum_{j=1}^{5} j^2$ **11.** $\displaystyle\sum_{j=1}^{\infty} j(j+1)$ **13.** $\displaystyle\sum_{j=1}^{\infty} \dfrac{j+1}{j}$ **15.** 63   **17.** 806   **19.** -6

21. 1092   **23.** $\dfrac{31}{32}$ **25.** $\dfrac{364}{729}$ **27.** 168   **29.** 196   **31.** $\dfrac{3}{4}, \dfrac{7}{8}, \dfrac{15}{16}, \dfrac{31}{32}; 1$

33. $p = 4, q = -3$

35. Since $s_n = a + (n-1)d$ and $S_{n+1} = S_n + s_n$, it follows that **a.** For $n=1: S_n = s_n = s_1$
$= a; \dfrac{n}{2}(a + s_n) = \dfrac{1}{2}(2a) = a.$ **b.** For $n=k: S_k = \dfrac{k}{2}(a + s_k), s_k = a + (k-1)d$; now
$s_{k+1} = a + [(k+1) - 1]d = a + kd$; adding produces $S_k + s_{k+1} = S_{k+1}$
$= \dfrac{k}{2}(a + s_k) + (a + kd) = \dfrac{k}{2}[a + a + (k-1)d] + \dfrac{2}{2}(a + kd)$
$= \dfrac{2ka + k^2d - kd + 2a + 2kd}{2} = \dfrac{2ka + 2a + k^2d + kd}{2} = \dfrac{(k+1)2a + k(k+1)d}{2}$
$= \dfrac{(k+1)}{2}[2a + kd] = \dfrac{k+1}{2}[a + (a + kd)] = \dfrac{k+1}{2}[a + s_{n+1}].$

37. Let $s_1, s_2, s_3, \ldots, s_n$ and $t_1, t_2, t_3, \ldots, t_n$ be two sequences with terms in arithmetic progression; then, by Problem 36 above, $(s_1 + t_1), (s_2 + t_2), \ldots, (s_n + t_n)$ is also a sequence with terms in arithmetic progression. $S_1 = s_1 + s_2 + s_3 + \cdots + s_n = \displaystyle\sum_{j=1}^{n} s_j,$
$S_2 = t_1 + t_2 + t_3 + \cdots + t_n = \displaystyle\sum_{j=1}^{n} t_j,$ and $S_1 + S_2 = \displaystyle\sum_{j=1}^{n} s_j + \sum_{j=1}^{n} t_j$; however,
$(S_1 + S_2) = (s_1 + t_1) + (s_2 + t_2) + \cdots + (s_n + t_n),$ and $S_1 + S_2 = \displaystyle\sum_{j=1}^{n} (s_j + t_j);$
hence $\displaystyle\sum_{j=1}^{n} s_j + \sum_{j=1}^{n} t_j = \sum_{j=1}^{n} (s_j + t_j).$

Exercise 12.4 (page 379)

1. $\lim_{n \to \infty} s_n = 0$ **3.** $\lim_{n \to \infty} s_n = 1$ **5.** $\lim_{n \to \infty} s_n$ is undefined. **7.** $\lim_{n \to \infty} s_n = 0$

9. Convergent, $\lim_{n \to \infty} \left| 0 - \dfrac{1}{2^n} \right| = 0$ **11.** Divergent, $\lim_{n \to \infty} n$ is undefined.

13. Convergent, $\lim_{n \to \infty} \left| 0 - (-1)^{n+1} \dfrac{1}{2^{n-1}} \right| = 0$ **15.** 24 **17.** No sum **19.** 2

21. $\dfrac{31}{99}$ **23.** $2\dfrac{410}{999}$ **25.** $\dfrac{29}{225}$ **27.** 20 cm. **29.** 0.980 **31.** 0.920 **33.** 0.296

35. 1 **37.** -1

Exercise 12.5 (page 385)

1. $8 \cdot 7 \cdot 6 \cdot 5 \cdot 4 \cdot 3 \cdot 2 \cdot 1$ **3.** $6 \cdot 5 \cdot 4 \cdot 3 \cdot 2 \cdot 1$ **5.** $5 \cdot 4 \cdot 3 \cdot 2 \cdot 1 = 120$ **7.** $\dfrac{9 \cdot 8 \cdot 7!}{7!} = 72$

9. $\dfrac{5 \cdot 4 \cdot 3 \cdot 2 \cdot 1 \cdot 7!}{8 \cdot 7!} = 15$ **11.** $\dfrac{8 \cdot 7 \cdot 6!}{2 \cdot 1 \cdot 6!} = 28$ **13.** $3!$ **15.** $\dfrac{6!}{2!}$ **17.** $\dfrac{8!}{5!}$

19. $\dfrac{6!}{5! 1!} = 6$ **21.** $\dfrac{3!}{3! 0!} = 1$ **23.** $\dfrac{7!}{0! 7!} = 1$ **25.** $\dfrac{5!}{2! 3!} = 10$

27. $(n)(n-1)(n-2) \cdots \cdots 3 \cdot 2 \cdot 1$ **29.** $(3n)(3n-1)(3n-2) \cdots \cdots 3 \cdot 2 \cdot 1$

31. $(n-2)(n-3)(n-4) \cdots \cdots 3 \cdot 2 \cdot 1$ **33.** $(n+2)(n+1)$ **35.** $\dfrac{n+1}{n+3}$ **37.** $\dfrac{2n-1}{2n-2}$

39. $x^5 + 15x^4 + 90x^3 + 270x^2 + 405x + 243$

41. $x^4 - 12x^3 + 54x^2 - 108x + 81$ **43.** $8x^3 - 6x^2y + \dfrac{3}{2}xy^2 - \dfrac{1}{8}y^3$

45. $\dfrac{1}{64}x^6 + \dfrac{3}{8}x^5 + \dfrac{15}{4}x^4 + 20x^3 + 60x^2 + 96x + 64$

47. $x^{20} + 20x^{19}y + \dfrac{20 \cdot 19}{2!}x^{18}y^2 + \dfrac{20 \cdot 19 \cdot 18}{3!}x^{17}y^3$, or

$\dbinom{20}{0}x^{20} + \dbinom{20}{1}x^{19}y + \dbinom{20}{2}x^{18}y^2 + \dbinom{20}{3}x^{17}y^3$

49. $a^{12} + 12a^{11}(-2b) + \dfrac{12 \cdot 11}{2!}a^{10}(-2b)^2 + \dfrac{12 \cdot 11 \cdot 10}{3!}a^9(-2b)^3$, or

$\dbinom{12}{0}a^{12} + \dbinom{12}{1}a^{11}(-2b) + \dbinom{12}{2}a^{10}(-2b)^2 + \dbinom{12}{3}a^9\left(-2b\right)^3$

51. $x^{10} + 10x^9(-\sqrt{2}) + \dfrac{10 \cdot 9}{2!}x^8(-\sqrt{2})^2 + \dfrac{10 \cdot 9 \cdot 8}{3!}x^7(-\sqrt{2})^3$, or

$$\binom{10}{0}x^{10} + \binom{10}{1}x^9(-\sqrt{2}) + \binom{10}{2}x^8(-\sqrt{2})^2 + \binom{10}{3}x^7(-\sqrt{2})^3$$

53. 1.22   **55.** \$1480   **57.** $-3003a^{10}b^5$ **59.** $3360x^6y^4$

61. *a.* $1 - x + x^2 - x^3 + \cdots$,   **b.** $1 - x + x^2 - x^3 + \cdots$

Exercise 13.1 (page 394)

1. 1, 5, 8   **3.** 2, 6, 16   **5.** 2, 2, 4   **7.** 4   **9.** 24   **11.** 16   **13.** 64

15. 24   **17.** 216   **19.** 375   **21.** 30   **23.** 10   **25.** 48   **27.** $\dfrac{5!}{2!}$, or 60

29. $\dfrac{8!}{3!}$, or 6720   **31.** $P_{5,3} = \dfrac{5!}{2!} = \dfrac{5 \cdot 4!}{2!} = 5\left(\dfrac{4!}{2!}\right) = 5(P_{4,2})$

33. $P_{n,3} = \dfrac{n!}{(n-3)!} = \dfrac{n(n-1)!}{(n-3)!} = n\left[\dfrac{(n-1)!}{(n-3)!}\right] = n(P_{n-1,2})$ **35.** 9   **37.** 24

39. 48

Exercise 13.2 (page 398)

1. 7   **3.** 15   **5.** $\binom{52}{5}$ **7.** $\binom{13}{5} \cdot \binom{13}{5} \cdot \binom{13}{3}$ **9.** $4 \cdot \binom{13}{5}$ **11.** 164

13. 10   **15.** 210   **17.** 12   **19.** $\binom{10}{10}a^{10} + \binom{10}{9}a^9b + \binom{10}{8}a^8b^2 + \binom{10}{7}a^7b^3 + \cdots$

Exercise 13.3 (page 403)

1. $\{1, 2, 3, 4, 5, 6\}, \{3, 4, 5, 6\}, \dfrac{2}{3}$

3. $\{(H, H), (H, T), (T, H), (T, T)\}, \{(H, H), (T, T)\}, \dfrac{1}{2}$ **5.** $\dfrac{1}{6}$ **7.** $\dfrac{1}{18}$ **9.** $\dfrac{5}{9}$

11. $\dfrac{1}{52}$ **13.** $\dfrac{3}{26}$ **15.** $\dfrac{1}{17}$ **17.** $\dfrac{11}{221}$ **19.** $\dfrac{1}{190}$ **21.** $\dfrac{3}{38}$

Exercise 13.4 (page 406)

1. $\dfrac{1}{6}$ **3.** $\dfrac{7}{18}$ **5.** $\dfrac{13}{18}$ **7.** $\dfrac{5}{33}$ **9.** $\dfrac{1}{11}$ **11.** $\dfrac{5}{22}$ **13.** $\dfrac{14}{33}$ **15.** $\dfrac{15}{22}$

17. $\dfrac{3}{13}$ **19.** $\dfrac{15}{77}, \dfrac{16}{77}, \dfrac{46}{77}$ **21.** \$2.14; less   **23.** 5.3 cents   **25.** \$3.29

Exercise 13.5 (page 410)

1. $\dfrac{20}{91}$; no   **3. a.** $\dfrac{2}{45}$ **b.** $\dfrac{28}{75}$ **c.** $\dfrac{4}{225}$ **d.** $\dfrac{1}{9}$ **5. a.** $\dfrac{11}{36}$ **b.** $\dfrac{5}{36}$ **c.** $\dfrac{2}{11}$ **d.** No

7. a. $\dfrac{1}{2}$ **b.** $\dfrac{1}{2}$ **c.** $\dfrac{1}{4}$ **d.** $\dfrac{1}{4}$ **e.** $\dfrac{1}{8}$ **9. a.** $\dfrac{71}{72}$ **b.** $\dfrac{5}{9}$ **c.** $\dfrac{5}{36}$ **d.** $\dfrac{61}{72}$

11. a. $\dfrac{1}{210}$ **b.** $\dfrac{29}{210}$ **c.** $\dfrac{29}{70}$ **d.** $\dfrac{29}{30}$; yes   **13.** $\dfrac{1}{4}$

Index

Analytical geometry supplemental index